Advances in the Theory of Fréchet Spaces

NATO ASI Series

Advanced Science Institutes Series

A Series presenting the results of activities sponsored by the NATO Science Committee, which aims at the dissemination of advanced scientific and technological knowledge, with a view to strengthening links between scientific communities.

The Series is published by an international board of publishers in conjunction with the NATO Scientific Affairs Division

A Life Sciences **B Physics**	Plenum Publishing Corporation London and New York
C Mathematical and Physical Sciences **D Behavioural and Social Sciences** **E Applied Sciences**	Kluwer Academic Publishers Dordrecht, Boston and London
F Computer and Systems Sciences **G Ecological Sciences** **H Cell Biology**	Springer-Verlag Berlin, Heidelberg, New York, London, Paris and Tokyo

Series C: Mathematical and Physical Sciences - Vol. 287

Advances in the Theory of Fréchet Spaces

edited by

T. Terzioğlu
Middle East Technical University,
Ankara, Turkey

Kluwer Academic Publishers

Dordrecht / Boston / London

Published in cooperation with NATO Scientific Affairs Division

Proceedings of the NATO Advanced Research Workshop on
Advances in the Theory of Fréchet Spaces
Istanbul, Turkey
August 15–19, 1988

Library of Congress Cataloging in Publication Data

NATO Advanced Research Workshop Advances in the Theory of Fréchet Spaces (1988 : Bosphorus University)
Advances in the theory of Fréchet spaces : proceedings of the NATO Advanced Research Workshop Advances in the Theory of Fréchet Spaces, Bosphorus University, Istanbul, August 15-19, 1988 / edited by T. Terzioğlu.
p. cm. -- (NATO ASI series. Series C, Mathematical and physical sciences ; vol. 287)

1. Fréchet spaces--Congresses. I. Terzioğlu, T., 1942- . II. Title. III. Series: NATO ASI series. Series C, Mathematical and physical sciences ; no. 287.
QA322.N395 1988
515'.73--dc20 89-36646

ISBN-13: 978-94-010-7608-1 e-ISBN-13: 978-94-009-2456-7
DOI: 10.1007/978-94-009-2456-7

Published by Kluwer Academic Publishers,
P.O. Box 17, 3300 AA Dordrecht, The Netherlands.

Kluwer Academic Publishers incorporates the publishing programmes of
D. Reidel, Martinus Nijhoff, Dr W. Junk and MTP Press.

Sold and distributed in the U.S.A. and Canada
by Kluwer Academic Publishers,
101 Philip Drive, Norwell, MA 02061, U.S.A.

In all other countries, sold and distributed
by Kluwer Academic Publishers Group,
P.O. Box 322, 3300 AH Dordrecht, The Netherlands.

Printed on acid free paper

Softcover reprint of the hardcover 1st edition 1989

Table of Contents

PREFACE

Fréchet spaces have been studied since the days of Banach. These spaces, their inductive limits and their duals played a prominent role in the development of the theory of locally convex spaces. Also they are natural tools in many areas of real and complex analysis.

The pioneering work of Grothendieck in the fifties has been one of the important sources of inspiration for research in the theory of Fréchet spaces. A structure theory of nuclear Fréchet spaces emerged and some important questions posed by Grothendieck were settled in the seventies. In particular, subspaces and quotient spaces of stable nuclear power series spaces were completely characterized. In the last years it has become increasingly clear that the methods used in the structure theory of nuclear Fréchet spaces actually provide new insight to linear problems in diverse branches of analysis and lead to solutions of some classical problems.

The unifying theme at our Workshop was the recent developments in the theory of the projective limit functor. This is appropriate because of the important role this theory had in the recent research. The main results of the structure theory of nuclear Fréchet spaces can be formulated and proved within the framework of this theory. A major area of application of the theory of the projective limit functor is to decide when a linear operator is surjective and, if it is, to determine whether it has a continuous right inverse. Using the recent developments one can tackle successfully such questions in the case of linear partial differential operators and convolution equations. This is an area of contact with the more classical fields of analysis.

Recent research in the structure theory of some special classes of Fréchet spaces, like various spaces of continuous functions, Köthe spaces and spaces of analytic functions of several variables as well as infinite-dimensional holomorphy constitute the subject matter of several articles. Progress in some outstanding open problems is also discussed. By presenting the recent developments in these proceedings in a coherent and unified fashion, we aim to stimulate further research in the theory of Fréchet spaces. We also hope that the considerable wealth of ideas and applications of the theory exhibited in this volume will attract the attention of mathematicians working in related fields.

We are very grateful to the NATO Science Division for providing funds for this Advanced Research Workshop. Some support was also received from the Turkish Mathematical Society. The Workshop was held at Bosphorus University in Istanbul. The efforts of the local organizing committee, A. Aytuna, A. Erkip, A. Fevzioğlu and A. Ülger as well as the cooperation of the Director of the Graduate School of Basic Sciences of Bosphorus University, Prof. Dr. M. Dikmen, contributed to the success of our meeting. Many colleagues and friends also helped in one way or the other, but I would like to mention especially the dedicated assistance of Zehra Tuğlu. I thank them all.

It is indeed a pleasure for me to acknowledge the essential contributions of the co-organizers, Ed Dubinsky, R. Meise and D. Vogt. Without their efforts and the interest of all the participants, the Workshop could not have taken place.

When preparations of these proceedings were nearly finished, we learned that Professor G. Köthe had died on April 30, 1989. Professor Köthe holds a prominent place in the development of Functional Analysis as one of the early pioneers of this field. With his powerful research and his book *Topological Vector Spaces* he has influenced many mathematicians throughout the world. However his loss was especially deeply felt by the participants of our Workshop, since most of us had the privilege of knowing him personally; we had benefitted from his guidance and generous encouragement. With these thoughts and on behalf of all the participants, I would like to dedicate this volume to the memory of Prof. Dr. G. Köthe.

Ankara - May 11, 1989

T. Terzioğlu

LIST OF PARTICIPANTS

A. AKSOY
Claremont McKenna College
Department of Mathematics,
California 91711,
USA

E. AMAR
Université de Bordeaux 1,
351 de la Libération,
33405 Talence
FRANCE

J.M. ANSEMIL
Universidad Complutense,
Facultad de Ciencias Matematicas,
Departamento de Analisis Matematico
28040 Madrid - SPAIN

A. ATZMON
Tel Aviv University,
Raymond and Beverly Sackler-Faculty of Exact Sci.
School of Mathematical Sciences,
Ramat Aviv, 69978 Tel Aviv - ISRAEL

Y. AVCI
İstanbul University,
Mathematics Department,
Vezneciler, İstanbul - TURKEY

A. AYTUNA
Department of Mathematics
Middle East Technical University
06531 Ankara - TURKEY

T. BASKAYA
Department of Mathematics
Middle East Technical University
06531 Ankara - TURKEY

K.D. BIERSTEDT
Universität Gesamthochschule Padernborn,
Fachbereich 17 - Mathematik - Informatik
4790 Paderborn, F.R. GERMANY

J. BONET
Departamento de Matematica Aplicade,
E.T.S. Arquitectura,
Universitat Politecnica de Valencia
E-46071 Valencia - SPAIN

R.W. BRAUN
Mathematisches Institut der
Universität Düsseldorf
Universitätstrasse 1
4000 Düsseldorf - F.R. GERMANY

L. CATTABRIGA
Universita Degli Studi di Bologna
Dipartimento di Matematica
Piazza di Porta S. Donato
5 40126 Bologna - ITALY

J. CNOPS
Seminar of Algebra and Functional Analysis
State University of Ghent-RUG
Galglaan 2, 9000 Gent
BELGIUM

S. DIEROLF
Universität Trier
FB IV - Mathematik
Postfach 3825
5500 Trier - F.R. GERMANY

S. DINEEN
Department of Mathematics
University of College
Belfield, Dublin 4
IRELAND

E. DUBINSKY
Mathematical Department
Purdue University, W. Lafayette
Indiana 47907 - USA

A. ERKIP
Department of Mathematics
Middle East Technical University
06531 Ankara
TURKEY

M. FRAGOULOPOULOU
University of Athens
Mathematical Institute
57 Solonos St.
Athens 143
GREECE

A. KANEKO
Department of Mathematics
College of General Education
University of Tokyo
Komaba, Meguro, Tokyo
JAPAN

M. KOCATEPE
Bilkent University
Mathematics Department
Ankara - TURKEY

H. KOMATSU
Department of Mathematics
Faculty of Science, University of Tokyo,
Tokyo, 113 JAPAN

J. KRONE
Bergische Universität
Gesamthochschule Wuppertal
Gauss-Strasse 20
5600 Wuppertal 1
F.R. GERMANY

A. KYRIAZIS
University of Athens,
Mathematical Institute
57 Solonos St.
Athens 143
GREECE

M. LANGENBRUCH
Universität Münster
Fachbereich 15
Mathematisches Institut
4400 Münster
F.R. GERMANY

R. MEISE
Mathematisches Institut der
Universität Düsseldorf
Universitätsstrasse 1
4000 Düsseldorf
F.R. GERMANY

G. METAFUNE
Universita Degli Studi di Lecce
Facolta di Scienze
Dipartimento di Matematica
Via Arnesano, 73100 Lecce - ITALY

P.P. NARAYANASWAMI
Memorial University of Newfoundland
St. John's, Newfoundland
Canada A1C 5S7
CANADA

Z. NURLU
Department of Mathematics
Middle East Technical University
06531 Ankara - TURKEY

S. ÖNAL
Department of Mathematics
Middle East Technical University
06531 Ankara - TURKEY

H.J. PETZSCHE
Universität Dortmund
Fachbereich Mathematik
4600 Dortmund 50
F.R. GERMANY

J. PRADA
Seccion' de Matematicas
Universidad de Salamanca
37008 Salamanca
SPAIN

J. SCHMETS
Université de Liege
Institut de Mathématique
Avenue des Tilleuls, 15
B-4000 Liege - BELGIUM

S.P. SINGH
Memorial University of Newfoundland
St. John's, Newfoundland
Canada A1C 5S7
CANADA

B.A. TAYLOR
University of Michigan
Department of Mathematics
347 West Engineering Building
Ann Arbor, Michigan 48109-1092, USA

T. TERZİOĞLU
Department of Mathematics
Middle East Technical University
06531 Ankara
TURKEY

M. TIDTEN
Fachbereich 7 - Mathematik
Bergische Universität
Gesamthochschule, Wuppertal
Gaussstrasse 20, D 5600 Wuppertal
F.R. GERMANY

A. ÜLGER
Boğazici University
Mathematics Department
Bebek
İstanbul - TURKEY

A. ÜNAL
The University of Michigan-Flint
Department of Mathematics
Flint,
Michigan 48502-2186
USA

M. VALDIVIA
Universidad de Valencia
Facultad de Matematicas
Depto. de Teoria de Functiones
Doctor Moliner, 50, 46100 Burjasot-Valen
SPAIN

D. VOGT
Bergische Universität
Gesamthochschule Wuppertal
Fachbereich 7 Mathematik
Gauss-Strasse 20, D 5600 Wuppertal 1
F.R. GERMANY

M. YURDAKUL
Department of Mathematics
Middle East Technical University
06531 Ankara
TURKEY

LIST OF CONTRIBUTORS

J.M. ANSEMIL
Universidad Complutense
Facultad de Ciencias Matematicas
Departamento de Analisis Matematico
28040 Madrid - SPAIN

A. AYTUNA
Department of Mathematics
Middle East Technical University
06531 Ankara
TURKEY

K.D. BIERSTEDT
Universität Gesamthochschule Paderborn
Fachbereich 17 - Mathematik - Informatik
4790 Paderborn
F.R. GERMANY

J. BONET
Departamento de Matematica Aplicade
E.T.S. Arquitectura
Universitat Politecnica de Valencia
E-46071 Valencia
SPAIN

R.W. BRAUN
Mathematisches Institut der
Universität Düsseldorf
Universitätstrasse 1
4000 Düsseldorf
F.R. GERMANY

L. CATTABRIGA
Universita Degli Studi di Bologna
Dipartimento di Matematica
Piazza di Porta S. Donato
5 40126 Bologna
ITALY

S. DIEROLF
Universität Trier
FB IV - Mathematik
Postfach 3825
5500 Trier
F.R. GERMANY

S. DINEEN
Department of Mathematics
University College
Belfield,
Dublin 4
IRELAND

E. DUBINSKY
Mathematical Department
Purdue University
W. Lafayette
Indiana 47907
USA

M. FRAGOULOPOULOU
University of Athens
Mathematical Institute
57 Solonos St.
Athens 143
GREECE

A. KANEKO
Department of Mathematics
College of General Education
University of Tokyo
Komaba, Meguro, Tokyo
JAPAN

M. KOCATEPE
Bilkent University
Mathematics Department
Ankara
TURKEY

J. KRONE
Bergische Universität
Gesamthochschule Wuppertal
Gauss-Strasse 20
5600 Wuppertal 1
F.R. GERMANY

A. KYRIAZIS
University of Athens
Mathematical Institute
57 Solonos St.
Athens 143
GREECE

M. LANGENBRUCH
Universität Münster
Fachbereich 15
Mathematisches Institut
4400 Münster
F.R. GERMANY

M. MAESTRE
Departamento de Analisis Matematico
Facultad de Farmacia
Universidad de Valencia
46010 Valencia
SPAIN

R. MEISE
Mathematisches Institut der
Universität Düsseldorf
Universität-Strasse 1
4000 Düsseldorf
F.R. GERMANY

G. METAFUNE
Universita Degli Studi di Lecce
Facolta di Scienze
Dipartimento di Matematica
Via Arnesano
73100 Lecce
ITALY

V.B. MOSCATELLI
Dipartimento di Matematica
Universita - C.P. 193
73100 Lecce
ITALY

Z. NURLU
Department of Mathematics
Middle East Technical University
06531 Ankara
TURKEY

S. ÖNAL
Department of Mathematics
Middle East Technical University
06531 Ankara
TURKEY

B.A. TAYLOR
University of Michigan
Department of Mathematics
347 West Engineering Building
Ann Arbor, Michigan 48109-1092
USA

T. TERZİOĞLU
Department of Mathematics
Middle East Technical University
06531 Ankara
TURKEY

D. VOGT
Bergische Universität
Gesamthochschule Wuppertal
Fachbereich 7 - Mathematik
Gauss-Strasse 20, D 5600 Wuppertal 1
F.R. GERMANY

M. J. WAGNER
Weinheimer Str. 54
6703 Limburgerhof
F.R. GERMANY

APPROXIMATION PROPERTIES OF NUCLEAR FRECHET SPACES

ED DUBINSKY
Mathematics Department
Purdue University
W. Lafayette, IN 47907
USA

To celebrate the growth of nuclear Fréchet Spaces with this NATO Workshop attended by mathematicians from so many countries,
To thank our hosts for successful organizing efforts,
To apologize for having nothing new to say about Fréchet Spaces,

WHAT I HAVE DECIDED TO DO

Think about 25 years of studying Fréchet spaces,
Reminisce on some of the things I and others thought about,
Share with you some philosophical points of view that guided some of our research,
Begin to suggest a perspective in which to place all of these results.

1 Grothendièck's philosophy in creating nuclear Fréchet spaces

A. Properties of function spaces from abstract considerations. The function spaces in question included spaces of analytic functions, C^∞ functions, distributions and test functions, solution spaces of certain linear partial differential equations, solutions of systems of infinitely many linear equations in infinitely many unknowns.

Investigations of these objects have been carried on by the school of D. Vogt and profound contributions have been made by him and other mathematicians such as R. Meise, B. Taylor, and others.

T. Terzioğlu (ed.), Advances in the Theory of Fréchet Spaces, 1–10.

B. Approximation of an infinite dimensional space by its finite dimensional subspaces.

This includes bases, finite dimensional decompositions and the bounded approximation property. Questions that have been studied include the following.

1. Does a space necessarily have such a structure?
2. To what extent is the existence or absence of such a structure related to subspaces, quotient spaces, and complemented subspaces?

In this lecture I will recall some definitions, discuss the totally negative answer to the first question, and indicate the richness of investigations that have been made into the second.

2 Definitions

We refer to [14] for basic definitions about nuclear Fréchet spaces. A *basis* in a nuclear Fréchet space E is a sequence $(x_n)_n$ in E such that for each $x \in E$ there is a unique sequence $(t_n)_n$ of scalars such that

$$x = \sum_{n=1}^{\infty} t_n x_n$$

A *finite dimensional decomposition* (*FDD*) in E is a sequence $(A_n)_n$ of continuous linear operators $A_n : E \to E$ such that $\dim(A_n(E)) < \infty$, $A_n A_m = \delta_{mn} A_n$ (here δ_{mn} is the Kronecker symbol) and

$$x = \sum_{n=1}^{\infty} A_n x, \quad x \in E$$

A *k-FDD*, $k = 1, 2, \ldots$ is an *FDD* in which $(A_n)_n$ can be chosen such that the range of each operator has dimension $\leq k$. A *strong FDD* is an *FDD* which is a *k-FDD* for some k.

We say that E has the *bounded approximation property* (*BAP*) if there is a sequence $(A_n)_n$ of continuous linear operators $A_n : E \to E$ such that for each n $\dim(A_n(E)) < \infty$, and

$$x = \lim_{n \to \infty} A_n x, \quad x \in E$$

If E is a nuclear Fréchet space with a basis (x_n) and $(||\cdot||_k)$ is an increasing sequence of seminorms that defines the topology of E then we have an infinite matrix $a = (a_n^k)$ where

$$a_n^k = ||x_n||_k, \quad k, n = 1, 2, \ldots$$

We define the corresponding *Köthe space* $K(a)$ to be

$$K(a) = \{\xi = (\xi_n) : |\xi|_k = \sum_n |\xi_n| a_n^k < \infty, k = 1, 2, \ldots\}$$

with topology defined by the sequence of seminorms $(|\cdot|_k)$

We say that a Köthe space $K(a)$ is *regular* if

$$\frac{a_{n+1}^k}{a_{n+1}^{k+1}} \leq \frac{a_n^k}{a_n^{k+1}}$$

and we say that a basis is *regular* if there is a system of seminorms for which the corresponding Köthe space is regular. The Köthe space $K(a)$ is nuclear if and only if the seminorms can be chosen so that the matrix a satisfies

$$\sum_n \frac{a_n^k}{a_n^{k+1}} < \infty$$

An increasing sequence of positive numbers $\alpha = (\alpha_n)$ is called a *nuclear exponent sequence of infinite type* if

$$\sup_n \frac{\log n}{\alpha_n} < \infty$$

In this case we define *the infinite type power series space* $\Lambda_\infty(\alpha)$ to be the Köthe space $K(a)$ where

$$a_n^k = e^{k\alpha_n}$$

An increasing sequence of positive numbers $\alpha = (\alpha_n)$ is called a *nuclear exponent sequence of finite type* if

$$\lim_n \frac{\log n}{\alpha_n} = 0$$

In this case we define *the finite type power series space* $\Lambda_1(\alpha)$ to be the Köthe space $K(a)$ where

$$a_n^k = e^{-\frac{1}{k}\alpha_n}$$

We say that an exponent sequence (of finite or infinite type) is *stable* if

$$\sup_n \frac{\alpha_{2n}}{\alpha_n} < \infty$$

and *unstable* if

$$\lim_n \frac{\alpha_{n+1}}{\alpha_n} = \infty$$

Finally we have the *limit point condition* on an exponent sequence α which is that the set of finite limit points of the set

$$\{\frac{\alpha_m}{\alpha_n} : m, n = 1, 2, \ldots\}$$

is bounded.

We will denote by ω the space of all sequences of scalars. With the standard product topology, this is a nuclear Fréchet space.

3 Non-existence of finite dimensional approximations

In 1975 and 1976 B.S. Mitiagin, N. M. Zobin, and P. B. Djakov gave the first examples of nuclear Fréchet spaces without basis [17,3]. Their approach was to "paste" together infinitely many 2-dimensional subspaces. In each subspace they used Hilbert norms and considered the geometry of the ellipses formed by the unit ball. It was possible to twist these ellipses in such a way that in order to find a system of norms in which they were realigned (which would be possible if there was a basis), it was necessary to construct a transformation whose modulus of continuity varied with the subspace. This variation could be designed so as to have insufficient uniformity to be pasted together into a transformation that was continuous over the whole space.

In 1980, Moscatelli [13] gave a quite different kind of example. Back in 1967, I had observed that any Fréchet space with an unconditional basis was isomorphic to a countable product of spaces with continuous norm. This can be explained in a very rough analogy with the "lining up" of the ellipses in the Mitiagin-Zobin-Djakov approach. In any locally convex space, the kernel of a continuous seminorm is a closed subspace and the induced seminorm on the quotient by this kernel is a norm so that the completion of the quotient is a Banach space. If the same thing is done with a larger seminorm then one gets a second Banach space and the identity map induces a mapping between these two Banach spaces. It is the kernels of these linking maps that must "line up" in order for the space to have a representation as a countable product of spaces with continuous norm. Moscatelli constructs a fairly simple example in which the kernels are so badly "twisted" that this is impossible and the space can not have a basis.

In 1979, I constructed a nuclear Fréchet space which did not have *BAP*. A Fréchet space is *countably normed* if it has a fundamental sequence of seminorms for which the linking maps described above are one-to-one. It is immediate that countably normed implies the existence of a continuous norm so the question arises about the converse. If a Fréchet space has *BAP* and a continuous norm then, from a theorem of Pełczyński, it is a complemented subspace of a Fréchet space with basis. From this it follows easily that the original space is countably normed. Thus, one can obtain the counterexample by constructing a nuclear Fréchet space which has a continuous norm but is not countably normed. This is done with yet a different kind of twisting of the linking maps.

An interesting point arises in connection with the result of Pełczyński that was just mentioned. All of the examples of Mitiagin, Zobin and Djakov have *BAP*, they even possess *FDDs*. Hence they are complemented subspaces of Fréchet spaces with bases. Thus it is possible to have a complemented subspace of a Fréchet space with basis fail to have a basis, even if the subspace is nuclear. What is still an open question, since the construction of Pełczyński can lead to a non nuclear space, is whether a complemented subspace of a nuclear Fréchet space with basis must have a basis.

4 Subspaces, quotient spaces, complemented subspaces

4.1 Search for subspaces and quotients of a single space

For several years, a number of researchers engaged in a search for simple characterizations of all subspaces and quotients of a particular space. We were particularly interested in doing this when the space was a power series space of infinite or finite type.

I was able to obtain reasonable characterizations, for the most important power series spaces, of subspaces and quotient spaces with bases. The approach of D. Vogt and his school, using the (DN) and (Ω) conditions was much more powerful. It solved this problem without the assumption of a basis and obtained characterizations that easily reduced to mine when there was a basis. All of this work went on fairly simultaneously.

4.2 The method of sup-norms on a Köthe space

Consider a Köthe space $K(a)$ and recall the definition

$$K(a) = \{\xi = (\xi_n) : |\xi|_k = \sum_n |\xi_n| a_n^k < \infty, k = 1, 2, \ldots\}$$

If the space is nuclear, then an equivalent definition is

$$K(a) = \{\xi = (\xi_n) : ||\xi||_k = \sup_n |\xi_n| a_n^k < \infty, k = 1, 2, \ldots\}$$

Moreover, again because of the nuclearity it follows that for each k, ξ,

$$||\xi||_k = a_q^k |\xi_q|$$

where $q = q^k(\xi)$ depends on k and ξ.

If (x_n) is a basic sequence in $K(a)$, that is, a basis for the closed subspace it generates, then we can write,

$$x_n = \sum_i t_i^n e_i$$

where the t_i^n are scalars and the e_i are the "coordinate sequences", that is e_i is 0 except for its ith term which is 1. We can then write

$$||x_n||_k = |t^n_{q_n^k}| a^k_{q_n^k}$$

and study the behavior of q_n^k as k, n vary.

The first important result has been called the *fundamental inequality* (see [7] for the proofs of all these) which says that for all n, k, j,

$$\frac{a^j_{q_n^k}}{a^k_{q_n^k}} \leq \frac{||x_n||_j}{||x_n||_k} \leq \frac{a^j_{q_n^j}}{a^k_{q_n^j}}$$

and

$$q_n^k \leq q_n^{k+1}$$

These properties hold whenever (x_n) is a basic sequence in a regular nuclear Köthe space $K(a)$. Conversely, any given set of values which satisfy these conditions can be realized as the values of $(||x_n||_k)$ for an appropriate sequence (x_n). Everything would be extremely simple except for the fact that it is not easy to choose (x_n) so that it is a basic sequence.

4.3 Power series subspaces of power series subspaces

One of the earliest results in this structure theory is the construction of S. Rolewicz [15] who gave many examples of $\Lambda_\infty(\beta)$ subspaces of $\Lambda_1(\alpha)$ in the case when α is stable. On the other hand, V. P. Zahariuta showed [16] that there is no $\Lambda_1(\beta)$ subspace of any $\Lambda_\infty(\alpha)$ space.

The situation is quite different if α is unstable [4]. In this case, there are no infinite type power series subspaces of $\Lambda_1(\alpha)$ and a finite type power series space $\Lambda_1(\beta)$ is isomorphic to a subspace of $\Lambda_1(\alpha)$ if and only if there is a subsequence γ of α such that the ratio $\frac{\beta_n}{\gamma_n}$

is bounded above and bounded away from 0. This latter condition is equivalent to $\bigwedge_1(\beta)$ being isomorphic to a complemented subspace of $\bigwedge_1(\alpha)$.

The proof of the second statement is fairly simple. One direction is obvious and if $\bigwedge_1(\beta)$ is isomorphic to a subspace of $\bigwedge_1(\alpha)$, then we can consider the basic sequence (x_n) in $\bigwedge_1(\alpha)$ corresponding to the coordinate basis in $\bigwedge_1(\beta)$ and analyze the corresponding set of indices (q_n^k). It follows in a straightforward manner that , for each n, q_n^k is essentially independent of k.

4.4 Subspaces and quotients of arbitrary nuclear Fréchet spaces

There is quite a bit known about this, but still a number of open questions. In the following list, the statements in roman type are known and a reference is given. The statements in italics are open questions.

Every nuclear Fréchet space not isomorphic to ω has

a subspace with a basis [2].

a quotient with a basis and not isomorphic to ω [2] .

a complemented subspace with basis?

a complemented subspace?

a subspace without a basis [6].

a quotient without a basis [9].

a subspace without strong FDD [1].

a quotient without strong FDD [1].

a subspace without FDD?

a quotient without FDD [1].

a subspace without BAP?

a quotient without BAP [10] .

We can make some comments on these. One can show that a nuclear Fréchet space with a 2-FDD has a complemented subspace with basis [8] . This seems to be the most general result known on this question.

Regarding the question of existence of basis in any complemented subspace, the known results along with some new ones are presented by J. Krone in another paper given at this conference.

The existence of a subspace without basis in an arbitrary nuclear Fréchet space not isomorphic to ω is an interesting synthesis of the construction of Djakov/Mitiagin and the general constructions of the structure theory. Using the latter, it was not very difficult to simply embed the Djakov/Mitiagin construction in a given space.

Regarding the existence of bases in complemented supspaces of spaces of a nuclear Fréchet space with basis, the most general results are the following.

THEOREM. In a finite type power series space, every complemented subspace has a basis. [12]

THEOREM. In an infinite type power series space in which the exponent sequence satisfies the limit point condition, every complemented subspace has a basis. [11]

The question for general infinite type power series spaces remains open.

4.5 Special subspaces

All of the results in this section are contained in [5] . They are not, perhaps, the most important facts in the structure theory, but they are my personal favorites.

The general question that was investigated here concerns the possiblity of finding a small class of subspaces from which all spaces are made up in some reasonable sense. If that is not possible, then perhaps it is at least possible to find a small class that is found everywhere. The results show that if "made up" involves subspaces, and if the small class is restricted to power series spaces, or to some countable collection of spaces, then the answer is no. The last result shows, however, that the various nuclear Fréchet spaces are not completely different from each other.

THEOREM. Every nuclear Fréchet space has a subspace which has no power series subspaces.

THEOREM. Given any countable collection of nuclear Fréchet spaces, then every nuclear Fréchet space not isomporphic to ω has a subspace which has no subspaces isomorphic to a space in the collection.

THEOREM. given any two nuclear Fréchet spaces not isomporphic to ω, there is a nuclear Fréchet space which is isomorphic to a subspace of both.

5 Open Questions

It is perhaps reasonable to close this personal retrospective with a brief statement of the status of the "classical" problems of the structure theory of nuclear Fréchet spaces. The question of existence of bases and approximation properties seems to be fairly well settled and perhaps the only important remaining one is the following which is important for the question of subspaces with BAP.

Does every countably normed nuclear Fréchet space have BAP?

On the other hand, the following two questions have been open and the subject of (at most partially successful) investigation for many years.

BESSAGA'S CONJECTURE. Is a complemented subspace of a nuclear Fréchet space with basis necessarily isomorphic to the subspace generated by a subsequence of the basis?

QUASIEQUIVALENCE. Given two bases in a nuclear Fréchet space, are they necessarily quasiequivalent in the sense that there is a permutation of one of them and sequence of scalars by which each may be multiplied so that the resulting sequence is carried onto the other basis by some isomorphism of the space into itself?

REFERENCES

1. C. Bessaga and Ed Dubinsky, *Nuclear Fréchet spaces without bases*, **Arch. der Math. 31**, 6 (1978), 597-604.
2. C. Bessaga and A. Pełczyński, *Własności baz w przestrzeniach typu B*, **Prace. Mat. III** (1959), 123-142.
3. P.B. Djakov and B.S. Mitiagin, *Modified construction of a nuclear Fréchet space without basis*, **J. Funct. Anal. 23** 4 (1976), 415-423.
4. Ed Dubinsky, *Infinite type power series subspaces of finite type power series spaces*, **Israel J. Math. 15** (1973), 257-281.
5. Ed Dubinsky, *Concrete subspaces of nuclear Fréchet spaces*, **Stud. Math. 52** (1975), 209-219.
6. Ed Dubinsky, *Subspaces without bases in nuclear Fréchet spaces*, **J. Funct. Anal. 26**, 2 (1977), 121-130.

7. Ed Dubinsky, *The structure of nuclear Fréchet spaces*, **LNIM** 720, Berlin:Springer-Verlag, (1979).

8. Ed Dubinsky, *Complemented basic sequences in nuclear Fréchet spaces with finite dimensional decomposition*, **Arch. der Math. 38**, 2 (1982), 138-150.

9. Ed Dubinsky and B.S. Mitiagin, *Quotient spaces without bases in nuclear Fréchet spaces*, **Can. J. Math. 30**, 6 (1978), 1296-1305.

10. Ed Dubinsky and Dietmar Vogt, *Fréchet spaces with quotients failing the bounded approximation property*, **Stud. Math. 81** (1985), 71-77.

11. Ed Dubinsky and Dietmar Vogt, *Bases in complemented subspaces of power series spaces*, **Bull. Polon. Acad. Sci. Math. 34**, 1-2, (1986), 65-67.

12. B.S. Mitiagin, *Equivalence of bases in Hilbert scales* (Russian), **Stud. Math. 37** (1971), 111-137.

13. V. B. Moscatelli, *Fréchet spaces without continuous norms and without bases*, **Bull. London Math. Soc. 12** (1980), 63-66.

14. A. Pietsch, **Nukleare lokalkonvexe Räume**, Berlin (1965).

15. S. Rolewicz, *On spaces of holomorphic functions*, **Studia Math. 21** (1961), 135-160.

16. V. P. Zahariuta, *On the isomorphism of Cartesian products of locally convex spaces*, **Studia Math. 46** (1973), 201-221.

17. N.M. Zobin and B.S. Mitiagin, *Examples of nuclear linear metric spaces without a basis*, **Funct. Anal. and Appl. 8**, 4 (1975), 304-313.

TOPICS ON PROJECTIVE SPECTRA OF (LB)–SPACES

DIETMAR VOGT
Fachbereich Mathematik
Bergische Universität–GH Wuppertal
Gaußstr. 20
D–5600 Wuppertal
Fed. Rep. of Germany

Abstract. The paper first gives a new introduction to Palamodov's theory of the projective limit functor avoiding categorical and abstract homological concepts. Then Retakh's condition for $\mathrm{Proj}^1 \mathcal{X} = 0$ for a spectrum $\mathcal{X}$ of (LB)–spaces is discussed. Conditions are derived which are accessible for evaluation. In §3 these conditions are connected to certain topological properties of the projective limit and finally the case of sequence spaces is presented, where we have a complete characterization in terms of the defining matrices.

Introduction.

The present paper, based on the author's lectures during the conference and on [13], first gives a brief introduction into the theory of the projective limit functor as developed by Palamodov in [9] and [10]. We do not use tools of abstract homological algebra and category theory, however define directly and explicitly Proj^0 and Proj^1 and obtain in a very natural way Palamodov's exact sequence of six spaces ([9, p. 542]). Then we turn to projective limits of regular (LB)–spaces. We present Retakh's result ([11, Theorem 3]) and derive from it conditions (P_1^*) and (P_2^*), which are estimates in terms of the dual norms of a given setting and can be used (see [1], [2], [3]) for the solution of solvability problems in analysis.

The §3 essentially gives a survey on results of [13] which connect the conditions for $\mathrm{Proj}^1 \mathcal{X} = 0$ to topological properties of the projective limit as barrelledness, reflexivity, etc.. In §4 we present a result of [13], which extends results of [7], and gives in the case of sequence spaces a necessary and sufficient condition for $\mathrm{Proj}^1 \mathcal{X} = 0$ and for a lot of topological properties of the projective limit and its dual, which in this case are equivalent. This in particular contains results of Grothendieck on topological properties of (F)– and (DF)–spaces ([4, II, §4]).

Finally we discuss how far all this depends on the given spectrum and what exactly has to be shown in the typical application (as eg. [1]).

T. Terzioğlu (ed.), Advances in the Theory of Fréchet Spaces, 11–27.

1. Projective spectra of linear spaces

In the following projective or injective spectrum will always mean countable projective or injective spectrum of linear spaces.

A projective spectrum is a sequence $\mathcal{X} = (X_n, \iota^n_{n+1})$ of linear spaces X_n and linear maps $\iota^n_{n+1} : X_{n+1} \to X_n$. We put $\iota^m_m = \mathrm{id}_{X_m}$ and $\iota^n_m = \iota^n_{n+1} \circ \cdots \circ \iota^{m-1}_m$ for all $m, n < m$.

If $\mathcal{X} = (X_n, \iota^n_{n+1})$ and $\mathcal{Y} = (Y_n, \iota^n_{n+1})$ are projective spectra, then a map $\Phi : \mathcal{X} \to \mathcal{Y}$ is a sequence $\varphi^n_{k(n)} : X_{k(n)} \to Y_n$ of linear maps, where $k(n) \leq k(n+1)$ and $\varphi^n_{k(n)} \circ \iota^{k(n)}_{k(n+1)} = \iota^n_{n+1} \circ \varphi^{n+1}_{k(n+1)}$ for all n. We put $\varphi^n_m = \varphi^n_{k(n)} \circ \iota^{k(n)}_m$ for all $k(n) \leq m$ and obtain linear maps $X_m \to Y_n$ satisfying $\varphi^n_m \circ \iota^m_M = \iota^n_\mu \circ \varphi^\mu_M$ whenever all maps are defined.

Two maps $\Phi = \left(\varphi^n_{k(n)}\right)$ and $\tilde{\Phi} = \left(\tilde{\varphi}^n_{l(n)}\right)$ are called equivalent $\left(\Phi \sim \tilde{\Phi}\right)$ if for every n there is $m(n) \geq \max(k(n), l(n))$ such that $\varphi^n_{m(n)} = \tilde{\varphi}^n_{m(n)}$ for all n. This defines an equivalence relation.

For two maps $\Phi : \mathcal{X} \to \mathcal{Y}, \Psi : \mathcal{Y} \to \mathcal{Z}$ the composition $\Psi \circ \Phi : \mathcal{X} \to \mathcal{Z}$ is defined by $\left(\psi^n_{l(n)} \circ \varphi^{l(n)}_{k(l(n))}\right)_n$, where $\Phi = \left(\varphi^n_{k(n)}\right)_n$, $\Psi = \left(\psi^n_{l(n)}\right)_n$. Composition respects equivalence. A map $\Phi : \mathcal{X} \to \mathcal{Y}$ is called an equivalence map if there is a map $\Phi^{-1} : \mathcal{Y} \to \mathcal{X}$ such that $\Phi^{-1} \circ \Phi \sim \mathrm{id}_{\mathcal{X}}$, $\Phi \circ \Phi^{-1} \sim \mathrm{id}_{\mathcal{Y}}$. $\mathcal{X}$ and $\mathcal{Y}$ are called equivalent if there is an equivalence map $\mathcal{X} \to \mathcal{Y}$ (resp. $\mathcal{Y} \to \mathcal{X}$).

Example. Let $I = \left(\iota^n_{k(n)}\right)_n$, then $I \sim \mathrm{id}$. Therefore $\left(X_{k(n)}, \iota^{k(n)}_{k(n+1)}\right)_n \sim \mathcal{X}$. $\Phi \sim \tilde{\Phi}$ if and only if there exists $\tilde{I}$ such that $\Phi \circ I = \tilde{\Phi} \circ \tilde{I}$.

Definition. For $\mathcal{X} = (X_n, \iota^n_{n+1})_n$ we set

$$\mathrm{Proj}^0\, \mathcal{X} = \left\{(x_n)_n \in \prod_n X_n : \iota^n_{n+1} x_{n+1} = x_n \text{ for all } n\right\}$$

$$\mathrm{Proj}^1\, \mathcal{X} = \left(\prod_n X_n\right) / \mathrm{B}(\mathcal{X})$$

where

$$\mathrm{B}(\mathcal{X}) = \Big\{(a_n)_n \in \prod_n X_n : \text{there exists } (b_n)_n \in \prod_n X_n, \text{ such that } a_n = \iota^n_{n+1} b_{n+1} - b_n \text{ for all } n\Big\}.$$

There is a natural exact sequence

$$0 \to \mathrm{Proj}^0\, \mathcal{X} \hookrightarrow \prod_n X_n \xrightarrow{\sigma} \prod_n X_n \xrightarrow{q} \mathrm{Proj}^1\, \mathcal{X} \to 0$$

where $\sigma : (x_n)_n \mapsto \left(\iota^n_{n+1} x_{n+1} - x_n\right)_n$ and q is the quotient map.

For $\Phi : \mathcal{X} \to \mathcal{Y}$, $\Phi = \left(\varphi^n_{k(n)}\right)_n$ we define

$$^\pi\Phi\, x = \left(\varphi^n_{k(n)} x_{k(n)}\right)_n$$

$$\Phi^\pi x = \left(\sum_{m=k(n)}^{k(n+1)-1} \varphi^n_m x_m\right)_n$$

These are linear maps from $\prod_n X_n$ to $\prod_n Y_n$, satisfying $\Phi^\pi \circ \sigma = \sigma \circ {}^\pi\Phi$. This means we have the following commutative diagram:

$$\begin{array}{ccccccccc} 0 \to & \mathrm{Proj}^0 \mathcal{X} & \to & \prod_n X_n & \xrightarrow{\sigma} & \prod_n X_n & \to & \mathrm{Proj}^1 \mathcal{X} & \to 0 \\ & \Phi^0 \downarrow & & {}^\pi\Phi \downarrow & & \downarrow \Phi^\pi & & \downarrow \Phi^1 & \\ 0 \to & \mathrm{Proj}^0 \mathcal{Y} & \to & \prod_n Y_n & \xrightarrow{\sigma} & \prod_n Y_n & \to & \mathrm{Proj}^1 \mathcal{Y} & \to 0 \end{array}$$

where Φ^0 and Φ^1 are the maps induced by ${}^\pi\Phi$ and Φ^π. We used the following definition:

Definition. (1) For $x = (x_n)_n \in \mathrm{Proj}^0 \mathcal{X}$ we set

$$\Phi^0 x = \left(\varphi^n_{k(n)} x_{k(n)}\right)_n$$

(2) For $a = (a_n)_n + \mathrm{B}(\mathcal{X}) \in \mathrm{Proj}^1 \mathcal{X}$ we set

$$\Phi^1 a = \left(\sum_{m=k(n)}^{k(n+1)-1} \varphi^n_m a_m\right)_n + \mathrm{B}(\mathcal{Y}).$$

We have the following:

1.1 Proposition. *Let $\Phi, \tilde{\Phi} : \mathcal{X} \to \mathcal{Y}$, $\Psi : \mathcal{Y} \to \mathcal{Z}$ be maps, $I : \mathcal{X} \to \mathcal{X}$ as above. Then*
(1) $(\Psi \circ \Phi)^0 = \Psi^0 \circ \Phi^0$, $(\Psi \circ \Phi)^1 = \Psi^1 \circ \Phi^1$
(2) $I^0 = \mathrm{id}_{\mathrm{Proj}^0 \mathcal{X}}$, $I^1 = \mathrm{id}_{\mathrm{Proj}^1 \mathcal{X}}$
(3) *if $\Phi \sim \tilde{\Phi}$ then $\Phi^0 = \tilde{\Phi}^0$, $\Phi^1 = \tilde{\Phi}^1$.*

The Proof of (1) is obvious. (2) requires some calculation. (3) is an immediate consequence of (1) and (2) since $\Phi \sim \tilde{\Phi}$ means the existence of I and $\tilde{I}$ such that $\Phi \circ I = \tilde{\Phi} \circ \tilde{I}$.

1.2 Corollary. *If $\mathcal{X} \sim \mathcal{Y}$ and $\Phi : \mathcal{X} \to \mathcal{Y}$ is an equivalence map then*

$$\Phi^0 : \mathrm{Proj}^0 \mathcal{X} \to \mathrm{Proj}^0 \mathcal{Y},\ \Phi^1 : \mathrm{Proj}^1 \mathcal{X} \to \mathrm{Proj}^1 \mathcal{Y}$$

are isomorphisms.

Let $\Phi : \mathcal{X} \to \mathcal{Y}$, $\Psi : \mathcal{Y} \to \mathcal{Z}$ be maps, $\Phi = \left(\varphi^n_{k(n)}\right)_n$, $\Psi = \left(\psi^n_{l(n)}\right)_n$.

Definition. $\mathcal{X} \xrightarrow{\Phi} \mathcal{Y} \xrightarrow{\Psi} \mathcal{Z}$ is called exact in $\mathcal{Y}$ if

(1) $\Psi \circ \Phi \sim 0$
(2) for every $n, N \geq k(n)$ there are $\mu, m \geq \max(n, l(\mu))$ such that $\mathrm{im}\, \varphi^n_N \supset \iota^n_m \ker \psi^\mu_m$.

It is not difficult to see that this is invariant under equivalence. More precise:

Remark. If

$$\begin{array}{ccccc} \mathcal{X} & \to & \mathcal{Y} & \to & \mathcal{Z} \\ \updownarrow\sim & & \updownarrow\sim & & \updownarrow\sim \\ \tilde{\mathcal{X}} & \to & \tilde{\mathcal{Y}} & \to & \tilde{\mathcal{Z}} \end{array}$$

is a diagram, commutative up to equivalence, and one row is exact, then also the other.

The definition of exactness of $\mathcal{X} \to \mathcal{Y} \to \mathcal{Z}$ means that by just taking subsequences we can find equivalent spectra such that (using the same notation after the change) $\Phi = (\varphi_n^n)_n$, $\Psi = (\psi_n^n)_n$ and

$$\iota_{n+1}^n \ker \psi_{n+1}^{n+1} \subset \operatorname{im} \varphi_n^n \subset \ker \psi_n^n$$

for all n.

1.3 Proposition. *(1) If $0 \to \mathcal{X} \to \mathcal{Y} \to \mathcal{Z}$ is exact, then*

$$0 \to \operatorname{Proj}^0 \mathcal{X} \to \operatorname{Proj}^0 \mathcal{Y} \to \operatorname{Proj}^0 \mathcal{Z}$$

is exact.
(2) If $\mathcal{X} \to \mathcal{Y} \to \mathcal{Z} \to 0$ is exact, then

$$\operatorname{Proj}^1 \mathcal{X} \to \operatorname{Proj}^1 \mathcal{Y} \to \operatorname{Proj}^1 \mathcal{Z} \to 0$$

is exact.

Let $0 \to \mathcal{X} \xrightarrow{\Phi} \mathcal{Y} \xrightarrow{\Psi} \mathcal{Z} \to 0$ be a short exact sequence of spectra. Then by going to equivalent spectra by taking subsequences we first may assume the standardization described before Proposition 1.3. From there it is again easy to see, that the spectra $(\ker \psi_n^n, \iota_{n+1}^n)_n$, $(\operatorname{im} \psi_n^n, \iota_{n+1}^n)_n$ are equivalent to $\mathcal{X}$ (resp. $\mathcal{Z}$). Hence we obtained spectra $\tilde{\mathcal{X}}$, $\tilde{\mathcal{Y}}$, $\tilde{\mathcal{Z}}$ and maps $\tilde{\Phi}$, $\tilde{\Psi}$ such that

$$\text{(ST)} \qquad \begin{array}{ccccccc} 0 \to & \mathcal{X} & \xrightarrow{\Phi} & \mathcal{Y} & \xrightarrow{\Psi} & \mathcal{Z} & \to 0 \\ & \sim\downarrow \Phi_{\mathcal{X}} & & \sim\downarrow \Phi_{\mathcal{Y}} & & \sim\downarrow \Phi_{\mathcal{Z}} & \\ 0 \to & \tilde{\mathcal{X}} & \xrightarrow{\tilde{\Phi}} & \tilde{\mathcal{Y}} & \xrightarrow{\tilde{\Psi}} & \tilde{\mathcal{Z}} & \to 0 \end{array}$$

commutes up to equivalence and the lower sequence is of the form

$$0 \to \tilde{\mathcal{X}}_n \hookrightarrow \tilde{\mathcal{Y}}_n \to \tilde{\mathcal{Z}}_n \to 0\,,$$

where this sequence is exact, for all n.

A short exact sequence of spectra with these properties is called *exact sequence in standard form*. If we have the diagramm (ST) then $0 \to \tilde{\mathcal{X}} \to \tilde{\mathcal{Y}} \to \tilde{\mathcal{Z}} \to 0$ is called a *standard representation* of $0 \to \mathcal{X} \to \mathcal{Y} \to \mathcal{Z} \to 0$.

For an exact sequence $0 \to \mathcal{X} \xrightarrow{\Phi} \mathcal{Y} \xrightarrow{\Psi} \mathcal{Z} \to 0$ in standard form we have

$$\begin{aligned} \Phi^0 x &= (\varphi_n^n x_n)_n \\ \Phi^1 a &= (\varphi_n^n a_n)_n + \mathrm{B}(\mathcal{Y}) \end{aligned}$$

and likewise for Ψ.

In this case we define for $z = (z_n)_n \in \operatorname{Proj}^0 \mathcal{Z}$

$$\delta^* z = \left(\iota^n_{n+1} y_{n+1} - y_n\right)_n + \mathrm{B}(\mathcal{X})$$

where $y_n \in Y_n$ is chosen such that $\psi^n_n y_n = z_n$. Note that, because of

$$\psi^n_n \left(\iota^n_{n+1} y_{n+1} - y_n\right) = \iota^n_{n+1} z_{n+1} - z_n = 0,$$

we have $\iota^n_{n+1} y_{n+1} - y_n \in X_n$ for all n. δ^* is a well defined linear map from $\operatorname{Proj}^0 \mathcal{Z}$ to $\operatorname{Proj}^1 \mathcal{X}$.

1.4 Lemma. *If*

$$\begin{array}{ccccccc} 0 \to & \mathcal{X}^1 & \to & \mathcal{Y}^1 & \to & \mathcal{Z}^1 & \to 0 \\ & \downarrow \Phi_{\mathcal{X}} & & \downarrow \Phi_{\mathcal{Y}} & & \downarrow \Phi_{\mathcal{Z}} & \\ 0 \to & \mathcal{X}^2 & \to & \mathcal{Y}^2 & \to & \mathcal{Z}^2 & \to 0 \end{array}$$

is a commutative (up to equivalence) diagram with exact rows in standard form then

$$\begin{array}{ccc} \operatorname{Proj}^0 \mathcal{Z}^1 & \xrightarrow{\delta^*} & \operatorname{Proj}^1 \mathcal{X}^1 \\ \downarrow \Phi^0_{\mathcal{Z}} & & \downarrow \Phi^1_{\mathcal{X}} \\ \operatorname{Proj}^0 \mathcal{Z}^2 & \xrightarrow{\delta^*} & \operatorname{Proj}^1 \mathcal{X}^2 \end{array}$$

commutes.

From this and Proposition 1.1 (3) we conclude that the following definition is unique.

Definition. For any exact sequence we define $\delta^* : \operatorname{Proj}^0 \mathcal{Z} \to \operatorname{Proj}^1 \mathcal{X}$ by $\delta^* = {\Phi^1_{\mathcal{X}}}^{-1} \circ \delta^* \circ \Phi^0_{\mathcal{Z}}$, where we used a standard representation and (ST).

We obtain and admit without proof (Palamodov [9, p. 542]).

1.5 Theorem. *For every exact sequence*

$$0 \to \mathcal{X} \xrightarrow{\Phi} \mathcal{Y} \xrightarrow{\Psi} \mathcal{Z} \to 0$$

of spectra the sequence

$$0 \to \operatorname{Proj}^0 \mathcal{X} \xrightarrow{\Phi^0} \operatorname{Proj}^0 \mathcal{Y} \xrightarrow{\Psi^0} \operatorname{Proj}^0 \mathcal{Z} \xrightarrow{\delta^*} \operatorname{Proj}^1 \mathcal{X} \xrightarrow{\Phi^1} \operatorname{Proj}^1 \mathcal{Y} \xrightarrow{\Psi^1} \operatorname{Proj}^1 \mathcal{Z} \to 0$$

is exact. All maps in this exact sequence depend only on the equivalence class of Φ and Ψ. Moreover this exact sequence depends functorially on the exact sequence of spectra.

The last sentence means the following: If

$$\begin{array}{ccccccc} 0 \to & \mathcal{X}^1 & \to & \mathcal{Y}^1 & \to & \mathcal{Z}^1 & \to 0 \\ & \downarrow \Phi_{\mathcal{X}} & & \downarrow \Phi_{\mathcal{Y}} & & \downarrow \Phi_{\mathcal{Z}} & \\ 0 \to & \mathcal{X}^2 & \to & \mathcal{Y}^2 & \to & \mathcal{Z}^2 & \to 0 \end{array}$$

is a commutative (up to equivalence) diagram with exact rows, then

$$\begin{array}{ccccccccccccc} 0 \to & \mathrm{Proj}^0\,\mathcal{X}^1 & \to & \mathrm{Proj}^0\,\mathcal{Y}^1 & \to & \mathrm{Proj}^0\,\mathcal{Z}^1 & \xrightarrow{\delta^*} & \mathrm{Proj}^1\,\mathcal{X}^1 & \to & \mathrm{Proj}^1\,\mathcal{Y}^1 & \to & \mathrm{Proj}^1\,\mathcal{Z}^1 & \to 0 \\ & \downarrow & & \downarrow & & \downarrow & & \downarrow & & \downarrow & & \downarrow & \\ 0 \to & \mathrm{Proj}^0\,\mathcal{X}^2 & \to & \mathrm{Proj}^0\,\mathcal{Y}^2 & \to & \mathrm{Proj}^0\,\mathcal{Z}^2 & \xrightarrow{\delta^*} & \mathrm{Proj}^1\,\mathcal{X}^2 & \to & \mathrm{Proj}^1\,\mathcal{Y}^2 & \to & \mathrm{Proj}^1\,\mathcal{Z}^2 & \to 0 \end{array}$$

is a commutative diagram with exact rows.

2. Projective spectra of LB–spaces

In this section we assume $\mathcal{X} = (X_n, \iota^n_{n+1})$ to be a spectrum of (LB)–spaces, i.e. every X_n has the form $X_n = \bigcup_k X_{n,k}$, where $X_{n,k}$ is a Banach space with a norm $\|\ \|_{n,k}$ and X_n carries the locally convex inductive limit topology of the $X_{n,k}$. ι^n_{n+1} is assumed to be continuous.

Hence we may assume without loss of generality that $X_{n,k} \subset X_{n,k+1}$ and

$$\| x \|_{n,k} \geq \| x \|_{n,k+1}$$

$$\| \iota^n_{n+1} x \|_{n,k} \leq \| x \|_{n+1,k}$$

for all n, k and $x \in X_{n,k}$ (resp. $x \in X_{n,k+1}$). We put $B_{n,k} = \{x \in X_{n,k} : \| x \|_{n,k} \leq 1\}$ and assume that the $B_{n,k}$, $k = 1, 2, \ldots$ are a fundamental system of bounded sets in X_n. By X (resp. Y, Z) we denote always $\mathrm{Proj}^0\,\mathcal{X}$ (resp. $\mathrm{Proj}^0\,\mathcal{Y}$, $\mathrm{Proj}^0\,\mathcal{Z}$) equipped with the projective topology and by $\iota^n : X \to X_n$ the canonical map.

In this case a necessary and sufficient condition for $\mathrm{Proj}^0\,\mathcal{X} = 0$ has been given by V.S. Retakh [11, Theorem 3].

2.1 Theorem (Retakh). $\mathrm{Proj}^1\,\mathcal{X} = 0$ *if and only if the following holds: For every μ the space X_μ contains a bounded Banach ball B_μ such that*
(1) $\iota^\mu_{\mu+1} B_{\mu+1} \subset B_\mu$ for all μ
(2) for every μ there is $k \geq \mu$ such that $\iota^\mu_k(X_k) \subset \iota^\mu X + B_\mu$.

Remark. Then for every $\varepsilon > 0$ we have even

$$\iota^\mu_k(X_k) \subset \iota^\mu X + \varepsilon B_\mu.$$

We want to put this into a form that makes it ready for evaluation in concrete cases and for closer investigation in general.

2.2 Lemma. *In Theorem 2.1 condition (2) can be replaced by*

(2)′ for every μ there is $k \geq \mu$ such that for every $K \geq k$ and $\varepsilon > 0$

$$\iota^\mu_k(X_k) \subset \iota^\mu_K X_K + \varepsilon B_\mu.$$

PROOF. Since (2)⇒(2)′ is clear we have to prove the converse . Given (1), (2)′ we obtain a sequence $k(m)$, $k(0)=\mu$ such that for every m we have

$$\iota^{k(m)}_{k(m+1)}X_{k(m+1)} \subset \iota^{k(m)}_{k(m+2)}X_{k(m+2)} + 2^{-m}B_{k(m)}.$$

We proceed inductively starting with $x = x_1 \in X_{k(1)}$. Let $X_{k(m+1)}$ be chosen. We find $x_{k(m+2)} \in X_{k(m+2)}$ such that

$$\iota^{\nu}_{k(m+1)}x_{k(m+1)} - \iota^{\nu}_{k(m+2)}x_{k(m+2)} \subset 2^{-m}B_\nu$$

for $1 \le \nu \le k(m)$.

Then $\lim\limits_{m\to\infty} \iota^{\nu}_{k(m)}x_{k(m)} =: \xi_\nu$ exists for all ν and clearly $\xi = (\xi_\nu)_\nu \in \mathrm{Proj}^0\,\mathcal{X}$. We have with $k = k(1)$

$$\begin{aligned}\| \iota^{\nu}_{k}x - \iota^{\nu}\xi \|_\nu &\le \sum_{m=1}^{\infty} \| \iota^{\nu}_{k(m)}x_{k(m)} - \iota^{\nu}_{k(m+1)}x_{k(m+1)} \|_\nu \\ &\le \sum_{m=1}^{\infty} \| x_{k(m)} - \iota^{k(m)}_{k(m+1)}x_{k(m+1)} \|_{k(m)} \\ &\le 1\end{aligned}$$

where $\| \;\; \|_\nu$ denotes the Minkowski norm of B_ν. This proves the assertion.

2.3 Lemma. *Condition (2)′ in Lemma 2.2 implies*

(2)″ for every μ there is $k \ge \mu$ such that for every $K \ge k$ and m there are N and S such that

$$\iota^{\mu}_{k}B_{K,m} \subset S\,(\iota^{\mu}_{K}B_{k,N} + B_\mu)\ .$$

PROOF. This follows from Grothendieck's factorization theorem [4, Théorème A, p.16] and 2.2.

2.4 Lemma. *Condition (2)′ in Lemma 2.2 is implied by*

(2)‴ for every μ there is $k \ge \mu$ such that for every $K \ge k, m$ and ε there are N and S such that

$$\iota^{\mu}_{k}B_{k,m} \subset S\iota^{\mu}_{K}B_{K,N} + \varepsilon B_\mu\ .$$

The proof is obvious. For the interpretation of this condition see [13, Theorem 4.9], which is a dual version of [11, Theorem 2].

For the following Lemma cf. the Remark after Theorem 2 in [11]. We use the following notation.

Definition. $\mathcal{X}$ is called a (DFS)–spectrum if for every k and m there exists M such that the inclusion $X_{k,m} \hookrightarrow X_{k,M}$ is compact.

2.5 Lemma. *If $\mathcal{X}$ is a (DFS)-spectrum then (2) implies (2)‴.*

PROOF. Since (2) implies also (2)″ we may assume with suitable quantifiers

(i) $\iota_k^\mu X_k \subset \iota_K^\mu X_K + \frac{\varepsilon}{2} B_\mu$

(ii) $\iota_k^\mu B_{k,M} \subset S\,(\iota_K^\mu B_{K,N} + B_\mu)$

(iii) $X_{k,m} \hookrightarrow X_{k,M}$ compact.

We put

$$E = \iota_K^\mu X_{K,N} + \operatorname{span} B_\mu .$$

This is in a natural way a Banach space. Because of (ii), (iii) $\iota_k^\mu B_{k,m}$ is relatively compact in E. We consider the sets

$$U_L = \{\iota_K^\mu x + y \ : \| x \|_{K,L} < L,\ \| y \|_\mu < \varepsilon\}$$

for $L \geq N$. Because of (i) they are an open covering of the compact set

$$\overline{\iota_k^\mu B_{k,m}} = \iota_k^\mu \overline{B_{k,m}}^{\| \|_{k,M}} .$$

This proves the assertion.

2.6 Proposition. *If $\mathcal{X}$ is a (DFS)-spectrum then* $\operatorname{Proj}^1 \mathcal{X} = 0$ *if and only if the following holds: For every μ the space X_μ contains a bounded Banach ball B_μ such that*

(1) $\iota_{\mu+1}^\mu B_{\mu+1} \subset B_\mu$ *for all* μ

(2) $\forall \mu\ \exists k\ \forall K, m\ \exists N, S\ :\ \iota_k^\mu B_{k,m} \subset S\,(B_{K,N} + B_\mu)$

This condition is certainly fulfilled if there is n such that (2) holds with $B_\mu = B_{\mu,n}$. On the other hand (1) and (2) remain true if we replace B_μ by $\varepsilon_\mu B_\mu$, $\varepsilon_\mu \geq \varepsilon_{\mu+1}$ for all μ. Therefore we may assume $B_\mu \subset B_{\mu,n(\mu)}$ for all μ with suitable $n(\mu)$. We obtain the following conditions.

Definition.

(P_1) $\exists n\ \forall \mu\ \ \exists k\ \forall K, m\ \exists N, S\ :\ \iota_k^\mu B_{k,m} \subset S\,(B_{K,N} + B_{\mu,n})$

(P_2) $\forall \mu\ \exists n, k\ \forall K, m\ \exists N, S\ :\ \iota_k^\mu B_{k,m} \subset S\,(B_{K,N} + B_{\mu,n})$

And we obtained:

2.7 Theorem. *If $\mathcal{X}$ is a (DFS)-spectrum, then* $(P_1) \Rightarrow \operatorname{Proj}^1 \mathcal{X} = 0 \Rightarrow (P_2)$.

The advantage of these conditions is that by means of dualization they can be turned into inequalities.

To formulate that we use the following notation $\jmath_\mu^k : X_\mu' \to X_k'$ is the transpose of ι_k^μ for $\mu \le k$. For $y \in X_\mu'$ we set

$$\| y \|_{\mu,n}^* = \sup \{|y(x)| : \| x \|_{\mu,n} \le 1\}.$$

This is an extended real valued "norm". We have

$$\begin{aligned} \| y \|_{\mu,n}^* &\le \| y \|_{\mu,n+1}^* \\ \| y \|_{\mu,n}^* &\ge \| \jmath_\mu^{\mu+1} y \|_{\mu+1,n}^* \end{aligned}$$

for all μ, n and $y \in X_\mu'$.

Definition.

(P_1^*) $\exists n \, \forall \mu \, \exists k \, \forall K, m \, \exists N, S \, \forall y \in X_\mu' : \| \jmath_\mu^k y \|_{k,m}^* \le S \left(\| \jmath_\mu^K y \|_{K,N}^* + \| y \|_{\mu,n}^* \right)$

(P_2^*) $\forall \mu \, \exists n, k \, \forall K, m \, \exists N, S \, \forall y \in X_\mu' : \| \jmath_\mu^k y \|_{k,m}^* \le S \left(\| \jmath_\mu^K y \|_{K,N}^* + \| y \|_{\mu,n}^* \right)$

By means of classical duality theory, in particular the bipolar theorem, (P_1^*) and (P_2^*) are equivalent to (P_1) and (P_2) respectively. Therefore we obtain:

2.8 Theorem. *If $\mathcal{X}$ is a (DFS)-space, then $(P_1^*) \Rightarrow \operatorname{Proj}^1 \mathcal{X} = 0 \Rightarrow (P_2^*)$.*

Remark. It should be noticed that the condition for $\operatorname{Proj}^1 \mathcal{X} = 0$ in Proposition 2.6 can always be turned into (P_1). One just has to change the fundamental systems $(B_{\mu,k})$ in each X_μ by setting $B_{\mu,0} = B_\mu$, etc. This, however, is not very helpful if the $X_{n,k}$ have a concrete meaning as in most examples.

3. Topological properties of projective limits of (LB)–spaces

We keep the notations and assumptions of §2. We assume additionally, that

$$\bigcup_m B_{k,m} = X_k \text{ for all } k.$$

Then for every bounded set $B \subset X_k$ there is m such that $B \subset B_{k,m}$. We achieve that e.g. by replacing $B_{k,m}$ by $mB_{k,m}$.

We set $B_{k,m}^\infty = (\iota^k)^{-1} B_{k,m}$. Then the sets

$$\mathrm{B} = \bigcap_k B_{k,m(k)}^\infty ,$$

where $(m(k))_k$ runs through all sequences of integers, are a basis of bounded sets in X. This means every bounded set in X is contained in such a set. Clearly these sets are Banach balls.

Remark. X has a basis of bounded sets consisting of Banach balls.

The uniform boundedness principle implies that in X' every weak*– bounded set is bounded in the strong topology.

We call the spectrum $\mathcal{X}$ *reduced* if $X_k = \overline{\iota^k X}$ for all k. From Theorem 2.1 (Retakh's theorem) follows easily:

Remark. If $\mathrm{Proj}^1 \mathcal{X} = 0$ then $\mathcal{X}$ is equivalent to a reduced spectrum.

For a reduced spectrum $\mathcal{X}$ we identify X_k' with $X_k^* := \jmath^k X_k' \subset X'$. Then $X_k^* \subset X_{k+1}^*$ for all k and we obtain an imbedding spectrum of Fréchet spaces. By X^* we denote the dual space X' equipped with the inductive topology. X_b' denotes X' equipped with the strong topology. $\mathrm{id} : X^* \to X_b'$ is continuous.

Remark. Let $\mathcal{X}$ be reduced. For an absolutely convex bounded set $\mathrm{B} \subset X'$ the following are equivalent:

(1) B is equicontinuous

(2) B is relatively weak*–compact

(3) B is contained in a bounded Banach ball

(4) B is contained in some X_k^* and bounded there.

PROOF. (1) ⇒ (2) ⇒ (3) is clear, (3) ⇒ (4) follows from Grothendieck's factorization theorem, (4) ⇒ (1) since X_k is barrelled.

In particular X is a Mackey space, i.e. carries its Mackey topology. The following Lemma is an immediate consequence of the previous Remarks.

3.1 Lemma. *Let $\mathcal{X}$ be reduced. The following are equivalent:*

(1) X barrelled

(2) X quasibarrelled

(3) X_b' sequentially complete (quasi–complete)

(4) X' weak–sequentially complete (quasi–complete)*

(5) Every bounded set $\mathrm{B} \subset X'$ is contained in some X_k^* and bounded there.

From [13] we take the following two crucial Lemmas.

3.2 Lemma. ([13, 5.6]) *If* $\operatorname{Proj}^1 \mathcal{X} = 0$ *then* X *is bornological.*

3.3 Lemma. ([13, 5.9, 5.10]) *If* $\mathcal{X}$ *is reduced and does not satisfy* (P_2^*) *then there is* μ *and a sequence* $(y_n)_n$ *in* X_μ *such that* $\lim_n (2^n \jmath_\mu y_n) = 0$ *in* X^* *but* $\left(\frac{1}{n}\jmath_\mu^k y_n\right)_n$ *does not converge to* 0 *for all* $k \geq \mu$.

Collecting the previous information we obtain the following theorem.

3.4 Theorem. *Let* $\mathcal{X}$ *be reduced. Then (1)* $\Rightarrow$ *(2)* $\Rightarrow$ *(3)* $\Rightarrow$ *(4)* $\Rightarrow$ *(5)* $\Rightarrow$ *(6), where:*

(1) $\operatorname{Proj}^1 \mathcal{X} = 0$

(2) X *bornological*

(3) X_b' *complete*

(4) X *barrelled*

(5) X^* *regular*

(6) Condition (P_2^*).

Here the inductive limit X^* is called *regular* if every bounded set is contained in some X_k^* and bounded there.

If all X_k are reflexive then $X^{*\prime} = X$, hence $X^* = X_b{}'$ (because the topology of X^* is admissible). If $\mathcal{X}$ is a (DFS)–spectrum, then every X_k is reflexive and $X^* = X_b'$ is a Schwartz space. According to a theorem of L. Schwartz ([12, p.43]) completeness of $X^* = X_b'$ implies that $(X_b')_b'$ is bornological (=ultrabornological in our case). Since completeness of X_b' also implies that X is barrelled (see Lemma 3.1) we have X bornological.

Remark. If $\mathcal{X}$ is a (DFS)–spectrum then $X^* = X_b'$ and X is bornological if and only if $X^* = X_b'$ is complete.

3.5 Theorem. *Let* $\mathcal{X}$ *be a reduced (DFS)–spectrum. Then* $X^* = X_b'$ *and (1)* $\Rightarrow$ *(2)* $\Longleftrightarrow$ *(3)* $\Rightarrow$ *(4)* $\Longleftrightarrow$ *(4)'* $\Longleftrightarrow$ *(5)* $\Longleftrightarrow$ *(5)'* $\Rightarrow$ *(6) where*

(1) $\operatorname{Proj}^1 \mathcal{X} = 0$

(2) X *bornological*

(3) X^* *complete*

(4) X *barrelled* — *(4)'* X *reflexive*

(5) X^* *regular* — *(5)'* X^* *reflexive*

(6) Condition (P_2^*).

Proof. According to the previous Remarks and Theorem 3.4 we have only to show (4) $\Longleftrightarrow$ (4)$'$ and (5) $\Longleftrightarrow$ (5)$'$.

In X every bounded set is relatively compact, hence X is semireflexive, so (4) implies (4)$'$. The converse is clear . If X^* is reflexive every bounded set in X'_b is relatively weak–compact, hence relatively weak*– compact. So (5)$'$ $\Rightarrow$ (5). If X^* is regular then $(X^*)'_b = X$, since all X_k are reflexive. This proves (5) $\Rightarrow$ (5)$'$.

4. The case of sequence spaces

Let $(a_{j;k,m})_{j,k,m\in\mathbb{N}}$ be an infinite matrix with

$$a_{j;k,m} > 0$$
$$a_{j;k,m} \geq a_{j;k,m+1}$$
$$a_{j;k,m} \leq a_{j;k+1,m}$$

for all j,k,m. For fixed $1 \leq p < +\infty$ we define

$$X_{k,m} = \{x = (x_1, x_2, \ldots) : \| x \|^p_{k,m} = \sum_j |x_j|^p a^p_{j;k,m} < \infty\},$$

and for $p = +\infty$

$$X_{k,m} = \{x = (x_1, x_2, \ldots) : \lim_j |x_j| a_{j;k,m} = 0\}$$

with the norm $\| x \|_{k,m} = \sup_j |x_j| a_{j;k,m}$.

We put $X_k = \bigcup_m X_{k,m}$, equipped with the inductive topology. Then $X_{k+1} \subset X_k$ and we consider this as a projective spectrum $\mathcal{X}$ with the inclusions as connecting maps. As always in this paper $X = \operatorname{Proj}^0 \mathcal{X}$ equipped with the projective topology.

Dually we define for $1 < p \leq +\infty$

$$X^*_{k,m} = \{y = (y_1, y_2, \ldots) : \| y \|^{*q}_{k,m} = \sum_j |y_j|^q a^{-q}_{j;k,m} < +\infty\}$$

where $\frac{1}{p} + \frac{1}{q} = 1$ and for $p = 1$

$$X^*_{k,m} = \{y = (y_1, y_2, \ldots) : \| y \|^*_{k,m} = \sup_j |y_j| a^{-1}_{j;k,m} < +\infty\}.$$

We put $X^*_k = \bigcap_m X^*_{k,m}$ with the projective topology, i.e. X^*_k is the Köthe space with the matrix $(a_{j;k,m})_{j,m\in\mathbb{N}}$. Then $X^*_k \subset X^*_{k+1}$ and we consider this as an inductive spectrum $\mathcal{X}^*$ with the inclusions as connecting maps. We set $X^* = \bigcup_k X^*_k$ equipped with the inductive topology.

Obviously $X^*_{k,m} = X'_{k,m}$, $X^*_k = X'_k$ by canonical identification. The topologies are the strong topologies (see [6, p. 406 ff]). Also $X^* = X'$ by the same identification, and the topology is the strong topology for $1 < p < +\infty$, due to reflexivity (see §3).

For $p = 1$ this needs not to be the case. Let $a_{j;k,m} = a_{j,k}$ for all j,k,m, where $(a_{j,k})_{j,k}$ is the matrix of a non–distinguished Köthe space (see Köthe [6, p. 438]). Notice that in this case $\operatorname{Proj}^1 \mathcal{X} = 0$, however $X^* \neq X'_b$.

For $1 \leq p < +\infty$ the $B_{k,m}$ are closed in X_k, hence the $B_{k,m}$ are a fundamental system of bounded sets in X_k (see [6, p. 406 f]). For $p = +\infty$ this needs not to be the case (see [6, p. 437 f]).

To avoid these difficulties we assume for $p = 1, \infty$

$$\forall k, m \, \exists M \; : \; \lim_j \frac{a_{j;k,m}}{a_{j;k,M}} = 0 .$$

Hence we are in the (DFS)–case , so $X^* = X_b'$ (see §3) and $\overline{B_{k,m}}^{X_k} \subset C B_{k,M}$, so the $B_{k,m}$ are a fundamental system of bounded sets in X_k.

Using the ideas of [7] we obtain (see [13, Lemma 6.1]):

4.1 Lemma. *Under the assumptions of this section* $\mathrm{Proj}^1 \mathcal{X} = 0$ *and* (P_2^*) *coincide with :*

(P) $\forall \mu \, \exists n, k \, \forall m, K \, \exists N, S \, \forall j \; : \; \frac{1}{a_{j;k,m}} \leq S \max \left(\frac{1}{a_{j;K,N}}, \frac{1}{a_{j;\mu,n}} \right).$

By use of 3.4 and 3.5 this yields (see [13, Theorem 6.]).

4.2 Theorem. *Under the assumptions of this section* $X^* = X_b'$ *and the following are equivalent:*

(1) $\mathrm{Proj}^1 \mathcal{X} = 0$

(2) X bornological

(3) X^* *complete*

(4) X barrelled — *(4)' X reflexive*

(5) X^* *regular* — *(5)'* X^* *reflexive*

(6) Condition (P).

Remark. We mention that another condition is equivalent to (1),...,(6) (see [13, Theorem 6.4]):

(7) $X^* = \{ y = (y_1, y_2, \ldots) \; : \; \sum_j |x_j y_j| < +\infty$ for all $x \in X \}$ ("Köthe dual").

This Theorem generalizes results of Grothendieck [4, II, §4] and Krone–Vogt [7]. The following special case generalizes results of Vogt–Wagner [14], [15], Hebbecker [5], Nyberg [8]. For the case (ii), $\rho = 0$ and $r = +\infty$ see Braun–Meise–Vogt [1].

We assume that $\alpha = (\alpha_j)_j$, $\beta = (\beta_j)_j$ are nonnegative numerical sequences, $\alpha_j + \beta_j \to +\infty$, and $r, \rho \in \mathbb{R} \cup \{+\infty\}$. For $r_k \nearrow r$, $\rho_k \nearrow \rho$ we consider the matrix

$$a_{j;k,m} = e^{r_k \alpha_j - \rho_m \beta_j} .$$

It is easy to see that $\mathcal{X}$ depends, up to equivalence, only on α, β, r, ρ.

4.3 Theorem. *(P) is satisfied if and only if (i) or (ii), where*

(i) $\rho = +\infty$

(ii) $\mathbb{N} = J_1 \dot{\cup} J_2$ *such that* $\inf_{j \in J_1} \frac{\alpha_j}{\beta_j} > 0$ *and* $\lim_{j \in J_2} \frac{\alpha_j}{\beta_j} = 0$.

PROOF. After taking logarithms and dividing by β_j condition (P) takes the form

$$\forall \mu\ \exists n, k\ \forall m, K\ \exists N, S\ \forall j\ :\ \rho_m - r_k \frac{\alpha_j}{\beta_j} \le \frac{S}{\beta_j} + \max\left(\rho_N - r_K \frac{\alpha_j}{\beta_j}, \rho_n - r_\mu \frac{\alpha_j}{\beta_j}\right)$$

or, taking into account that $\alpha_j + \beta_j \to \infty$, with different k, n and N (using $\frac{S}{\beta_j} - (r_{k+1} - r_k)\frac{\alpha_j}{\beta_j} \le \varepsilon$ for $\alpha_j + \beta_j$ large)

$\forall \mu\ \exists n, k\ \forall m, K\ \exists N$ such that for all but finitely many j

$$\rho_m - r_k \frac{\alpha_j}{\beta_j} \le \max\left(\rho_N - r_K \frac{\alpha_j}{\beta_j}, \rho_n - r_\mu \frac{\alpha_j}{\beta_j}\right).$$

The inequality we can write as

$$\rho_m - \rho_n \le (r_k - r_\mu)\frac{\alpha_j}{\beta_j} \quad \text{or} \quad \rho_N - \rho_m \ge (r_K - r_k)\frac{\alpha_j}{\beta_j}.$$

If $\rho = +\infty$ and $m > n$, $K > k > \mu$ we choose N such that

$$\frac{\rho_N - \rho_n}{\rho_m - \rho_n} \ge \frac{r_K - r_k}{r_k - r_\mu}$$

and obtain for j with $\rho_m - \rho_n > (r_k - r_\mu)\frac{\alpha_j}{\beta_j}$

$$(r_K - r_k)\frac{\alpha_j}{\beta_j} \le \frac{r_K - r_k}{r_k - r_\mu}(\rho_m - \rho_n) \le \rho_N - \rho_m.$$

If $\rho < +\infty$ and (ii) holds, then for given μ we choose $k = \mu + 1$ and n so large that

$$\rho - \rho_n \le (r_k - r_\mu)\frac{\alpha_j}{\beta_j} \quad \text{for all } j \in J_1.$$

If $N > m$ the second possibility will be satisfied for all $j \in J_2$ up to finitely many.

Now assume $\rho < +\infty$ and (P). For $\mu = 1$ we choose k and n. We put

$$J_1 = \{j\ :\ \rho - \rho_n \le (r_k - r_1)\frac{\alpha_j}{\beta_j}\}$$

and $J_2 = \mathbb{N} \setminus J_1$. For every m the inequality

$$\rho - \rho_m \ge (r_{k+1} - r_k)\frac{\alpha_j}{\beta_j}$$

holds for all $j \in J_2$ up to finitely many. This implies the assertion.

For $J \subset \mathbb{N}$, $1 \leq p < +\infty$, α and r we set (different notation for *)

$$\Lambda_r(\alpha, J) = \{(x_i)_{i\in J} \ : \| x \|_t^p = \sum_{j\in J} |x_j|^p e^{pt\alpha_j} < +\infty \text{ for all } t < r\}$$

$$\Lambda_r^*(\alpha, J) = \{(x_j)_{j\in J} \ : \| x \|_t^p = \sum_{j\in J} |x_j|^p e^{pt\alpha_j} < +\infty \text{ for some } t < r\}$$

and obtain (see Braun–Meise–Vogt [1]):

4.4 Corollary. *If $\rho = 0$ then (P) is satisfied if and only if $\mathbb{N} = J_1 \dot\cup J_2$ such that $X = \Lambda_r(\alpha, J_1) \oplus \Lambda_0^*(\beta, J_2)$.*

Important special cases are those used in Braun–Meise–Vogt [1] (cf. [2]), to solve the problem of solvability of convolution equations in Gevrey classes on $\mathbb{R}$, and the cases of matrices

$$a_{\nu,j;k,m} = e^{r_k\alpha_\nu - \rho_m\beta_j}$$

where $\alpha_\nu \to +\infty$, $\beta_j \to +\infty$. They occur in connection with the investigation of tensor-products of (F)– and (DF)–spaces and of Ext^1 (see remarks above).

Restricting us to the nuclear case, i.e. to sequences α, β with

$$\limsup_\nu \frac{\log \nu}{\alpha_\nu} < +\infty \text{ for } r = +\infty$$

$$\lim_\nu \frac{\log \nu}{\alpha_\nu} = 0 \text{ for } r < +\infty$$

and analogous for β, we have $X \cong L(\Lambda_\rho(\beta), \Lambda_r(\alpha))$. The decomposition in Corollary 4.4 gives a decomposition in a "lower triangular" and "upper triangular" part. One maps a fixed neighborhood of zero into a (variable) bounded set, the other some neighborhood of zero into a fixed bounded set. All maps are compact, i.e. $L(\Lambda_\rho(\beta), \Lambda_r(\alpha)) = LB(\Lambda_\rho(\beta), \Lambda_r(\alpha))$ (see Nyberg [8]).

5. Final remarks

The usual way of application of the previous results is the following. We are given an exact sequence

$$0 \to \mathcal{X} \xrightarrow{\Phi} \mathcal{Y} \xrightarrow{\Psi} \mathcal{Z} \to 0$$

of spectra of complete (LB)–spaces. X, Y, Z are the projective limits, $\varphi = \Phi^0$, $\psi = \Psi^0$. We have the exact sequence

$$0 \to X \xrightarrow{\varphi} Y \xrightarrow{\psi} Z$$

and we want to know whether ψ is surjective.

In many of these applications we have $\mathrm{Proj}^1 \mathcal{Y} = 0$. In this case we have

5.1 Theorem. *If* $\mathrm{Proj}^1 \mathcal{Y} = 0$, *then* ψ *is surjective if and only if* $\mathrm{Proj}^1 \mathcal{X} = 0$.

PROOF. Immediate consequence of Theorem 1.5.

If one investigates surjectivity of a map ψ by this method (see e.g. Braun–Meise–Vogt [2]) then Theorem 4.5 suggests the question whether this is a property of $\mathcal{X}$ or of the space X. This means whether we can replace the investigation of $\mathcal{X}$ by that of any other spectrum of (LB)–spaces generating X.

By means of Grothendieck's factorization theorem one proves (cf. [13, Proposition 3.2])

5.2 Proposition. *If* $\mathcal{X}, \mathcal{Y}$ *are projective spectra of complete (LB)–spaces,* X, Y *the projective limits,* $\varphi : X \to Y$ *continuous linear and* $\mathcal{X}$ *reduced, then there exists* $\Phi : \mathcal{X} \to \mathcal{Y}$ *such that* $\varphi = \Phi^0$.

An immediate consequence is:

5.3 Corollary. *Any two reduced spectra of complete (LB)–spaces generating* X *are equivalent.*

This answers to some extend the question above. However $\mathcal{X}$ occurring as spectrum of kernels of Ψ needs not to be reduced. So the surjectivity of ψ depends in fact on $\mathcal{X}$.

Example. Let $1 \leq p \leq +\infty$ and

$$\begin{array}{ll} X_n = \ell_p & , \ \iota^n_{n+1} x = (0, x_1, x_2, \ldots) \text{ for all } n \\ Y_n = \ell_p & , \ \iota^n_{n+1} = \mathrm{id} \text{ for all } n \\ Z_n = \mathbb{K}^n & , \ \iota^n_{n+1}(x_1, \ldots, x_{n+1}) = (x_1, \ldots, x_n) \text{ for all } n\,. \end{array}$$

We put $\varphi^n_n x = (0, \ldots, 0, x_1, x_2, \ldots)$ writing n zeros at the beginning, $\psi^n_n x = (x_1, \ldots, x_n)$ and obtain an exact sequence

$$0 \to \mathcal{X} \xrightarrow{\Phi} \mathcal{Y} \xrightarrow{\Psi} \mathcal{Z} \to 0$$

of projective spectra of Banach spaces.

Since $X = \{0\}, Y = \ell_p, Z = \omega := \mathbb{K}^{\mathbb{N}}$, the space X has all good properties, but $\psi : \ell_p \hookrightarrow \omega$ is obviously not surjective.

If we combine the results of this section with the 2. Remark of §3, then we obtain for the question at the beginning of this section.

5.4 Theorem. *If* $\mathrm{Proj}^1 \mathcal{Y} = 0$, *then* ψ *is surjective if and only if* $\mathcal{X}$ *is reduced and* X *admits a spectrum* $\tilde{\mathcal{X}}$ *of complete (LB)–spaces with* $\mathrm{Proj}^1 \tilde{\mathcal{X}} = 0$.

References

[1] Braun R., Meise R., Vogt D., 'Existence of fundamental solutions and surjectivity of convolution operators on classes of ultradifferentiable functions', preprint.

[2] Braun R., Meise R., Vogt D.,'Application of the projective limit functor to convolution and partial differential equations', these proceedings.

[3] Braun R., Meise R., Vogt D., 'Characterization of the linear partial differential operators which are surjective on non–quasianalytic classes of Roumien type', manuscript.

[4] Grothendieck A., 'Produits tensoriels topologiques et espaces nucléaires', *Mem. Amer. Math. Soc.*,**16** (1953).

[5] Hebbecker J., 'Auswertung der Splittingbedingungen (S_1^*) und (S_2^*) für Potenzreihenräume und L_f–Räume', Diplomarbeit Wuppertal 1984.

[6] Köthe G., 'Topologische lineare Räume', *Springer* 1960.

[7] Krone J., Vogt D., 'The splitting relation for Köthe spaces', *Math. Z.* 190 (1985),387-400.

[8] Nyberg K., 'Tameness of pairs of nuclear power series spaces and related topics', *Trans. Amer. Math. Soc.* **283** (1984), 645-660.

[9] Palamodov V.P., 'The projective limit functor in the category of linear topological spaces' (Russian), *Mat. Sbornik* **75** (1968), 567-603, English transl. *Math. USSR–Sb.* **4** (1968), 529-558.

[10] Palamodov V.P., 'Homological methods in the theory of locally convex spaces' (Russian), *Usp. Mat. Nauk* **26** 1 (1971), 3-66, English transl. Russian Math. Surveys **26** 1(1971), 1-64.

[11] Retakh V.S., 'Subspaces of a countable inductive limit', *Dokl. Akad. Nauk SSSR* **194** (1970), No. 6, English transl. *Soviet Math. Dokl.* **11** (1970), 1384-1386.

[12] Schwartz L., 'Théorie des distributions a valeurs vectorielles', *Ann. Inst. Fourier* **7** (1957), 1-141.

[13] Vogt D., 'Lectures on projective spectra of (DF)–spaces', Seminar lectures, AG Funktionalanalysis Düsseldorf/Wuppertal 1987.

[14] Vogt D., 'Charakterisierung der Unterräume von s', *Math. Z.* **155** (1977), 109-117.

[15] Vogt D., Wagner M. J., 'Charakterisierung der Quotientenräume von s und eine Vermutung von Martineau', *Studia Math.* **68** (1980), 225-240.

APPLICATIONS OF THE PROJECTIVE LIMIT FUNCTOR TO CONVOLUTION AND PARTIAL DIFFERENTIAL EQUATIONS

RÜDIGER W. BRAUN
Mathematisches Institut
Heinrich-Heine-
Universität Düsseldorf
Universitätsstr. 1
D–4000 Düsseldorf
Fed. Rep. of Germany

REINHOLD MEISE
Mathematisches Institut
Heinrich-Heine-
Universität Düsseldorf
Universitätsstr. 1
D–4000 Düsseldorf
Fed. Rep. of Germany

DIETMAR VOGT
FB Mathematik
Bergische Universität
Gesamthochschule Wuppertal
Gaußstr. 20
D–5600 Wuppertal
Fed. Rep. of Germany

Abstract

Let $\mathcal{E}_{\{\omega\}}(\mathbb{R}^N)$ denote the non-quasianalytic class of all $\{\omega\}$-ultradifferentiable functions on $\mathbb{R}^N$. This notion is an extension of the classical Gevrey classes $\Gamma^{\{d\}}(\mathbb{R}^N)$, $d > 1$. Reporting on our work [5] and [6], we explain how the projective limit functor introduced by Palamodov [21] can be used to characterize the surjectivity of

(1) convolution operators T_μ on $\mathcal{E}_{\{\omega\}}(\mathbb{R})$ and

(2) linear partial differential operators $P(D)$ on $\mathcal{E}_{\{\omega\}}(\mathbb{R}^N)$.

In both cases we apply recent results of Vogt [28], [29] on the projective limit functor. In (1) we also use a sequence space representation of $\ker T_\mu$ given in Meise [15], and in (2) we use basic results on Phragmén-Lindelöf conditions on algebraic varieties of Meise, Taylor and Vogt [19]. We remark that (1) and (2) extend previous results of Chou [10], Cattabriga [7], [8], and Zampieri [31].

Introduction

At the time when Meise and Vogt [20] was written, we noticed that no characterization of the surjectivity of convolution operators T_μ on the classical Gevrey classes $\Gamma^{\{d\}}(\mathbb{R})$, $d > 1$, could be found in the literature. The results of Ehrenpreis [11] on convolution operators on $C^\infty(\mathbb{R}^N)$ suggested the guess that the surjectivity of T_μ should be equivalent to the existence of a fundamental solution for T_μ. However, $\Gamma^{\{d\}}(\mathbb{R})$ is a projective limit of (DFN)-spaces and it was open how to extend the classical methods to this case.

The solution of the problem was obtained in the following way: By results of Palamodov [21], it was easy to see that the surjectivity of T_μ is equivalent to the "local surjectivity" of T_μ and $\mathrm{Proj}^1(\ker T_\mu) = 0$. For a convolution operator T_μ admitting a fundamental solution, Meise [15] had obtained a sequence space representation for $\ker T_\mu$. To this representation

T. Terzioğlu (ed.), Advances in the Theory of Fréchet Spaces, 29–46.

one could apply a result of Vogt [28], who had characterized $\mathrm{Proj}^1 \Lambda(A) = 0$ for countable projective spectra $\Lambda(A)$ of (DFS)-sequence spaces. From this and some auxiliary considerations, we obtained that T_μ is surjective on $\Gamma^{\{d\}}(\mathbb{R})$ if and only if T_μ admits a fundamental solution and the zeros of the Fourier-Laplace transform $\widehat{\mu}$ of μ satisfy a certain condition concerning their location in the complex plane.

Knowing this, we suspected that a similar result should hold for linear partial differential operators $P(D)$ with constant coefficients on $\Gamma^{\{d\}}(\mathbb{R}^N)$. Later we learned that this effect was known already. In fact, Cattabriga [8] has observed that for $N \geq 3$ there are operators $P(D)$ which are not surjective on $\Gamma^{\{d\}}(\mathbb{R}^N)$. In [7] he gave sufficient conditions for the surjectivity of $P(D)$, and Zampieri [31] used methods of Hörmander [12] to derive sufficient conditions for the surjectivity of $P(D)$, phrased in terms of a Phragmén-Lindelöf condition on the zero variety of P.

To obtain the desired characterization one has to overcome the following difficulty: For arbitrary projective spectra $\mathcal{X} = (X_n, \iota^n_{n+1})_{n\in\mathbb{N}}$ of (DFS)-spaces no evaluable characterization of $\mathrm{Proj}^1 \mathcal{X} = 0$ is known, except in the presence of a basis. In Vogt [28], [29] a result of Retakh [24] is used to derive an evaluable sufficient condition and also a necessary one, but in general they are not equivalent. However, for the present problem, their equivalence can be shown in the following way: Both characterizations can be translated to Phragmén-Lindelöf conditions for analytic functions on the zero variety of P. By a result of Meise, Taylor, and Vogt [19], these conditions are equivalent to the corresponding conditions for plurisubharmonic functions. The equivalence of these can be seen very easily. Thus the surjectivity of $P(D)$ on $\Gamma^{\{d\}}(\mathbb{R}^N)$ is equivalent to a Phragmén-Lindelöf condition on the zero variety of P.

1 Preliminaries

In this section we introduce the classes $\mathcal{E}_{\{\omega\}}(\mathbb{R}^N)$ and convolution operators T_μ on these classes. Moreover, we show how to use the projective limit functor in order to characterize the surjectivity of T_μ.

1.1 Definition. An increasing continuous function $\omega: [0,\infty[\to [0,\infty[$ is called a weight function if it has the following properties:

(α) there exists $K \geq 1$ with $\omega(2t) \leq K(1+\omega(t))$ for all $t \geq 0$,

(β) $\displaystyle\int_0^\infty \frac{\omega(t)}{1+t^2}\,dt < \infty$,

(γ) $\log t = o(\omega(t))$ for $t \to \infty$,

(δ) $\varphi: t \mapsto \omega(e^t)$ is convex on $\mathbb{R}$.

The Young conjugate $\varphi^*: [0,\infty[\to \mathbb{R}$ of φ is defined by $\varphi^*(y) := \sup_{x\geq 0}(xy - \varphi(x))$.

1.2 Definition. Let ω be a weight function and let $B_k := \left\{x \in \mathbb{R}^N \;\middle|\; |x| \le k\right\}$. Then we define

$$\mathcal{E}_{\{\omega\}}(\mathbb{R}^N) := \Big\{f \in C^\infty(\mathbb{R}^N) \;\Big|\; \forall k \in \mathbb{N}\, \exists m \in \mathbb{N} : \\ \|f\|_{k,m} := \sup_{x\in B_k} \sup_{\alpha\in\mathbb{N}_0^N} |f^{(\alpha)}(x)| \exp(-\frac{1}{m}\varphi^*(m|\alpha|)) < \infty\Big\},$$

and we endow $\mathcal{E}_{\{\omega\}}(\mathbb{R}^N)$ with the topology of the projective limit over k of the inductive limits over m. Furthermore, we define

$$\mathcal{D}_{\{\omega\}}(B_k) := \left\{f \in \mathcal{E}_{\{\omega\}}(\mathbb{R}^N) \;\middle|\; \operatorname{Supp}(f) \subset B_k\right\}$$

and

$$\mathcal{D}_{\{\omega\}}(\mathbb{R}^N) := \operatorname*{ind}_{k\rightarrow} \mathcal{D}_{\{\omega\}}(B_k).$$

1.3 Remark. (a) Using a variation of the original idea of Beurling [2] (see also Björck [3] and Petzsche and Vogt [23]) it is shown in Braun, Meise, and Taylor [4] that $\mathcal{D}_{\{\omega\}}(\mathbb{R}^N)$ is a non-trivial (DFN)-space. Hence the class $\mathcal{E}_{\{\omega\}}(\mathbb{R}^N)$ is non-quasianalytic.
(b) For a weight function ω and each $k \in \mathbb{N}$ we have the topological identity

$$\mathcal{D}_{\{\omega\}}(B_k) = \Big\{f \in C(\mathbb{R}^N) \;\Big|\; \operatorname{Supp}(f) \subset B_k \quad \text{and there exists } m \in \mathbb{N}: \\ \int_{\mathbb{R}^N} |\hat{f}(t)| \exp(\frac{1}{m}\omega(|t|))\, dt < \infty\Big\},$$

where $\hat{f}$ denotes the Fourier transform of f.
(c) Two functions $\omega_1, \omega_2 \colon [0,\infty[\rightarrow [0,\infty[$ are called equivalent if there exists $C \ge 1$ so that $\omega_1 \le C(1+\omega_2)$ and $\omega_2 \le C(1+\omega_1)$. It is easy to see that each weight function is equivalent to a weight function satisfying $\omega(1) = 0$. Since the definitions in 1.2 do not change if ω is replaced by an equivalent weight function, we can, therefore, assume in the sequel that φ^* has values in $[0,\infty[$.

1.4 Example. It is easy to check that the following functions $\omega\colon [0,\infty[\rightarrow [0,\infty[$ are equivalent to weight functions:

(1) $$\omega(t) = t^\beta, \quad 0 < \beta < 1,$$

(2) $$\omega(t) = (\log(1+t))^\beta, \quad \beta > 1,$$

(3) $$\omega(t) = t(\log(e+t))^{-\beta}, \quad \beta > 1.$$

In case (1) we get $\varphi^*(x) = (x/\beta)\log(x/(e\beta))$ and hence with $d := 1/\beta$

$$\mathcal{E}_{\{\omega\}}(\mathbb{R}^N) = \Gamma^{\{d\}}(\mathbb{R}^N) = \left\{f \in C^\infty(\mathbb{R}^N) \;\middle|\; \forall k \in \mathbb{N}\, \exists m \in \mathbb{N} : \sup_{x\in B_k} \sup_{\alpha\in\mathbb{N}_0^N} \frac{|f^{(\alpha)}(x)|}{m^{|\alpha|}(|\alpha|!)^d} < \infty\right\}.$$

1.5 Convolution operators. For $\mu \in \mathcal{E}_{\{\omega\}}(\mathbb{R}^N)'$ and $f \in \mathcal{E}_{\{\omega\}}(\mathbb{R}^N)$ the function

$$\mu * f : x \mapsto \langle \mu_y, f(x-y) \rangle, \quad x \in \mathbb{R}^N,$$

is in $\mathcal{E}_{\{\omega\}}(\mathbb{R}^N)$, as an easy computation shows. Moreover,

$$T_\mu : \mathcal{E}_{\{\omega\}}(\mathbb{R}^N) \to \mathcal{E}_{\{\omega\}}(\mathbb{R}^N), \quad T_\mu(f) := \mu * f,$$

is a continuous linear operator, which is called the convolution operator induced by μ.

For $\nu \in \mathcal{D}_{\{\omega\}}(\mathbb{R}^N)'$ we define $\mu * \nu \in \mathcal{D}_{\{\omega\}}(\mathbb{R}^N)'$ by

$$\mu * \nu : g \mapsto \langle \nu, \check{\mu} * g \rangle,$$

where $\check{\mu} : h \mapsto \langle \mu_x, h(-x) \rangle$. An ultradistribution $\nu \in \mathcal{D}_{\{\omega\}}(\mathbb{R}^N)'$ with $\mu * \nu = \delta$ is called a fundamental solution for T_μ.

The simplest examples of convolution operators are the constant coefficient partial differential operators $P(D) = T_\mu$, for $\mu = \sum_{|\alpha| \le m} a_\alpha i^{|\alpha|} \delta^{(\alpha)}$. By Cattabriga [8], the operator

$$P(D) = \left(\frac{\partial}{\partial x}\right)^2 + \left(\frac{\partial}{\partial y}\right)^2 : \Gamma^{\{d\}}(\mathbb{R}^3) \to \Gamma^{\{d\}}(\mathbb{R}^3)$$

is, for each $d > 1$, not surjective (see 3.3 below for an alternative proof). Therefore the following question is reasonable.

Question. For which $\mu \in \mathcal{E}_{\{\omega\}}(\mathbb{R}^N)'$ is $T_\mu : \mathcal{E}_{\{\omega\}}(\mathbb{R}^N) \to \mathcal{E}_{\{\omega\}}(\mathbb{R}^N)$ surjective?

In the next two sections we will answer this question for $N = 1$ in the case of convolutions and for $N > 1$ in the case of partial differential operators. To prove our answer, we use the projective limit functor of Palamodov [21], which we now introduce briefly, following Vogt [29], section 1.

1.6 The projective limit functor. A sequence $\mathcal{X} = (X_n, \iota^n_{n+1})_{n \in \mathbb{N}}$ of linear spaces X_n and linear maps $\iota^n_{n+1} : X_{n+1} \to X_n$ is called a projective spectrum. For a projective spectrum $\mathcal{X} = (X_n, \iota^n_{n+1})_{n \in \mathbb{N}}$ we define

$$\begin{aligned} \operatorname{Proj}^0 \mathcal{X} &:= \left\{ (x_n)_{n \in \mathbb{N}} \in \prod_{n \in \mathbb{N}} X_n \;\middle|\; \iota^n_{n+1}(x_{n+1}) = x_n \text{ for all } n \in \mathbb{N} \right\} \\ \operatorname{Proj}^1 \mathcal{X} &:= \left(\prod_{n \in \mathbb{N}} X_n \right) \Big/ B(\mathcal{X}) \end{aligned}$$

where

$$B(\mathcal{X}) := \left\{ (a_n)_{n \in \mathbb{N}} \in \prod_{n \in \mathbb{N}} X_n \;\middle|\; \text{there exists } (b_n)_{n \in \mathbb{N}} \in \prod_{n \in \mathbb{N}} X_n \text{ with } a_n = \iota^n_{n+1}(b_{n+1}) - b_n \text{ for all } n \in \mathbb{N}. \right\}$$

1.7 Example. For a given weight function ω and $n \in \mathbb{N}$ we put

$$\mathcal{E}_{\{\omega\}}(n) := \left\{ f \in \mathcal{E}_{\{\omega\}}(\mathbb{R}^N) \;\middle|\; f|_{B_n} \equiv 0 \right\},$$

and we define $\mathcal{E}_n := \mathcal{E}_{\{\omega\}}(\mathbb{R}^N)/\mathcal{E}_{\{\omega\}}(n)$. The corresponding quotient map is denoted by $q_n : \mathcal{E}_{\{\omega\}}(\mathbb{R}^N) \to \mathcal{E}_n$. It is easy to check that the identity map on $\mathcal{E}_{\{\omega\}}(\mathbb{R}^N)$ induces a continuous linear map $\iota^n_{n+1} : \mathcal{E}_{n+1} \to \mathcal{E}_n$ satisfying $q_n = \iota^n_{n+1} \circ q_{n+1}$. Thus we have defined a projective spectrum $\mathcal{E} := (\mathcal{E}_n, \iota^n_{n+1})_{n \in \mathbb{N}}$.

Note that $\mathcal{E}_n$ is a (DFN)-space. This follows from the known fact that for $s > 0$ the space $\mathcal{E}^{s-\mathrm{per}}_{\{\omega\}}(\mathbb{R}^N)$ of those functions in $\mathcal{E}_{\{\omega\}}(\mathbb{R}^N)$ which are s-periodic in each variable is a complemented (DFN)-subspace of $\mathcal{E}_{\{\omega\}}(\mathbb{R}^N)$ (see 2.8 and 2.12 below) and that $\mathcal{E}_n$ is a quotient space of $\mathcal{E}^{s-\mathrm{per}}_{\{\omega\}}(\mathbb{R}^N)$ if $s > n$.

1.8 Lemma. *For the projective spectrum $\mathcal{E} = (\mathcal{E}_n, \iota^n_{n+1})_{n \in \mathbb{N}}$ defined in 1.7 we have*

(1)
$$\mathrm{Proj}^0\, \mathcal{E} \cong \mathcal{E}_{\{\omega\}}(\mathbb{R}^N)$$

and

(2)
$$\mathrm{Proj}^1\, \mathcal{E} = 0.$$

PROOF. (1) It is easy to check that $A : \mathcal{E}_{\{\omega\}}(\mathbb{R}^N) \to \prod_{n \in \mathbb{N}} \mathcal{E}_n$, $A(f) := (q_n(f))_{n \in \mathbb{N}}$, is a linear bijection between $\mathcal{E}_{\{\omega\}}(\mathbb{R}^N)$ and $\mathrm{Proj}^0\, \mathcal{E}$.

(2) To show $B(\mathcal{E}) = \prod_{n \in \mathbb{N}} \mathcal{E}_n$, let $(a_n)_{n \in \mathbb{N}} \in \prod_{n \in \mathbb{N}} \mathcal{E}_n$ be given. Then, for each $n \in \mathbb{N}$ there exists $f_n \in \mathcal{E}_{\{\omega\}}(\mathbb{R}^N)$ with $q_n(f_n) = a_n$. We put $b_n := q_n(\sum_{j=1}^{n-1} f_j)$ for $n \in \mathbb{N}$ and get

$$\iota^n_{n+1}(b_{n+1}) - b_n = \iota^n_{n+1} \circ q_{n+1}\Big(\sum_{j=1}^{n} f_j\Big) - q_n\Big(\sum_{j=1}^{n-1} f_j\Big) = q_n(f_n) = a_n.$$

1.9 Example. Let ω be a weight function and let $\mu \in \mathcal{E}_{\{\omega\}}(\mathbb{R}^N)'$ be given with $\mathrm{Supp}(\mu) \subset B_k$ for some $k \in \mathbb{N}$. Then we have

$$T_\mu(\mathcal{E}_{\{\omega\}}(n+k)) \subset \mathcal{E}_{\{\omega\}}(n) \quad \text{for all } n \in \mathbb{N},$$

since

$$\mathrm{Supp}(\mu * f) \subset \mathrm{Supp}(\mu) + \mathrm{Supp}(f) \quad \text{for all } f \in \mathcal{E}_{\{\omega\}}(\mathbb{R}^N).$$

Hence there are continuous linear maps $\tau^n_{n+k} : \mathcal{E}_{n+k} \to \mathcal{E}_n$, $n \in \mathbb{N}$, which are uniquely determined by the condition

$$\tau^n_{n+k}(q_{n+k}(f)) = q_n(T_\mu(f)) \quad \text{for all } f \in \mathcal{E}_{\{\omega\}}(\mathbb{R}^N),\ n \in \mathbb{N}.$$

Then $\mathcal{T}_\mu = (\tau^n_{n+k})_{n \in \mathbb{N}}$ is a map $\mathcal{T}_\mu : \mathcal{E} \to \mathcal{E}$ between projective spectra. If we put $K_n := 0$ and $j^n_{n+1} = 0$ for $n \le k$ and

$$K_{n+k} := \ker \tau^n_{n+k} \subset \mathcal{E}_{n+k}, \quad j^{n+k}_{n+k+1} := \iota^{n+k}_{n+k+1}|_{K_{n+k+1}} \quad \text{for } n \in \mathbb{N},$$

then $\mathcal{K}(\mu,\omega) := (K_n, j_{n+1}^n)_{n\in\mathbb{N}}$ is a projective spectrum. Moreover, for each $n \in \mathbb{N}$ we have the following exact sequence

$$(*) \qquad 0 \longrightarrow K_{n+k} \xrightarrow{i_{n+k}^{n+k}} \mathcal{E}_{n+k} \xrightarrow{\tau_{n+k}^{n}} \mathcal{E}_n,$$

where $i_n^n \colon K_n \hookrightarrow \mathcal{E}_n$ denotes the inclusion.

In order to discuss the surjectivity of T_μ (resp. of τ_{n+k}^n), we introduce the following notation.

1.10 Definition. A convolution operator T_μ on $\mathcal{E}_{\{\omega\}}(\mathbb{R}^N)$ is called locally surjective if for each $n \in \mathbb{N}$ and each $g \in \mathcal{E}_{\{\omega\}}(\mathbb{R}^N)$ there exists $f \in \mathcal{E}_{\{\omega\}}(\mathbb{R}^N)$ with $T_\mu(f)|_{B_n} = g|_{B_n}$.

1.11 Remark. (a) Each surjective convolution operator is locally surjective.
(b) If T_μ admits a fundamental solution then T_μ is locally surjective.
(c) Each non-trivial linear partial differential operator is locally surjective since it admits a fundamental solution.
(d) Local surjectivity does not imply surjectivity, as the examples of Cattabriga [8] show (see also 3.3 below).

1.12 Proposition. *A convolution operator T_μ on $\mathcal{E}_{\{\omega\}}(\mathbb{R}^N)$ is surjective if and only if it satisfies (1) and (2):*

(1) T_μ is locally surjective,

(2) $\operatorname{Proj}^1 \mathcal{K}(\mu,\omega) = 0$.

PROOF. Necessity: If T_μ is surjective then we have (1) by 1.11(a). Moreover, from 1.9(*) we get that

$$(3) \qquad 0 \longrightarrow K_{n+k} \xrightarrow{i_{n+k}^{n+k}} \mathcal{E}_{n+k} \xrightarrow{\tau_{n+k}^{n}} \mathcal{E}_n \longrightarrow 0,$$

is an exact sequence for each $n \in \mathbb{N}$. Therefore, the properties of the projective limit functor (see Palamodov [21], Vogt [29], 1.5) together with 1.8(2) imply that we have the following exact sequence

$$(4) \qquad 0 \longrightarrow \operatorname{Proj}^0 \mathcal{K}(\mu,\omega) \longrightarrow \operatorname{Proj}^0 \mathcal{E} \xrightarrow{T_\mu^0} \operatorname{Proj}^0 \mathcal{E} \xrightarrow{\delta^*} \operatorname{Proj}^1 \mathcal{K}(\mu,\omega) \xrightarrow{T_\mu^1} 0.$$

By 1.8(1) we can identify $\operatorname{Proj}^0 \mathcal{E}$ with $\mathcal{E}_{\{\omega\}}(\mathbb{R}^N)$. If we do this then the map T_μ^0 corresponds to T_μ. Hence T_μ^0 is surjective. By the exactness of the sequence this implies $\delta^* = 0$ and

$$\operatorname{Proj}^1 \mathcal{K}(\mu,\omega) = \ker T_\mu^1 = \operatorname{im} \delta^* = 0.$$

Sufficiency: It is easy to see that the local surjectivity of T_μ implies the surjectivity of τ_{n+k}^n in 1.9(*). Hence (1) implies that the sequence in (3) is exact. Therefore we have the exact sequence (4) also in the present case. Because of (2) we have $\delta^* = 0$ and hence

$$\operatorname{im} T_\mu^0 = \ker \delta^* = \operatorname{Proj}^0 \mathcal{E}.$$

Identifying $\operatorname{Proj}^0 \mathcal{E}$ with $\mathcal{E}_{\{\omega\}}(\mathbb{R}^N)$ and T_μ^0 with T_μ again, this proves the surjectivity of T_μ.

2 Convolution operators on $\mathcal{E}_{\{\omega\}}(\mathbb{R})$

In this section we indicate how one can use Fourier analysis in order to evaluate the conditions (1) and (2) of Proposition 1.12 for convolution operators on $\mathcal{E}_{\{\omega\}}(\mathbb{R})$. In doing this, we report on our article [5].

2.1 Definition. Let $A(\mathbb{C}^N)$ denote the space of all entire functions on $\mathbb{C}^N$. For a weight function ω and for $k, m \in \mathbb{N}$ we put

$$A_{k,m} := \left\{ f \in A(\mathbb{C}^N) \;\middle|\; \sup_{z\in\mathbb{C}^N} |f(z)| \exp(-k|\operatorname{Im} z| - \frac{1}{m}\omega(|z|)) < \infty \right\},$$

and we define

$$A_{\{\omega\}}(\mathbb{C}^N) := \operatorname*{ind}_{k\to} \operatorname*{proj}_{\leftarrow m} A_{k,m}.$$

It is easy to check that the following holds:

2.2 Proposition. *For each weight function ω, the Fourier-Laplace transform*

$$\mathcal{F}\colon \mathcal{E}_{\{\omega\}}(\mathbb{R}^N)_b' \to A_{\{\omega\}}(\mathbb{C}^N), \quad \mathcal{F}(\mu) = \widehat{\mu}\colon z \mapsto \left\langle \mu_x, \exp(-i\langle x, z\rangle)\right\rangle,$$

is a linear topological isomorphism. Moreover, for each $\mu \in \mathcal{E}_{\{\omega\}}(\mathbb{R}^N)'$ the following diagram commutes:

$$\begin{array}{ccc} \mathcal{E}_{\{\omega\}}(\mathbb{R}^N)_b' & \xrightarrow{T_\mu^t} & \mathcal{E}_{\{\omega\}}(\mathbb{R}^N)_b' \\ \mathcal{F}\downarrow & & \downarrow\mathcal{F} \\ A_{\{\omega\}}(\mathbb{C}^N) & \xrightarrow{M_{\widehat{\check{\mu}}}} & A_{\{\omega\}}(\mathbb{C}^N) \end{array}$$

where $M_{\widehat{\check{\mu}}}$ denotes the operator of multiplication by $\widehat{\check{\mu}} = \mathcal{F}(\check{\mu})$.

From Proposition 2.2 it follows that studying convolution operators on $\mathcal{E}_{\{\omega\}}(\mathbb{R}^N)$ is—up to duality—the same as investigating multiplication operators on $A_{\{\omega\}}(\mathbb{C}^N)$. By this classical observation, arguments and methods due to Ehrenpreis [11], Berenstein and Taylor [1], and Meise, Taylor, and Vogt [17] lead to the following characterization, which can be found in [5], section 2.

2.3 Theorem. *Let ω be a weight function. For $\mu \in \mathcal{E}_{\{\omega\}}(\mathbb{R})'$, $\mu \neq 0$, the following conditions are equivalent:*

(1) T_μ is locally surjective,

(2) T_μ admits a fundamental solution,

(3) for each $n \in \mathbb{N}$ there exists $x_n > 0$ so that for each $x \in \mathbb{R}$ with $|x| \geq x_n$ there exists $t \in \mathbb{R}$ with $|t - x| \leq \frac{1}{n}\omega(|x|)$ and

$$|\widehat{\mu}(t)| \geq \exp(-\frac{1}{n}\omega(|t|)),$$

(4) for each bounded set B in $\mathcal{D}_{\{\omega\}}(\mathbb{R})$ the set $T_\mu^{-1}(B)$ is bounded in $\mathcal{D}_{\{\omega\}}(\mathbb{R})$,

(5) the operator $S_\mu: \mathcal{D}_{\{\omega\}}(\mathbb{R})'_b \to \mathcal{D}_{\{\omega\}}(\mathbb{R})'_b$, $S_\mu(\lambda) := \mu * \lambda$, is surjective,

(6) there exists a weight function s with $s = o(\omega)$ satisfying (i) and (ii)

(i) there exists $L > 0$ with $\sup_{z\in\mathbb{C}} |\widehat{\mu}(z)| \exp\Big(-L\big(|\operatorname{Im} z| + s(|z|)\big)\Big) < \infty$,

(ii) there exist $\varepsilon, C, D > 0$ so that for each component S of the set

$$S(\widehat{\mu}, \varepsilon, C) := \Big\{z \in \mathbb{C} \;\Big|\; |\widehat{\mu}(z)| < \varepsilon \exp\Big(-C\big(|\operatorname{Im} z| + s(|z|)\big)\Big)\Big\}$$

we have

$$\sup_{z\in S}(|\operatorname{Im} z| + s(|z|)) \leq D\Big(1 + \inf_{z\in S}\big(|\operatorname{Im} z| + s(|z|)\big)\Big)$$

and

$$\sup_{z\in S}(|\operatorname{Im} z| + \omega(|z|)) \leq D\Big(1 + \inf_{z\in S}\big(|\operatorname{Im} z| + \omega(|z|)\big)\Big).$$

Theorem 2.3 explains the significance of the local surjectivity condition. However, 2.3(6) can also be used to decide whether condition 1.12(2) holds. To explain this, we observe first that by classical duality theory and Proposition 2.2 for each $\mu \in \mathcal{E}_{\{\omega\}}(\mathbb{R}^N)'$ we have the following isomorphisms

$$(\ker T_\mu)'_b \cong \mathcal{E}_{\{\omega\}}(\mathbb{R}^N)'_b/(ker T_\mu)^\perp = \mathcal{E}_{\{\omega\}}(\mathbb{R}^N)'_b/\overline{\operatorname{im} T_\mu^t} \cong A_{\{\omega\}}(\mathbb{C}^N)/\overline{I(\mathcal{F}(\check{\mu}))},$$

where $I(\mathcal{F}(\check{\mu}))$ denotes the principal ideal generated by $\mathcal{F}(\check{\mu})$ in the algebra $A_{\{\omega\}}(\mathbb{C}^N)$. Since all these isomorphisms are explicit, we get

2.4 Proposition. *For each weight function ω and each $\mu \in \mathcal{E}_{\{\omega\}}(\mathbb{R}^N)'$ the map*

$$\Phi: (\ker T_\mu)'_b \to A_{\{\omega\}}(\mathbb{C}^N)/\overline{I(\mathcal{F}(\check{\mu}))}, \quad \Phi(\lambda) := \widehat{\lambda} + \overline{I(\mathcal{F}(\check{\mu}))},$$

is a linear topological isomorphism.

From 2.3(6) and the maximum principle we get easily:

2.5 Corollary. *Assume that for $\mu \in \mathcal{E}_{\{\omega\}}(\mathbb{R})'$ one of the equivalent conditions in 2.3 holds. Then $I(\mathcal{F}(\check{\mu}))$ is closed in $A_{\{\omega\}}(\mathbb{C})$. Hence*

$$(\ker T_\mu)'_b \cong A_{\{\omega\}}(\mathbb{C})/I(\mathcal{F}(\check{\mu})).$$

In Meise [15] it was shown that $A_{\{\omega\}}(\mathbb{C})/I(\mathcal{F}(\check{\mu}))$ is isomorphic to a sequence space. To describe this result, we need some more notations.

2.6 Sequence spaces. (1) Let α and β be sequences in $[0,\infty[$ with $\lim_{j\to\infty}\beta_j = \infty$. For $k,m \in \mathbb{N}$ we put

$$\lambda_{k,m}(\alpha,\beta) := \left\{x \in \mathbb{C}^{\mathbb{N}} \;\middle|\; \sum_{j=1}^{\infty} |x_j| \exp(k\alpha_j + \frac{1}{m}\beta_j) < \infty\right\}.$$

Then for each $k \in \mathbb{N}$ the space

$$\lambda_k(\alpha,\beta) := \operatorname*{ind}_{m\to} \lambda_{k,m}(\alpha,\beta)$$

is a (DFS)-space. Obviously, the inclusion $\iota_{k+1}^k : \lambda_{k+1}(\alpha,\beta) \hookrightarrow \lambda_k(\alpha,\beta)$ is continuous. Hence $\Lambda(\alpha,\beta) := (\lambda_k(\alpha,\beta), \iota_{k+1}^k)_{k\in\mathbb{N}}$ is a projective spectrum of (DFS)-spaces. We put

$$\lambda(\alpha,\beta) := \operatorname*{proj}_{\leftarrow k} \lambda_k(\alpha,\beta).$$

(2) It is easy to check (see Meise [15], 1.6) that

$$\lambda(\alpha,\beta)_b' = \operatorname*{ind}_{k\to}\operatorname*{proj}_{\leftarrow m} \chi_{k,m}(\alpha,\beta),$$

where

$$\chi_{k,m}(\alpha,\beta) = \left\{y \in \mathbb{C}^{\mathbb{N}} \;\middle|\; \sup_{j\in\mathbb{N}} |y_j| \exp(-k\alpha_j - \frac{1}{m}\beta_j) < \infty\right\}.$$

(3) For an increasing unbounded sequence γ in $[0,\infty[$ and $0 < R \leq \infty$ we define the power series space

$$\Lambda_R(\gamma) := \left\{x \in \mathbb{C}^{\mathbb{N}} \;\middle|\; \sum_{j=1}^{\infty} |x_j| r^{\gamma_j} < \infty \text{ for } 0 < r < R\right\}.$$

Obviously, $\Lambda_R(\gamma)$ is a Fréchet-Schwartz space.
(4) Note that for γ and δ as in (3) there exist α and β as in (1) so that $\Lambda_\infty(\gamma)\hat{\otimes}\Lambda_1(\delta)_b' \cong \lambda(\alpha,\beta)$.

2.7 Theorem (Meise [15], 3.6). *Let T_μ be a convolution operator on $\mathcal{E}_{\{\omega\}}(\mathbb{R})$ which admits a fundamental solution and for which $\ker T_\mu$ is infinite dimensional. Then $\ker T_\mu$ is isomorphic to $\lambda(\alpha,\beta)$, where $\alpha = (|\operatorname{Im} a_j|)_{j\in\mathbb{N}}$, $\beta = (\omega(|a_j|))_{j\in\mathbb{N}}$, and where $(a_j)_{j\in\mathbb{N}}$ counts the zeros of $\widehat{\mu}$ with multiplicities.*

We briefly indicate the idea of the proof. For the sake of simplicity we assume that all zeros of $\widehat{\mu}$ are simple. Then we define

$$\rho: A_{\{\omega\}}(\mathbb{C}) \to \mathbb{C}^{\mathbb{N}}, \quad \rho(f) := (f(a_j))_{j\in\mathbb{N}}.$$

From 2.3(6) it follows by the arguments used in the proof of 2.5 that $\ker\rho = I(\widehat{\mu})$. Hence ρ induces a continuous linear bijection $\overline{\rho}: A_{\{\omega\}}(\mathbb{C})/I(\widehat{\mu}) \to \operatorname{im}\rho$. For α and β as in the hypothesis it is easy to check that $\operatorname{im}\rho$ is contained in the set

$$\left\{y \in \mathbb{C}^{\mathbb{N}} \;\middle|\; \exists k \in \mathbb{N}\, \forall m \in \mathbb{N} : \sup_{j\in\mathbb{N}} |y_j| \exp(-k\alpha_j - \frac{1}{m}\beta_j) < \infty\right\} = \lambda(\alpha,\beta)'$$

since the defining estimates for $f \in A_{\{\omega\}}(\mathbb{C})$ also hold on $\{a_j | j \in \mathbb{N}\}$. If $\operatorname{im}\rho = \lambda(\alpha,\beta)'_b$ then the open mapping theorem for (LF)-spaces implies that $\overline{\rho}$ is an isomorphism, i.e., $A_{\{\omega\}}(\mathbb{C})/I(\widehat{\mu}) \cong \lambda(\alpha,\beta)'_b$. Hence we get from Corollary 2.5

$$(\ker T_\mu)'_b \cong (\ker T_{\check{\mu}})'_b \cong \lambda(\alpha,\beta)'_b.$$

It turns out that even in the present special case $\operatorname{im}\rho$ can be strictly contained in $\lambda(\alpha,\beta)'$. Therefore, one has to determine $\operatorname{im}\rho \subset \mathbb{C}^{\mathbb{N}}$, which amounts to an interpolation problem. Furthermore one has to overcome the difficulty that $\ker T_\mu$ and $\lambda(\alpha,\beta)$ need not be reflexive. Using 2.3(6), Hörmander's solution of the $\overline{\partial}$-problem as well as ideas and results of Berenstein and Taylor [1], Meise [14], Meise and Taylor [16], and Taylor [25], these technical problems can be solved. For the details, we refer to Meise [15].

2.8 Example. Let ω be a weight function and put $\mu := \delta_\pi - \delta_{-\pi}$. Then we have $\widehat{\mu}(z) = -2i\sin\pi z$. From Berenstein and Taylor [1], Theorem 4 and Corollary 2, it follows that μ satisfies condition 2.3(6). Therefore we get from 2.7 that for $\beta(\omega) := (\omega(j))_{j\in\mathbb{N}}$ we have

$$\mathcal{E}^{2\pi-\text{per}}_{\{\omega\}}(\mathbb{R}) = \ker T_\mu \cong \Lambda_1(\beta(\omega))'_b.$$

This can also be obtained in different ways (see Petzsche [21], sect. 3, and Vogt [26], 7.7).

Using Vogt [29], 4.1, the following proposition is proved in [5].

2.9 Proposition. *Let $\Lambda(\alpha,\beta)$ be the projective spectrum defined in 2.6(1). Then*

$$\operatorname{Proj}^1 \Lambda(\alpha,\beta) = 0$$

if and only if there exists a partition $N_0 \cup N_1$ of $\mathbb{N}$ so that

$$\lim_{j\in N_0} \frac{\alpha_j}{\beta_j} = 0 \quad \textit{and} \quad \liminf_{j\in N_1} \frac{\alpha_j}{\beta_j} > 0.$$

Now the desired characterization of the surjectivity of a convolution operator T_μ on $\mathcal{E}_{\{\omega\}}(\mathbb{R})$ will follow from Proposition 1.12 by Theorem 2.3 and Proposition 2.9 as soon as we have shown that $\operatorname{Proj}^1 \mathcal{K}(\mu,\omega) = 0$ is equivalent to $\operatorname{Proj}^1 \Lambda(\alpha,\beta) = 0$. This holds by [5], 3.6. Hence we get:

2.10 Theorem. *Let ω be a weight function. The convolution operator T_μ on $\mathcal{E}_{\{\omega\}}(\mathbb{R})$ is surjective if and only if (1) and (2) hold:*

(1) T_μ admits a fundamental solution,

(2) the zero set $V(\widehat{\mu})$ of $\widehat{\mu}$ admits a partition $V = V_0 \cup V_1$ such that

$$\lim_{z\in V_0} \frac{|\operatorname{Im} z|}{\omega(|z|)} = 0 \quad \textit{and} \quad \liminf_{z\in V_1} \frac{|\operatorname{Im} z|}{\omega(|z|)} > 0.$$

2.11 Remark. In Theorem 2.10 condition (2) can be replaced by each one of the following conditions:

(3) $\ker T_\mu$ is reflexive (quasi-barrelled, barrelled, bornological),

(4) $(\ker T_\mu)'_b$ is complete (sequentially complete),

(5) $\ker T_\mu \cong X \times Y$ for some (FN)-space X and some (DFN)-space Y.

The equivalence of (3) and (4) with (2) follows from Vogt [29], 4.2. From (2) and Theorem 2.7 we get (5). Since (5) implies (3), it also implies (2).

As the following two propositions show, the conditions in 2.10(2) have an analytic significance of their own.

2.12 Proposition. *Let T_μ be a locally surjective convolution operator on $\mathcal{E}_{\{\omega\}}(\mathbb{R})$. Then the following conditions are equivalent*

(1) $T_\mu \colon \mathcal{E}_{\{\omega\}}(\mathbb{R}) \to \mathcal{E}_{\{\omega\}}(\mathbb{R})$ admits a continuous linear right inverse,

(2) $S_\mu \colon \mathcal{D}_{\{\omega\}}(\mathbb{R})'_b \to \mathcal{D}_{\{\omega\}}(\mathbb{R})'_b$ admits a continuous linear right inverse,

(3) there exist fundamental solutions ν_+ and ν_- of T_μ and $a_+, a_- \in \mathbb{R}$ with $\operatorname{Supp} \nu_+ \subset [a_+, \infty[$ *and* $\operatorname{Supp} \nu_- \subset]-\infty, a_-]$,

(4) $\displaystyle \lim_{z \in V(\widehat{\mu})} \frac{|\operatorname{Im} z|}{\omega(|z|)} = 0.$

To formulate the second proposition, we introduce for a weight function ω the space

$$\mathcal{E}_{(\omega)}(\mathbb{R}) := \left\{ f \in C^\infty(\mathbb{R}) \,\middle|\, \forall k, n \in \mathbb{N} : \|f\|_{k,1/n} < \infty \right\}.$$

Under its natural topology, $\mathcal{E}_{(\omega)}(\mathbb{R})$ is a nuclear Fréchet space, which is continuously embedded in $\mathcal{E}_{\{\omega\}}(\mathbb{R})$.

2.13 Proposition. *Let T_μ be a locally surjective convolution operator on $\mathcal{E}_{\{\omega\}}(\mathbb{R})$. Then the following conditions are equivalent:*

(1) $\ker S_\mu = \{\nu \in \mathcal{D}_{\{\omega\}}(\mathbb{R})' \mid \mu * \nu = 0\} \subset \mathcal{E}_{(\omega)}(\mathbb{R})$,

(2) $\ker T_\mu \subset \mathcal{E}_{(\omega)}(\mathbb{R})$,

(3) $\ker T_\mu$ *is a Fréchet space under the topology induced by* $\mathcal{E}_{\{\omega\}}(\mathbb{R})$,

(4) $\displaystyle \liminf_{z \in V(\widehat{\mu})} \frac{|\operatorname{Im} z|}{\omega(|z|)} > 0.$

Remark. Theorem 2.10 gives rise to the question whether every surjective convolution operator T_μ on $\mathcal{E}_{\{\omega\}}(\mathbb{R})$ can be factored as $T_\mu = T_\lambda \circ T_\nu$ where the zeros of $\widehat{\lambda}$ satisfy condition 2.12(4) while those of $\widehat{\nu}$ satisfy condition 2.13(4). This, however, is false even for the Gevrey-classes as is shown in [5], 4.11. On the other hand, the factorization of ultradifferential operators is possible under a technical condition, which holds, e.g., in the Gevrey-classes (see [5], 4.10).

3 Partial differential operators on $\mathcal{E}_{\{\omega\}}(\mathbb{R}^N)$

Reporting on our forthcoming article [6] we now indicate how Proposition 1.12 can be used to characterize the surjectivity of linear partial differential operators with constant coefficients on $\mathcal{E}_{\{\omega\}}(\mathbb{R}^N)$.

For $N > 1$ and $P \in \mathbb{C}[z_1, \ldots, z_N]$ with $P(z) = \sum_{|\alpha| \le m} a_\alpha z^\alpha$, we define

$$P(D): \mathcal{E}_{\{\omega\}}(\mathbb{R}^N) \to \mathcal{E}_{\{\omega\}}(\mathbb{R}^N), \quad f \mapsto \sum_{|\alpha| \le m} a_\alpha \left(\frac{1}{i}\right)^{|\alpha|} \frac{\partial^\alpha}{\partial x_1^{\alpha_1} \cdots \partial x_N^{\alpha_N}} f.$$

Moreover, the polynomial $P_m: z \mapsto \sum_{|\alpha| = m} a_\alpha z^\alpha$ is called principal part of P provided m is the degree of P. The set

$$V = V(P) := \{z \in \mathbb{C}^N \mid P(-z) = 0\}$$

is called the zero variety of P.

Since $P(D)$ admits a fundamental solution for every $P \neq 0$ in $\mathbb{C}[z_1, \ldots, z_N]$, we get from 1.12 and 1.11(b):

3.1 Proposition. *Let ω be a weight function and let $P \in \mathbb{C}[z_1, \ldots, z_N]$ with $P \neq 0$ be given. Then $P(D): \mathcal{E}_{\{\omega\}}(\mathbb{R}^N) \to \mathcal{E}_{\{\omega\}}(\mathbb{R}^N)$ is surjective if and only if* $\mathrm{Proj}^1 \mathcal{K}(P(D), \omega) = 0$.

Since $\mathcal{K}(P(D), \omega)$ is a projective spectrum of (DFN)-spaces, we get from Vogt [29], 3.5:

3.2 Corollary. *If $P(D): \mathcal{E}_{\{\omega\}}(\mathbb{R}^N) \to \mathcal{E}_{\{\omega\}}(\mathbb{R}^N)$ is surjective, then the kernel of this operator is bornological and barrelled.*

Corollary 3.2 can be used to get the following examples of non-surjective operators $P(D)$ on $\mathcal{E}_{\{\omega\}}(\mathbb{R}^N)$, which extend previous examples of Cattabriga [8].

3.3 Example. Let $P \in \mathbb{C}[z_1, \ldots, z_N]$ be an elliptic polynomial, let $M \in \mathbb{N}$ be given, and define $\widetilde{P} \in \mathbb{C}[z_1, \ldots, z_{N+M}]$ by $\widetilde{P}(z_1, \ldots, z_{N+M}) = P(z_1, \ldots, z_N)$. Then the operator $\widetilde{P}(D): \mathcal{E}_{\{\omega\}}(\mathbb{R}^{N+M}) \to \mathcal{E}_{\{\omega\}}(\mathbb{R}^{N+M})$ is not surjective. To prove this, we first note that the ellipticity of P implies that $\ker P(D)$ does not depend on ω and that by Wiechert [30], 2.2, Satz 9

(1) $$\ker P(D) \cong \Lambda_\infty(j^{1/(N-1)}) \cong A(\mathbb{C}^{N-1}).$$

Next note that

(2) $$\ker \widetilde{P}(D) = (\ker P(D)) \hat{\otimes} \mathcal{E}_{\{\omega\}}(\mathbb{R}^M).$$

For the space

$$X = \left\{ f \in \ker \widetilde{P}(D) \;\middle|\; f \text{ is } 2\pi\text{-periodic in } x_{N+1}, \ldots, x_{N+M} \right\}$$

we have

(3) $$X = (\ker P(D)) \hat{\otimes} \mathcal{E}^{2\pi-\mathrm{per}}_{\{\omega\}}(\mathbb{R}) \hat{\otimes} \cdots \hat{\otimes} \mathcal{E}^{2\pi-\mathrm{per}}_{\{\omega\}}(\mathbb{R}).$$

Therefore 2.12 and (2) imply that X is a complemented subspace of $\ker \widetilde{P}(D)$. On the other hand, we get from 2.8, (1) and (3)

$$X \cong \Lambda_\infty(j^{1/(N-1)}) \hat{\otimes} \Lambda_1(\alpha(\omega))'_b \hat{\otimes} \cdots \hat{\otimes} \Lambda_1(\alpha(\omega))'_b$$

where $\alpha(\omega) = (\omega(j))_{j\in\mathbb{N}}$ is a stable exponent sequence by 1.1(α). Therefore Krone and Vogt [13], 2.1, in connection with Krone and Vogt [13], 1.1, and Vogt [27], 4.2, show that

$$\Lambda_\infty(j^{1/(N-1)}) \hat{\otimes} \Lambda_1(\alpha(\omega))'_b \cong L_b\Big(\Lambda_1(\alpha(\omega)), \Lambda_\infty(j^{1/(N-1)})\Big)$$

is not bornological. Hence $\ker \widetilde{P}(D)$ is not bornological since it contains a complemented subspace which is not bornological. Now Corollary 3.2 shows that $\widetilde{P}(D)$ is not surjective.

Remark. From 3.3 we get in particular that the operator

$$P(D) = \Big(\frac{\partial}{\partial x}\Big)^2 + \Big(\frac{\partial}{\partial y}\Big)^2 : \mathcal{E}_{\{\omega\}}(\mathbb{R}^3) \to \mathcal{E}_{\{\omega\}}(\mathbb{R}^3)$$

is, for each weight function ω, not surjective. This is the example of Cattabriga [8] which we have mentioned in section 1.

To characterize the surjective operators $P(D)$ on $\mathcal{E}_{\{\omega\}}(\mathbb{R}^N)$ we first remark:

3.4 Remark. (a) For $Q, R \in \mathbb{C}[z_1, \dots, z_N]$ put $P = QR$. Then $P(D)$ is surjective if and only if $Q(D)$ and $R(D)$ are surjective.
(b) Because of (a) it suffices to consider only irreducible polynomials.

For an irreducible polynomial P we can again use Fourier analysis to get an appropriate representation of $(\ker P(D))'_b$ where $P(D)$ acts on $\mathcal{E}_{\{\omega\}}(\mathbb{R}^N)$. To describe this, we introduce the following notation.

3.5 Definition. Let ω be a weight function, let $P \in \mathbb{C}[z_1, \dots, z_N]$ be given, put $V = V(P)$ and denote by $A(V)$ the space of all analytic functions on the variety V. For $k, m \in \mathbb{N}$ define

$$A_{k,m}(V) := \Big\{ f \in A(V) \;\Big|\; \sup_{z\in V} |f(z)| \exp\Big(-k|\operatorname{Im} z| - \frac{1}{m}\omega(|z|)\Big) < \infty \Big\},$$

$$A_k(V) := \operatorname*{proj}_{\leftarrow m} A_{k,m}(V),$$

and

$$A_{\{\omega\}}(V) := \operatorname*{ind}_{k\rightarrow} A_k(V) = \operatorname*{ind}_{k\rightarrow} \operatorname*{proj}_{\leftarrow m} A_{k,m}(V).$$

We remark that $A_k(V)$ is a nuclear Fréchet space for each $k \in \mathbb{N}$ and that $A_{\{\omega\}}(V)$ is an (LF)-space. Using some steps out of the proof of Ehrenpreis' fundamental principle and a technical argument due to Braun, Meise, and Taylor [4], one derives from 2.4:

3.6 Proposition. *For each weight function ω and each irreducible polynomial P in $\mathbb{C}[z_1, \dots, z_N]$ the Fourier-Laplace transform*

$$\mathcal{F}: (\ker P(D))'_b \to A_{\{\omega\}}(V), \quad \mathcal{F}(\mu) := \widehat{\mu}: z \mapsto \langle \mu_x, e^{-i\langle x,z\rangle}\rangle, \; z \in V,$$

is a linear topological isomorphism.

For a general projective spectrum Vogt [29], 2.8, gives a necessary condition for $\mathrm{Proj}^1 \mathcal{X} = 0$, called (P_2^*), and a sufficient one, called (P_1^*). We will not go into these. In our case, they correspond to certain properties of the variety V, which we define now.

3.7 Phragmén-Lindelöf conditions. For ω, P, and V as in 3.6 and $f \in A(V)$ consider the following inequalities (1)–(3), where μ, n, L, M, k, m, and C are natural numbers:

$$(1) \qquad \text{for all } z \in V: \quad \log|f(z)| \le \mu|\mathrm{Im}\, z| + \frac{1}{n}\omega(|z|)$$

$$(2) \qquad \text{for all } z \in V: \quad \log|f(z)| \le L|\mathrm{Im}\, z| + \frac{1}{M}\omega(|z|)$$

$$(3) \qquad \text{for all } z \in V: \quad \log|f(z)| \le k|\mathrm{Im}\, z| + \frac{1}{m}\omega(|z|) + C$$

We say that P satisfies the analytic $(\mathrm{PL})_\omega$-condition (resp. $(\mathrm{PL})'_\omega$-condition) if the following holds:

$(\mathrm{PL})_\omega$: $\forall \mu\, \exists n, k\, \forall L, m\, \exists M, C\, \forall f \in A_\mu(V)$: (1) and (2) imply (3),

$(\mathrm{PL})'_\omega$: $\exists n\, \forall \mu\, \exists k\, \forall L, m\, \exists M, C\, \forall f \in A_\mu(V)$: (1) and (2) imply (3).

We say that P satisfies the plurisubharmonic $(\mathrm{PL})_\omega$-condition (resp. $(\mathrm{PL})'_\omega$-condition) if we can replace $\log|f|$ in the above by any plurisubharmonic function u on V (provided the condition $f \in A_\mu(V)$ is replaced by the corresponding estimates for u).

From 3.6, 1.8(2), and Vogt [29], 5.1 and 2.8 we get similarly as in section 2:

3.8 Proposition. *For each weight function ω and each irreducible P we have*

(1) If P satisfies the analytic $(\mathrm{PL})'_\omega$-condition, then $\mathrm{Proj}^1 \mathcal{K}(P(D), \omega) = 0$,

(2) If $\mathrm{Proj}^1 \mathcal{K}(P(D), \omega) = 0$, then P satisfies the analytic $(\mathrm{PL})_\omega$-condition.

In general, no evaluable characterization of $\mathrm{Proj}^1 \mathcal{X} = 0$ is known, except in the presence of a basis. However, in our case we can show that the analytic $(\mathrm{PL})_\omega$-condition implies the analytic $(\mathrm{PL})'_\omega$-condition. To do so, we use that both of these analytic conditions are equivalent to their respective plurisubharmonic counterparts by Meise, Taylor and Vogt [19] (see also [18], section 5). Since the equivalence of the plurisubharmonic $(\mathrm{PL})_\omega$-condition with the plurisubharmonic $(\mathrm{PL})'_\omega$-condition is trivial, this proves:

3.9 Theorem. *For each weight function ω and each $P \in \mathbb{C}[z_1, \dots, z_N]$, $P \neq 0$, the following conditions are equivalent:*

(1) $P(D)\colon \mathcal{E}_{\{\omega\}}(\mathbb{R}^N) \to \mathcal{E}_{\{\omega\}}(\mathbb{R}^N)$ is surjective,

(2) P satisfies the analytic $(\mathrm{PL})_\omega$-condition,

(3) P satisfies the plurisubharmonic $(\mathrm{PL})_\omega$-condition.

Remark. Theorem 3.9 extends previous results of Cattabriga [7] and Zampieri [31], who gave sufficient conditions for the surjectivity of $P(D)$ on the Gevrey classes $\Gamma^{\{d\}}(\mathbb{R}^N)$, $d > 1$ (see also Cattabriga [9]).

It is easy to see that in the limit case $d = 1$ the Gevrey-class $\Gamma^{\{1\}}(\mathbb{R}^N)$ equals $A(\mathbb{R}^N)$, the space of all complex-valued real-analytic functions on $\mathbb{R}^N$. The surjectivity of partial differential operators on $A(\mathbb{R}^N)$ has been characterized already by Hörmander [12]. For the sake of comparison we recall one of his results.

3.10 Theorem (Hörmander [12]). *Let $P \in \mathbb{C}[z_1, \ldots, z_N]$, $P \neq 0$, be given. A partial differential operator $P(D)\colon A(\mathbb{R}^N) \to A(\mathbb{R}^N)$ is surjective if and only if P satisfies*

(PL) *there exists $A \geq 1$ such that for each plurisubharmonic function u on $\mathbb{C}^N$ the conditions (1) and (2) imply (3)*

(1) $u(z) \leq |z|$ for all $z \in \mathbb{C}^N$,

(2) $u(z) \leq 0$ for all $z \in V(P_m) \cap \mathbb{R}^N$,

(3) $u(z) \leq A|\operatorname{Im} z|$ for all $z \in V(P_m)$.

Note that the (PL)-condition depends only on the principle part of the operator $P(D)$. This is no longer true in the non-quasianalytic classes $\Gamma^{\{d\}}(\mathbb{R}^N)$, as the following example shows.

3.11 Example. The heat operator

$$P(D) = \left(\frac{\partial}{\partial x}\right)^2 + \left(\frac{\partial}{\partial y}\right)^2 - \frac{\partial}{\partial z} : \Gamma^{\{d\}}(\mathbb{R}^3) \to \Gamma^{\{d\}}(\mathbb{R}^3)$$

is not surjective for $1 < d < 2$ (see Cattabriga [8]) and is surjective for $d \geq 2$ (see Cattabriga [7] and 3.13 below). However, its principle part is not surjective by 3.3.

The following result can be used to derive examples from Hörmander [12].

3.12 Proposition. *Let $P \in \mathbb{C}[z_1, \ldots, z_N]$, $P \neq 0$, be a homogeneous polynomial.*

(1) If there exists a weight function ω_0 so that P satisfies $(PL)_{\omega_0}$ then P satisfies $(PL)_\omega$ for all weight functions ω.

(2) If P satisfies condition (PL) of 3.10, then P satisfies $(PL)_\omega$ for all weight functions ω.

Using the classical estimates for the distance between the points in the variety V of a hypoelliptic polynomial and $\mathbb{R}^N$, it is easy to see that Theorem 3.9 implies:

3.13 Proposition. *If $P \in \mathbb{C}[z_1, \ldots, z_N]$, $P \neq 0$, is hypoelliptic, then there exists $d_0 \geq 1$ so that $P(D)$ is surjective on $\Gamma^{\{d\}}(\mathbb{R}^N)$ for each $d > d_0$.*

The algebraic subvarieties of $\mathbb{C}^2$ can be described rather explicitely in terms of Puiseux expansion. Therefore, condition 3.9(3) can be evaluated almost completely in dimension 2:

3.14 Proposition. *For every polynomial P in two variables there is a finite set $\{p_1, \ldots, p_k\}$ of rational numbers in $]0,1[$ such that the operator $P(D)\colon \mathcal{E}_{\{\omega\}}(\mathbb{R}^2) \to \mathcal{E}_{\{\omega\}}(\mathbb{R}^2)$ is surjective whenever ω satisfies for $1 \le j \le k$ either*

$$\lim_{r\to\infty} \frac{r^{p_j}}{\omega(r)} = 0$$

or

$$\liminf_{r\to\infty} \frac{r^{p_j}}{\omega(r)} > 0.$$

3.15 Corollary. *For each $P \in \mathbb{C}[z_1, z_2]$, $P \neq 0$, and each $d > 1$ the operator $P(D)$ is surjective on $\Gamma^{\{d\}}(\mathbb{R}^2)$.*

Remark. (a) 3.13 (resp. 3.14) extend Cattabriga [7], 1.2 ii) (resp. [8], 4.1)(see also [9]).
(b) As the research on the subject matter of the present section is still going on, we refer to our forthcoming article [6] for proofs and further results.

References

[1] C. A. Berenstein, B. A. Taylor, 'A new look at interpolation theory for entire functions of one variable', *Advances in Math.* **33** (1979), 109–143.

[2] A. Beurling, 'Quasi-analyticity and general distributions', *Lectures 4. and 5. AMS Summer institute* (Stanford, 1961).

[3] G. Björck, 'Linear partial differential operators and generalized distributions', *Ark. Math.* **6** (1965), 351–407.

[4] R. W. Braun, R. Meise, B. A. Taylor, 'Ultradifferentiable functions and Fourier analysis', preprint.

[5] R. W. Braun, R. Meise, D. Vogt, 'Existence of fundamental solutions and surjectivity of convolution operators on classes of ultradifferentiable functions', preprint.

[6] R. W. Braun, R. Meise, D. Vogt, 'Characterization of the linear partial differential operators with constant coefficients which are surjective on non-quasianalytic classes of Roumieu type', manuscript.

[7] L. Cattabriga, 'Solutions in Gevrey-spaces of partial differential equations with constant coefficients', *Astérisque* **89/90** (1981), 129–151.

[8] L. Cattabriga, 'On the surjectivity of differential polynomials on Gevrey-spaces', *Atti del Convegno 'Linear partial and pseudodifferential operators'* (30 Sett. – 2 Ott. 1982).

[9] L. Cattabriga, 'Remarks on the existence of solutions of partial differential equations in Gevrey spaces', these Proceedings.

[10] C. C. Chou, *La Transformation de Fourier Complexe et l'Equation de Convolution* (Springer LNM 325, 1973).

[11] L. Ehrenpreis, 'Solutions of some problems of division IV', *Amer. J. Math.* **82** (1960), 522–588.

[12] L. Hörmander, 'On the existence of real analytic solutions of partial differential equations with constant coefficients', *Invent. Math.* **21** (1973), 151–183.

[13] J. Krone, D. Vogt, 'The splitting relation for Köthe spaces', *Math. Z.* **190** (1985), 387–400.

[14] R. Meise, 'Sequence space representations for (DFN)-algebras of entire functions modulo closed ideals', *J. reine angew. Math.* **363** (1985), 59–95.

[15] R. Meise, 'Sequence space representations for zero-solutions of convolution equations on ultradifferentiable functions of Roumieu type', *Studia Math.*, to appear.

[16] R. Meise, B. A. Taylor, 'Sequence space representations for (FN)-algebras of entire functions modulo closed ideals', *Studia Math.* **85** (1987), 203–227.

[17] R. Meise, B. A. Taylor, D. Vogt, 'Equivalence of slowly decreasing conditions and local Fourier expansions', *Indiana Univ. Math. J.* **36** (1987), 729–756.

[18] R. Meise, B. A. Taylor, D. Vogt, 'Partial differential operators with continuous linear right inverse', these Proceedings.

[19] R. Meise, B. A. Taylor, D. Vogt, 'Equivalence of analytic and plurisubharmonic Phragmén-Lindelöf principles', manuscript.

[20] R. Meise, D. Vogt, 'Characterization of convolution operators on spaces of C^∞-functions admitting a continuous linear right inverse', *Math. Ann.* **279** (1987), 141–155.

[21] V. P. Palamodov, 'The projective functor in the category of linear topological spaces', *Math. USSR-Sbornik* **4** (1968), 529–559.

[22] H. J. Petzsche, 'Die Nuklearität der Ultradistributionsräume und der Satz vom Kern I', *Manuscripta Math.* **24** (1978), 133–174.

[23] H. J. Petzsche, D. Vogt, 'Almost analytic extension of ultradifferentiable functions and the boundary values of holomorphic functions', *Math. Ann.* **267** (1984), 17–35.

[24] V. S. Retakh, 'Subspaces of a countable inductive limit', *Dokl. Akad. Nauk SSSR* **194** (1970), No. 6, English transl. *Soviet Math. Dokl.* **11** (1970), 1384–1386.

[25] B. A. Taylor, 'On weighted polynomial approximation of entire functions', *Pacific J. Math.* **36** (1971), 523–539.

[26] D. Vogt, 'Sequence space representations of spaces of test functions and distributions', pp. 405–443 in: *Functional Analysis, Holomorphy, and Approximation Theory,* (ed. G. Zapata), Lect. Notes in Pure and Appl. Math. 83, (Dekker 1983).

[27] D. Vogt, 'On the functors $\mathrm{Ext}^1(E, F)$ for Fréchet spaces', *Studia Math.* **85** (1987), 163–197.

[28] D. Vogt, 'Lectures on projective spectra of (DF)-spaces', Seminar lectures, AG Funktionalanalysis Düsseldorf/Wuppertal 1987.

[29] D. Vogt, 'Topics on projective spectra of (LB)-spaces', these Proceedings.

[30] G. Wiechert, 'Dualitäts- und Strukturtheorie der Kerne linearer Differentialoperatoren', Dissertation Wuppertal (1982).

[31] G. Zampieri, 'An application of the fundamental principle of Ehrenpreis to the existence of global Gevrey solutions of linear partial differential equations', *Boll. U.M.I.* **6** (1986), 361–392.

PARTIAL DIFFERENTIAL OPERATORS WITH CONTINUOUS LINEAR RIGHT INVERSE

R. MEISE
Mathematisches Institut
Universität Düsseldorf
Universitätsstr 1
D-4000 Düsseldorf
F.R. Germany

B.A. TAYLOR
Department of Mathematics
University of Michigan
Ann Arbor, MI 48109
USA

D. VOGT
Universität-GHS-Wuppertal
Fachbereich Mathematik
Gaussstr. 20
D-5600 Wuppertal
F.R. Germany

§1. Introduction.

For Ω an open subset of $\mathbf{R}^N$, let $\mathcal{E}(\Omega)$ and $\mathcal{D}'(\Omega)$ denote respectively the space of all infinitely differentiable functions and the space of all distributions on Ω. For a polynomial $P(z) = \sum a_\alpha z^\alpha$ on $\mathbf{C}^N$, we denote by $P(D)$ the linear operator from $\mathcal{E}(\Omega)$ to $\mathcal{E}(\Omega)$, or from $\mathcal{D}'(\Omega)$ to $\mathcal{D}'(\Omega)$ defined by

$$P(D)f = \sum a_\alpha (i)^{-|\alpha|} f^{(\alpha)}$$

where $i = \sqrt{-1}$ and $f^\alpha = \frac{\partial^{|\alpha|} f}{\partial x^\alpha}$. It is well known that $P(D)$ is a continuous linear map that is surjective on both $\mathcal{D}'(\Omega)$ and $\mathcal{E}(\Omega)$ when Ω is an open convex subset of $\mathbf{R}^N$.

Professor L. Schwartz has posed the problem of determining when the operator $P(D)$ has a continuous linear right inverse; that is, when does there exist a continuous linear map R such that

$$P(D)Rf = f \tag{1.1}$$

for all $f \in \mathcal{E}(\Omega)$ or $f \in \mathcal{D}'(\Omega)$? For example, when Ω is $\mathbf{R}^N$ and $P(D)$ is hyperbolic in a direction, such an operator exists; one can take Rf to be the unique solution to the Cauchy problem $P(D)u = f$ with zero initial data. Negative results for the most important special classes of operators have been given by several authors. For $N \geq 2$, Grothendieck has shown that no elliptic operator can have a right inverse on $\mathcal{E}(\Omega)$ (see e.g. [TR]). The same result holds for hypoelliptic operators as Vogt has proved [V]. D. Cohoon [CO] has proved that parabolic operators (and an even more general class) have no right inverse on $\mathcal{E}(\Omega)$.

T. Terzioğlu (ed.), Advances in the Theory of Fréchet Spaces, 47–62.

In these lectures, we outline a complete solution of L. Schwartz's problem for open convex sets in $\mathbf{R}^N$, along with related work on the Phragmén-Lindelöf inequality and applications to various classes of partial differential operators. It is not possible to give complete proofs in the time available, so we will outline the most important parts of our work.

Theorem 1. Let Ω be an open convex set in $\mathbf{R}^N$, and let P be a polynomial on $\mathbf{C}^N, P \not\equiv 0$. Then the following are equivalent:

(1) $P(D): \mathcal{E}(\Omega) \to \mathcal{E}(\Omega)$ has a continuous linear right inverse.

(2) for each compact subset K of Ω there exists a compact subset K' of Ω such that for all $\varphi \in \mathcal{E}(\Omega)$ such that $\varphi(x) = 0$ for all $x \in K'$, there exists $u \in \mathcal{E}(\Omega)$ such that $P(D)u = \varphi$ and $u(x) = 0$ for all $x \in K$.

(3) for each compact subset K of Ω, there exists a compact convex subset K' of Ω such that if $\xi \in \Omega \setminus K'$, the equation

$$P(D)T = \delta_\xi$$

has a solution $T \in \mathcal{D}'(\Omega)$ with support outside of K. (Here δ_ξ denotes the point evaluation at ξ).

(4) $P(D): \mathcal{D}'(\Omega) \to \mathcal{D}'(\Omega)$ has a continuous linear right inverse.

(5) The Phragmén-Lindelöf principle, $PL(\Omega)$, described in Definition 1.1, is valid for the variety $V = \{P(-z) = 0\}$

Thus, the description of the operators $P(D)$ with a continuous linear right inverse can be made, according to parts (2) and (3) of Theorem 1, in terms of the supports of solutions of the equation. We will show in Section 6 how this interacts with hyperbolicity to show it is quite rare for operators on bounded domains to have a right inverse. For positive results the most interesting case is $\Omega = \mathbf{R}^N$ where there are lots of examples of operators that admit a continuous linear right inverse. In this case, it turns out that the Phragmén-Lindelöf condition is much easier to evaluate than the conditions (2), (3) of Theorem 1. We shall discuss the Phragmén-Lindelöf condition in Section 5, and relate it to a similar condition studied by Hörmander [H1].

The Phragmén-Lindelöf condition we study here is the following. Recall that the support function of a compact convex set K in $\mathbf{R}^N$ is

$$h_K(\xi) = \sup\{x \cdot \xi \; : \; x \in K\}, \quad \xi \in \mathbf{R}^N \tag{1.2}$$

(here $x \cdot \xi = x_1\xi_1 + \cdots + x_n\xi_n$.)

Definition 1.1. Let V be an algebraic variety in $\mathbf{C}^N$ and Ω an open convex set in $\mathbf{R}^N$. We say that the Phragmén-Lindelöf condition $PL(\Omega)$ is valid provided that for every compact convex subset K of Ω, there exists a compact convex subset K' of Ω such that, given any compact convex subset K'' of Ω, there exists a constant $B = B(K'')$ such that the following implication holds for every plurisubharmonic function u on $\mathbf{C}^N$: if

$$(i) \quad u(z) \leq h_K(Im\, z) + O\left(\log(1+ \mid z \mid)\right), \quad z \in \mathbf{C}^N$$

$$(ii) \qquad u(z) \leq h_{K''}(Im\, z), \qquad z \in V$$

then

$$(iii) \quad u(z) \leq h_{K'}(Im\, z) + B \log(2+ \mid z \mid), \quad z \in V.$$

Theorem 1 is not the first time that a Phragmén-Lindelöf inequality has been shown to be equivalent to properties of constant coefficient partial differential operators. The first result in this direction known to us is the work of Hörmander [H1] who proved that a certain, different Phragmén-Lindelöf principle is equivalent to the surjectivity of the operator on the space of real-analytic functions. See section 5 for a description of this principle (called (HPL) here). Other results of this type are given in Braun, Meise, and Vogt [BMV]. A. Kaneko has also shown that continuation properties of pde's can be described in terms of a Phragmén-Lindelöf condition. For $\Omega = \mathbf{R}^N$, Theorem 1.1 was announced in [MTV].

§2. Characterization for $\mathcal{D}'(\Omega)$.

It is possible to give a simple characterization of when there exists a right inverse for $P(D)$ on $\mathcal{D}'(\Omega)$ in terms of supports of solutions of the equation $P(D)u = f$ in case Ω is an arbitrary open set in $\mathbf{R}^N$. It is convenient to introduce the following notation. First, for $\varepsilon > 0$ let

$$\Omega_\varepsilon = \left\{x \in \Omega \mid \; dist\left(x, \mathbf{R}^N \setminus \Omega\right) > \varepsilon \text{ and } \mid x \mid < \frac{1}{\varepsilon}\right\},$$

and

$$\mathcal{D}'(\Omega, \Omega_\varepsilon) = \left\{\mu \in \mathcal{D}'(\Omega) \mid \; \mu\mid_{\Omega_\varepsilon} \equiv 0\right\}.$$

Further, let

$$\mathcal{N}(\Omega) = \{\mu \in \mathcal{D}'(\Omega) \mid \; P(D)\mu = 0\}$$

denote the null space of $P(D)$. In the following theorem, we consider $\mathcal{D}'(\Omega)$ as being equipped with its usual strong topology, and denote this space $\mathcal{D}'_b(\Omega)$.

Theorem 2.1. For each open set $\Omega \subset \mathbf{R}^N$ and each polynomial P on $\mathbf{C}^N$, $P \not\equiv 0$, the following are equivalent:

(1) $P(D) : \mathcal{D}'_b(\Omega) \to \mathcal{D}'_b(\Omega)$ has continuous linear right inverse.

(2) for each $\varepsilon > 0$, there exists $0 < \delta < \varepsilon$ such that for every $\mu \in \mathcal{N}(\Omega_\delta)$, there exists $\nu \in \mathcal{N}(\Omega)$ such that $\nu \mid_{\Omega_\varepsilon} = \mu \mid_{\Omega_\varepsilon}$.

(3) for each $\varepsilon > 0$, there exists $0 < \delta < \varepsilon$ such that for every $f \in \mathcal{D}'(\Omega, \Omega_\delta)$ there exists $g \in \mathcal{D}'(\Omega, \Omega_\varepsilon)$ such that $P(D)g = f$.

(4) for each $\varepsilon > 0$, there exists $0 < \delta < \varepsilon$ such that for every $f \in \mathcal{E}'(\mathbf{R}^N, \Omega_\delta)$, there exists $g \in \mathcal{D}'(\Omega, \Omega_\varepsilon)$ such that $P(D)g = f$.

Proof: That (1) $\Rightarrow$ (2) and (1) $\Rightarrow$ (3) are immediate consequences of the existence of a continuous linear right inverse. In (3), the desired solution of the equation is given by Rf where R is the right inverse while in (2), $\nu = \mu - R\mu$. This relation also shows that (2) and (3) are equivalent. The implication (3) $\Rightarrow$ (4) is trivial, so we only have to prove that (4) $\Rightarrow$ (1). This is done by a linearized Mittag-Leffler argument. To simplify technical detail, we will only give here the case $\Omega = \mathbf{R}^N$.

Choose a partition of unity $\{\varphi_j\}$ on $\mathbf{R}^N$ that is locally finite and so that each φ_j has compact support in $\mathbf{R}^N$ inside a ball about a point x_j of radius 1. Then $\mid x_j \mid \to \infty$ and by (4), there exists a solution T_j of $P(D)T_j = \delta_{x_j}$, where δ_{x_j} is the Dirac measure at x_j, and a sequence $R_j \to +\infty$ such that T_j has no support in $\{\mid x \mid \leq R_j\}$. Thus, if $\tau_j(x) = x + x_j$ denotes translation by $-x_j$, then $E_j(x) = \tau_* T_j(x) = T_j(x + x_j)$ is a fundamental solution of $P(D)$ i.e. $P(D)E_j = \delta_0$. Then $P(D)(E_j * f) = f$ for all distributions f with compact support. Further, if f is supported in $\{\mid x - x_j \mid \leq 1\}$, then $E_j * f$ has no support in $\{\mid x \mid \leq R_j - 1\}$. Thus,

$$Rf = \sum_j E_j * (\varphi_j f) \tag{2.1}$$

is a convergent sum in $\mathcal{D}'_b(\Omega)$, since it is locally finite. Thus, R is a right inverse for $P(D)$.

Corollary 2.2. If $P(D)$ has a continuous linear right inverse on $\mathcal{D}'_b(\Omega)$, then it also has one on $\mathcal{E}(\Omega)$.

Proof: The formula (2.1) shows that R also maps $\mathcal{E}(\Omega)$ into $\mathcal{E}(\Omega)$.

Corollary 2.3. If $\{\Omega_i\}$ is a family of open sets in $\mathbf{R}^N$, if $P(D)$ has a right inverse on $\mathcal{D}'_b(\Omega_i)$ for all i, and if $\Omega = \cap\Omega_i$ is open, then $P(D)$ has a right inverse on Ω.

Proof: This follows from Theorem 2.1 (4), using a suitable partition of unity.

§3. Characterization for $\mathcal{E}(\Omega)$.

The conditions analogous to (2) and (3) of Theorem 2.1 for the existence of a right inverse on $\mathcal{E}(\Omega)$ are almost the same as for $\mathcal{D}'(\Omega)$. They may well be the same, but we have not yet been able to establish this except with an additional hypothesis on Ω. In this section, we will explain these results.

For K a compact subset of Ω, we let $C^\infty(K)$, (respectively $C^m(K)$) denote the space of infinitely differentiable Whitney functions on K (respectively m-times continuously differentiable functions) equipped with the usual seminorms

$$(3.1) \qquad |\,\varphi\,|_{m,K} = sup\{|\,D^\alpha\varphi(x)\,| : \; x \in K, \; |\,\alpha\,| \leq m\}$$

where $D^\alpha = i^{-|\alpha|}\frac{\partial^{|\alpha|}}{\partial x^\alpha}$. We also set

$$\begin{aligned}
\mathcal{N}^\infty(\Omega) &= \{f \in \mathcal{E}(\Omega) \; : \; P(D)f = 0\} = \mathcal{N}(\Omega) \cap \mathcal{E}(\Omega). \\
\mathcal{N}^\infty(K) &= \{f \in C^\infty(K) \; : \; P(D)f = 0\}. \\
H^0(\Omega) &= \{f \in L^2(\Omega) \; : \; P(D)f = 0 \text{ on } \Omega\}. \\
\mathcal{E}(\Omega, K) &= \{f \in \mathcal{E}(\Omega) \; : \; f = 0 \text{ on } K\}.
\end{aligned}$$

Lemma 3.1. For an open set $\Omega \subset \mathbf{R}^N$ and P a polynomial on $\mathbf{C}^N$, $P \not\equiv 0$ consider the following conditions.

(1) $P(D) : \mathcal{E}(\Omega) \to \mathcal{E}(\Omega)$ has a continuous linear right inverse.

(2) for every $\varepsilon > 0$, there exists $0 < \delta < \varepsilon$ and a continuous linear map $\Phi \; : C^\infty(\overline{\Omega}_\delta) \to N^\infty(\Omega)$ such that for all $u \varepsilon N^\infty(\overline{\Omega}_\delta)$, $\Phi(u)\,|_{\Omega_\varepsilon} = u\,|_{\Omega_\varepsilon}$.

(3) for every $\varepsilon > 0$, there exists $0 < \delta < \varepsilon$ such that for all $0 < \eta < \delta$, there exists $m \geq 0$ and $D > 0$ so that for each $\mu \in C^\infty(\overline{\Omega}_\varepsilon)'$ with

$$|\,\mu(\chi)\,|^2 \leq A^2 \int_{\Omega_\eta} |\,\chi\,|^2, \quad \chi \in H^0(\Omega_\eta)$$

we have

$$|\,\mu(\chi)\,| \leq AD\,|\,\chi\,|_{m,\overline{\Omega}_\delta}, \quad \chi \in \mathcal{N}^\infty(\overline{\Omega}_\delta).$$

(4) for every $\varepsilon > 0$, there exists $\delta > 0$ such that for all $0 < \eta < \delta$ there exists $m \geq 0$ such that for all $\mu \in C^m(\Omega_\eta)$ with $P(D)\mu = 0$ on Ω_δ, there exists $\nu \in H^0(\Omega_\eta)$ such that $\mu = \nu$ on Ω_ε.

(5) for every $\varepsilon > 0$, there exists $0 < \delta < \varepsilon$ such that for every $0 < \eta < \delta$, there exists $m \geq 0$ such that for all $f \in C^m(\Omega_\eta)$ such that $f = 0$ on Ω_ε, there exists $g \in L^2(\Omega_\eta)$ such that $P(D)g = f$ in Ω_η and $g = 0$ in Ω_ε.

Then we have the following implications.

$$(1) \Longleftrightarrow (2) \Rightarrow (3) \Rightarrow (4) \Rightarrow (5)$$

We omit the proof of this lemma. Note, however, that condition (5) is almost exactly like that of Theorem 2.1 (4). The only difference is in the regularity of the right hand side of the equation f. In one case, $f \in \mathcal{E}'(\Omega)$ is a distribution while in the other case f has m continuous derivatives. For a large class of open sets, these conditions are equivalent.

Definition 3.2. An open set $\Omega \subset \mathbf{R}^N$ has the *integration property* (IP) if the following condition holds.

(IP) for each $\varepsilon > 0$, there exists $0 < \delta < \varepsilon$ such that for all $0 < \eta < \delta$, $\mu \in \mathcal{E}'(\Omega, \Omega_\delta)$, $n \geq 0$, there exists $m \in \mathbf{N}$, $f_j \in C^n(\overline{\Omega}_\eta, \Omega_\varepsilon)$ and multi-indices α_j for $1 \leq j \leq m$ such that

$$\mu = \sum_{j=1}^{n} f_j^{(\alpha_j)} \quad \text{on } \Omega_\eta.$$

The integration property (IP) can be verified to hold if Ω is convex or, more generally, if Ω is starshaped with respect to one of its points. Note also that if (IP) holds, then condition (5) of Lemma 3.1 implies condition (4) of Theorem 2.1. To see this, let $f \in \mathcal{E}'(\mathbf{R}^N, \Omega_\delta)$. Then by (IP), with a suitable choice of ε, δ, m, we can write $f = F^{(\alpha)}$. If α is a suitable sufficiently high order multi-index, then F will have continuous partial derivatives of order m over a large subset of Ω. Further, we can, by property (IP), preserve the fact that F vanishes on a large subset Ω_δ (possibly a different δ) of Ω. By property (5), we can then solve $P(D)G = F$ with G vanishing on Ω_ε, so that $g = G^{(\alpha)}$ is the distribution required in Theorem 2.1. (4), since $P(D)$ commutes with the differentiation operator. Thus we obtain the following result.

Theorem 3.3. For a convex open set in $\mathbf{R}^N$, the following are equivalent.

(1) $P(D) : \mathcal{E}(\Omega) \to \mathcal{E}(\Omega)$ has a right inverse.

(2) $P(D) : \mathcal{D}'_b(\Omega) \to \mathcal{D}'_b(\Omega)$ has a right inverse.

Further, (1) and (2) are equivalent to conditions (2)-(5) of Lemma 3.1.

§4. The Phragmén-Lindelöf condition.

In this section, we show the equivalence of the Phragmén-Lindelöf condition $PL(\Omega)$ with the existence of a right inverse for $P(D)$ on an open convex set in $\mathbf{R}^N$. That is, we explain the equivalence of the conditions (5) and (1) of Theorem 1.

The proof is based on the so-called Fundamental Principle for solutions of the equation $P(D)u = 0$. For K a compact subset of Ω, let $\mathcal{E}'(K)$ denote the space of all entire functions $F(z)$ satisfying the growth estimate

$$|F(z)| \leq C_1 (1+|z|)^{C_2} exp(h_K(Im\, z)), \quad z \in \mathbf{C}^N \tag{4.1}$$

where h_K, defined in (1.2), denotes the support function of the compact convex set K. The Paley-Wiener-Schwartz theorem (see e.g. [H2] or [E]) implies that the Fourier transform

$$\hat{\mu}(z) = \mu(x \to exp(-ix \cdot z))$$

gives an isomorphism between $C^\infty(K)'$, the space of distributions supported on K and $\hat{\mathcal{E}}'(K)$. Thus the dual space of $\mathcal{E}(\Omega)$ is identified with the space $\hat{\mathcal{E}}'(\Omega)$ of all entire functions that satisfy (4.1) for some $C_1, C_2 > 0$ and some $K \subset\subset \Omega$.

If P is a polynomial on $\mathbf{C}^N$, then the null space $\mathcal{N}^\infty(\Omega)$ of $P(D)$ in $\mathcal{E}(\Omega)$ therefore has a dual space that is identified with the quotient space $\mathcal{E}'(\Omega)/\mathcal{N}^\infty(\Omega)^\perp$, where

$$\mathcal{N}^\infty(\Omega)^\perp = \{\mu \in \mathcal{E}'(\Omega) : \mu(f) = 0 \text{ whenever } P(D)f = 0, \quad f \in \mathcal{E}(\Omega)\}.$$

Under Fourier transform $\mathcal{N}^\infty(\Omega)^\perp$ is identified with the submodule of $\hat{\mathcal{E}}'(\Omega)$ (over the polynomials),

$$\mathcal{J}(P) = P(-z)\hat{\mathcal{E}}'(\Omega) = \{P(-z)\hat{\nu}(z) : \quad \hat{\nu} \in \hat{\mathcal{E}}'(\Omega)\}.$$

Since if $\hat{\mu} \in \hat{\mathcal{E}}'(K)$ satisfies (4.1) and is divisible by $P(-z)$ as an analytic function, it is also divisible by $P(-z)$ in $\hat{\mathcal{E}}'(K)$, the space $\hat{\mathcal{E}}'(\Omega)/\mathcal{N}^\infty(\Omega)^\perp$ is identified via Fourier transform as a space of functions on the multiplicity variety defined by the submodule $\mathcal{J}(P)$.

For our purposes here, we do not have to worry about the notion of higher order multiplicities.

Lemma 4.1. Let $P = P_1 \cdots P_\ell$ where $P_1, \cdots, P_\ell$ are polynomials on $\mathbf{C}^N$. Then $P(D)$ has a right inverse on $\mathcal{E}(\Omega)$ (resp. $\mathcal{D}'(\Omega)$) if and only if each $P_i(D)$ has a right inverse on $\mathcal{E}(\Omega)$ (resp.$\mathcal{D}'(\Omega)$).

Thus, to study the existence of a right inverse, we can assume that P is irreducible. We will make this simplifying assumption throughout the rest of this section. Under this assumption, the quotient space $\hat{\mathcal{E}}'(K)/P(-z) \cdot \hat{\mathcal{E}}'(K)$ can be identified with

$$\hat{\mathcal{E}}'(V) = \{F(z) : F \text{ is analytic on V and satisfies (4.1) for } z \in V\}$$

where

$$V = \{z \in \mathbf{C}^N \; : P(-z) = 0\}.$$

Definition 4.2. Let V be an algebraic variety in $\mathbf{C}^N$ and Ω an open convex set in $\mathbf{R}^N$. We say that the analytic Phragmén-Lindelöf condition, $APL(\Omega)$, holds on V, provided that the conditions of Definiton 1.1 hold for psh functions u of the form $u(z) = log \mid f(z) \mid$, f entire on $\mathbf{C}^N$.

Theorem 4.3. For Ω an open convex set in $\mathbf{R}^N$, the following are equivalent.

(1) $P(D) : \mathcal{E}(\Omega) \to \mathcal{E}(\Omega)$ has a continuous linear right inverse.

(2) The variety $V = \{z \in \mathbf{C}^N \; : P(-z) = 0\}$ satisfies the analytic Phragmén-Lindelöf condition.

Proof: We already know that (1) is equivalent to the condition (3) of Lemma 3.1. It turns out that $APL(\Omega)$ is the direct translation of (3) via Fourier transform. For, let $K = \overline{\Omega}_\varepsilon$, $K' = \overline{\Omega}_\delta$, $K'' = \overline{\Omega}_\eta$. The condition $\mu \in C^\infty(\overline{\Omega}_\varepsilon)'$ means exactly the estimate (i) of Definition 1.1 by the Paley-Wiener- Schwartz theorem. The condition

$$\mid \mu(\chi) \mid^2 \leq A^2 \int_{\Omega_\eta} \mid \chi \mid^2$$

for χ satisfying $P(D)\chi = 0$ implies the estimate (ii) of Definition 1.1. And the last condition

$$\mid \mu(\chi) \mid \leq A \cdot D \parallel \chi \parallel_{m,\overline{\Omega}_\delta}$$

of (3) of Lemma 3.1 implies an estimate of the form (iii) in Definition 1.1. Except for the exact value of the constants it is easy to verify that these steps are all reversible, which gives the equivalence of $APL(\Omega)$ with (1)

§5. General Remarks on Phragmén-Lindelöf Principles.

There are several natural questions that arise when one considers Phragmén-Lindelöf conditions on algebraic varieties. We will discuss six of them in this section.

(I) Does the validity of $PL(\Omega)$ depend on the class of psh functions u considered?

(II) Does the validity of $PL(\Omega)$ depend on whether (i) of the definition is enforced globally ($z \in \mathbf{C}^N$) or just on V?

(III) Does the validity of the condition $PL(\Omega)$ depend on Ω?

(IV) Does the validity of $PL(\Omega)$ depend on $P(z)$, or just on the leading homogeneous term of $P(z)$?

(V) Can the function $log(1+ \mid z \mid)$ be replaced by other growth rates?

(VI) Is there a geometric characterization of the varieties V for which the Phragmén-Lindelöf condition holds?

First, to discuss questions I and II. We already showed in Theorem 4.3 that the relevant condition for the equivalence with the existence of a right inverse is to have the conditions hold only for $u = log \mid f \mid$ where f is entire on $\mathbf{C}^N$. On the other hand, one could require that the conditions $(i) + (ii) \Rightarrow (iii)$ hold not just for psh functions u on $\mathbf{C}^N$, but also for weakly psh functions u on V (recall that u is weakly psh on V if u is locally bounded above and psh on the regular points of V). In this latter case, one can require that (i) hold only for $z \in V$. It turns out that all these conditions are equivalent. One can give a precise result that implies this equivalence. Unfortunately, it is quite technical. However, it implies that for a large class of Phragmén-Lindelöf principles, including the one studied by Kaneko and that studied by Braun, Meise and Vogt in [BMV], the principle is independent of whether it is required to hold for all weakly psh functions u on V or just for those of the form $u(z) = log \mid f(z) \mid$, f entire on $\mathbf{C}^N$.

The theorem that gives the equivalence of $APL(\Omega)$ and $PL(\Omega)$ is as follows. First, for the algebraic variety V of codimension 1 in $\mathbf{C}^N$, choose coordinates $s \in \mathbf{C}, w \in \mathbf{C}^{N-1}$, such that the projection map $z = (s, w) \to w$ is a proper map on V. Then we can write

$$V = \left\{(s_j(w), w) : w \in \mathbf{C}^{N-1},\ 1 \le j \le m\right\}.$$

There exists a polynomial $D(w)$ so that projection is a covering map that is branched only over $\{D(w) = 0\}$. Coordinates can also be chosen so that

$$| s_j(w) | \le C_0(1 + | w |), \quad w \in \mathbf{C}^{N-1}.$$

Then it is possible to choose constants $\varepsilon, C, A, B > 0$ such that the following conditions hold:

(α) $\mathcal{O} = \left\{z \in \mathbf{C}^N : log\, | P(z) | + A\, log(2 + | z |) + B\, log\frac{1}{|D(w)|} < 0\right\}$
is pseudoconvex.

(β) $S_0 = \left\{z = (s, w) \in \mathbf{C}^N : | D(w) | < \varepsilon(1 + | w |)^{-C}\right\}$
has the property that $z \in V \cap S_0$ implies for all u psh on $\mathbf{C}^N$,
$u(z) \le max\{u(\zeta) : \zeta \in V \setminus S_0, | \zeta - z | \le 1\}$.

Theorem 5.1. Let $\rho > 0$ and $\Theta > 1$ be given. Then there exists a constant $D > 0$ and constants $\varepsilon_1, C_1 > 0$, depending only on V, ρ, Θ, and the constants in α, β such that for all weakly psh functions u on V, $0 \le u \le \rho | z |$, $z \in V$, and for all $z \in V \setminus S_0$, there exists a subset $E(z)$ of the set

$$B(z) = \left\{\zeta = (s, w) \in V : | \zeta - z | < \varepsilon_1(1 + | z |)^{-C_1}\right\}$$

such that

(a) $| B(z) \setminus E(z) | \cdot max_{\zeta \in B(z)} | u(\zeta) | \le D \cdot | B(z) |$ where $| F |$ denotes the Lebesgue measure of the set of $w \in \mathbf{C}^{N-1}$ with $(\zeta, w) \in F$; and

(b) given $\zeta \in E(z)$, there exists F entire on $\mathbf{C}^N$ such that

(i) $log\, | F(\zeta) | \ge \frac{1}{\Theta} u(\zeta) - D log(2 + | \zeta |)$; and

(ii) $log\, | F(\tau) | \le max_{|\tilde{\tau} - \tau| \le 1} u(\tilde{\tau}) + D log(2 + | \tau |)$ for $\tau, \tilde{\tau} \in V$.

It is a direct consequence of this result that $APL(\Omega)$ implies $PL(\Omega)$.

Corollary 5.2. The conditions $APL(\Omega)$ and $PL(\Omega)$ are equivalent.

Regarding question III it turns out that the condition $PL(\Omega)$ does depend a great deal on the set Ω. As we will point out in the next section, if Ω has boundary points with normals that are non- characteristic directions for $P(D)$, then P must be hyperbolic in these directions.

Question IV is suggested because of the work of Hörmander [H1] who was the first to consider such Phragmén-Lindelöf inequalities. He studied the Phragmén-Lindelöf condition, denoted (HPL) defined as follows.

Condition (HPL)

For each compact convex set $K \subset \Omega$, there exists a compact convex set $K' \subset \Omega$ such that for every compact convex set $K'' \subset \Omega$ and every $\delta > 0$, there exists $0 < \delta' < \delta$ and a constant $C > 0$ such that, for all entire functions f on $\mathbf{C}^N$ such that

$$(i) \qquad log \mid f(z) \mid \leq h_K(Im\ z) + \delta' \mid z \mid \quad , \ z \in \mathbf{C}^N$$

$$(ii) \qquad log \mid f(z) \mid \leq h_{K''}(Im\ z) \quad , \ z \in V$$

we have

$$(iii) \quad log \mid f(z) \mid \leq h_{K'}(Im\ z) + \delta \mid z \mid + C \quad , \ z \in V.$$

Hörmander proved that this condition on $V = \{P(-z) = 0\}$ is equivalent to the following condition on the homogeneous variety

$$V_h = \{P_m(-z) = 0\}, \ \ P(z) = P_m(z) + \text{lower order terms}$$

where P_m is the principal part of P. For every compact convex set $K \subset \Omega$, there exists $K' \subset \Omega$ and a constant $\delta > 0$ such that for every psh function u on $\mathbf{C}^N$, if

$$(i) \quad u(z) \leq h_K(Imz) + \delta \mid z \mid \quad , \ z \in \mathbf{C}^N$$

$$(ii) \qquad u(z) \leq 0 \qquad , \ z \in V_h \cap \mathbf{R}^N$$

then

$$(iii) \qquad u(z) \leq h_{K'}(Imz) \qquad , \ z \in V_h.$$

Hörmander also showed how the conditions could be localized.

For the conditions $PL(\Omega)$, there is no such equivalence with the principle holding for the top order homogeneous part. A simple example is provided in $N = 2$ by the operator $\frac{\partial^2}{\partial x^2}$ on $\{(x,t) \in \mathbf{R}^2\}$. This has a right inverse, while the heat operator $\frac{\partial}{\partial t} - \frac{\partial^2}{\partial x^2}$ does not.

It is possible to give a much more detailed analysis of this general question and we will do so in future work.

The question V is quite interesting and turns out to be closely related to question VI. First of all, we can consider other Phragmén-Lindelöf conditions, $PL(\Omega,\omega)$ obtained by replacing the functions $log(1+ \mid z \mid)$ in (i) and (iii) of Definition (1.1) by another psh function $\omega(z) = \omega(\mid z \mid)$ such that $\omega(\mid z \mid) = o(\mid z \mid), \mid z \mid\rightarrow \infty$. With only minor technical conditions on ω, it turns out that $PL(\Omega,\omega)$ also holds. More generally, if the Phragmén-Lindelöf condition holds with one weight function ω, then it is true for larger weight functions $\tilde{\omega}$. The converse is false. On $\Omega = \mathbf{R}^2$, the condition $PL(\Omega) = PL(\Omega, log(2+ \mid z \mid))$ fails for the heat operator $\frac{\partial}{\partial t} - \frac{\partial^2}{\partial x^2}$. On the other hand, the condition $PL\left(\Omega, \sqrt{\mid z \mid}\right)$ is valid.

There is an interesting connection between the validity of the condition $PL(\Omega,\omega)$ and the geometry of the variety, at least when $\Omega = \mathbf{R}^N$. Again let $V_h = \{P_m(-z) = 0\}$ where $P(z) = P_m(z) +$ lower order terms, so that V_h is the tangent cone of the variety V at infinity. We say that V_h has *order of contact* $\alpha < 1$ with V at infinity if, given $z_j \in V, \mid z_j \mid\rightarrow \infty$, there exists $\tilde{z}_j \in V_h$ with $\mid z_j - \tilde{z}_j \mid = 0(\mid z_j \mid^\alpha)$, and, further, α is the smallest positive number with this property.

Theorem 5.3. Suppose that V satisfies $PL(\mathbf{R}^N,\omega)$. Then

(a) V_h satisfies $PL(\mathbf{R}^N)$; and

(b) $\mid z \mid^\alpha = O(\omega(\mid z \mid))$, where $0 \leq \alpha < 1$ is the order of contact of V_h with V at infinity.

The proof of Theorem 5.3 is too long for us to give here. It seems likely that the converse of the theorem is also true.

Finally, we mention some results about homogeneous varieties that are extremely useful in checking examples.

Definition 5.4. A homogeneous algebraic variety V_h in $\mathbf{C}^N$ is said to satisfy the *dimension condition* if and only if every irreducible component W of V_h satisfies

$$dim_{\mathbf{R}} W \cap \mathbf{R}^N = dim_{\mathbf{C}} V_h. \tag{5.1}$$

We say that V_h satisfies the *strong dimension condition* if (5.1) also holds for each local irreducible component W of V_h near a point $z_0 \in V_h \cap \mathbf{R}^N, z_0 \neq 0$.

Definition 5.5. A homogeneous algebraic variety V_h in $\mathbf{C}^N$ is said to satisfy the local Phragmén-Lindelöf condition (local (PL)) near a point $\xi \in \mathbf{R}^N \cap V_h$ if and only if there exists $0 < \varepsilon_1 < \varepsilon_2 < \varepsilon_3$ and a constant $A > 0$ such that every function u psh on $\{\mid \zeta - \xi \mid < \varepsilon_3\}$

which satisfies $0 \le u \le 1$, there and $u(\zeta) \le 0$ for $\zeta \in \mathbf{R}^N \cap V_h \cap \{|\zeta - \xi| < \varepsilon_2\}$, also satisfies $u(\zeta) \le A \mid Im\zeta \mid$ for all $\zeta \in V_h \cap \{|\zeta - \xi| < \varepsilon_1\}$.

Theorem 5.6. If V_h is a homogeneous algebraic variety in $\mathbf{C}^N$, then V_h satisfies $PL(\mathbf{R}^N)$ if and only if

(i) V_h satisfies the dimension condition; and

(ii) V_h satisfies local (PL) at each $\xi \in V_h \cap \mathbf{R}^N, |\xi| = 1$.

Further, in this case V_h satisfies the strong dimension condition.

Remark The ideas of Theorem 5.5 are contained in §5 of [H1]. See especially Theorem 6.3, 6.5, and 6.6. There is some difference in the results, however. For example, if $P(D) = P_m(D)$ is an elliptic operator, then $\{0\} = \{\xi \in \mathbf{R}^N : P_m(\xi) = 0\}$, so (i) fails. On the other hand, every elliptic operator satisfies $HPL(\mathbf{C}^N)$.

§6. Hyperbolicity and the existence of a right inverse.

It is well-known that for a polynomial P in N-variables hyperbolic in a direction $\vec{n} \in \mathbf{R}^N$, $P(D)$ has a continuous linear right inverse on the following spaces:

$$\mathcal{E}(\mathbf{R}^N),\ \mathcal{D}'(\mathbf{R}^N),\ \mathcal{E}\left(H_\pm(\vec{n})\right),\ \text{and}\ \mathcal{D}'\left(H_\pm(\vec{n})\right),$$

where $H_\pm(\vec{n})$ denotes the half space $\{x \in \mathbf{R}^N : x \cdot \vec{n} > 0 \text{ (resp. } < 0)\}$. This is due to the fact that $P(D)$ has fundamental solutions supported in proper closed convex cones $\Gamma_\pm$ in the half space $H_\pm(\vec{n})$. The following Lemma gives information in the converse direction.

Lemma 6.1. Let Ω be an open subset of $\mathbf{R}^N$ with C^1-boundary; and assume that $P(D)$ has a continuous linear right inverse on $\mathcal{D}'(\Omega)$ or $\mathcal{E}(\Omega)$. For $\xi \in \partial\Omega$, let n_ξ denote the unit outer normal to $\partial\Omega$ at ξ. Then n_ξ is either a characteristic vector for P or P is hyperbolic with respect to n_ξ.

The proof of Lemma 4.1 is based on Holmgren's uniqueness theorem. In case Ω is convex, one can also give a proof based on the Phragmén- Lindelöf principle of Section 4. Based on Lemma 6.1, once can prove that there are very few operators that have a continuous linear right inverse on a smoothly bounded open set.

Theorem 6.2. For a polynomial P on $\mathbf{C}^N$, $P \not\equiv 0$, the following are equivalent:

(1) there exists an open bounded set $\Omega \subset \mathbf{R}^N$, with C^1 boundary so that $P(D)$ has a right inverse on $\mathcal{E}(\Omega)$.

(2) $P(D)$ has a right inverse on $\mathcal{E}(\Omega)$ and on $\mathcal{D}'(\Omega)$ for each open convex subset Ω of $\mathbf{R}^N$.

(3) P and its principal part P_m are equally strong (see [H2] for the definition) and P_m is a product of m linear factors with real coefficients.

(4) P is hyperbolic and P_m, the principal part of P, is a product of m real linear factors.

For domains that do not have smooth boundary, the condition can be quite different. Note that if Ω is a polygon whose faces are noncharacteristic for $P(D)$ and such that the normal vectors to the faces are hyperbolic directions for $P(D)$, then $P(D)$ has a right inverse on $\mathcal{E}(\Omega)$ and $\mathcal{D}'(\Omega)$ by Corollaries 2.2 and 2.3.

§7. Examples for the case $\Omega = \mathbf{R}^N$.

We have seen from the results of §6 that the condition that $P(D)$ have a right inverse is quite stringent when Ω is a bounded open set. Many examples are explained by Theorem 6.2. However, when $\Omega = \mathbf{R}^N$ there are many more operators that have right inverses. In this case, it seems easier to use the Phragmén-Lindelöf condition to check examples. We will mention here a few nonhyperbolic operators that have continuous linear right inverses. A large class of examples is given by the following special case.

Theorem 7.1. Let $P = P_m$ be a homogeneous polynomial that has real coefficients and satisfies

(i) $dim_{\mathbf{R}}\{Q = 0\} \cap \mathbf{R}^N = n - 1$ for every irreducible factor Q of P_m; and

(ii) $\nabla P_m \neq 0$ on $\{P_m = 0\} \cap \mathbf{R}^N \cap \{|\,\xi\,| = 1\}$.

Then $P(D) = P_m(D)$ has a continuous linear right inverse on $C^\infty(\mathbf{R}^N)$ and $\mathcal{D}'(\mathbf{R}^N)$.

Remark: Note that this includes examples like $P(z) = z_1^3 + z_2^3 + z_3^3$ on $\mathbf{C}^3$, and the ultrahyperbolic operator, $P(z) = z_1^2 + z_2^2 - z_3^2 - z_4^2$ on $\mathbf{C}^4$.

Proof: This follows easily from Theorem 5.5. The condition (i) is the dimension condition for $V_h = \{P_m(-z) = 0\}$. The condition (ii) of Theorem 5.5 is implied by (ii) of Theorem

7.1. This is readily verified by using the implicit function theorem and the classical local version of the Phragmén-Lindelöf inequality. We also comment that, as in Theorem 6.5 of [H1], the condition (ii) of Theorem 7.1 can be relaxed to the assumption that $P(z) = P_m(z)$ is locally hyperbolic at each $\xi \in \{P(z) = 0\} \cap \mathbf{R}^N$, $\xi \neq 0$.

It is also easy to give examples in the negative direction for large classes of operators. For example, if $P(\xi)$ is elliptic or hypoelliptic then the dimension condition fails for $P_m(\xi)$, the principal part of P. Thus, by Theorem 5.2, P cannot have a right inverse on $\mathcal{E}(\mathbf{R}^N)$. If $P(D)$ is a parabolic operator, then $P_m(\xi)$ also fails the dimension condition, so that again $P(D)$ has no right inverse on $\mathcal{E}(\mathbf{R}^N)$. Finally, when $N = 2$, one can completely answer the question of which operators have a right inverse. Namely, only those that are products of hyperbolic operators.

Theorem 7.2. For P a polynomial in two variables, $P \not\equiv 0$, the following are equivalent.

(1) $P(D)$ has a continuous linear right inverse on $\mathcal{E}(\mathbf{R}^N)$ or $\mathcal{D}'(\mathbf{R}^N)$.

(2) each irreducible factor of P is hyperbolic.

(3) P is hyperbolic.

(4) P is hyperbolic with respect to each noncharacteristic direction.

(5) P has a right inverse on $\mathcal{E}(\Omega)$ and on $\mathcal{D}'(\Omega)$ for each open convex subset Ω of $\mathbf{R}^2$.

(6) there exists an open half space $H \subset \mathbf{R}^2$ such that $P(D)$ has a right inverse on $\mathcal{E}(H)$ or $\mathcal{D}'(H)$.

Proof: We won't verify all parts of this theorem. But, for example, to see that (1) $\Rightarrow$ (2), choose coordinates $z = (s, \omega)$ on C^2 so that the Puisseux expansion of the roots of $P(-z) = 0$ are

$$s_j(\omega) = a_j\omega^{r_j} + b_j\omega^{r_{j+1}} + \cdots$$

where r_j is a rational number, $0 \leq r_j \leq 1$, $r_{j+1} < r_j$ and b_j, $a_j \neq 0$. If $0 \leq r_j < 1$, then the contact condition (b) of Theorem 5.3 is violated. Thus, we must have $r_j = 1$. Then if $r_{j+1} > 0$, the contact condition is again violated, so we must have $r_{j+1} \leq 0$. If a_j is not real, then the dimension condition (i) of Theorem 5.6 is violated by V_h, so a_j must be real. This means that P is hyperbolic.

REFERENCES

[BMV] Braun, R., Meise, R., and Vogt, D., 'Applications of the projective limit functor to convolution and partial differential equations', these Proceedings.

[CO] Cohoon, D.K., 'Nonexistence of a continuous right inverse for linear partial differential operators with constant coefficients', *Math. Scand.* **29**, (1971), 337-342.

[E] Ehrenpreis, L., *Fourier Analysis in Several Complex Variables*, Wiley-Interscience, 1970.

[H1] Hörmander, L., 'On the existence of real analytic solutions of partial differential equations with constant coefficients', *Inv. Math.* **21**, (1973), 151-182.

[H2] Hörmander, L., *The Analysis of Linear Partial Differential Operators,* Vol. I, II, Springer-Verlag, 1983.

[MTV] Meise, R., Taylor, B.A., and Vogt, D., 'Characterisation des opérateurs linéaires aux'derivées partielles avec coefficients constants sur $\mathcal{E}(\mathbf{R}^N)$ admettant un inverse a droite qui est linéaire et continu', *C.R. Acad. Sci. Paris,* **307**, (1988), 239-242.

[TR] Treves, F., *Locally Convex Spaces and Linear Partial Differential Equations,* Springer-Verlag, 1967.

[V] Vogt, D., 'Some results on continuous linear maps between Fréchet spaces', pp.349-381 in Functional Analysis: Surveys and Recent Results III, K.D. Bierstedt and B. Fuchssteiner (Eds.), North-Holland Mathematics Studies 90, (1984).

HARTOGS TYPE EXTENSION THEOREM OF REAL ANALYTIC SOLUTIONS OF LINEAR PARTIAL DIFFERENTIAL EQUATIONS WITH CONSTANT COEFFICIENTS

A. KANEKO
Department of Mathematics
College of General Education
University of Tokyo
Tokyo - Japan

Abstract

A necessary and sufficient condition is given for extension of real analytic solutions of $P(D)u = 0$ in a situation analogous to Hartogs's theorem in several complex variables. It is expressed in terms of a Phragmén-Lindelöf type principle on $N(P)$.

§1. In this talk we present results on extension of the real analytic solutions of a linear partial differential equation with constant coefficients. The problem has its mental origin in the theory of functions of several complex variables. First we briefly review it in general.

Riemann's theorem on removable isolated singularity is a model of the results on the extension of solutions under additional growth conditions at the singularity. Here *singularity* means the set of points where the solution does not satisfy the equation (in this model the Cauchy-Riemann equation). For the development in this line see the survey article by Polking [10]. On the other hand, Hartogs's continuity theorem is a model of the results where the solutions may be extended to an obstacle without any condition if they satisfy *many* equations, constituting an over-determined system, as clarified by Ehrenpreis and Malgrange. In 1966, however, Grušin [1] presented a result where sufficiently smooth solutions may be unconditionally extended to a sufficiently thin obstacle even if they satisfy only one equation. Stimulated by this, we gave the following complete answer to the extension problem of real analytic solutions in the case of a compact obstacle:

T. Terzioğlu (ed.), Advances in the Theory of Fréchet Spaces, 63–72.

Theorem. ([3]). *Let $K \subset U$ be a pair of compact set and its neighborhood such that $U \setminus K$ is connected. Then the following are equivalent:*

(1) *Every real analytic solution of $P(D)u = 0$ on $U \setminus K$ can be extended to a real analytic solution on U.*

(2) *$P(D)$ has no elliptic factor.*

The following result for the case of systems of equations was suggested by a result of Palamodov [9] on the removability of weak isolated singularity of C^∞ solutions, which in turn generalized another work of Grušin:

Theorem. ([4]). *Let $K \subset U$ be as above. The following are equivalent:*

(1) *Every real analytic solution of $P(D)u = 0$ can be extended to a hyperfunction solution on U.*

(2) *$Ext^1(Coker^t P, \mathbf{C}[D])$ has no elliptic component.*

Here, choosing a free resolution over the ring $\mathbf{C}[D]$ of polynomials of D,

$$0 \leftarrow Coker^t P \leftarrow \mathbf{C}[D]^{\ell_0} \overset{{}^tP(D)}{\longleftarrow} \mathbf{C}[D]^{\ell_1} \overset{{}^tQ(D)}{\longleftarrow} \mathbf{C}[D]^{\ell_2},$$

we have by definition

$$Ext^1(Coker^t P, \mathbf{C}[D]) = \text{Ker}\ \{Q(D) : \mathbf{C}[D]^{\ell_1} \to \mathbf{C}[D]^{\ell_2}\}/P(D)\mathbf{C}[D]^{\ell_0}. \qquad (1)$$

If we require the extended solution also to be real analytic on the whole U, then the condition (2) should be supplemented with another one. Hom($Coker^t P$, $\mathbf{C}[D]$) = 0, i.e. the system should be *determined.* Recall that the condition for the extendability of any hyperfunction solution on $U \setminus K$ was the *overdeterminedness* $Ext^1(Coker^t P, C[D]) = 0$. The above weakened condition has its reason in the unique continuation property of the solution. In fact, Abramczuk [1] has shown that entirely the same assertion holds for any quasi-analytic class solutions.

Hartogs's continuity theorem has another form, i.e. the case where K, U in the above are replaced by their intersection with a half space, which will be assumed to be of the form $x_n < const.$ throughout to fix the idea. We shall call hereafter such a pair $K \subset U$ a *Hartogs pair.* Recently we obtained the following abstract answer to the extendability of real analytic solutions in this case:

Theorem. ([5]). *The following are equivalent:*

(1) *Every real analytic solution on $U \setminus K$ can be continued to a real analytic solution on U for any Hartogs pair $K \subset U$.*

(2) *For any $A, a > 0$ we can find $B \geq A$, $b < a$, $\delta > 0$ such that the following Phragmén-Lindelöf type principle holds for global holomorphic functions on every irreducible component of the variety $P(\zeta) = 0$: If $F(\zeta)$ satisfies, for any $\epsilon > 0$,*

$$| F(\zeta) | \leq C_\epsilon \exp(\epsilon | \zeta | + A | \operatorname{Im} \zeta' | + a(\operatorname{Im} \zeta_n)_-), \qquad \text{for } \forall \zeta \in N(P),$$

and, for any $\epsilon > 0$, (2)

$$| F(\zeta) | \leq C_\epsilon \exp(\epsilon | \zeta | + A | \operatorname{Im} \zeta' | + b | \operatorname{Im} \zeta |), \text{ for } | \operatorname{Im} \zeta | \leq \delta(| \operatorname{Re} \zeta | + 1),$$

then it satisfies, for any $\epsilon > 0$, (3)

$$| F(\zeta) | \leq C_\epsilon \exp(\epsilon | \zeta | + B | \operatorname{Im} \zeta' | + b | \operatorname{Im} \zeta |), \quad \text{for} \quad \forall \zeta \in N(P). \tag{4}$$

Here $(t)_- := \max\{-t, 0\}$. This is related to the following choice of the corresponding notions:

$$D = \frac{1}{i}\partial, \quad (\mathcal{F}u)(\zeta) = \int e^{-ix\zeta} u(x) dx.$$

§2. In the rest of this talk we shall give an outline of the proof of this last theorem. First of all remark that we can make a reduction so that $P(D)$ is irreducible and that

$$U = \{x_n < 0\}, \quad K = \{| x' | \leq A, \ -a \leq x_n < 0\}. \tag{5}$$

For the details of this reduction see [6]. In the sequel, for an open set U we denote in general by $\mathcal{A}(U)$, $\mathcal{B}(U)$ the space of real analytic functions and of hyperfunctions on U respectively, and, for a closed subset $K \subset U$, by $\mathcal{B}[K]$ the subspace of $\mathcal{B}(U)$ of those with supports contained in K. Note that $\mathcal{B}[K]$ is determined only by K and does not depend on the choice of its neighborhood U.

Let u be a hyperfunction solution of $P(D)u = 0$ on $U \setminus K$. Let $[u] \in \mathcal{B}(U)$ be any extension of u. (We can take such u thanks to the flabbiness of the sheaf of hyperfunctions.) Then $P(D)[u] \in \mathcal{B}[K]$. Let $[[P(D)[u]]] \in \mathcal{B}[\overline{K}]$ denote a further extension of $P(D)[u]$ with compact support. This is determined modulo $P(D)\mathcal{B}[\overline{K}] + \mathcal{B}[\partial K]$, where somewhat abusively we are denoting $\partial K := \overline{K} \setminus K = \overline{K} \cap \{x_n = 0\}$. Thus applying the Fourier transform and restricting to the variety $N(P) := \{\zeta \in \mathbf{C}^N; \ P(\zeta) = 0\}$, we obtain a globally defined holomorphic function

$F(\zeta) = \mathcal{F}[[P(D)[u]]] \mid_{N(P)}$ on $N(P)$ which satisfies, in view of a Paley-Wiener type theorem,

$$| F(\zeta) | \leq C_\epsilon \exp(\epsilon | \zeta | + H_{\overline{K}}(\text{ Im } \zeta)). \tag{6}$$

Here $H_{\overline{K}}(\text{ Im } \zeta)$ denotes the supporting function of $\overline{K}$. For K as in (5) it becomes

$$H_{\overline{K}}(\text{ Im } \zeta) = A | \text{ Im } \zeta' | + a(\text{ Im } \zeta_n)_-.$$

Next, in order to extract information from the real analyticity of u, we take an "extension" $v \in \mathcal{B}(\mathbf{R}^n) \cap \mathcal{A}(\mathbf{R}^n \setminus \overline{K})$ of u preserving its singularity modulo $\mathcal{A}$ at K. That is, we consider $[u] \in \Gamma_K(U, \mathcal{B}/\mathcal{A})$, and employing the flabbiness of the quotient sheaf $\mathcal{B}/\mathcal{A}$, we now take an extension $v \in \Gamma_{\overline{K}}(\mathbf{R}^n, \mathcal{B}/\mathcal{A})$, just as we took the extension $[[P(D)[u]]]$ from $P(D)[u]$. By the vanishing of the first cohomology groups for the sheaf $\mathcal{A}$, we can take a global representative of this element in $\mathcal{B}(\mathbf{R}^n)$, which we will denote by the same letter v as above. We can further make this procedure in a more refined way so that v becomes an exponentially decreasing holomorphic function on a conical neighborhood of the real axis toward the infinity. We shall give in the Appendix an elementary way of constructing such a modification. To prepare notation, however, we recall here the standard argument for such a construction: This is to use the same procedure as above employing, instead of $\mathcal{O}$ and $\mathcal{B}$, the sheaf $\mathcal{O}_\approx$ of rapidly decreasing holomorphic functions of modified type and the sheaf $\tilde{\mathcal{Q}}$ of modified Fourier hyperfunctions, respectively. These are sheaves on $\mathbf{D}^n + i\mathbf{R}^n$ and $\mathbf{D}^n$, respectively, where $\mathbf{D}^n := \mathbf{R}^n \coprod \mathbf{S}^{n-1}_\infty$ denotes the directional compactification of $\mathbf{R}^n$. Intuitively, a section of $\mathcal{O}_\approx$ is a holomorphic function which decreases exponentially on a conical neighborhood of every point at infinity contained in the domain of definition of the section. If the exponential decay is replaced by the infra-exponential growth, it gives the definition of the sheaf $\tilde{\mathcal{O}}$ of slowly increasing holomorphic functions of modified type. The sheaf $\tilde{\mathcal{Q}}$ of modified Fourier hyperfunctions are constructed from this latter just as $\mathcal{B}$ from $\mathcal{O}$ is, namely, as the boundary values of the sections of $\tilde{\mathcal{O}}$ to the "real axis" $\mathbf{D}^n$. By the definition, $\tilde{\mathcal{O}}$, $\mathcal{O}_\approx$ agree with $\mathcal{O}$ on $\mathbf{C}^n$, and $\tilde{\mathcal{Q}}$ with $\mathcal{B}$ on $\mathbf{R}^n$. For further details of these sheaves, see [8], [11]. Anyway, we can thus take $v \in \tilde{\mathcal{Q}}(\mathbf{D}^n)$ such that $v \mid_{\mathbf{D}^n \setminus \overline{K}} \in \mathcal{O}_\approx(\mathbf{D}^n \setminus \overline{K})$ and that $[u] - v \mid_U \in \mathcal{A}(U)$. In the sequel we shall express this decreasing condition for v simply as $v \in \mathcal{B}(\mathbf{R}^n) \cap \mathcal{O}_\approx(\mathbf{D}^n \setminus \overline{K})$.

The utility of these sheaves is explained by the following Paley-Wiener-Ehrenpreis type assertion given in Kawai [8]:

Lemma. *Let K be a convex compact set. The following are equivalent:*

(1) $v(x) \in \mathcal{B}(\mathbf{R}^n) \cap \mathcal{O}_{\approx}(\mathbf{D}^n \setminus K)$.

(2) $(\mathcal{F}v)(\zeta)$ *is a real analytic function which, for any $\delta > 0$, can be holomorphically extended to a conic neighborhood*

$$| \operatorname{Im} \zeta | \leq c_\delta(| \operatorname{Re} \zeta | + 1)$$

and there satisfies, for any $\epsilon > 0$,

$$|(\mathcal{F}v)(\zeta)| \leq C_\epsilon \exp(\epsilon | \zeta | + H_K(\operatorname{Im} \zeta) + \delta | \operatorname{Im} \zeta |).$$

Put $w = [[P(D)[u]]] - P(D)v$. We have $w \in \mathcal{B}(\mathbf{R}^n) \cap \mathcal{O}_{\approx}(\mathbf{D}^n \setminus \overline{K})$. On the other hand, from the assumption we have $w |_U \in \mathcal{A}(U)$. Hence $w \in \mathcal{B}(\mathbf{R}^n) \cap \mathcal{O}_{\approx}(\mathbf{D}^n \setminus \partial K)$, and we have

$$F(\zeta) = \mathcal{F}[[P(D)[u]]] |_{N(P)} = \mathcal{F}w |_{N(P)}$$

on some conical neighborhood $| \operatorname{Im} \zeta | \leq \delta(| \operatorname{Re} \zeta | + 1)$ of the real axis. Namely, $F(\zeta)$ satisfies there another estimate

$$| F(\zeta) | = | (\mathcal{F}w)(\zeta) | \leq C_\epsilon \exp(\epsilon | \zeta | + H_{\partial K}(\operatorname{Im} \zeta) + b | \operatorname{Im} \zeta |).$$

Thus by the above lemma we have

$$| F(\zeta) | \leq C_\epsilon \exp(\epsilon | \zeta | + B | \operatorname{Im} \zeta' | + b | \operatorname{Im} \zeta |), \text{ for } \forall \zeta \in N(P).$$

By the Fundamental Principle corresponding to the growth order of Fourier images of hyperfunctions with compact support (see [4]), we can find an entire holomorphic function with the same estimate, hence a hyperfunction g with support in the set $\{| x' | \leq B, \ | x_n | \leq b\}$ such that

$$F(\zeta) = \mathcal{F}g |_{N(P)} \cdot$$

Thus $\mathcal{F}[[P(D)[u]]] - \mathcal{F}g$ is divisible by $P(\zeta)$ as an entire function and by Malgrange's inequality we can find another hyperfunction h with support in $\{| x' | \leq B, \ -a \leq x_n \leq b\}$ such that

$$[[P(D)[u]]] = g + P(D)h,$$

hence

$$P(D)([u] - h) = 0 \quad \text{in} \quad x_n < -b.$$

This means that we can extend u as a hyperfunction solution to $x_n < -b$ at the cost of modifying it in the set $\{| x' | \leq B, \ x_n \geq -a\}$. By a result on the propagation of real analyticity (see [8], Theorem 5.1.1), the extended solution is real analytic. Repeating this procedure (employing reduction by absurdity, to be more precise), we obtain a required extension to $x_n < 0$.

Conversely, suppose that there exists a global holomorphic function $F(\zeta)$ on $N(P)$ satisfying the estimates (2), (3) of the theorem. First of all, from (2) with use of the Fundamental Principle and the Paley-Wiener theorem we can find a hyperfunction $v(x)$ with support in $\{|x'| \leq A, \quad -a \leq x_n \leq 0\}$ such that

$$F(\zeta) = \mathcal{F}v\,|_{N(P)}\,.$$

Next, from (3), employing a localized form of the Fundamental Principle whose proof is similar to the one given in [4], we can find a holomorphic function in a smaller conic neighborhood of the type $|\ \mathrm{Im}\ \zeta\ | \leq \delta'(|\ \mathrm{Re}\ \zeta\ | + 1)$, satisfying the same estimate as (3). Then by the above lemma of Kawai such a function is the Fourier transform of a hyperfunction $w(x) \in \mathcal{B}(\mathbf{R}^n) \cap \mathcal{O}_{\approx}(\mathbf{D}^n \setminus \{|x'| \leq A+b,\ |x_n| \leq b\})$. Now the difference $v - w$ is divisible by $P(D)$ and again by the Malgrange inequality and Kawai's lemma we can find a hyperfunction $u(x) \in \mathcal{B}(\mathbf{R}^n) \cap \mathcal{O}_{\approx}(\mathbf{D}^n \setminus \{|x'| \leq A+b,\ -a \leq x_n \leq b\})$ such that

$$P(D)u(x) = v(x) - w(x).$$

Let $\gamma > 0$ be a small constant and let $h(x)$ be a real analytic solution of $P(D)h = w$ on $x_n < -b - \gamma$. Such an h exists because v is holomorphic on a convex complex neighborhood of the real set $\{x_n < -b - \gamma\}$ by the definition of $\mathcal{O}_{\approx}$. Then

$$P(D)(u + h) = v \text{ in } x_n < -b - \gamma$$

and we have obtained a real analytic solution on $U \setminus K$, where now $U = \{x_n < -b - \gamma\}$ and $K = \{|x'| \leq A + b,\ -a \leq x_n \leq -b - \gamma\}$. (Note that our operator is translation invariant.) If this is extendable as a hyperfunction solution to the whole U, then modulo $P(D)\mathcal{B}[\overline{K}]$, v will reduce to one with support in $\{|x'| \leq A + b, \quad -b - \gamma \leq x_n \leq 0\}$. Thus after application of Fourier transform and restriction to $N(P)$, we will conclude that $F(\zeta) = \mathcal{F}v\,|_{N(P)}$ will actually satisfy

$$|F(\zeta)| \leq C_{\gamma,\epsilon} \exp(\epsilon\,|\zeta| + (A + b)\,|\ \mathrm{Im}\ \zeta'\,| + (b + \gamma)\,|\ \mathrm{Im}\ \zeta_n\,|).$$

Since $\gamma > 0$ is arbitrary, we have obtained (4).

Remark 1: The above proof shows the existence of the following natural injection

$$\mathcal{A}_P(U \setminus K)/\mathcal{A}_P(U) \to \widetilde{\mathcal{B}[\overline{K}]}\{N(P)\}/\widetilde{\mathcal{B}[\partial K]}\{N(P)\}. \tag{7}$$

Here in general $\mathcal{A}_P(U)$ denotes the space of real analytic solutions of $P(D)u = 0$ on U. (Hence the left-hand side quotient space represents the obstruction for extension of real analytic solutions.) For a compact set L in general $\widetilde{\mathcal{B}[L]}\{N(P)\}$ denotes the space of holomorphic functions on $N(P)$ with the same growth condition as $\widetilde{\mathcal{B}[L]} = \mathcal{F}(\mathcal{B}[L])$. Furthermore, any representative of an element in the image of this mapping satisfies, for any $\epsilon > 0$,

$$| F(\zeta) | \leq C_\epsilon \exp(\epsilon | \zeta | + H_{\overline{K}}(\operatorname{Im} \zeta)) \quad \text{for} \quad \forall \zeta \in N(P), \tag{8}$$

and, for any $\delta > 0$ there exists $c_\delta > 0$ such that for any $\epsilon > 0$,

$$| F(\zeta) | \leq C_{\epsilon,\delta} \exp(\epsilon | \zeta | + H_{\partial K}(\operatorname{Im} \zeta') + \delta | \operatorname{Im} \zeta |) \quad \text{for} \quad | \operatorname{Im} \zeta | \leq c_\delta(| \operatorname{Re} \zeta | + 1). \tag{9}$$

We expect that these two properties characterize the image of (7) completely. If this is true, then introducing the another condition: for any $\epsilon > 0$

$$| F(\zeta) | \leq C_\epsilon \exp(\epsilon | \zeta | + H_{\partial K}(\operatorname{Im} \zeta')) \quad \text{for} \quad \forall \zeta \in N(P), \tag{10}$$

the Phragmén-Lindelöf type principle "(8)+(9) $\Rightarrow$ (10)" will give a necessary and sufficient condition for $\mathcal{A}_P(U \setminus K)/\mathcal{A}_P(U) = 0$ for a fixed convex obstacle K. This is, however, not so clear because in the latter half of the above proof, we assumed the existence of a global real analytic solution, and this requires to diminish U a little. For the moment a precise condition for a fixed K can be formulated only by means of some projective limit spaces. See [5].

Remark 2: Similar assertion holds for a system of equations. The condition is stated on the *multiplicity variety* $N(Ext^1(Coker^t P, \mathbf{C}[D]))$, with use of the *Noether operator*. Note that for the extension of hyperfunction solutions in the situation of general Hartogs pair, the overdeterminedness is sufficient but is not necessary in general. A necessary and sufficient condition is written in terms of a kind of hyperbolicity of the module $Ext^1(Coker^t P, \mathbf{C}[D])$. See [6].

§3. The problem of paraphrasing the above abstract condition in terms of algebraic one for $P(\zeta)$ is still open. For the moment we have the following partial results (see [5]):

Theorem. *Each of the following conditions are sufficient for the extension of real analytic solutions to hold for any Hartogs pair:*

(1) *There exists a sequence of directions ϑ_k tending to $(0, \cdots, 0, 1)$ with respect to which $P(D)$ is hyperbolic.*
(2) *$P(D)$ has the principal part $P_m(D)$ which does not contain D_n and which is elliptic with respect to the variables actually contained in $P_m(D)$.*
(3) *$P(D)$ is of second order and has real principal part independent of D_n.*
(4) *The operator $P(D)$ itself does not contain D_n.*

Condition (1) is obtained directly by a totally different argument based on the hyperbolicity. (Here hyperbolicity means the weakest one.) It also gives the following necessary condition:

Theorem: *If* $(0,\cdots,0,1)$ *is a non-characteristic direction of* $P(D)$ *and the extension of real analytic solutions holds for any Hartogs pair, then* $P(D)$ *must be hyperbolic to this direction. Hence if* $(0,\cdots,0,1)$ *is an interior point in the set of directions with the extension property, then we must have the situation (1) in the preceding theorem.*

Thus interesting situations are rather degenerate. A typical example is

$$D_1^2 \pm \cdots \pm D_k^2 + cD_n, \quad \text{with } k < n,$$

which contains the heat equation, or the Schrödinger equation. (Concerning the heat equation in the direction of time evolution we have an explanation based on the theory of holomorphic semi-groups. See [5]. But the other cases do not seem trivial at all for such a simple operator.)

Remark 3: There are researches on the extension of holomorphic solutions. See e.g. references in [12]. It seems that the problem of extension of real analytic solutions is not directly connected with that of holomorphic solutions.

Remark 4: When the set K is thin and is contained in a non-characteristic hypersurface for the operator, we can apply a completely different method based on the boundary value theory which is (micro-)localizable and valid to equations with variable coefficients. For this topic see [7].

§4. Appendix: We explain here a most elementary way to obtain an exponentially decreasing modification of $u \in \mathcal{B}(U) \cap \mathcal{A}(U \setminus K)$. It is to use the following variant of Kashiwara's twisted Radon decomposition of $\delta(x)$:

$$\delta(x) = \int_{S^{n-1}} W(x,\omega)d\omega,$$

where $W(x,\omega)$ denotes the boundary value of

$$W(z,\omega) = \frac{(n-1)!}{(-2\pi i)^n} \frac{J(z,\omega)e^{-z^2}}{\{z\omega + i(z^2 - (z\omega)^2)/\sqrt{1+z^2}\}^n}$$

and $J(z,\omega)$ is a polynomial in $z, \omega, (1+z^2)^{-\frac{1}{2}}$, of which the explicit form is not important. Let $\{\Gamma_j\}_{j=1}^N$ be convex open cones whose duals Γ_j^0 give a decomposition of S^{n-1}. For $z \in \mathbf{R}^n + i\Gamma_j^0$, put

$$W(z,\Gamma_j^0) = \int_{\Gamma_j^0 \cap S^{n-1}} W(z,\omega)d\omega.$$

Given $u(x) \in \mathcal{B}(U) \cap \mathcal{A}(U \setminus K)$, we first represent u as a hyperfunction by the boundary-values

$$u(x) = \sum_{j=1}^{N} F_j(x + i\Gamma_j^0)$$

with defining functions $F_j(z)$ analytically continuable to $U \setminus K$. Then, choosing a domain D such that $K \subset\subset D \subset\subset U$ and small vectors $a_j \in \Gamma_j$, we put

$$G_k(z) = \sum_{j=1}^{N} \int_{D+ia_j} W(z - w, \Gamma_k^0) F_j(w) dw.$$

It is clear that $G_k(x + i\Gamma_k^0) \in \mathcal{B}(\mathbf{R}^n) \cap \mathcal{O}_{\approx}(\mathbf{D}^n \setminus K)$ and that the difference

$$\sum_{k=1}^{N} G_k(x + i\Gamma_k^0) - u(x)$$

is real analytic on a neighborhood of K.

REFERENCES

[1] Abramczuk, W.: 'On continuation of quasi-analytic solutions of partial differential equations to compact convex sets,' *J. Austral. Math. Soc.* **39** (1985), 306-316.

[2] Grušin, V.V.: 'On solutions with isolated singularities for partial differential equations with constant coefficients', *Trans. Moscow Math. Soc.* **15** (1966), 262-278.

[3] Kaneko, A.: 'Prolongement des solutions régulières de l'équation aux dérivées partielles *à* coefficients constants', *Séminaire Goulaouic-Schwartz* 1976-7, Ecole Polytech., Paris, 1977, Exposé No.18.

[4] Kaneko, A.: 'On continuation of regular solutions of partial differential equations to compact convex sets II', *J. Fac. Sci. Univ. Tokyo Sec.* 1A **18** (1972), 415-433.

[5] Kaneko, A.: 'On Hartogs type continuation theorem for regular solutions of linear partial differential equations with constant coefficients', *ibid* **35** (1988), 1-26.

[6] Kaneko, A.: 'On continuation of regular solutions of partial differential equations with constant coefficients', *J. Math. Soc. Japan* **26** (1974), 92-123.

[7] Kaneko, A.: 'On continuation of real analytic solutions of linear partial differential equations', *Astérisque* **89-90**, *Soc. Math.* France, 1981, pp.11-44.

[8] Kawai, T.: 'On the theory of Fourier hyperfunctions and its applications to partial differential equations with constant coefficients', *J. Fac. Sci. Univ. Tokyo Sec.* 1A **17** (1970), 467-517.

[9] Palamodov, V.P.: *Linear Differential Equations with Constant Coefficients*, Moscow, 1967 (in Russian; English translation from Springer, 1970).

[10] Polking, J.C.: 'A survey of removable singularities', *Seminar on non-linear partial differential equations,* Ed. by S.S. Chern, Springer, 1984, pp.261-292.

[11] Saburi, Y.: 'Fundamental properties of modified Fourier hyperfunctions', *Tokyo J. Math.* **8** (1985), 231-273.

[12] Tajima, S.: 'Analyse microlocale sur les variétés de Cauchy-Riemann et problème du prolongement des solutions holomorphes des équations aux dérivées partielles', *Publ. RIMS Kyoto Univ.* **18** (1982), 911-945.

REMARKS ON THE EXISTENCE OF SOLUTIONS OF PARTIAL DIFFERENTIAL EQUATIONS IN GEVREY SPACES

L. CATTABRIGA
Dipartimento di Matematica
Università di Bologna
Piazza di Porta S. Donato, 5
40127 Bologna - ITALY

Let $x = (x_1, \cdots, x_n) \epsilon \mathbf{R}^n$; $D = (D_1, \cdots, D_n)$, $D_j = -i\partial \backslash \partial x_j$, $j = 1, \cdots, n$ and for $\alpha \epsilon \mathbf{Z}_+^n$ let $| \alpha | = \alpha_1 + \cdots + \alpha_n$ and $D^\alpha = D_1^{\alpha_1} \cdots D_n^{\alpha_n}$. Let Ω be an open set in $\mathbf{R}^n, \tilde{\Omega}$ a relatively compact open subset of Ω and for $\sigma \geq 1$, $A > 0$

$$\mathcal{E}^{\{\sigma\},A}(\tilde{\Omega}) = \left\{ f \epsilon C^\infty(\Omega);\ sup_{\alpha \epsilon \mathbf{Z}_+^n} sup_{x \epsilon \tilde{\Omega}} A^{-|\alpha|} \alpha!^{-\sigma} \mid D^\alpha f(x) \mid < +\infty \right\},$$

$$\mathcal{E}^{\{\sigma\},A}(\Omega) = \lim_{\substack{\longleftarrow \\ \tilde{\Omega} \to \Omega}} \mathcal{E}^{\{\sigma\},A}(\tilde{\Omega}),$$

$$\mathcal{E}^{\{\sigma\}}(\Omega) = \lim_{\substack{\longleftarrow \\ \tilde{\Omega} \to \Omega}} \lim_{\substack{\longrightarrow \\ A \to +\infty}} \mathcal{E}^{\{\sigma\},A}(\tilde{\Omega}),$$

$$\mathcal{E}^{(\sigma)}(\Omega) = \lim_{\substack{\longleftarrow \\ \tilde{\Omega} \to \Omega}} \lim_{\substack{\longleftarrow \\ A \to 0}} \mathcal{E}^{\{\sigma\},A}(\tilde{\Omega}).$$

For a given $P(D) = \sum_{|\alpha| \leq m} c_\alpha D^\alpha, c_\alpha$ complex valued constants, m a positive integer, the problem of its surjectivity in $\mathcal{E}^{\{\sigma\}}(\Omega)$ has drawn considerable attention in recent years. Limiting our consideration here to the case when $\Omega = \mathbf{R}^n$, we recall that the equality

$$P(D)\mathcal{E}^{\{\sigma\}}(\mathbf{R}^n) = \mathcal{E}^{\{\sigma\}}(\mathbf{R}^n) \tag{1}$$

was first proved to hold for any $P(D)$ when $\sigma = 1, n = 2$ by E. DeGiorgi - L. Cattabriga [9] and that sufficient conditions on $P(D)$ for (1) to hold when $\sigma = 1, n > 2$ were subsequently given by K.G. Andersson [1] and L. Cattabriga [3]. Also in the case when $\sigma = 1$, necessary and sufficient conditions on $P(D)$ for (1) were obtained by L. Hörmander [10]. When σ is a rational number ≥ 1 sufficient conditions for (1) has been proved by L. Cattabriga [5]

T. Terzioğlu (ed.), Advances in the Theory of Fréchet Spaces, 73–78.

and [6][1]. In the last of these papers equality (1) is proved to hold for every $\sigma \geq 1$ and for every $P(D)$, when $n = 2$, whereas some counter examples to (1) for $n > 2$ are exhibited[2]. Sufficient conditions for (1) analogous to those by Hörmander [10] has been proved by G. Zampieri [12] and necessary and sufficient conditions of the same type have been presented by R. Meise [2] at this workshop.

Finally we recall that for every $P(D)$ and any $n \geq 2, \sigma \geq 1, A > 0$

$$P(D)\mathcal{E}^{(\sigma)}(\mathbf{R}^n) = \mathcal{E}^{(\sigma)}(\mathbf{R}^n), \quad P(D)\mathcal{E}^{\{\sigma\},A}(\mathbf{R}^n) = \mathcal{E}^{\{\sigma\},A}(\mathbf{R}^n) \ ^3. \tag{2}$$

In this talk we consider differential polynomials $P(D)$ for which (1) does not hold and show some conditions on $P(D)$ and on $f \epsilon \mathcal{E}^{\{\sigma\}}(\mathbf{R}^n)$, $\sigma > 1$, in order for the equation

$$P(D)u = f$$

has a solution $u \epsilon \mathcal{E}^{\{\sigma\}}(\mathbf{R}^n)$.

First we recall that the surjectivity results of [5] follows from the following lemmas.

Lemma 1. Let $\sigma = r/s, r$ and s positive integers and let

$$Q(\xi,\tau) = \left(\sum_{j=1}^{n} \xi_j^2\right)^s + \left(\sum_{h=1}^{\overline{n}} \tau_h^2\right)^r, \quad \xi \epsilon \mathbf{R}^n, \quad \tau \epsilon \mathbf{R}^{\overline{n}},$$

$$E(x,t,\nu) = \mathcal{F}^{-1}_{(\xi,\tau)}(exp(-\nu Q(\xi,\tau))\,(x,t), \quad x \epsilon \mathbf{R}^n, \quad t \epsilon \mathbf{R}^{\overline{n}},$$

where $\mathcal{F}^{-1}_{(\xi,\tau)}$ denotes the inverse Fourier transformation with respect to the variables (ξ,τ) and $\overline{n}$ is greater than 2r.

For a given

$$P(D) = D_n^{\ell} + \sum_{j=1}^{\ell} a_j(D')D_n^{\ell-j}, \quad D' = (D_1, \cdots, D_{n-1}),$$

where $a_j(D')$ are differential operators in D' of order $\leq m - \ell + j$, $m \geq \ell$, with constant coefficients, assume that

a) $\exists\, k > 1, \ \rho > \sigma, \ c_1, \ c_2 > 0$ such that:

$$(\xi', \lambda) \epsilon \mathbf{R}^{n-1} \times \mathbf{C}, \quad |\xi'| > k, \quad P(\xi', \lambda) = 0 \Rightarrow$$

[1]See also [7]. The restriction to rational σ's can eventually be removed with suitable changes in the proof of Lemma 1.

[2]See [8] for $\sigma = 1$

[3]See for example F. Treves [11], Chapter VII

$$\text{either} \quad Im\,\lambda \geq -c_1 \mid \xi' \mid^{1/\rho} \quad \text{or} \quad Im\,\lambda \leq -c_2(\mid \xi' \mid + \mid Re\,\lambda \mid)^{1/\sigma}.$$

If $f_1 \epsilon \mathcal{E}^{\{\sigma\}}(\mathbf{R}^n)$ can be represented as

$$f_1(x) = \int_{\mathbf{R}^n} dy \int_0^\infty G_1(x-y,\eta)\, g_1(y,\eta) d\eta, \quad x \epsilon \mathbf{R}^n,$$

where g_1 is a C^∞ function on $\mathbf{R}^n \times \mathbf{R}^+$ such that

(i) supp $g_1 \subset \{(y,\eta)\epsilon\mathbf{R}^n \times \mathbf{R}^+;\ y_n \geq c_0,\ \ \eta \geq \chi(y_n)\}$
for a suitable $c_0 \epsilon \mathbf{R}$ and a positdive function $\chi \epsilon C(\mathbf{R})$;

(ii) $\int_{\mathbf{R}^n} exp(c' \mid y \mid) dy \int_0^\infty \mid g_1(y,\eta) \mid d\eta < +\infty$,
where c' is a positive constant that depends on P; and

$$G_1(x,\eta) = \int_0^{\nu_0} E(x,t,\nu) \mid_{|t|=\eta} d\nu, \quad x\epsilon\mathbf{R}^n, \quad t\epsilon\mathbf{R}^{\overline{n}}, \quad \nu_0 > 0,$$

then there exists $u_1 \epsilon \mathcal{E}^{\{\sigma\}}(\mathbf{R}^n)$ such that $P(D)\, u_1 = f_1$.

Lemma 2. Let

$$f_2(x) = \int_{\mathbf{R}^n} dy \int_0^{+\infty} G_2(x-y,\eta)\, g_2(y,\eta) d\eta, \quad x\epsilon\mathbf{R}^n,$$

where $g_2 \epsilon C^\infty(\mathbf{R}^n \times \mathbf{R}^+)$ is such that

$$\text{supp } g_2 \subset \left\{(y,\eta)\epsilon\mathbf{R}^n \times \mathbf{R}^+;\ \ \Psi(\mid y \mid^2) \leq \eta \leq 2\Psi(\mid y \mid^2)\right\}$$

for a positive non increasing function $\Psi \epsilon C^\infty(\mathbf{R})$ and

$$G_2(x,\eta) = \int_{\nu_0}^{+\infty} E(x,t,\nu) \mid_{|t|=\eta} d\nu, \quad \nu_0 > 0,$$

then $f_2 \epsilon \mathcal{E}^{(\sigma)}(\mathbf{R}^n)$ and for the first equality in (2), there exists $u_2 \epsilon \mathcal{E}^{\{\sigma\}}(\mathbf{R}^n)$ such that $P(D) u_2 = f_2$.

Remark 3. Examples of $P(D)$ on $\mathbf{R}^3$ satisfying **a)** are

(1) $(-D_1^2 - D_2^2 - 1)(-D_1^2 - D_2^2 + D_3^2)$

(2) $(-D_1^2 - D_2^2 - iD_3)(-D_1^2 - D_2^2 + D_3^2)$.

However, according to what has been proved in [6], 1) and 2) do not satisfy (1), for any $\sigma \geq 1$ and for any $\sigma \epsilon [1,2)$ respectively.

By using Lemma 1 and 2 and the representation formula for functions in $\mathcal{E}^{\{\sigma\}}(\mathbf{R}^n)$ proved in [5] it can be proved

Theorem 4. Let $\sigma > 1$ and suppose that $P(D)$ satisfies **a**). Assume that

b) $\quad f \epsilon \mathcal{E}^{\{\sigma\}}(\mathbf{R}^n) \cap \mathcal{E}^{\{\sigma\},A}(\mathbf{R}^n \setminus \Delta)$
for $A > 0$ and $\Delta = \{x \epsilon \mathbf{R}^n;\ | x | \leq \gamma x_n,\ x_n \geq \gamma_0 > 0\}$, then there exists $u \epsilon \mathcal{E}^{\{\sigma\}}(\mathbf{R}^n)$ such that $P(D)u = f$.

Remark 5. When in condition **b)** Δ is not convex we suppose that Δ can be covered by a finite number of convex cones Δ_j, each of which containing a $N_j \neq 0$ such that $P(D)$ satisfies **a**) after a rotation that sends N_j on $(0, \cdots, 0, 1)$. When $\Delta = \mathbf{R}^n$ we obtain, as a consequence, the sufficient condition for (1) proved in [5].

We give now the definition of σ-wave front set at infinity, $WF_\sigma^\infty f$, for a function $f \epsilon \mathcal{E}^{\{\sigma\}}(\mathbf{R}^n)$.

Definition 6. Let $f \epsilon \mathcal{E}^{\{\sigma\}}(\mathbf{R}^n)$. We denote by $WF_\sigma^\infty f$ the complement in $\mathbf{R}^n \times \mathbf{R}^n \setminus \{0\}$ of the set of (x^0, ξ^0) such that there exist a conic neighborhood Λ of ξ^0 and positive constants r and c such that for every $\Psi \epsilon \mathcal{E}^{(\sigma)}(\mathbf{R}^n) \cap C_0^\infty(B(0,r))$, $\Psi(0) \neq 0$

$$| \mathcal{F}(\Psi_{tx^0} f)(\xi) | \leq C\, exp(-c \,| \xi |^{1/\sigma}) \quad \forall \xi \epsilon \Lambda, \quad t > 0,$$

where $\Psi_{tx^0}(x) = \Psi(x - tx^0)$ and $C > 0$ depends on Ψ and t.

By suitably adapting the proof of theorem 2.4 of [5] it can be proved

Theorem 7. Let $\sigma > 1$ and suppose that $P(D)$ satisfies

a_Γ) there exist an open cone $\Gamma \subset \mathbf{R}^{n-1}$ and constants $k > 1$, $\rho > \sigma$, $c_1 > 0$, $c_2 > 0$ such that:

$$(\xi', \lambda) \epsilon \Gamma \times \mathbf{C}, \ | \xi' | > k, \ P(\xi', \lambda) = 0 \Rightarrow$$

$$\text{either} \quad Im\, \lambda \geq -c_1 | \xi' |^{1/\rho} \quad \text{or} \quad Im\, \lambda \leq -c_2(| \xi' | + | Re\, \lambda |)^{1/\sigma}.$$

Assume that $f \epsilon \mathcal{E}^{\{\sigma\}}(\mathbf{R}^n)$ satisfies

c) $\quad WF_\sigma^\infty f \subset \Delta \times (\Gamma' \times \mathbf{R})$,
where Δ is as in **b)** Theorem 4 and Γ' is an open cone with its closure contained in Γ.

Then there exists $u \epsilon \mathcal{E}^{\{\sigma\}}$ such that $P(D)u = f$.

As a final remark we observe that following [4] the results stated in Theorems 4 and 7 can be proved to hold also when $P(D)$ is replaced by a convolution operator on $\mathcal{E}^{\{\sigma\}}(\mathbf{R}^n)$ of the form

$$P = \sum_{j=0}^{m} A_j \otimes D_n^j \delta,$$

where $A_j \epsilon \mathcal{E}^{\{\sigma\}'}(\mathbf{R}^n), j = 0, \cdots, m$.

Actually we have to replace $P(\xi', \lambda)$ with

$$\sum_{j=0}^{m} \mathcal{F}(A_j)(\xi')\lambda^j$$

in **a)** and a_Γ) and assume furthermore that:

d) there exist positive constants c_3, c_4 such that

$$\sum_{j=0}^{m} | \mathcal{F}(A_j)(\xi') | \geq c_3\, exp(-c_4 | \xi' |^{1/\rho}), \quad \xi' \epsilon \mathbf{R}^{n-1}, \quad | \xi' | > k.$$

Complete proofs of the results stated above will appear elsewhere.

REFERENCES

[1] K.G. Andersson, 'Global solvability of partial differential equations in the space of real analytic functions', *Coll. on Analysis*, Rio de Janiero 1972, Analyse fonctionelle, Hermann 1974.

[2] R.W. Braun, R. Meise, D. Vogt, 'Applications of the projective limit functor to convolution and partial differential equations', these Proceedings.

[3] L. Cattabriga, 'Sull'esistenza di soluzioni analitiche reali di equazioni a derivate parziali a coefficienti costanti', *Boll. Un. Mat. Ital.* (4) **12** (1975), 221-234.

[4] L. Cattabriga, 'Construction of a fundamental solution for certain convolution operators', *Atti Acc. Scienze Ist. Bologna* (13) **4** (1977), 59-68.

[5] L. Cattabriga, 'Solutions in Gevrey spaces of partial differential equations with constant coefficients', *Astérisque* **89-90** (1981), 129-151.

[6] L. Cattabriga, 'On the surjectivity of differential polynomials on Gevrey spaces', Meeting on 'Linear partial and pseudo-differential operators', Torino 1982, *Rend. Sem. Mat. Univers. Politecn.* Torino, fascicolo speciale 1983, 81-89.

[7] L. Cattabriga, *Alcuni problemi per equazioni differenziali lineari con coefficienti costanti*, Pitagora ed. Bologna, 1983.

[8] E. DeGiorgi, 'Solutions analytiques des équations aux dérivées partielles à coéfficients constant,' Ecole Polytechnique. *Séminaire Goulaouic-Schwartz* 1971-72, ex.29.

[9] E. DeGiorgi - L. Cattabriga, 'Una dimostrazione diretta dell'esistenza di soluzioni analitiche nel piano reale di equazioni a derivate parziali a coefficienti costanti', *Boll. Un. Mat. Ital.* (4) **4** (1971), 1015-1027.

[10] L. Hörmander, 'On the existence of real analytic solutions of partial differential equations with constant coefficients', *Inventiones Math.* **21** (1973), 151-182.

[11] F. Treves, *Locally Convex Spaces and Linear Partial Differential Equations*, Springer-Verlag 1967.

[12] G. Zampieri, 'An application of the fundamental principle of Ehrenpreis to the existence of global Gevrey solutions of linear differential equations', *Boll. Un. Mat. Ital.* (6) **5-B** (1986), 361-392.

TAME RIGHT INVERSES FOR PARTIAL DIFFERENTIAL EQUATIONS

M. LANGENBRUCH
Mathematical Institute
Einsteinstr. 62
D-4400 Münster
F.R. Germany

Abstract

The existence of continuous linear right inverses for partial differential equations is proved in certain weighted spaces of C^∞-functions and distributions. Continuity estimates are given for the right inverses. This makes the problem accessible to a perturbation argument and may be used to prove the existence of right inverses for certain partial differential operators with C^∞-coefficients, which are "constant at ∞". These right inverses satisfy the same continuity estimates as in the case of constant coefficients. The main tools are the *-tame splitting of the $\bar\partial$-complex and the *-tame sequence space representations and splitting theory of D. Vogt for power series spaces of infinite type.

Introduction.

The existence or nonexistence of continuous linear right inverses for partial differential operators with constant coefficients has recently been studied in many different spaces of distributions or ultradistributions ([4-6], [10]). In this paper some of the results of [6] will be extended into two directions:

On one hand, rather precise continuity estimates will be given for any of the continuous linear mappings used in this paper (including the right inverses). To handle several types of continuity estimates simultaneously, we will use the notion of *-tameness, which is a common notation for tameness, almost 1-tameness and ℓ-tameness (see section 1). Naturally, the necessary assumptions (collected in (*), see section 1) will depend on the type of continuity estimates to be proved. Essentially, we will have to use o-type conditions instead of O-type conditions to improve on the continuity estimates. In their weakest form these conditions are weaker than those of [6], extending the set of examples (e.g. by the

T. Terzioğlu (ed.), Advances in the Theory of Fréchet Spaces, 79–114.

spaces $S(\pm M)$, see 1.1 and 1.10) and leading to ℓ-tame estimates instead of continuous right inverses only ([6]).

As a second improvement on the results of [6], the precise continuity estimates of this paper can be used to prove the existence of *-tame right inverses for partial differential operators with C^∞- coefficients of the form

$$P(x,D) = P_0(D) + \sum_{k=1}^{\ell} c_k(x) Q_k(D),$$

where P_0 is hypoelliptic, each of the Q_k is strictly weaker than P_0 and (c_k) are C^∞-functions, tending to 0 at a certain rate, depending on the weighted space, where the right inverse is to be constructed.

The paper is divided into four parts:

The first part contains the basic results and definitions from Fourier theory and structure theory of power series spaces of infinite type: A *-tame Paley-Wiener-type theorem is proved for the weighted spaces used in the subsequent sections. As an immediate consequence, any of these spaces is (ℓ-tamely) isomorphic to (s), the space of rapidly decreasing sequences (see [6], §1 for a different proof.)

The second section contains a *-tame splitting theorem for the $\overline{\partial}$-complex in certain weighted spaces, which are associated to the growth conditions coming from the Paley-Wiener-Theorem above.

Using this *-tame right inverse for the $\overline{\partial}$-operator, a left inverse for the multiplication with $P(-z)$ is explicitly given in section 3, which gives a *-tame right inverse for the associated partial differential operator $P(D)$ via Fourier transformation and transposition (see [6] for a different approach via the fundamental principle of Ehrenpreis). This implies, that the kernel of $P(D)$ is *-tamely complemented in (s), especially, it has a basis ([3]), and it is indeed isomorphic to (s) in many cases. A similar result could not be obtained in the continuous category, since the basis problem in complemented subspaces of (s) is still open.

The final section is devoted to the proof of the existence of *-tame right inverses for operators $P(x,D)$ with C^∞-coefficients, which were mentioned above. The main point is the existence of a right inverse for $P_0(D)$, which has "optimal regularity", such that the part with variable coefficients causes a compact perturbation of the identity on the step spaces, defining our weighted spaces. So one has the usual Riesz-Schauder-theory on these step spaces, leading to finite dimensional exceptional spaces, which indeed do not depend on the step spaces. The right inverse of $P(x,D)$ constructed in this way on a space of finite codimension is also *-tame. So it may be used in the Nash-Moser- theorem to prove existence results for certain nonlinear equations in the weighted spaces of this paper.

§1. Basic Structural Results.

This section is concerned with the study of Fourier transformation and sequence space representation for the spaces, which are considered in this paper. These were already studied in [6] in the continuous category, while special emphasis is laid on precise continuity estimates in this paper. We will only deal with weighted spaces of C^∞-functions and distributions here, and skip the case of ultradifferentiable functions and ultradistributions for the sake of simplicity.

The weighted spaces are defined as follows: Let

$$M(x) := \sum_{i=1}^{N} M_i(|x_i|) \text{ and } V(x) = \sum_{i=1}^{N} V_i(|x_i|)$$

with $M_i,\ V_i \in C^1(\mathbf{R}_+)$.

1.1. Definition: For positive V let

$$C^\infty(M,V) := \{f \in C^\infty(\mathbf{R}^N) \mid \ \|f\|_{kn} := \sup_{|j|_\infty \le k} \|f^{(j)} e^{M+nV}\|_\infty < \infty \ \forall\, k,n \in \mathbf{N}\}$$

$C^\infty(M, \sum_i \ell n(1+|x_i|))$ is denoted by $S(M)$.
The quotients of the weights defining $S(M)$ are polynomials like for the Schwartz space $S(\mathbf{R}^N)$, justifying the terminology.

We will assume later, that M is positive and dominates V, and we will consider right inverses for partial differential operators in
$C^\infty(M,V)'_b$, a weighted space of distributions, and in
$C^\infty(-M,V)$, a weighted space of C^∞-functions.

To obtain readable statements on continuity estimates, we have to introduce some more notation, which has proved to be very useful:

The first is the concept of a graded (F)-space $(E, |\ |_n)$, that is, we fix an increasing system $\{|\ |_n \mid n \ge 1\}$ of semi-norms defining the topology of E.

The second is the modul of continuity $s_T(n)$ of a continuous mapping $T : (E, |\ |_n) \to (F, |\ |_n)$, which is defined by

$$s_T(n) := \inf\{t \ge 1 \mid \exists\, C_t : |\, T(e)\, |_n \le C_t \,|\, e\, |_t \text{ for any } e \in E\}.$$

Finally, we introduce three types of continuity estimates, which will be considered simultaneously in this paper: T is called

(i) tame, iff $s_T(n) \le n + a$ for some a and large n,

(ii) ℓ-tame, iff $s_T(n) \le an$ for some a and large n.

The seminorms used in this paper usually will depend on two or three indices (k,n) (resp. (ℓ,k,n)), where the first always denotes the order of differentiation, and the last one denotes the dependence on the weights M and V. We will then use $\{|\ \ |_{n,n}|\ n \geq 1\}$ (resp. $\{|\ \ |_{n,n,n}|\ n \geq 1\}$) as a grading, when dealing with tameness or 1-tameness.

This is different for the third type of tameness: A mapping $T : (E, |\ \ |_{k,n}) \to (F, |\ \ |_{k,n})$ is called almost 1-tame iff

$$\forall a > 1\ \exists K_0\ \forall K \geq K_0 :\ s_T(n) \leq an \text{ for large } n,$$

where $s_T(n)$ is calculated w.r.t. the grading $\{|\ \ |_{n,Kn}|\ n \geq 1\}$. For the norms $|\ \ |_{\ell,k,n}$ we similarly use $|\ \ |_{n,Kn,K^2n}$ in connection with almost 1-tameness.

A graded space E is called almost 1-tamely isomorphic to F iff for any $a > 1$ and large K there is a bijection T from E onto F such that $s_T(n)$ and $s_{T^{-1}}(n)$ are bounded by an, for the grading $\{|\ \ |_{n,Kn}|\ n \geq 1\}$.

Similarly, for almost 1-tamely complemented subspaces and almost 1-tame right inverses the projection (and the right inverse, resp.) may depend on the tameness constant a to be proved.

Almost 1-tameness is especially useful as far as the so called "hard implicit function theorem" of Nash and Moser is concerned ([8]). In fact, ℓ-tame estimates are needed with $a < 2$ for the right inverses of the derivatives of the nonlinear function. To give these, one may use any number of compositions of almost 1-tame mappings.

*-tameness is used as a common notation for any of the three types of tameness.

T is a *-tame isomorphism iff T is bijective and both T and T^{-1} are *-tame.

Gradings and *-tame mappings in (DFS)-spaces F are easily defined by duality (using a decreasing set of dual norms defining the bounded sets in F etc.). Duals of graded (FS)-spaces are always considered with the dual grading. Subspaces and quotients are endowed with their canonical induced gradings. Power series spaces of infinite type $\Lambda_\infty(\alpha_n)$ are graded spaces via

$$|\,(c_k)\,|_n := \|(c_k e^{n\alpha_k})\|_1^2.$$

To obtain unified and short statements, we collect the assumptions used in this paper already at this stage: Let $m_i := M_i'$ and $v_i := V_i'$ for $i \leq N$. We consider the following assumptions:

$$m_i(t) + nv_i(t) \uparrow \infty \text{ for } n \in \mathbf{Z} \text{ and large } t \quad v_i(t) = o(m_i(t)). \tag{1}$$

This means, that the weights $M + nV$ are convex functions and M is the dominating term.

$$\ell n\, m_i(t) + \ell n\, t = 0(V_i(t)) \tag{2}$$

$$\text{resp.}\quad o(V_i(t))) \tag{2$_o$}$$

$$\exists \hat{C} > 1, C\ (\forall C > 1\ \exists \hat{C} > 1) : V_i \circ m_i^{-1}(\hat{C}t) \leq CV_i \circ m_i^{-1}(t). \tag{3}$$

(resp. $(3)_o$)

(3) and $(3)_o$ are stability assumptions for the functions $V_i \circ m_i^{-1}$. They may be substituted by similar conditions on V_i and v_i, if v_i is increasing or decreasing (see Remark 1.7).

$$t = 0(m_i(t)^\alpha V_i(t)) \text{ for some } \alpha. \tag{4}$$

The following conditions are used to prove
ℓ-tameness: (1) - (4)
almost 1-tameness: (1), (2), $(3)_o$, (4) (*)
tameness: (1), $(2)_o$, $(3)_o$, (4).
This assignment of the conditions to the type of continuity estimates is denoted by (*). In 1.10 we will give some examples showing that (*) is easily verified in many cases. In [6] we used (1)-(3) and instead of (4) the stronger assumption $t = 0(V_i(t))$. Especially, $S(M)$ could not be treated there, since $V_i(t) = \ell n(1+t)$ in this case (for the significance of slowly increasing V_i see also Theorem 4.4).

The general form of the weights often allows to consider the case of one variable only (i.e. $N = 1$). The index i in M_i and V_i is then deleated, and the case of several variables is left to the reader.

Let $N = 1$ and

$$U(t) := \int_0^t (|\, m(\tau)\,| + |\, v(\tau)\,| + 1) d\tau.$$

Let the sequence $(x_r)_{r\in\mathbf{Z}}$ be defined by $x_0 = 1$.

$$U(x_{r+1}) := U(x_r) + 1 \text{ and } x_{-r} := x_{|r|} \text{ for } r \in \mathbf{N}. \tag{5}$$

(x_r) is defined, since U is continuous, strictly increasing and unbounded on $\mathbf{R}_+$. Let $\epsilon_r := x_{|r|+1} - x_{|r|}$. For $r \in \mathbf{N}$ we get:

$$0 < \epsilon_r \leq \int_{x_r}^{x_{r+1}} (|\, m(\tau)\,| + |\, v(\tau)\,| + 1) d\tau = U(x_{r+1}) - U(x_r) = 1$$

by the definition of x_r. (x_r) is unbounded, since $U(x_r) \geq r$ for $r \in \mathbf{N}$.

Let the Fourier transform be defined by $\hat{f}(z) := \langle f, e^{iz\cdot}\rangle$. Let $I_r := (x_{|r|-1}, x_{|r|+1})sgn r$ for $|\, r\,| \geq 1$ and $I_0 := (-x_1, x_1)$.

1.2. Lemma: Let $N = 1$ and let (1) and (2) (resp. $(2)_o$) be valid. Then there is a C^∞-partition of unity $\{\phi_r \mid r \in \mathbf{Z}\}$ subordinate to $\{I_r \mid r \in \mathbf{Z}\}$ such that

$$\exists\, A > 0\ (\forall A > 0)\, \forall k \in \mathbf{N}\ \exists C_1\ \forall f \in C^\infty(\mathbf{R}),\ \ z \in \mathbf{C} :$$

$$|\widehat{f\phi_r}(z)|(1+|z|)^k \le C_1 \sup_{\substack{j\le k\\ x\in I_r}} |f^{(j)}(x)| e^{A(k+1)V(x_{|r|-1})} \int_{I_r} e^{x\cdot Im\, z} dx \quad \text{for large } |r|.$$

Proof: a) Let $r \in \mathbf{N}$ and $x, y \in I_r$.

$$|M(x)-M(y)| + |V(x)-V(y)| \le \int_{x_{r-1}}^{x_{r+1}} (|m(\tau)| + |v(\tau)| + 1)d\tau = U(x_{r+1}) - U(x_{r-1}) = 2 \tag{6}$$

(1) and (2) (resp. $(2)_o$) imply:

$$\exists A > 0 (\forall A > 0): \ 2\ell n\, t + \ell n(|m(t)| + |v(t)|) \le AV(t) + C_A. \tag{7}$$

$$\begin{aligned} 1/\epsilon_r = (U(x_{r+1}) - U(x_r))/(x_{r+1} - x_r) &= |m(\tau)| + |v(\tau)| + 1 \le \\ &\le e^{A+AV(\tau)} \le e^{3A+AV(x_{r-1})} \text{ for large } r \end{aligned} \tag{8}$$

by the mean value theorem and (6).

b) Choose $0 \le \psi \in D(\mathbf{R})$ such that $\|\psi\|_1 = 1$ and $\psi(t) = 0$ for $|t| \ge 1/16$. For $r \in \mathbf{N}_0$ let χ_r be the characteristic function of $(-\infty, x_r + \epsilon_r/2)$ and $\psi_r(t) := \chi_r * \psi(\cdot/\epsilon_r)/\epsilon_r$. Let $\phi_0(t) := \psi_0(|t|)$ and $\phi_r := (\psi_r - \psi_{r-1})\chi_{\mathbf{R}_+}$ and $\phi_{-r} = \check{\phi}_r$ for $r \in \mathbf{N}$. This choice implies (for $k \ge 0$):

$$supp\, \phi_r \subset I_r \text{ and } \sum \phi_r = 1$$

$$supp\, \phi_r \subset [x_r - 9\epsilon_{r-1}/16,\ x_r + 9\epsilon_r/16] \text{ for } |r| \ge 1 \tag{9}$$

$$\begin{aligned} |(f\phi_r)\hat{}(z)z^k| &= |(D^k(f\phi_r))\hat{}(z)| \\ &\le C_k \sup_{\substack{j\le k\\ x\in I_r}} |f^{(j)}(x)| \sup_{j\le k} \|\phi_r^{(j)}\|_\infty \int_{I_r} e^{x\cdot Im\, z} dx \text{ for } |r| \ge 1. \end{aligned}$$

The Lemma follows from (8).

Let $\langle x, y\rangle := \sum x_i y_i$ and let $F^*(y) := \sup_{x\in\mathbf{R}^N} \langle x, y\rangle - F(x)$ be the Young conjugate of the convex function F on $\mathbf{R}^N$.

1.3. Definition: Assume (1).

$$\begin{aligned} H_+ := \{f \in H(\mathbf{C}^N) \mid |f|_{k,n} &:= \sup_{z\in\mathbf{C}^N} |f(z)| e^{-(M+nV)^*(Im\, z)} \prod_{i\le N} (1+|z_i|)^k \\ &< \infty \text{ for some } k, n \ge 1\}. \\ H_- := \{f \in H(\mathbf{C}^N) \mid |f|^-_{k,n} &:= \sup_{z\in\mathbf{C}^N} |f(z)| e^{-(M-nV)^*(Im\, z)} \prod_{i\le N} (1+|z_i|)^{-k} \\ &< \infty \text{ for some } k, n \ge 1\}. \end{aligned}$$

Let $\| \|_{k,n}$ and $\| \|_{k,n}^{-}$ define the grading of $C^\infty(M,V)$ (and $C^\infty(-M,V)$, resp.) according to 1.1. The following theorem is basic for this paper:

1.4. Theorem: Assume (*). Then the Fourier transform is a *-tame isomorphism from $C^\infty(M,V)$ onto H_+ (and from $C^\infty(-M,V)'_b$ onto H_-).

Proof: Let $N = 1$.

a) We first consider $C^\infty(M,V)$. (7) implies:

$$\begin{aligned} | \hat{f}(z)z^k | &\leq \int | f^{(k)}(x)e^{x\, Im\, z} | dx \\ &\leq \|f\|_{k,A+n} \int e^{-AV(x)}dx \sup_x e^{x\, Im,z-(M+nV)(x)} \\ &\leq C_1 \|f\|_{k,A+n} e^{(M+nV)^*(Im\, z)} \end{aligned}$$

$C^\infty(M,V)$ is contained in $S(\mathbf{R}^N)$ by (2). The Fourier inversion formula and Cauchy's integral formula imply for any $y \in \mathbf{R}$:

$$\begin{aligned} | f^{(k)}(x) | &= \frac{1}{2\pi} | \int \widehat{f^{(k)}}(\zeta + iy)e^{ix(\zeta+iy)}d\zeta | \\ &\leq C_1 \int (1+ | \zeta + iy |)^{-2} d\zeta \sup_\zeta | \hat{f}(\zeta + iy) | (1+ | \zeta + iy |)^{k+2} e^{-xy} \\ &\leq C_2 | f |_{k+2,n}\, e^{-xy+(M+nV)^*(y)}. \end{aligned}$$

a) follows, if we choose y such that

$$-(xy - (M+nV)^*(y)) = -((M+nV)^*)^*(x) = -(M+nV)(x).$$

b) The grading in $C^\infty(-M,V)'_b$ is defined by

$$\|T\|_{k,n}^{-*} := \sup\{| \langle T, \psi \rangle | \mid \psi \in C^\infty(-M,V),\ \|\psi\|_{k,n}^{-} \leq 1\} \text{ for } T \in C^\infty(-M,V)'_b.$$

i) $\hat{T}$ is entire, since locally $\hat{T}$ is the uniform limit of entire functions of the form $(T\psi)^\wedge$ with $\psi \in D(\mathbf{R})$.

$$| \hat{T}(z) | \leq \|T\|_{k,n}^{-*} \|e^{-iz\cdot}\|_{k,n}^{-} \leq C_1 \|T\|_{k,n}^{-*} (1+ | z |)^k e^{(M-nV)^*(Im\, z)}$$

ii) For $f \in H_-$ there is a unique $T \in S'(\mathbf{R})$ such that $\mathcal{F}_{S'(\mathbf{R})}(T) = f |_{\mathbf{R}}$. If $T |_{D(\mathbf{R})}$ is continuous for the topology of $C^\infty(-M,V)$, then $f |_{\mathbf{R}} = \mathcal{F}_{S'}(T) = \hat{T} |_{\mathbf{R}}$ and hence $\hat{T} = f$, since $\hat{T} \in H_-$ by i). For $\psi \in D(\mathbf{R})$ and $y \in \mathbf{R}$ we obtain:

$$\langle T, \check{\psi} \rangle = \frac{1}{2\pi} \langle \hat{T}, \hat{\psi} \rangle = \frac{1}{2\pi} \int f(x)\hat{\psi}(x)dx = \frac{1}{2\pi} \int f(x+iy)\hat{\psi}(x+iy)dx,$$

since $\mathcal{F}_{S'}(T) = f\,|_{\mathbf{R}}$ and $f \in H_-$. This implies for $\psi \in D(I_r)$:

$$\begin{aligned} |\langle T, \check{\psi}\rangle| &\leq C_1 \,|\, f\,|_{k,n}^{-} \sup_{j\leq k+2} |\,\widehat{\psi^{(j)}}(x+iy)\,|\, e^{(M-nV)^*(y)} \\ &\leq C_2 \,|\, f\,|_{k,n}^{-} \,\|\psi\|_{k+2,n+A}^{-} \int e^{-AV(\zeta)}d\zeta \sup_{\zeta\in I_r} e^{\zeta y+(M-nV)(\zeta)+(M-nV)^*(y)} \end{aligned}$$

By (6) we may choose y such that

$$\begin{aligned} \sup_{\zeta\in I_r} \zeta y + (M-nV)(\zeta) &= \sup_{\zeta\in I_r} -|\,\zeta\,|\,|\,y\,| + (M-nV)(\zeta) \\ &\leq -x_{|r|-1}\,|\,y\,| + (M-nV)(x_{|r|+1}) \\ &\leq C_3 - x_{|r|-1}\,|\,y\,| + (M-nV)(x_{|r|-1}) = C_3 - (M-nV)^*(y). \end{aligned}$$

This choice and (7) then show

$$|\langle T, \check{\psi}\rangle| \leq C_4 \,|\, f\,|_{k,n}^{-} \,\|\psi\|_{k+2,n+A}^{-} \text{ for } \psi \in D(I_r).$$

With $\{\phi_r\}$ as in Lemma 1.2 we get by (8) for $\psi \in D(\mathbf{R})$:

$$\|\check{\psi}\phi_r\|_{k+2,n+A}^{-} \leq C_5 \|\check{\psi}\|_{k+2,n+(k+5)A}^{-} \epsilon_r e^{-AV(x_{|r|-1})} \sum_r \epsilon_r\, e^{-AV(x_{|r|-1})} \leq \int e^{-AV(x)}dx < \infty. \tag{10}$$

by (7). So we get:

$$\begin{aligned} |\langle T, \check{\psi}\rangle| \leq \sum_r |\langle T, \check{\psi}\phi_r\rangle| &\leq C_6 \,|\, f\,|_{k,n}^{-} \sum_r \|\psi\phi_r\|_{k+2,n+A}^{-} \\ &\leq C_7 \,|\, f\,|_{k,n}^{-} \,\|\psi\|_{k+2,\,n+(k+5)A}^{-}. \end{aligned}$$

Since $\{f \in D(\mathbf{R}) \mid \|f\|_{k,n}^{-} \leq C\}$ is dense in $\{f \in C^\infty(-M,V) \mid \|f\|_{k,n+1}^{-} \leq 1\}$ w.r.t. $\|\ \|_{k,n}^{-}$, we get:

$$\|T\|_{k+2,n+(k+5)A+1}^{-*} \leq C_7 \,|\, f\,|_{k,n}^{-} = C_7 \,|\, \hat{T}\,|_{k,n}^{-}\,.$$

Tameness (and ℓ-tameness) of $\mathcal{F}$ and $\mathcal{F}^{-1}$ are now clearly a consequence of (7). Almost 1-tameness is seen as follows:

$$\|T\|_{a(n,Kn)}^{-*} \leq \|T\|_{n+2,\,Kn+(n+5)A+1}^{-*} \leq C_7 \|\hat{T}\|_{n,Kn}^{-},$$

if $a \geq (n+2)/n$ and $a \geq 1 + (1+5/n)A/K + 1/(Kn)$. This is always true for large K and n.

The above shift in the estimates from the index corresponding to differentiation to that corresponding to the weights is a typical feature of many estimates in this paper, and it is the reason for choosing the grading $\{|\ |_{n,\,Kn,\,K^2n}\}$ in connection with almost 1-tameness.

1.5. Remark: Let $H_{k,n}^{\pm} := \{f \in H(\mathbf{C}^N) \mid |\, f\,|_{k,n}^{\pm} < \infty\}$ and assume (*). With the constant A from (7) we get:

(a) $H_\pm$ is dense in $H_{k+2,n+A+1}$ for the topology of $H_{k,n}$.

(b) $H_{0,0}^-$ is dense in $H_{k,n}^-$ for the topology of $H_{k+3,n+(k+5)A+2}^-$.

Proof: a) Let $N = 1$ and $E_{k,n} := \{f \in C^k(\mathbf{R}) \mid \|f\|_{k,n} < \infty\}$. Then $D(\mathbf{R})$ (and hence $C^\infty(M,V)$) is dense in $E_{k,n+1}$ for the topology of $E_{k,n}$. In the proof of 1.4a) we indeed have shown, that

$$\mathcal{F}^{-1} : H_{k+2,n+A+1} \to E_{k,n+A+1} \text{ and } \mathcal{F} : E_{k,n+A} \to H_{k,n}$$

are continuous. This shows a).

b) For $N = 1$ let $E_{k,n}^- := \{T \in C^\infty(-M,V)^* \mid \|T\|_{k,n}^{-*} < \infty\}$, where F^* denotes the algebraic dual of the linear space F. $E'(\mathbf{R}) \cap E_{k,n}^-$ is dense in $E_{k,n}^-$ for the topology of $E_{k,n+1}^-$. Hence $D(\mathbf{R})$ and also $E_{0,0}^-$ are dense in $E_{k,n}^-$ w.r.t. $\| \; \|_{k+1,n+1}^{-*}$ (use convolution). The conclusion follows from the proof of 1.4b), since $\mathcal{F} : E_{k,n}^- \to H_{k,n}^-$ and

$$\mathcal{F}^{-1} : H_{k,n}^- \to E_{k+2,n+(k+5)A+1}^-$$

are continuous.

To get a *-tame sequence representation from Theorem 1.4, we first need a simple lemma:

1.6. Lemma: Assume (*) (and $N = 1$). Then $H_\pm$ may be defined by the grading

$$\{p_{k,n}^\pm(f) := \sup_{z \in \mathbf{C}} | f(z) | (1+ | z |)^{\pm k} e^{-M^*(Im\, z) \pm nV \circ m^{-1}(Im\, z)}\}.$$

This change of grading is a *-tame isomorphism.

Proof: We may assume, that m and $(m \pm v)$ are increasing on $\mathbf{R}_+$.

i) $M^*(t) - (M+V)^*(t) = \int_0^t (m^{-1}(\tau) - (m+v)^{-1}(\tau))d\tau \leq \int_0^{m^{-1}(t)} v(\tau)d\tau = V \circ m^{-1}(t) \leq (M-V)^*(t) - M^*(t)$.

ii) $M^*(t) - (M+V)^*(t) \geq V \circ (m+v)^{-1}(t) \geq V \circ m^{-1}(t/\hat{C})$
$(M-V)^*(t) - M^*(t) \leq V \circ (m-v)^{-1}(t) \leq V \circ m^{-1}(t/\hat{C})$
for any $\hat{C} > 1$ and large t, since $v = o(m)$ by (1). Applying these for estimates to nV instead of V, we get 1-tame isomorphisms from (3) (choose $\hat{C} = 2$) and tame (and a forteriori almost 1-tame) estimates from $(3)_o$.

1.7. Remark: If v is increasing (resp. decreasing), then (3) and $(3)_o$ in (*) may be substituted by the following conditions:

$$\exists C>1,\ \hat{C}>1\ (\forall C>1\ \exists \hat{C}>1):\ \ v\circ V^{-1}(Ct)\geq \hat{C}v\circ V^{-1}(t) \tag{11}$$

(resp. $(11)_o$)

$$(\text{ resp. } \exists C>1,\ \hat{C}>1\ (\forall C>1\ \ \exists \hat{C}>1):\ \ v\circ V^{-1}(t/C)\geq \hat{C}v\circ V^{-1}(t) \tag{12}$$

(resp. $(12)_o$)

$$\exists \alpha,\ C\ (\exists \alpha\, \forall C>1):\ \ V\circ m^{-1}(t+t^{-\alpha})\leq CV\circ m^{-1}(t) \tag{13}$$

(resp. $(13)_o$)

Proof: We have to show, that

$$V\circ(m+v)^{-1}(t)\geq V\circ m^{-1}(t)/C \text{ and } V\circ(m-v)^{-1}(t)\leq CV\circ m^{-1}(t).$$

Both inequalities follow from

$$m\circ V^{-1}(Ct)-m\circ V^{-1}(t)\geq \begin{cases} v\circ V^{-1}(Ct), & \text{if } v \text{ is increasing}\\ v\circ V^{-1}(t), & \text{if } v \text{ is decreasing}\end{cases}$$

Since $m\pm kv$ is increasing by (1), we get

$$m\circ V^{-1}(Ct)-m\circ V^{-1}(t)\geq k\mid v\circ V^{-1}(Ct)-v\circ V^{-1}(t)\mid$$

and the right hand side is larger than $v\circ V^{-1}(Ct)$ (resp. $v\circ V^{-1}(t)$) by (11) (and (12), resp.).

(13) will be needed in Proposition 1.9.

For $V(t)=t$ one can use (14) instead of (11) or (12):

$$\exists C>1\ (\forall C>1):\ \ m(Ct)-m(t)\geq 1 \text{ for large } t \tag{14}$$

(resp. $(14)_o$).

We can now prove the announced *-tame sequence space representation for $H_{\pm}$, using the results of D. Vogt on sequence space representations with continuity estimates ([13]). The somewhat complicated statement is due to the fact, that there are non tamely equivalent gradings on (s).

1.8. Theorem: Assume (*). Then H_+ and H_- are *-tamely isomorphic to some $\Lambda_\infty(\alpha_n)$ (resp. $\Lambda_\infty(\alpha_n)'_b$).$\Lambda_\infty(\alpha_n)$ is ℓ-tamely isomorphic to (s).

Proof: i) We first show, that $C^\infty(M,V)$ is *-tamely isomorphic to a *-tamely complemented subspace of $C^\infty(-M,V)$.

Proof: Let ψ and ϕ_r be chosen as in Lemma 1.2. Let $\Psi_r := \tilde{\chi}_r * \psi(\cdot/\tilde{\epsilon}_r)/\tilde{\epsilon}_r$ and $\Psi_{-r} := \check{\Psi}_r$ for $r \geq 1$, where $\tilde{\epsilon}_r := min(\epsilon_r, \epsilon_{r-1})$ and $\tilde{\chi}_r$ is the characteristic function of

$$[x_r - 3\epsilon_{r-1}/8,\ x_r + 3\epsilon_r/8].$$

If Ψ_0 is chosen suitably, we get:

$$\Psi_j = 1 \text{ on } supp\,\phi_j(2\cdot -x_j) \quad supp\,\Psi_j \cap supp\,\phi_r(2\cdot -x_r) = \emptyset \text{ for } r \neq j, (14')$$

since $supp\,\phi_r \subset [x_r - 9\epsilon_{r-1}/16,\ x_r + 9\epsilon_r/16]$ by (9).
Let $\kappa_1 : C^\infty(M,V) \to C^\infty(-M,V)$ and $\kappa_2 : C^\infty(-M,V) \to C^\infty(M,V)$ be defined by

$$\kappa_1(f)(x) := \sum_r (\phi_r f)(2x - x_r)\, e^{2M(x_r)}$$

$$\kappa_2(g)(x) := \sum_j (\Psi_j g)((x + x_j)/2)\, e^{-2M(x_j)}$$

$\kappa_2 \circ \kappa_1(f) = f$ by (14').

$$\begin{aligned} \|\kappa_1(f)\|^-_{k,n} &\leq C_1 \sum_r \sup_{\substack{j\leq k\\ x\in I}} |\, (\phi_r f)^{(j)}(x)\,|\, e^{M(x_r)+nV(x_r)} \\ &\leq C_2 \sum_r \sup_{\substack{j\leq k\\ x\in I_r}} |\, f^{(j)}(x)\,|\, e^{M(x_r)+(n+A(k+2))V(x_r)}\epsilon_r \leq C_3\|f\|_{k,n+A(k+3)} \end{aligned}$$

where we have used (6)-(8) and (10). A similar estimate follows for κ_2. So $\kappa_1 \circ \kappa_2$ is a *-tame projection in $C^\infty(-M,V)$ onto a subspace E, which is *-tamely isomorphic to $C^\infty(M,V)$ via $\kappa_2\,|_E$ and κ_1.

ii) We now use Theorem 1.7 in [13] to prove an ℓ-tame and almost 1-tame sequence space representation for $C^\infty(M,V)$ (and hence for H_+ by Theorem 1.4). Tame estimates and H_- are treated similarly. We fix $K \geq 1$ and get for $\tilde{n} > n$ and $\gamma(\tilde{n},n) := \tilde{n}/n - 1$ the following trivial estimate:

$$p^{\pm}_{n,Kn}(f)^{\gamma+1} \leq p^{\pm}_{0,0}(f)^{\gamma} p^{\pm}_{\tilde{n},K\tilde{n}}(f) \text{ for } f \in H_\pm.$$

The *-tame isomorphisms from Theorem 1.4 and Lemma 1.6 imply:

$$\|f\|^{\mu+1}_{n,Kn} \leq C_1 \|f\|^{\mu}_{n_0 K n_0} \|f\|_{\hat{n},K\hat{n}} \text{ for } f \in C^\infty(M,V) \text{ and large } \hat{n}$$

with $\mu(\hat{n},n) = \hat{n}/(an) - 1$.

$$\|T\|^{-*\nu+1}_{n,Kn} \leq C_1 \|T\|^{-*\nu}_{n_0,Kn_0} \|T\|^{-*}_{\hat{n},K\hat{n}} \text{ for } T \in C^\infty(-M,V)' \text{ and large } \hat{n}$$

with

$$\nu(\hat{n}, n) = \hat{n}\tilde{a}/n - 1 \tag{15'}$$

Here a and $\tilde{a}$ are the product of the ℓ-tameness constants from 1.4 and 1.6. $C^\infty(M,V)_b'$ is *-tamely isomorphic to a subspace of $C^\infty(-M,V)_b'$ by i). So (15') holds with new $\tilde{a}$ in $C^\infty(M,V)_b'$. Using the notation of [13] we choose $q = q(k)$ and $\delta > 0$ such that $(1-\delta)q(k) = ka\tilde{a}$. Then $\mu(Q,k)+1 = Q/(ak)$ and $\nu(Q,q) = Q\tilde{a}/q$ and

$$\lim_{Q\to\infty} \nu(Q+2, q(k))/(\mu(Q,k)+1) = a\tilde{a}k/q(k) = 1-\delta < 1.$$

This is condition (τ) of [13]. By Theorem 1.7 in [13] (for $A = id, \tau(k) = k, \delta(k) = 1 - a\tilde{a}k/q(k)$) and the remark after that theorem we get some $\Lambda_\infty(\alpha_n)$ and an isomorphism $T : C^\infty(M,V) \to \Lambda_\infty(\alpha_n)$ satisfying the estimates

$$\|T(f)\|_{(1-\delta/2)q(k)} \le C_k \|f\|_{q(k),Kq(k)} \text{ for } f \in C^\infty(M,V)$$

$$\|T^{-1}((c_j))\|_{k,Kk} \le C_k \|(c_j)\|_{(1-\delta/2)q(k)}.$$

So $C^\infty(M,V)$ is ℓ-tamely (resp. almost 1-tamely) isomorphic to $\Lambda_\infty(\alpha_n)$.

iii) The diametral dimension of $\Lambda_\infty(\alpha_n)$ is $\Delta((s))$, since $C^\infty(M,V)$ is nuclear by the Grothendieck/ Pietsch-criterion applied to H_+, and since $C^\infty(M,V)$ contains $D([0,1]) \cong (s)$ as a closed subspace. So α_n and $\ell n(1+n)$ are equivalent and $\Lambda_\infty(\alpha_n)$ and (s) are ℓ-tamely isomorphic.

To prepare the next section on the $\overline{\partial}$-equation, we now give a new description of $H_\pm$ using the following spaces:

$$C_+^\infty := \{f \in C^\infty(\mathbf{C}^N) \| \|f\|_{\ell,k,n} := \sup_{\substack{|j|\le\ell \\ z\in\mathbf{C}^N}} |f^{(j)}(z)| e^{-(M+nV)^*(Im\, z)} \prod_{i\le N}(1+|z_i|)^k < \infty$$

$$\text{for any } \ell,k,n \ge 1\}$$

$$C_-^\infty := \{f \in C^\infty(\mathbf{C}^N) \mid \|f\|_{\ell,k,n}^- := \sup_{|j|_\infty \le \ell} \int |f^{(j)}(z)| e^{(M-nV)^*(Im\, z)} \prod_{i\le N}(1+|z_i|)^k dz < \infty$$

$$\text{for any } \ell,k,n \ge 1\}.$$

1.9. Proposition: Assume (*).

(a) $C_\pm^\infty$ is *-tamely isomorphic to some $\Lambda_\infty(\alpha_k)$

(b) The identity mapping is a *-tame isomorphism

i) of H_+ onto $\ker \overline{\partial} \cap C_+^\infty$

ii) of H_- onto $\ker \overline{\partial} \cap (C_-^\infty)'_b$
where $\overline{\partial}$ is the Cauchy-Riemann operator.

Proof: Let $N = 1$. We may use $\mp M^* + nV \circ m^{-1}$ in the definition of $C_\pm^\infty$ instead of $\mp(M \pm nV)^*$ by Lemma 1.6. These norms are denoted by $p_{\ell,k,n}^\pm(f)$. The corresponding norms in $H_\pm$ are $p_{k,n}^\pm$.
Let y_r be defined by $y_0 = 1$ and

$$M^*(y_{r+1}) + y_{r+1}^{\alpha+1}/(\alpha+1) = M^*(y_r) + y_r^{\alpha+1}/(\alpha+1) \tag{15}$$

and $y_{-r} := y_r$ for $r \geq 1$, where α is larger than the constants from (13) and (4).

$$\epsilon_r := y_{|r|+1} - y_{|r|} \leq y_r^{-\alpha} \tag{16'}$$

$$1/\epsilon_r \leq m^{-1}(y_{r+1}) + y_{r+1}^\alpha \leq 2e^{\alpha A V \circ m^{-1}}(y_{r+1}) \leq 2e^{\alpha A C V \circ m^{-1}}(y_r) \tag{16}$$

by (7), (16') and (13) (resp. (3)).
Since $t = o(m(t)^\alpha V(t))$ by the choice of α, we get for any δ:

$$\begin{aligned} | M^*(x) - M^*(y) | &\leq | M^*(y_{r-1}) - M^*(y_{r-1} + y_{r-1}^{-\alpha}) | + | M^*(y_r) - M^*(y_r + y_r^{-\alpha}) | \leq \\ & 2m^{-1}(y_r + y_r^{-\alpha})/y_{r-1}^\alpha \leq \delta V \circ m^{-1}(y_r + y_r^{-\alpha}) \leq C\delta V \circ m^{-1}(y_r) \leq \\ & \delta C^2 V \circ m^{-1}(z) \text{ for } x, y, z \in I_r := [y_{|r|-1}, y_{|r|+1}] sgn\, r, \end{aligned} \tag{17}$$

where (4), (13) (or (3) were used.) This also shows, that

$$V \circ m^{-1}(x) \leq C^2 V \circ m^{-1}(y) \text{ for } x, y \in I_r. \tag{18}$$

So we may choose a partition of unity $\{\phi_r \mid r \in \mathbf{Z}\}$ subordinate to $\{I_r\}$ as in Lemma 1.2, and $\phi \in D(\mathbf{R})$ such that

$$\sum_{r,j \in \mathbf{Z}} \phi_r(Im\, z)\phi(Re\, z - j) = 1.$$

The partition of unity may now be used as $\{\phi_r\}$ in i) of the proof of Theorem 1.8 to show, that $C_\pm^\infty$ is *-tamely complemented in $(\tilde{S}, q_{\ell,k,n})$ for

$$\tilde{S} := \{f \in C^\infty(\mathbf{C}) \mid q_{\ell,k,n}(f) := \sup_{j \leq \ell} \int | f^{(j)}(z) | (1 + | z |)^k e^{nV \circ m^{-1}(Im\, z)} dz < \infty$$

for any $\ell, k, n \geq 1\}$.

$(\tilde{S}, q_{\ell,k,n})$ is *-tamely isomorphic to a *-tamely complemented subspace of (s) (use the partition of unity from above and Fourier series, see [6], §1). Finally, this also holds for $C_\pm^\infty$ and 1.9a) follows from Theorem 5.2 of [13].

b)i) $\ker \overline{\partial} \cap C_+^\infty = \{f \in H(\mathbf{C}^N) \mid \|f\|_{\ell,k,n} < \infty \; \forall \ell, k, n\} \subset H_+$

$$| f |_{k,n} \leq \|f\|_{0,k,n} \text{ for } f \in H(\mathbf{C}^N) \cap C_+^\infty.$$

For $f \in H_+$ and $z \in \mathbf{C}$ choose $\tilde{y}$ as in (15) for $y = Im\, z (\cong y_r)$. Let $\epsilon(y) := \tilde{y} - y$. Cauchy's inequality and (16) and (18) imply, since $\epsilon(y)$ is decreasing with y:

$$\begin{aligned} | f^{(j)}(z) | = | f^{(j_1+j_2)}(z) | &\leq C_1 | j |! \sup_{|z-\zeta|\leq\epsilon} | f(\zeta) | \epsilon(y)^{-|j|} \leq \\ &\leq C_2 | j |! e^{\alpha |j| C^2 A V om^{-1}(|Im\, z| - \epsilon(Im\, z))} \sup_{|z-\zeta|\leq\epsilon} | f(\zeta) | . \end{aligned}$$

Using (17) and (18) we get:

$$p_{\ell,k,n}(f) \leq C_3 p_{k, C^2 n + AC^3 \ell \alpha 2 + 1}(f).$$

This shows b)i).

(ii) $\ker \overline{\partial} \cap (C_-^\infty)'_b = \{f \in H(C) \mid \|f\|^{-*}_{\ell,k,n} < \infty \text{ for some } \ell, k, n\} \supset H_-$.

$$\|f\|^{-*}_{0,k,n} = \sup\{| \int f(z)\psi(z)dz | \mid \|\psi\|^{-}_{0,k,n} \leq 1\} \leq | f |^{-}_{k,n} \text{ for } f \in H_-$$

Let $q_{C,K}(\psi) := \sup_{x \in K, \ell} | \psi^{(\ell)}(x) | C^{-|\ell|_1 - 2\ell}$ and

$$\Gamma_0^2(\Omega) := \{f \in D(\Omega) \mid \exists C : \; q_{C,\Omega}(\psi) < \infty\}.$$

Then the topologies of $\Gamma_0^2(\Omega)'_b$ and $H(\Omega)$ coincide on $H(\Omega)$, since $H(\Omega)$ is a (FS)-space w.r.t. both of them. This implies for $f \in H(B_2(z))$:

$$\exists C_1 : | f(z) | \leq C_1 \sup\{| \int f(\zeta)\psi(\zeta)d\zeta | | supp\, \psi \subset B_1(z), \; q_{B_1(z), C_1}(\psi) \leq f\}$$

and therefore

$$\forall \ell \; \exists C_\ell \; \forall \epsilon \leq 1 \quad \forall \; f \in H(B_{2\epsilon}(z)) :$$

$$| f(z) | \leq C_1 \; \sup\{| \int f(\zeta)\psi(\zeta)d\zeta | \mid supp\, \psi \subset B_\epsilon(z), \; \sup_{|j|_\infty \leq \ell} \|\psi^{(j)}\|_1 \leq \epsilon^{-2\ell}\}.$$

If $\epsilon = \epsilon(Im\, z)$ is chosen as in b)i), such ψ satisfies:

$$p^{-}_{\ell,k,n}(\psi) \leq (1 + | z |)^k e^{M^*(Im\, z) + (nC^2 + 2AC^3 \ell \alpha + 1) V om^{-1}(Im\, z)}$$

So we get for $f \in H(\mathbf{C}) \cap (C_-^\infty)'_b$:

$$p^{-}_{k, C^2 n + 2AC^3 \ell \alpha + 1}(f) \leq C_\ell p^{-*}_{\ell,k,n}(f).$$

The Proposition is proved completely.

We conclude this section by discussing some simple examples satisfying condition (*).

1.10 Examples: Let $N = 1$.

a) Let $V(t) = \ell n\, t$ and let $m(t) + k/t \uparrow \infty$ for any $k \in \mathbf{N}$. (1)-(4) are equivalent to the single condition

$$\exists\, \gamma : t^{-\gamma} \leq m(t) \leq t^{\gamma} \text{ for large } t. \tag{19}$$

Indeed, the left estimate is equivalent to (4) and the right is equivalent to (2). (1) is trivial and (3) also follows from (19). $(13)_o$ is equivalent to

$m(t^C) - m(t) \geq 2m(t)^{\alpha}$ for some α and any $C > 1$. $\qquad (20)_o$

$(2)_o$ is not satisfied. So we get ℓ-tame estimates (resp. almost 1-tame estimates), if (19) (and $(20)_o$) holds. Tame estimates cannot be obtained in this case.

b) Let $V(t) = t$ and let $m(t)$ be increasing to ∞. Let

$$\exists\, C > 1\ (\forall C > 1) : m(Ct) \geq m(t) + 1 \text{ for large } t \tag{14}$$

(resp. $(14)_o$)

(1), (4) and $(13)_o$ are then trivial. (2) (resp. $(2)_o$) hold iff

$$\exists\, A > 0\ (\forall A > 0) : \ m(t) \leq e^{At}.$$

So we get tame estimates e.g. for $M(t) = t(\ell n\, t)^{\beta}$, $\beta > 1$; $M(t) = t^{\alpha}(\ell n\, t)^{\beta}$, $\alpha > 1$, $\beta \geq 0$; $M(t) = e^{(\ell n\, t)^{\beta}}$, $\beta > 1$; $M(t) = e^{t^{\alpha}}$, $0 < \alpha < 1$. L-tame estimates are obtained for $M(t) = e^{At}$, $A > 0$. In [6] instead of (14) the obviously stronger condition (3) was used, which means in this case:

$$\exists C > 1 : \ 2m(t) \leq m(Ct)$$

c) Let $M(t) = e^{\tilde{M}(t)}$ and $V(t) = \tilde{M}(t)$, where $\tilde{m} := \tilde{M}'$ is increasing to ∞. Then (1) and (4) are trivial. $(3)_o$ follows, since $m \circ V^{-1}(t) = e^t \tilde{m}(\tilde{M}^{-1}(t))$. (2) follows from

$$\ell n\, \tilde{m}(t) = 0(\tilde{M}(t)) \tag{21}$$

So almost 1-tame estimates follow from (21)

d) With M as in c) choose $V(t) = e^{\gamma \tilde{M}(t)}$, $0 < \gamma < 1$. Then (1) is trivial as well as (4). $(3)_o$ follows, since $m \circ V^{-1}(t) = t^{1/\gamma} \tilde{m}(\tilde{M}^{-1}(\ell n\, t^{1/\gamma}))$. $(2)_o$ holds if

$$\ell n\, \tilde{m}(t) = 0(e^{\gamma \tilde{M}(t)}). \tag{22}$$

So tame estimates follow from (22).

It was proved in [6], that any continuous function is bounded by some $\tilde{M}$ such that $\tilde{m} := \tilde{M}'$ is increasing to ∞ and $\tilde{m}(t) \leq \tilde{M}(t)^2$. Especially, by 1.10.c) or d) there is no a priori bound on the growth of the weight functions leading to any type of tame estimates considered in this paper.

In the literature it is usually assumed, that the weight systems in $C^\infty(\pm M, V)$ and H are stable for shifts. For M and V satisfying (1) this implies the a priori bounds $t \leq AV(t) \leq Ae^{At}$ for some A and $m(t) = 0(V(t)) = 0(exp(0(t)))$. So V and M are essentially bounded by e^{At} then.

§2. *-Tame Right Inverses for the $\overline{\partial}$-Equation

The $\overline{\partial}$-equation will be considered in the spaces C^∞_+ and $(C^\infty_-)'_b$ (see Proposition 1.9) in this section. The splitting of the $\overline{\partial}$-complex for (DFN)-algebras of entire functions was studied in [9]. The *-tame splitting of the $\overline{\partial}$-operator will be shown in the above spaces under assumption (*) (see the beginning of §1).

We recall, that the existence of an almost 1-tame right inverse means: For any $a > 1$ there is K_0 such that for $K \geq K_0$ there is a right inverse R with $s_R(n) \leq an$, where the modul of continuity s_R is calculated for the gradings $\{\|\,\|_{n,K^n,K^{2n}} \mid n \geq 1\}$ (resp. $\{\|\,\|^{-*}_{n,K^n,K^{2n}} \mid n \geq 1\}$). Let

$$K^{\pm}_N := \{(T_i) \in \left\{ \begin{array}{c} (C^\infty_+)^N \\ ((C^\infty_-)'_b)^N \end{array} \right| \overline{\partial}_i T_j = \overline{\partial}_j T_i \text{for} 1 \leq i,j \leq N\} \text{for} N \geq 2$$

and $K^+_1 := C^\infty_+$ and $K^-_1 := (C^\infty_-)'_b$.

2.1. Theorem: Assume (*). Then $\overline{\partial} : (C^\infty_-)'_b \to K^-_N$ has a *-tame right inverse.

Proof: a) We first consider the case $N = 1$, that is, the following sequence on $\mathbf{C}$:

$$O \to H_- \overset{id}{\to} (C^\infty_-)'_b \overset{\overline{\partial}}{\to} (C^\infty_-)'_b \to O \tag{23}$$

The gradings $|\ |'_{n,K^n}$ (and $\|\,\|^{-*}_{n,K^n,K^{2n}}$) are used in H_- (and in $(C^\infty_-)'_b$, resp.) with $K = 1$ to prove ℓ-tameness and tameness and with large K to prove almost 1-tameness.

(i) $id : H_- \to (C^\infty_-)'_b$ is a *-tame isomorphism by Proposition 1.9.

(ii) $\overline{\partial} : (C^\infty_-)'_b / id(H_-) \to (C^\infty_-)'_b$ is a *-tame isomorphism.

Proof: $\|\overline{\partial} T\|^{-*}_{\ell+1,k,n} \leq C_1 \|T\|^{-*}_{\ell,k,n}$ for $T \in (C^\infty_-)'$.

$$|\,\psi(z)\,| \leq C_2 \sup_{|j| \leq 2} \Big(\int_{|z-\zeta| \leq 1} |\,\psi^{(j)}(\zeta)\,|^2\, d\zeta\Big)^{\frac{1}{2}} \text{ for } \psi \in C^\infty(\mathbf{C})$$

by Sobolev's Lemma. This implies:

$$\forall \ell\, \exists C_\ell\, \forall \epsilon \leq 1\, \forall \psi \in C^\infty(\mathbf{C}) : \sup_{|j| \leq \ell} |\,\psi^{(j)}(z)\,| \leq C_\ell \epsilon^{-\ell-3} \sup_{|j| \leq \ell+2} \Big(\int_{|z-\zeta| \leq \epsilon} |\,\psi^{(j)}(\zeta)\,|^2\, d\zeta\Big)^{\frac{1}{2}}.$$

If $\epsilon = \epsilon(Im\, z)$ is chosen as in b)i) of the proof of Proposition 1.9, we get:

$$\sup_{|j|\le \ell} |\psi^{(j)}(z)| (1+|z|)^k e^{(M-nV)^*(Im\, z)} \le$$

$$\le C_\ell e^{AC^2(\ell+3)Vom^{-1}(Im\, z)} \sup_{|j|\le \ell+2} \left(\int_{|z-\zeta|\le\epsilon} |\psi^{(j)}(\zeta)|^2 d\zeta\right)^{\frac{1}{2}} e^{(M-nV)^*(Im\, z)}(1+|z|)^k$$

$$\le C_\ell' \sup_{|j|\le \ell+2} \left(\int |\psi^{(j)}(\zeta)|^2 e^{2(M-(nC+AC^2(\ell+3))V)^*(Im\,\zeta)}(1+|\zeta|)^{2k} d\zeta\right)^{\frac{1}{2}}$$

$$:= q^-_{\ell+2,k,n'}(\psi) C_\ell' \quad \text{for} \quad n' = Cn + AC^2(\ell+3)$$

by the proof of Lemma 1.6 and (17). This shows for $T \in (C^\infty_-)_b'$

$$q^{-*}_{\ell+2,k,n'}(T) \le C_\ell' \|T\|^{-*}_{\ell,k,n}.$$

Now if $\|T\|^{-*}_{\ell,k,n} \le 1$, then the Theorem of Hahn-Banach shows the existence of $f_\alpha \in L^2_{loc}$, $|\alpha| \le \ell+2$, such that

$$T = \sum_{|\alpha|\le \ell+2} D^\alpha f_\alpha \text{ and } \sum_{|\alpha|\le\ell+2} \left(\int |f_\alpha(z) e^{-(M-n'V)^*(Im\, z)}(1+|z|)^{-k}|^2 dz\right)^{\frac{1}{2}} \le C_\ell''.$$

$(M - n'V)^*(Im\, z) + k\,\ell n(1+|z|^2)$ is subharmonic. So there are $u_\alpha \in L^2_{loc}$ (by Theorem 15.1.2 of [1]) such that $\overline{\partial} u_\alpha = f_\alpha$ and

$$\sum_\alpha \left(\int |u_\alpha(z) e^{-(M-n'V)^*(Im\, z)}(1+|z|)^{-k-2}|^2 dz\right)^{\frac{1}{2}} \le \tilde{C}_\ell.$$

Let $u := \sum_\alpha D^\alpha u_\alpha$. Then $\overline{\partial} = T$ and

$$\|u\|^{-*}_{\ell+2,k+4,n'} \le \hat{C}_\ell \|T\|^{-*}_{\ell,k,n}$$

iii) We have shown in i) and ii), that the sequence (dual to (23))

$$O \to C^\infty_- \xrightarrow{-\overline{\partial}} C^\infty_- \xrightarrow{q} (H_-)_b' \to O \tag{24}$$

is exact and *-tame (i.e. $\overline{\partial} : C^\infty_- \to C^\infty_-$ and $q : C^\infty_-/(-\overline{\partial} C^\infty_-) \to (H_-)_b'$ are *-tame isomorphisms (into and onto, resp.)). (24) is *-tamely isomorphic to

$$0 \to \Lambda_\infty(\alpha_n) \xrightarrow{\tilde{p}} \Lambda_\infty(\alpha_n) \xrightarrow{\tilde{q}} \Lambda_\infty(\beta_n) \to O$$

for some $\Lambda_\infty(\alpha_n)$ and $\Lambda_\infty(\beta_n)$ by Proposition 1.9 and Theorem 1.8. So $\tilde{q}$ has a *-tame right inverse (see [11], [14], [15]).

Finally, $-\overline{\partial}$ (and $\overline{\partial}$) have *-tame left inverses (resp. right inverses) in C^∞_- (resp. $(C^\infty_-)_b'$).

b) The general case will be reduced to the case $N = 1$.
i) Let R_i be a *-tame right inverse of $\overline{\partial}_i = \overline{\partial}_{z_i}$ according to a).

Let $\tilde{B}_{1,1} := {}^tR_1$ and let

$$\tilde{B}_{N+1} := \prod_{i\le N+1} \tilde{B}_{N+1,i} : C^\infty_- \to (C^\infty_-)^{N+1}$$

be defined inductively for $N \geq 1$ and $z = (z', z_{N+1}) \in \mathbf{C}^N \times \mathbf{C}$ by

$$\tilde{B}_{N+1,i}(f)(z', z_{N+1}) := \tilde{B}_{N,i}(f(\cdot, z_{N+1}) + \overline{\partial}_{N+1} \circ {}^tR_{N+1}f(\cdot, z_{N+1}))(z')$$

$$\tilde{B}_{N+1,N+1}(f)(z', z_{N+1}) = {}^tR_{N+1}f(z', z_{N+1})$$

$\tilde{B}_{N+1,i}$ is defined and linear from C^∞_- into C^∞_- since

$$D^\alpha_{z'} L_{z''} f(\cdot, z'') = L_{z''} D^\alpha_{z'} f(\cdot, z'') \tag{25}$$

for $f \in C^\infty_-(\mathbf{C}^j \times \mathbf{C}^k), z = (z', z'') \in \mathbf{C}^j \times C^k = \mathbf{R}^{2j} \times \mathbf{R}^{2k}$, a linear and continuous operator $L_{z''}$ in $C^\infty_-(\mathbf{C}^k)$ and $\alpha \in \mathbf{N}^{2j}$.
This is evident for the single case to be proved, namely if $supp\, f \subset K' \times \mathbf{R}^{2k}$, where K' is compact in $\mathbf{R}^{2j}$. In general, the differential quotients of $f \in C^\infty_-$ need not even be contained in C^∞, though differentiation is continuous in C^∞_-. (25) also implies, that $\tilde{B}_{N+1,i}$ is *-tame by induction, since $\tilde{B}_{1,1} = {}^tR_1$ is *-tame by a). So

$$B_{N+1}(T_1, \cdots, T_{N+1}) := \sum {}^t\tilde{B}_{N+1,i}(T_i)$$

is a *-tame mapping from $((C^\infty_-)'_b)^{N+1}$ into $(C^\infty_-)'_b$.

ii) Let $L_N := (L_{N,i})$ be a continuous mapping from $C^\infty_-(\mathbf{C}^N)$ into $C^\infty_-(\mathbf{C}^N)^N$ such that $\langle (T_1, \cdots, T_N), L_N(f)\rangle = 0$ for any $f \in C^\infty_-(\mathbf{C}^N)$ and $(T_i) \in K_N^-$. Then

$$\langle (T_1, \cdots, T_N)_{,z'}\, L_N f(z', z_{N+1})\rangle = 0 \text{ for any } f \in C^\infty_-(\mathbf{C}^{N+1})$$

and $(T_1, \cdots, T_{N+1}) \in K^-_{N+1}$. Since $D(\mathbf{R}^{2N+2})$ (and hence $Lin\{\prod \psi_i(x_i) \mid \psi_i \in D(\mathbf{R})\}$) are dense in $C^\infty_-(\mathbf{C}^{N+1})$, this has to be proved for $f(z', z_{N+1}) = f_1(z')f_2(z_{N+1})$, $f_i \in D$, only.

$$\langle (T_1, \cdots, T_N)_{,z'}\, L_N f_1(z')f_2\rangle = \sum_{i\le N} \langle (T_i)_{f_2}, L_N f_1\rangle = 0, \text{ since } ((T_1)_{f_2}, \cdots, (T_N)_{f_2}) \in K_N^- :$$

$$\begin{aligned} \langle \overline{\partial}_j (T_i)_{f_2}, f\rangle &= \langle (T_i)_{f_2}, -\overline{\partial}_j f\rangle = \langle T_i, -\overline{\partial}_j f f_2\rangle \\ &= \langle \overline{\partial}_j T_i, f f_2\rangle = \langle \overline{\partial}_i T_j, f f_2\rangle = \langle \overline{\partial}_i (T_j)_{f_2}, f\rangle \end{aligned}$$

for $f \in C^\infty_-(\mathbf{C}^N)$ and $i, j \leq N$.

iii) $-\tilde{B}_N(\overline{\partial}_j f) = f e_j + L^j_N(f)$ with $\langle (T_i), L^j_N(f)\rangle = 0$ for $(T_i) \in K_N^-$, $f \in C^\infty_-(\mathbf{C}^N)$, $j \leq N$.

Proof: Let $N \geq 1$.

$$-\tilde{B}_{N+1,i}(\overline{\partial}_{N+1}f) = -\tilde{B}_{N,i}(\overline{\partial}_{N+1}f + \overline{\partial}_{N+1}({}^tR_{N+1}\overline{\partial}_{N+1})f) = 0 \text{ for } i \leq N$$

$$-\tilde{B}_{N+1,N+1}(\overline{\partial}_{N+1}f) = f,$$

since ${}^tR_{N+1} \circ \overline{\partial}_{N+1} = -id$. Hence iii) holds with $L_{N+1}^{N+1} = 0$. Let $j \leq N$.

$$-\tilde{B}_{N+1,i}(\overline{\partial}_j f) = -\tilde{B}_{n,i}\overline{\partial}_j(f + \overline{\partial}_{N+1} \circ {}^tR_{N+1}f) \text{ for } i \leq N$$

$$-\tilde{B}_{N+1,N+1}(\overline{\partial}_j f) = -\overline{\partial}_j\, {}^tR_{N+1}f$$

by (25). For $N = 1 = j = i$ we have: $-\tilde{B}_{1,1}\overline{\partial}_1 = -{}^tR_1\overline{\partial}_1 = id$ and hence iii) holds with $L_2^1(f) = (\overline{\partial}_2\, {}^tR_2 f, -\overline{\partial}_1\, {}^tR_2 f)$. For $N > 1$ we have by assumption:
$-\tilde{B}_N(\overline{\partial}_j f) = fe_j + L_N^j(f)$. So iii) holds again (by ii)) with

$$L_{N+1}^j(f) = (L_N^j(f + \overline{\partial}_{N+1}\, {}^tR_{N+1}f) + \overline{\partial}_{N+1}\, {}^tR_{N+1}fe_j, -\overline{\partial}_j\, {}^tR_{N+1}f).$$

iv) Let $(T_i) \in K_{N+1}^-$, $j \leq N+1$ and $f \in C_-^\infty$.

$$\langle \overline{\partial}_j B_{N+1}(T_i), f\rangle = -\sum_{i \leq N+1} \langle T_i, \tilde{B}_{N+1,i}\overline{\partial}_j f\rangle\rangle = \langle T_j, f\rangle + \langle (T_i), L_{N+1}^j(f)\rangle\rangle = \langle T_j, f\rangle$$

by iii). So B_{N+1} is a *-tame right inverse for $\overline{\partial} = (\overline{\partial}_1, \cdots, \overline{\partial}_{N+1})$ on K_{N+1}^-.

The corresponding proof in [9] ([9], Proposition 1.9) of the exactness of (23) used convolution, that is, C_-^∞ has to be invariant for shifts, which means that $t = O(V(t))$.

To prove the same result in C_+^∞, we need the following simple lemma on the existence of certain sh. functions:

2.2. Lemma: Assume (*) and $N = 1$. For any n and k there are sh. functions $\phi_{k,n}$ such that for A from (7)

$$-C_1 + (M + (n+1+Ak)V)^*(Im\, z) - (k+1)\ell n(1+|z|) \leq \phi_{k,n} \leq$$
$$\leq (M + nV)^*(Im\, z) - k\,\ell n(1+|z|) + C_1.$$

Proof: See also [6], Theorem 2.3c).

i) For $n' = n + 1 + Ak$ let

$$(M + n'V)^*(y_r) = (M + (n'+1)V)^*(y_{r+1}) \text{ for } r \geq 1 \text{ and } y_1 := 1.$$

We may assume, that $m+n'v$ and $m+(n'+1)v$ are increasing and that $V>0$ on $\mathbf{R}_+$. So y_r is increasing to ∞.
Let $x_0=0$ and $0\le x_r$ for $r\ge 1$ be chosen such that

$$(M+n'V)^*(y_r)=y_rx_r-(M+n'V)(x_r). \tag{26}$$

ii) Let $0\le\phi\in D((0,1/4))$ and $\|\phi\|_1=1$. For $r\ge 1$ let

$$\psi_r:=\phi((\cdot-x_r)/\epsilon_r)e^{-M(x_r)-n'V(x_r)}/\epsilon_r,$$

where $\epsilon_r=\tilde{x}_r-x_r\le 1$ and $\tilde{x}_r$ is defined by (5), i.e. $U(\tilde{x}_r)=U(x_r)+1$.
(1.6) and (1.8) imply:

$$\begin{aligned}|\hat{\psi}_r(z)|(1+|z|)^k &\le C_1\epsilon_r^{-k}e^{|\tilde{x}_r Im\, z|-M(\tilde{x}_r)-n'V(x_r)}\le\\ & C_2e^{|\tilde{x}_r Im\, z|-M(\tilde{x}_r)-nV(\tilde{x}_r)}\le C_3e^{(M+nV)^*(Im\, z)-V(x_r)}.\end{aligned}$$

iii) Let $|Re\, z|\le 1(\le 1/\epsilon_r)$, $y_{r+1}\ge Im\, z\ge y_r$ and $r\ge 1$. (26) and the definition of y_r show

$$\begin{aligned}|\hat{\psi}(z)|\ge e^{x_r y_r-M(x_r)-n'V(x_r)}cos(1/4) &= C_4e^{(M+n'V)^*(y_r)}=\\ &= C_4e^{(M+(n'+1)V)^*(y_{r+1})}\ge C_4e^{(M+(n'+1)V)^*(Im\, z)}.\end{aligned}$$

iv) $\psi_{rs}(z):=\ell n\,|\hat{\psi}_r(z-s)|-(k+1)\ell n(1+|s|)$

$$|\psi_{rs}(z)|\le C_5+(M+nV)^*(Im\, z)-k\,\ell n(1+|z|)-V(x_r)-\ell n(1+|s|) \tag{27}$$

by ii). Let $\tilde{\phi}_{k,n}:=\sup\{\psi_{rs}(z),\psi_{rs}(-z)\mid r\in\mathbf{N},\ s\in\mathbf{Z}\}$.
This supremum is locally a maximum by (27), so $\tilde{\phi}_{k,n}$ is continuous and sh., since ψ_{rs} is sh. $\tilde{\phi}_{k,n}$ has the desired properties for $|Im\, z|\ge 1$ xby iii). The conclusion follows for $\phi_{k,n}:=\max(\tilde{\phi}_{k,n},g_k)$, where g_k is sh. and

$$-C_6-(k+1)\ell n(1+|z|)\le g_k(z)\le C_6-k\,\ell n(1+|z|)+|Im\, z|.$$

2.3. Theorem: Assume (*). Then $\overline{\partial}:C_+^\infty\to K_N^+$ has a *-tame right inverse.

Proof: a) We again consider the case $N=1$ first, that is, we consider the following sequence on $\mathbf{C}$:

$$O\to H_+\to C_+^\infty\xrightarrow{\overline{\partial}}C_+^\infty\to O. \tag{28}$$

(i) $id:H_+\to C_+^\infty$ is a *-tame isomorphism by Proposition 1.9.

(ii) $\bar\partial : C_+^\infty / id(H_+) \to C_+^\infty$ is a *-tame isomorphism.

Proof: $\|\bar\partial f\|_{1,k,n} \le C_\ell \|f\|_{\ell+1,k,n}$ for $f \in C_+^\infty$.

$$q_{0,k,n'}(f) := \Big(\int |f(z)|^2 e^{-2(M+n'V)^*(Im\, z)}(1+|z|)^{2k} dz\Big)^{\frac12} \le C_1 \|f\|_{\ell,k+2,n'}.$$

By Lemma 2.2 there is a sh. function $\phi = \phi_{k,n}$ such that

$$\Big(\int |f(z)|^2 e^{-2\phi(z)} dz\Big)^{\frac12} \le C_2 q_{0,k+1,n'}(f)$$

with $n' = (n+1+Ak)$. Theorem 15.1.2 in [1] now gives a solution $u \in L^2_{loc}$ of $\bar\partial u = f$ such that (again by 2.2)

$$\Big(\int |u(z)|^2 e^{-2\phi(z)}(1+|z|^2)^{-2} dz\Big)^{\frac12} \le C_3 q_{0,k+1,n'}(f).$$

Using finally 2.2 again we get for $f \in C_+^\infty$ a solution u of $\bar\partial u = f$ such that

$$q_{0,k,n}(u) \le C_4 \|f\|_{\ell,k+5,n'}. \tag{29}$$

Sobolev's lemma implies for $\epsilon \le 1$ and any z:

$$\sup_{|j|\le \ell} |u^{(j)}(z)| \le \epsilon^{-1}\big(\|u\|_{L_2(B_\epsilon(z))} + \sup_{|j|\le \ell+1} \|f^{(j)}\|_{L_2(B_\epsilon(z))}\big).$$

We now choose $\epsilon = \epsilon(Im\, z)$ as in the proof of 1.9 and get

$$q_{\ell,k,n}(u) \le C_6\big(q_{0,k,Bn+B'\ell}(u) + q_{\ell+1,k,Bn+B'\ell}(f)\big),$$

where B and B' are monomials in C (and α, A, resp.; see 1.9). Using now (29) with $Bn+B'\ell$ instead of n we get a solution u such that

$$\|u\|_{\ell,k,n} \le C_7 \|f\|_{\ell+1,k+5,n''} \text{ with } n'' = nB + \tilde B + Ak + B'\ell.$$

Let $C_{\ell,k,n} := \{f \in C^\ell(\mathbf{R}^2) \mid \|f\|_{\ell,k,n} < \infty\}$. We have shown that $\bar\partial u = f$ has a solution $u \in C_{\ell,k,n}$ for every $f \in C_{\ell+1,k+5,n''}$. Now $\ker \bar\partial \cap C_{\ell,k,n}$ is contained in $H_{k,n} = \{f \in H(\mathbf{C}^2) \mid |f|_{k,n} < \infty\}$. $H_{k,Bn+B'\ell}$ is continuously embedded in $\ker\bar\partial \cap C_{\ell,k,n}$ by 1.9. Remark 1.5 implies, that H_+ is dense in $\ker\bar\partial \cap C_{\ell,k+2,n'}$ for the topology of $C_{\ell,k,n}$ (with $n' = Bn + B'\ell + A + 1$). We can now apply the Mittag-Leffler procedure to obtain a solution $u \in H_+$ such that

$$\|u\|_{\ell,k,n} \le C_8 \|f\|_{\ell+1,k+7,\tilde n}$$

where $\tilde n$ comes from n'' and n' above. This shows ii).

iii) (28) is a *-tamely exact sequence by i) and ii), which is *-tamely equivalent to a *-tamely exact sequence of power series spaces of infinite type by Proposition 1.9 and Theorem 1.8. So the sequence splits *-tamely by the results of [11], [14] and [15].

b) The general case will be proved by induction on N. Let B_N (and R_j) be *-tame right inverses for $\overline{\partial} : C_+^\infty \to K_N^+$ (and $\overline{\partial}_j : C_+^\infty(\mathbf{C}_j) \to C_+^\infty(\mathbf{C}_j)$, resp.). The induction hypothesis is satisfied for $B_1 = R_1$. Let

$$\begin{aligned} B_{N+1}(f_1,\cdots,f_{N+1})(z) := {}& {}_{z'}B_N(f_1(z',z_{N+1}),\cdots,f_N(z',z_{N+1}))+ \\ & -R_{N+1}({}_{z'}B_N(\overline{\partial}_{N+1}f_1(z',z_{N+1}),\cdots,\overline{\partial}_{N+1}f_N(z',z_{N+1})) - f_{N+1}(z',z_{N+1})) \quad (30) \\ & \text{for } z = (z',z_{N+1}) \in \mathbf{C}^N \times \mathbf{C}, (f_i) \in K_{N+1}^+ \text{ and } N \geq 1. \end{aligned}$$

$(\overline{\partial}^j_{N+1}f_1(\cdot,z_{N+1}),\cdots,\overline{\partial}^j_{N+1}f_N(\cdot,z_{N+1})) \in K_N^+$ for any z_{N+1} and $j = 1,0$. So the application of B_N in the definition of B_{N+1} is allowed and B_{N+1} is a *-tame mapping from K_{N+1}^+ into C_+^∞ (compare (2.2)). Similarly we get:

$$\begin{aligned} \overline{\partial}_{N+1}B_{N+1}(f_1,\cdots,f_{N+1}) &= B_N(\overline{\partial}_{N+1}f_1,\cdots,\overline{\partial}_{N+1}f_N)+ \\ &\quad -B_N(\overline{\partial}_{N+1}f_1,\cdots,\overline{\partial}_{N+1}f_N) + f_{N+1} = f_{N+1} \\ \overline{\partial}_j B_{N+1}(f_1,\cdots,f_{N+1}) &= f_j - R_{N+1}(\overline{\partial}_{N+1}f_j - \overline{\partial}_j f_{N+1}) = f_j \text{ for } j \leq N, \end{aligned}$$

since $(f_i) \in K_{N+1}^+$ and B_N is a right inverse for $\overline{\partial}$ by assumption.

The formulas defining B_N in 2.1 and 2.3 are essentially the same. The transposed version was chosen in 2.1 to deal with functions rather than distributions.

§3. *-Tame Right Inverses for Partial Differential Operators with Constant Coefficients

Using the *-tame right inverses for the $\overline{\partial}$-operator from the preceding section, we will show now the existence of a *-tame right inverse in $C^\infty(-M,V)$ and in $C^\infty(M,V)_b'$ for a general partial differential operator with constant coefficients. In contrast to the method of [6] these right inverses will be given by an explicit formula (and transposition) and the assumptions are weaker with improved conclusions compared with [6]. The main result is contained in Proposition 3.1 below. A similar application of a right inverse for to an extension problem for holomorphic functions is given in [12].

The *-tame right inverses will also give *-tame isomorphisms of the kernel of $P(D)$ in $C^\infty(-M,V)$ (and in $C^\infty(M,V)_b'$ to some $\Lambda_\infty(\alpha_n)$ (resp. $\Lambda_\infty(\alpha_n)_b'$), which in many cases is ℓ-tamely isomorphic to (s) (see [3, 5, 7] for related results in various types of generalized functions).

3.1. Proposition: Assume (*). Then $P(z)$ has a *-tame left inverse $L_\pm$ in $H_\pm$. To prove tame estimates, we also have to assume that P is hypoelliptic.

Proof: i) Let $d = \deg P$ and let J be an orthogonal matrix such that x_1 is noncharacteristic w.r.t. $P_J := P(J\cdot)$. We may assume, that $P_J(1,0) = 1$.
The following is shown in [1], p.293/294:
There is C_1 and $\gamma' < 1/6$ such that for any $\zeta \in \mathbf{C}^N$:

$$| P_J(s,z') | \geq C_1 \tilde{P}_J(\zeta)(1+|\zeta|)^{(1-\beta-d)d} \tag{31}$$

for $| s - \zeta_1 | = \gamma_1(1+|\zeta|)^{-\beta} =: \delta$ and $| z' - \zeta' | \leq \gamma'(1+|\zeta|)^{1-d-\beta}$, where $1/3 < \gamma_1 < 2/3$ and $\tilde{Q}(\zeta) := (\sum | Q^{(\alpha)}(\zeta) |^2)^{\frac{1}{2}}$. For $f \in H(\mathbf{C}^N)$ let

$$\Delta_\zeta(f)(z_1,z') := \frac{1}{2\pi i}\int_{|\zeta_1-s|=\delta} f(s,z')/(P_J(s,z')(s-z_1))ds.$$

Then $\Delta_\zeta(f)$ is defined on

$$B_{\gamma'}(\zeta) := \{z \mid | \zeta_1 - z_1 |^2 (1+|z|)^{2\beta} + | \gamma' - z' |^2 (1+|z|)^{2(\beta+d-1)} < \gamma'^2\}.$$

We obviously get from (31)

$$\Delta_\zeta : H(B_1(\zeta)) \to H(B_{\gamma'}(\zeta)) \tag{32}$$

is linear

$$\Delta_\zeta(P_J f) = f \text{ on } B_{\gamma'}(\zeta) \tag{33}$$

$$\sup_{z\in B_{\gamma'}(\zeta)} | \Delta_\zeta(f)^{(j)}(z) | \leq C_2 | j |! \sup_{\eta\in B_1(\zeta)} | f(\eta) | (1+|\zeta|)^{(d+\beta-1)d+\beta|j|}/\tilde{P}_J(\zeta). \tag{34}$$

So Δ_ζ is a local left inverse for P_J.

ii) By Theorem 1.4.10 of [1] we may now choose a partition of unity $\{\phi_r \mid r \in \mathbf{N}\}$ such that $supp\, \phi_r$ is contained in $B_{\gamma'}(\zeta_r)$ for some $\zeta_r \in \mathbf{C}^N$ and such that the number of mutual intersections of $supp\, \phi_r$ is globally bounded. Moreover,

$$\sup_{z\in B_{\gamma'}(\zeta_r)} | \phi_r^{(j)}(z) | \leq C_3(1+|\zeta_r|)^{d+\beta-1)|j|}. \tag{35}$$

Let $\tilde{C}_\mp^\infty$ be defined as $C_\mp^\infty$ but with the weights depending on Jz instead of z. For $f_+ \in \tilde{C}_+^\infty$ and $f_- \in (\tilde{C}_-^\infty)'_b$ let

$$\tilde{L}_\pm(f_\pm) := \sum \phi_r \Delta_{\zeta_r}(f_\pm)$$

$L_+ : \tilde{H}_+ \to \tilde{C}_+^\infty$ and $\tilde{L}_- : \tilde{H}_- : \tilde{H}_- \to (\tilde{C}_-^\infty)'_b \cap C^\infty(\mathbf{R}^{2N})$ are *-tame and $\tilde{L}_\pm(P_J f_\pm) = f_\pm$ for $f_\pm \in \tilde{H}_\pm$.

Proof: The sum defining $\tilde{L}_\pm$ is locally finite, so $\tilde{L}_\pm$ is defined and linear by (32).

$$\tilde{L}_\pm(P_J f_\pm) = \sum \phi_r f_\pm = f_\pm,$$

since (33) holds on $supp\,\phi_r$. Let $V_J := \{z \mid P_J(z) = 0\}$ and $\Omega_1 := \{z \mid d(z, V_J) > 3/2\}$. If $\zeta \in \Omega_1 \cap supp\,\phi_r$, then $d(supp\,\phi_r, V_J) > 7/6, f/P_J \in H(B_1(\zeta_r))$
and $\Delta_{\zeta_r}(f) = \Delta_{\zeta_r}(P_J(f/P_J)) = f/P_J$ on $B_{\gamma'}(\zeta_r)$ by (33). So

$$\tilde{L}_\pm(f_\pm)(\zeta) = f_\pm(\zeta)/P_J(\zeta) \text{ for } \zeta \in \Omega_1. \tag{36}$$

We obviously get tame estimates with constant 0 on Ω_1. On $\Omega_2 := \{z \mid d(z, V_J) \leq 2\}$ we get by (34) and (35) and Cauchy's estimate

$$\|\tilde{L}_+(f_+)\|^{\Omega_2}_{\ell,k,n} \leq C_4 \sup_{B_{\gamma'}(\zeta_r)\cap\Omega_2\neq\emptyset} (1+ \mid \zeta_r \mid)^{(d+2\beta-1)(2\ell N+1)} \sup_{\eta\in B_{2_{\gamma'}}(\zeta_r)} \mid f(\eta) \mid \;\; \|1\|^{B_{\gamma'}(\zeta_r)}_{0,k,n} \tag{37}$$

It was proved in 1.9, that the weights $(M+nV)^*(Im\,z)$ (and hence $(M+nV)^*(JIm\,z)$) do not change too much on $B_1(\zeta_r)$, (since this set is contained in $\{z \mid \mid \zeta_r - z \mid \leq (1+ \mid \zeta_r \mid)^{-\beta}\}$), if β is larger than α from (4). So we get ℓ-tame and almost 1-tame estimates. If P is hypoelliptic, then $(2)_o$ implies

$$\begin{gathered}(1+ \mid \zeta_r \mid)^{(d+2\beta-1)(2\ell N+1)} \leq C_5(1+ \mid J\,Im\,\zeta_r \mid)^{C_6\ell} \leq e^{V \circ m^{-1}(J\,Im\,\zeta_r)} \\ \text{for } B_{\gamma'}(\zeta_r) \cap \Omega_2 \neq \emptyset \text{ and large } \zeta_r.\end{gathered} \tag{38}$$

So we get tame estimates then.
Choose $\phi \in C^\infty(\mathbf{C}^N)$ such that $supp\,\phi \subset \Omega_1$ and $supp(1-\phi) \subset \Omega_2$ and $\|\phi^{(j)}\|_\infty < \infty$ for any j. Then

$$\langle \tilde{L}_-(f_-), \psi \rangle = \langle \tilde{L}_-(f_-), \phi\psi \rangle + \langle \tilde{L}_-(f_-), (1-\phi)\psi \rangle$$

and $\|\tilde{L}_-(f_-)\|^{-*}_{\ell,k,n}$ is estimated by treating the two terms on the right as above.

iii) Let $g_A(z) := g(Az)$ for $g \in C^\infty(\mathbf{C}^N)$ and and $A \in GL(n, \mathbf{R})$. Let

$$\hat{L}_\pm(f_\pm) := (\tilde{L}_\pm((f_\pm)_J))_{J^{-1}} \text{ for } f_\pm \in H_\pm.$$

Then $\hat{L}_+ : H_+ \to C^\infty_+$ and $\hat{L}_- : H_- \to (C^\infty_-)'_b$ are *-tame and

$$\hat{L}_\pm(Pf_\pm) = (\tilde{L}_\pm(P_J(f_\pm)_J))_{J^{-1}} = ((f_\pm)_J)_{J^{-1}} = f_\pm$$

by ii). Let $r_\pm$ be a *-tame right inverse for $\overline{\partial}$ in C^∞_+ (and $(C^\infty_-)'_b$, resp.) according to §2. Then

$$L_\pm(f_\pm) := (id - r_\pm \circ \overline{\partial}) \circ \hat{L}_\pm(f_\pm)$$

is a *-tame mapping from H_+ into C_+^∞ (and from H_- into $(C_-^\infty)_b'$).
$\overline{\partial} L_\pm = (\overline{\partial} - (\overline{\partial} \circ r_\pm) \circ \overline{\partial}) \circ \hat{L}_\pm = 0$, since $r_\pm$ is a right inverse for $\overline{\partial}$. So $L_\pm$ is *-tame from $H_\pm$ into $H_\pm$ by Proposition 1.9.

$$L_\pm \circ P = (id - r_\pm \circ \overline{\partial})(\hat{L}_\pm \circ P) = (id - r_\pm \circ \overline{\partial}) = id \text{ on } H_\pm,$$

since $\overline{\partial}(H_\pm) = 0$. This completes the proof of 3.1.

Condition (4) is related to the polynomial size of the radii of $supp\, \phi_r$ in the proof above. These are essentially invariant under orthogonal transformations, so we could choose x_1 to be noncharacteristic via J. If some coordinate is noncharcteristic from the beginning, then this transformation can be deleated and one can prove 3.1 without the assumption (4), choosing the radii by the formula $M^*(y + \epsilon_y) = M^*(y) + 1$. Indeed, also Proposition 1.9 can be proved assuming only (1) - (3).

3.2. Theorem: Same assumptions as in 3.1. Then $P(D)$ has *-tame right inverses R_+ in $C^\infty(M, V)_b'$ and R_- in $C^\infty(-M, V)$.

Proof: Let $L_\pm$ be a *-tame left inverse for $P(-\cdot)$ in $H_\pm$ as in 3.1. $R_\pm := (\mathcal{F}^{-1} \circ L_\pm \circ \mathcal{F})^t$ is *-tame in $C^\infty(M, V)_b'$ and in $C^\infty(-M, V)$ by Theorem 1.4, since the gradings of $C^\infty(M, V)_b'$ and $C^\infty(-M, V)$ are the dual gradings of $C^\infty(M, V)$ (and $C^\infty(-M, V)_b'$, resp.).

$$P(D) \circ R_\pm = (\mathcal{F}^{-1} \circ L_\pm \circ \mathcal{F} \circ P(-D))^t = (\mathcal{F}^{-1} \circ L_\pm \circ P(-\cdot) \circ \mathcal{F})^t = (\mathcal{F}^{-1}\mathcal{F})^t = id.$$

The conditions (1) - (3) and $t = O(V(t))$ (stronger than (4)) were used in [6] to prove the existence of continuous right inverses for $P(D)$ in $C^\infty(M, V)_b'$ and in $C^\infty(-M, V)$. So we get ℓ-tame estimates from weaker assumptions. Examples verifying condition (*) were given in 1.10.

Let $N_P^\pm(M, V)$ be the kernel of $P(D)$ in $C^\infty(M, V)_b'$ (and in $C^\infty(-M, V)$, resp.).

3.3. Corollary: Assume (1) - (4). Then N_P^+ and N_P^- are ℓ-tamely isomorphic to some $\Lambda_\infty(\alpha_n)_b'$ (and $\Lambda_\infty(\alpha_n)$, resp.), if $P \neq const$ (and $N \geq 2$, as is always assumed in this paper).

Proof: $N_P^\pm$ is ℓ-tamely isomorphic to a ℓ-tamely complemented subspace of $(s)'$ (resp. (s)) by Theorems 1.4, 1.8 and 3.2. So the conclusion follows from Theorem 5.2 in [13], since $\dim N_P^\pm = \infty$.

Especially, $N_P^\pm$ then has a basis. This could of course not be proved in the continuous category used in [6], since the corresponding basis problem in complemented subspaces of (s) is still open.

The space $\Lambda_\infty(\alpha_n)$ is often isomorphic to (s):

3.4. Theorem: Let $m_i(t) + k/t$ be increasing to ∞ for large t and any $k \in \mathbf{N}$ and let

$$t^{\frac{1}{\gamma}} \leq m_i(t) \leq t^{\gamma} \text{ for some } \gamma > 0 \text{ and large } t.$$

Then the kernels $S_P^{\pm}$ of $P(D)$ in $S(M)_b'$ (and $S(-M)$) are ℓ-tamely isomorphic to $(s)'$ (resp. (s)), if $P \neq const.$.

Proof: a) (1) - (4) are satisfied for $S(M)_b'$ and $S(-M)$ by Example 1.10a). So we only have to calculate $\Lambda_\infty(\alpha_n)$ i.e. the diametral dimension $\Delta(S_P^{\pm})$ of $S_P^{\pm}$. It is clear from 1.4 and 1.8, that S_P^+ and S_P^- are isomorphic to subspaces of $(s)'$ (and (s), resp.). Let $S_Q^{\pm}$ be the solutions of $Q(D) := P(D_1, D_2, 0)$ in $S(M)_b'$ (and $S(-M)$), which are 2π-periodical w.r.t. x_1. Periodical solutions of partial differential operators were considered in [2] in a different setting. We will show in b), that $S_Q^{\pm}$ contains $(s)'$ (resp. (s)) as a closed subspace. Since $S_Q^{\pm}$ is a subspace of $S_P^{\pm}$, the diametral dimensions of $(S_P^+)' \simeq \Lambda_\infty(\alpha_n) \simeq S_P^-$ coincide with $(s)'$, and α_n is equivalent to $\ell n\, n$. This shows our assertion.

b) $S_Q^{\pm}$ contains $(s)'$ (resp. (s)) as a closed subspace.

Proof: Since $P \neq const.$, we may assume that Q depends on x_2. Then we may also assume, that Q depends on x_1, since otherwise $f(S(M)_b') \simeq (s)'$ is a subspace of S_Q^+, if f is a solution of $Q(D)$ in $S(M)'$, depending only on x_2. (Similarly for S_Q^-). Now there is j_0 such that for $j \geq j_0$ $Q_j(\tau) = Q(j, \tau)$ is not the constant polynomial. Choose $\lambda_j \in \mathbf{C}$ such that $Q(j, \lambda_j) = 0$.

$$\Lambda^{\pm} := \{(c_j)_{j \geq j_0} \mid \mid (c_j) \mid_n^{\pm} := \sum_j \mid c_j \mid (1 + \mid (j, \lambda_j) \mid)^{\pm n} e^{(M \pm nV)^*(Im\, \lambda_j)} < \infty$$

for $\left\{ \begin{array}{l} some \\ any \end{array} \right. n \in \mathbf{N}\}$.

i) Let $\kappa((c_j)) := \sum c_j e^{i\langle (j,\lambda_j), x \rangle}$. Then

$$\|\kappa((c_j))\|_{n,n}^- \leq \sum \mid c_j \mid \quad \|e^{i\langle (j,\lambda_j), \cdot \rangle}\|_{n,n}^- \leq C_1 \mid (c_j) \mid_n^-$$

and for $\psi \in C^\infty(M, V)$ by Theorem 1.4a)

$$\mid \langle \kappa((c_j)), \psi \rangle \mid \leq \sum \mid c_j \mid \mid \hat{\psi}(-j, -\lambda_j) \mid \leq C_2 \|\psi\|_{n, A+n}^+ \mid (c_j) \mid_n^+ .$$

This implies:

$$\|\kappa((c_j))\|_{n,n+A}^* \leq C_2 \mid (c_j) \mid_n^+ .$$

So $\kappa : \Lambda^+ \to S(M)_b'$ and $\kappa : \Lambda^- \to S(-M)$ are continuous. $\kappa((c_j)$ is a 2π-periodical solution, since this holds for $e^{i\langle (j,\lambda_j, x \rangle}$.

ii) Let $G \in D(\mathbf{R})$ be chosen such that $\sum_{k \in \mathbf{Z}} G(\cdot - 2\pi k) = 1$. Choose $h_j^+ \in S(M)$ and $h_j^- \in S(-M)_b'$ as follows: For α from (4) let $\delta_j = (1 + \mid Im\, \lambda_j \mid)^{-\alpha}$.

Let $V_j := \{\lambda \mid Q(j,-\lambda) = 0\}$ and choose $\psi_j \in D(\mathbf{C})$ such that $\psi_j(z) = 1$ on $\{z \mid d(z,V_j) \le \delta_j/3\}$ and $supp\, \psi_j \subset \{z \mid d(z,V_j) \le \delta_j\}$. We may also get $\|D^k\psi_j\|_\infty \le C_1/\delta_j^{|k|}$. Let

$$\tilde{h}_j^{\pm} := \psi_j - Q(j,-\cdot)r_\pm(\overline{\partial}\psi_j/Q(j,-\cdot)),$$

where $r_\pm$ are the 1-tame right inverses of $\overline{\partial}$ in C_+^∞ (resp. $(C_-^\infty)_b'$).

$$\begin{aligned} \exists C_2\ \forall n\ \exists C_1\ \forall j:\ &|\ \tilde{h}_j^+\ |_{n,n}^+ \le C_1(1+|\ (j,\lambda_j)\ |)^{C_2 n} e^{-(M+nC_2V)^*(Im\,\lambda_j)} \\ &|\ \tilde{h}_j^-\ |_{C_2(n,n)}^- \le C_1(1+|\ (j,\lambda_j)\ |)^{-n} e^{-(M-nV)^*(Im\,\lambda_j)}. \end{aligned} \tag{39}$$

Here we have used Lemma 1.6, the proof of Proposition 1.9 and of course §2 and the fact, that $\overline{\partial}h_j^{\pm} = 0$.

iii) Let $\hat{h}_j^{\pm} = \tilde{h}_j^{\pm}$. For $f_\pm \in S_Q^{\pm}$ let

$$\nu_\pm(f_\pm) := \frac{1}{2\pi}(\langle_x f_\pm, G(x_1)e^{-ijx_1}h_j^{\pm}(x_2)\rangle)_{j\ge j_0}.$$

Theorem 1.4 and (39) imply:

$$\begin{aligned} \exists C_3\ \forall n:\ |\ \nu_+(f_+)\ |_{Cn}^+ &\le C'\|f_+\|_{n,n}^* \sum_j \|Ge^{ij\cdot}\|_{n,n}^+\ \ |\ \tilde{h}_j^+\ |_{C_3(n,n)}^+ \\ &\times\ (1+|\ (j,\lambda_j)\ |)^{Cn} e^{(M+CnV)^*(Im\,\lambda_j)} \\ &\le C''\|f_+\|_{n,n}^* \sum_j (1+|\ (j,\lambda_j)\ |)^{(-C+C_2C_3)n} \\ &\times\ e^{(M+CnV)^*(Im\,\lambda_j)-(M+C_3C_2nV)^*(Im\,\lambda_j)} \\ &\le C\|f_+\|_{n,n}^* \ \text{ for } C > C_2C_3+2. \end{aligned}$$

$$\begin{aligned} |\ \nu_-(f_-)\ |_n^- &\le C\|f_-\|_{CC_2C_3(n,n)}^- \sum_j |\ \hat{G}(j+\cdot)\tilde{h}_j^-\ |_{CC_2(n,n)}^- \\ &\times\ (1+|\ (j,\lambda_j)\ |)^n e^{(M-nV)^*(Im\,\lambda_j)} \\ &\le C''\|f\|_{CC_2C_3(n,n)}^- \end{aligned}$$

for large C. So $\nu_+ : S_Q^+ \to \Lambda^+$ and $\nu_- : S_Q^- \to \Lambda^-$ are continuous.

iv) $\nu_\pm \circ \kappa_\pm = id_{\Lambda^\pm}$.

Proof: Since $\nu_\pm \circ \kappa_\pm$ are continuous, we have to show this equation on the canonical unit vectors e_j only.

$$\begin{aligned} \nu_\pm \circ \kappa_\pm(e_j)_\ell = \nu_\pm(e^{i\langle(j,\lambda_j),\cdot\rangle})_\ell &= \frac{1}{2\pi}\langle e^{i\langle(j,\lambda_j),\cdot\rangle}, Ge^{-i\ell\cdot}h_\ell^{\pm}\rangle \\ &= \frac{1}{2\pi}\hat{G}(\ell-j)\tilde{h}_\ell^{\pm}(-\lambda_j) = \delta_{j\ell}\tilde{h}_\ell^{\pm}(-\lambda_\ell) = \delta_{j\ell}, \end{aligned}$$

since $\tilde{h}_j^{\pm}(-\lambda_j) = 1$ by construction.

v) Λ^+ and Λ^- are isomorphic to $(s)'$ (and (s), resp.).

Proof: $(1+ \mid j \mid) \leq (1+ \mid (j,\lambda_j) \mid) \leq C(1+ \mid j \mid)^{\tilde{d}}$ for some $\tilde{d}$, since $Q(j,\lambda_j) = 0$.

$$\mid Q(j,\tau) \mid = \mid \sum_{k=0}^{d} \mid Q_k(j)\tau^k \mid \geq \mid Q_d(j)\tau^d \mid /2 > 0$$

for large j and

$$\mid \tau \mid \geq H_Q(j) = \sum_{k=0}^{d-1} \mid Q_k(j)/Q_d(j) \mid^{1/(d-k)} 2d.$$

Notice, that $Q_d(j) \neq 0$ for large j and $H_Q(j) = O(j^{\tilde{d}})$.
$(M \pm nV)^*$ is equivalent to $M^* \pm nV \circ m^{-1}$ by Lemma 1.6.
$\ell n\, t \leq AV \circ m^{-1}(t)$ for large t by (1.2).
$V \circ m^{-1}(t) = \ell n(m^{-1}(t)) \leq \ell n\, t^\gamma = \gamma\, \ell n\, t$ for large t by assumption.

The main point in the above proof is that $V \circ m^{-1}(t)$ is equivalent to $\ell n\, t$ on a sufficiently large part of the zeroes of P. So we get the following

3.5. Remark: Assume (1) - (4). Then $N_P^+(M,V)$ and $N_P^-(M,V)$ are isomorphic to $(s)'$ (and (s), resp.) in each of the following cases:

a) The restriction of P to two variables is hyperbolic w.r.t. one variable.

b) $V_i = O(\ell n\, m_i(t))$ for some $i \leq N$. (40)

Proof: a) Let $Q(x_1,x_2) := P(x_1,x_2,0)$ be hyperbolic w.r.t. x_2. Then $Q(j,\tau) \neq 0$ for $\mid Im\, \tau \mid \geq \tau_0$ and hence (notation from 3.4) $\{\mid Im\, \lambda_j \mid \mid j \geq j_0\}$ is bounded and $V_2 \circ m_2^{-1}(Im\, \lambda_j)$ is equivalent to $\ell n(\mid Im\, \lambda_j \mid)$. The assertion follows from the proof of 3.4.

b) We may assume that $i = 2$. (40) means, that

$$V_2 \circ m_2^{-1}(t) \leq C\, \ell n\, t \text{ for large } t.$$

The opposite inequality follows from (1.2). So the proof of 3.4 applies again.

P need not be hyperbolic in case 3.5a), as is seen by the "ultrahyperbolic operator" $P(x) = x_1^2 - x_2^2 + x_3^2 - x_4^2$. Theorem 3.4 is contained in 3.5b). (40) is e.g. satisfied, if M and V are chosen as in Example 1.10c). So there are now a priori bounds for the growth of the weight conditions leading to spaces $N_P^+(M,V) \simeq (s)'$ and $N_P^-(M,V) \simeq (s)$ for any partial differential operator $P(D)$.

§4. Right Inverses for Partial Differential Operators with C^∞-Coefficients

Sofar we were only dealing with partial differential operators with constant coefficients. In this section, a perturbation argument is used to get *-tame right inverses for certain partial differential operators with C^∞-coefficients, which are "constant at ∞". As a main step in the proof we will show, that the left inverse of $P(z)$ constructed in Proposition 3.1 "looses no regularity" (see Proposition 4.2). if P is hypoelliptic.

Let P and Q be polynomials. Q is called strictly weaker than P $(Q << P)$ iff $\lim_{|x|\to\infty} \tilde{Q}(x)/\tilde{P}(x) = 0$, where

$$\tilde{P}(x) := (\sum \mid P^{(a)}(x) \mid^2)^{\frac{1}{2}}.$$

4.1. Remark: a) Let $Q << P$. Then there are $\epsilon, \delta > 0$ such that

$$\mid \tilde{Q}(x)/\tilde{P}(x) \mid \leq C(1+ \mid x \mid)^{-\epsilon} \text{ for any } x \in \mathbf{R}^N.$$

$$\mid Q(z) \mid \leq C(1+ \mid z \mid)^{-\epsilon} \mid P(z) \mid (1+ \mid Im\, z \mid)^{\delta}$$

for $z \in \Omega_1 := \{z \in C^N \mid d(z, V) \geq 3/2\}$, where $V := \{z \mid P(z) = 0\}$.

b) Let P be semielliptic with index $\tilde{d} := (d_1, \cdots, d_N)$, that is, $P(x) = \sum_{|a:\tilde{d}|\leq 1} c_a x^a$ and $P_{\tilde{d}}(x) := \sum_{|a:\tilde{d}|=1} c_a x^a \neq 0$ for $x \neq 0$, where $\mid a : \tilde{d} \mid = \sum a_i/d_i$. Then

$$Q << P \text{ iff } Q(x) = \sum_{|a:\tilde{d}|\leq\epsilon_0} \tilde{c}_a x^a \text{ for some } \epsilon_0 < 1.$$

Proof: a) Let $E := \{(\tau, y, x)\mathbf{R} \times \mathbf{R} \times \mathbf{R}^N \mid \tau^2 = \mid x \mid^2, \ \mid \tilde{P}(x) \mid^2 \ \ y = \mid \tilde{Q}(x) \mid^2\}$ and

$$f(\tau) := \sup\{y \mid (\tau, y, x) \in E\} = \sup\{\tilde{Q}(x)/\tilde{P}(x) \mid \ \mid x \mid = \tau\}.$$

Then E is semialgebraic, $f(\tau)$ is defined for $\tau > 0$ and $f(\tau) \neq 0$ for large τ. So Corollary A.2.6 of [1] implies, that $f(\tau) = A\tau^a(1 + O(1))$ with $A \neq 0$ and $a \in \mathbf{Q}$. Since f is positive and tends to 0 at ∞, we have $A > 0$ and $a > 0$. For any polynomials Q and P with $\deg Q, \deg P \leq d$ we have:

$$\mid Q(z) \mid = \mid \sum Q^{(a)}(Re\, z)(iIm\, z)^a/a! \mid \leq C_1(1+ \mid Im\, z \mid)^d \mid \tilde{Q}(Re\, z) \mid$$

$$\mid \tilde{P}(Re\, z) \mid \leq C_2(\sum_{a,b} \mid P^{(a+b)}(z)(-iIm\, z)^b/b! \mid^2)^{\frac{1}{2}} \leq C_3(1+ \mid Im\, z \mid)^d \mid \tilde{P}(z) \mid.$$

If $z \in \Omega_1$, then (11.1.3) of [1] shows

$$|\tilde{P}(z)| \leq C_4 \sup_{|\zeta|\leq 1} |P(z+\zeta)| \leq 2^d C_4 |P(z)|.$$

Putting these estimates together, we get for $z \in \Omega_1$ and $Q << P$:

$$|Q(z)| \leq C_5(1+|Im\, z|)^d \tilde{P}(Re\, z)(1+|Re\, z|)^{-\epsilon} \leq C_6(1+|Im\, z|)^{2d+\epsilon}(1+|z|)^{-\epsilon}|P(z)|.$$

b) " $\Rightarrow$ " If $\tilde{c}_a \neq 0$ for some $|a:\tilde{d}| \geq 1$, then there is $x \in \mathbf{R}^N$, such that for $t \cdot x := (t^{\frac{1}{d_i}} x_i)$ and large t

$$t \leq C|Q(t\cdot x)| \leq C|\tilde{Q}(t \cdot x)|.$$

Since $\tilde{P}(t\cdot x) \leq Ct$ and $|t \cdot x| \to \infty$ for $t \to \infty$, Q is not strictly weaker than P.
" $\Leftarrow$ " This easily follows, since
$|x^a| \leq (\sum |x_i|^{d_i})^{|a:\tilde{d}|}$ and $\sum |x_i|^{d_i} \leq C|P(x)| \leq C\tilde{P}(x)$.

4.2. Proposition: Assume (*), and let $Q << P$, where P is hypoelliptic. Let $L_\pm$ be the left inverses of $P(z)$ in $H_\pm$, which were constructed in Proposition 3.1.

a) $\exists \epsilon\ \forall k \geq k_0\ \forall n\ \exists n'$: $|L_+(Qf)|^+_{k,n} \leq C_1 |f|^+_{k-\epsilon,n'}$ for $f \in H_+$.

b) The following holds, if L_- is

$$\begin{aligned} tame\ &: \ \exists \epsilon, b\ \forall n \geq n_0 : |L_-(Qf)|^-_{n-\epsilon,n+b-\epsilon} \leq C_1 |f|^-_{n,0} \\ almost\,1-tame\ &: \ \exists \epsilon\ \forall n \geq n_0 : |L_-(Qf)|^-_{n-\epsilon,K(n-\epsilon)} \leq C_1 |f|^-_{n,0} \\ \ell-tame\ &: \ \exists \epsilon, b\ \forall n \geq n_0 : |L_-(Qf)|^-_{n-\epsilon,b(n-\epsilon)} \leq C_1 |f|^-_{n,0}. \end{aligned}$$

K is large and is the constant scaling the grading for almost 1-tameness.

Proof: We use the notation of the proof of Proposition 3.1.

$$\tilde{L}_\pm(f) := \sum \phi_r \Delta_{\zeta_r}(f) \text{ for } f \in \tilde{H}_\pm.$$

a) We first consider $f \in \tilde{H}_+$. Since $Q_J << P_J$, (36) and Remark 4.1a) imply:

$$|\tilde{L}_+(Q_J f)|^{\Omega_1}_{k,n} = |Q_J f/P_J|^{\Omega_1}_{k,n} \leq C_1 |f|^{\Omega_1}_{k-\epsilon,n'},$$

since

$$\ell n(1+|Im\, z|) \leq \ell n(1+|JIm\ |) \leq C_2 A \sum_i (1+V_i \circ m_i^{-1}(|Im(Jz)_i|))$$

by (2) and since $(M_i+nV_i)^* \circ J$ and $(M_i^* - nV_i \circ m_i^{-1}) \circ J$ are equivalent by Lemma 1.6. (37) and (38) imply:

$$\begin{aligned} \|\tilde{L}_+(Q_J f)\|^{\Omega_2}_{\ell,k,n} &\leq C_2 \sup_{B_{\gamma'}(\zeta_r)\cap\Omega_2\neq\emptyset}\ \sup_{\eta\in B_{2\gamma'}(\zeta_r)} |f(\eta)|\ \|1\|^{B_{\gamma'}(\zeta_r)}_{0,0,n'} \\ &\leq C_3 \|f\|^{\Omega_3}_{0,n''} \text{ for some } n'', \end{aligned}$$

where $\Omega_3 = \{z \mid d(z,V) \leq 3\}$.$\hat{L}_+(f) = (\tilde{L}_+((f)_J))_{J^{-1}}$ then satisfies the same estimates on H_+ (with Q instead of Q_J and Ω_i defined by P). For $f \in H_+$ we get:

$$\begin{aligned} \mid L_+(Qf) \mid_{k,n} &\leq \mid \hat{L}_+(Qf) \mid_{k,n} + \| r_+ \overline{\partial} \hat{L}_+(Qf) \|_{0,k,n} \\ &\leq C_4(\mid f \mid_{k-\epsilon,n_1} + \| \overline{\partial} \hat{L}_+(Qf) \|^{\Omega_2}_{n_2,n_2,n_2}) \\ &\leq C_5(\mid f \mid_{k-\epsilon,n_1} + \mid Qf \mid^{\Omega_3}_{0,n_3}) \leq C_6 \mid f \mid_{k-\epsilon,n_4}, \end{aligned}$$

since $\mid Q(z) \mid \leq C_7(1+ \mid z \mid)^d \leq C_8(1+ \mid Im\, z \mid)^{d\rho}$ on Ω_3.

b) The proof of Theorem 2.1 and section 1 show the existence of a such that for any $K \geq 1$ there is a ℓ-tame right inverse r_- for $\overline{\partial}$ in $((C^\infty_-)'_b, \; \| \; \|^{-*}_{n,Kn,K^2n})$ with ℓ-tameness constant a. So let r_- be almost ℓ-tame (resp. 1-tame) with tameness constant a w.r.t. $\| \; \|^{-*}_{n,Kn,K^2n}$. Let $f \in H_-$.

$$\begin{aligned} \mid L_-(Qf) \mid^-_{a(Kn-\epsilon,K^2(n-\epsilon)B_1+B_2(n-\epsilon))} &\leq C_1 \| L_-(Qf) \|^{-*}_{a(n-1,Kn-\epsilon,K^2(n-1))} \\ &\leq C_2 \mid \hat{L}_-(Qf) \mid^-_{a(Kn-\epsilon,\,K^2n)} \\ &\quad + C_2 \| r_-(\overline{\partial} \hat{L}_-(Qf)) \|^{-*}_{a(n-1,\,K(n-1),\,K^2(n-1))} \| r_-(\overline{\partial} \hat{L}_-(Qf)) \|^{-*}_{a(n-1,\,K(n-1),\,K^2(n-1))} \\ &\leq C_3 \| \overline{\partial} \hat{L}_-(Qf) \|^{-*}_{n-1,\,K(n-1),\,K^2(n-1)} \\ &\leq C_4 \mid \hat{L}_-(Qf) \mid^{-\Omega_2}_{K(n-1),\,K^2(n-1)} \\ &\leq C_5 \mid f \mid^-_{aKn,\,K^2(n-1)-B_3(aKn-K(n-1)-B_4} \\ &\leq C_5 \mid f \mid^-_{aKn,0}, \end{aligned}$$

since $K^2(n-1) \geq B_3((a-1)K(n-1)+aK)+B_4$ for large K.

$$\mid \hat{L}_-(Qf) \mid^-_{a(Kn-\epsilon,\,K^2n)} \leq C_6 \mid f \mid^-_{aKn,0}$$

by similar reasons as before. Rescaling shows the assertion. Tame estimates are obtained similarly.

4.3. Lemma: Assume (*) and let $h \in C^\infty(0,V)$.
a) $\mid \widehat{hg} \mid^+_{k,n} \leq C_1 \mid \hat{g} \mid^+_{k,0}$ for any $k,n \geq 1$ and $g \in C^\infty(M,V)$.
b) $\mid \widehat{hg} \mid^-_{k,0} \leq C_1 \mid \hat{g} \mid^-_{k,n}$ for any $k,n \geq C^\infty(-M,V)'$.

Proof: For $g \in C^\infty(M,V)$ or $g \in C^\infty(-M,V)'$ we have:

$$\widehat{hg}(z) = (2\pi)^{-N} \int \hat{g}(z-x)\hat{h}(x)dx \text{ for any } z \in \mathbf{C}^N. \tag{41}$$

This has to be proved for $z \in \mathbf{R}^N$ only, since both sides are entire functions of z. But $C^\infty(0,V)$ is contained is $S(\mathbf{R}^N)$ and $C^\infty(M,V) \cup C^\infty(-M,V)'$ is contained in $S'(\mathbf{R}^N)$. So (41) follows from the formula

$$\hat{h}\hat{g} = \widehat{h * g} \text{ for } h \in S(\mathbf{R}^N) \text{ and } g \in S'(\mathbf{R}^N).$$

Choose ϕ_r as in Lemma 1.2. Since $h \in C^\infty(0,V)$, the sum $\sum_{r\in\mathbf{Z}} h\phi_r$ converges to h in $S(\mathbf{R}^N)$ and therefore

$$\begin{aligned}(2\pi)^N \widehat{hg}(z) = \int \hat{g}(z-x)\hat{h}(x)dx &= \sum_r \int \hat{g}(z-x)\widehat{h\phi_r}(x)dx \\ &= \sum_r \int \hat{g}(Re\, z - iy_r - x)\widehat{h\phi_r}(x + i(y_r + Im\, z))dx\end{aligned}$$

for any $y_r \in \mathbf{R}^N$ by Cauchy's integral formula.

b) Let $g \in C^\infty(-M,V)'$.

$$\begin{aligned}| \widehat{hg}(z) | \leq C_1 | \hat{g} |^-_{k,n} \sum_r e^{(M-nV)^*(y_r)} \sup_x (1+ | x + i(y_r + Im\, z) |)^{k+N+1} \\ \times | \widehat{h\phi_r}(x + i(y_r + Im\, z)) | (1+ | z |)^k.\end{aligned}$$

Lemma 1.2 implies, that the sum is bounded by

$$\sum_r e^{(M-nV)^*(y_r)+A(k+N+2)V(x_{|r|-1})} \sup_{\substack{|j|\leq k+N+1 \\ x\in I_r}} | h^{(j)}(x) | \int_{I_r} e^{x(y_r+Im\, z)}dx. \tag{42}$$

We now choose y_r such that

$$\begin{aligned}| x_{|r|-1}y_r | + x\, Im\, z \leq -xy_r + | x_{|r|+1} Im\, z | \quad \text{for } x \in I_r \\ - | x_{|r|-1}y_r | + (M - nV)^*(y_r) = -(M - nV)(x_{|r|-1}).\end{aligned}$$

Then (42) is bounded by

$$\begin{aligned}\sum_r e^{-M(x_{|r|-1})+|x_{|r|+1} Im\, z|} \int_{I_r} 1\, dx \sup_{\substack{|j|\leq k+N+1 \\ x\in I_r}} | h^{(j)}(x)e^{(N+C)V(x)} | \\ \leq e^{M^*(Im\, z)} \|h\|_{k+N+1,n+C'} \text{ for suitable } C,\ C' \geq 1,\end{aligned}$$

where (6), (8) and (10) were used. This shows b).

a) Let $g \in C^\infty(M,V)$.

$$\begin{aligned}| \widehat{hg}(z)z^k | &\leq C_1 | \hat{g} |^+_{k,0} \sum_r e^{M^*(y_r)} \\ &\times \sup_x (1+ | x + i(y_r + Im\, z) |)^{k+N+1} | \widehat{h\phi_r}(x + i(y_r + Im\, z)) | .\end{aligned}$$

We now choose y_r as in b) (with $n = 0$) and get an estimate by

$$\begin{aligned}C_2 | \hat{g} |^+_{k,0} \sum_r e^{-M(x_{|r|-1})-nV(x_{|r|-1})+|x_{|r|+1} Im\, z|} \int_{I_r} 1dx \sup_{\substack{|j|\leq k+N+1 \\ x\in I_r}} | h^{(j)}(x) | e^{(n+C)V(x)} \\ \leq C_3 | \hat{g} |^+_{k,0} e^{(M+nV)^*(Im\, z)} \|h\|_{k+N+1,n+C'}.\end{aligned}$$

Now let $P(x,D) := \sum_{k=1}^{\ell} c_k(x)Q_k(D) + P_0(D)$, where P_0 is hypoelliptic, $Q_k << P_0$ and $c_k \in C^\infty(0,V)$ for any k.

4.4. Theorem: Assume (*) and let $P(x,D)$ be as above. Then $P(x,D)$ has *-tame right inverses on a subspace of finite codimension in $C^\infty(M,V)_b'$ (and in $C^\infty(-M,V)$, resp.). The right inverses are defined on $C^\infty(M,V)_b'$ (resp. $C^\infty(-M,V)$), if $\sup_{k\le\ell}\|c_k\|_{C,C}$ is sufficiently small.

Proof: Let

$$A_\pm(f_\pm) := \sum_{k=1}^{\ell} L_\pm Q_k(-\cdot)\mathcal{F}(c_k\mathcal{F}^{-1}(f_\pm)) \text{ for } f_\pm \in H_\pm,$$

where $L_\pm$ are the left inverses of $P(-z)$ in $H_\pm$. Proposition 4.2 and Lemma 4.3 imply:

$$| A_+(f_+) |^+_{n,Kn} \le C_1 \sup_{k\le\ell} | (c_k\mathcal{F}^{-1}(f_+))^\wedge |^+_{n-\epsilon,n'} \le C_2 | f_+ |^+_{n-\epsilon,K(n-\epsilon)} \sup_{k\le\ell} \|c_k\|_{n+N+1,n'+C'} \tag{43}$$

where $K=1$ for tameness and ℓ-tameness, and K is large for almost 1-tameness.

$$\begin{aligned} | A_-(f_-) |^-_{n-\epsilon,K(n-\epsilon)} &\le C_1 \sup_{k\le\ell} | (c_k\mathcal{F}^{-1}(f_-))^\wedge |^-_{n,0} \le \\ &\le C_2 | f_- |^-_{n,Kn} \sup_{k\le\ell} \|c_k\|_{n+N+1,Kn+C'} \text{ for large } K, \end{aligned} \tag{44}$$

where K is the constant from almost 1-tameness. For tameness and ℓ-tameness, Kn is to be substituted by $n+b$ and nb, resp.

a) We first consider H_-: Let $p_n^-(f) := | f |^-_{n,n+b}$ (resp. $| f |^-_{n,Kn}$, $| f |^-_{n,bn}$ for almost 1-tameness and ℓ-tameness, resp.). Let $\tilde{H}_n^- := \{f \in H(\mathbf{C}^N) \mid p_n^-(f) < \infty\}$ and let H_n^- be the closure of $\tilde{H}_{n_0}^-$ in H_n^- for $n \ge n_0$. Let $A_n^- := A_- |_{H_n^-}$. Then $A_n^- : H_n^- \to \tilde{H}_{n-\epsilon}^-$ is continuous by (44), moreover

$$\begin{gathered} A_n^-(H_n^-) = A_n(\overline{H_{n_0}}^{p_n}) \subset \overline{A_n^-(H_{n_0}^-)}^{p_{n-\epsilon}} = \overline{A_-(H_{n_0})}^{p_{n-\epsilon}} \subset \overline{H_{n_0}^-}^{p_{n-\epsilon}} = H_{n-\epsilon} \\ A_n^- : H_n^- \to H_n^- \text{ is compact,} \end{gathered} \tag{45}$$

since $id : \tilde{H}_{n-\epsilon}^- \to \tilde{H}_n^-$ is compact. For $f \in \ker(id + A) =: G$ we get $f = -Af = (-A)^k f$ for any k, and (44) implies:

$$\forall n\ \exists C_1 : p_{n_0}^-(f) \le C_1 p_n^-(f) \text{ for } f \in G.$$

So the (DFS)-topology of $ind(G \cap H_n^-)$ and the norm topology of $H_{n_0}^-$ coincide on G, therefore $\dim G =: k < \infty$ and G is contained in $H_{n_0}^-$ (without loss of generality). Let $I_n := (id + A_n^-)(H_n^-)$ and $I := (id + A_-)(H_-)$. Then I_n is closed by (45), I_{n_0} is contained in I_n and $q : H_{n_0}^-/I_{n_0} \to H_n^-/I_n$ is welldefined, continuous and has dense range, since $H_{n_0}^-$

is dense in H_n^-. Since $\dim(H_n^-/I_n) = k < \infty$ by (45), q is surjective and hence injective, since also $\dim(H_{n_0}^-/I_{n_0}) = k$. So a topological complement F of I_{n_0} in $H_{n_0}^-$ is an algebraic (and therefore topological) complement of I_n in H_n^-. Summarizing we have proved:
Let E be a topological complement of G in H_-. $id + A_- : E \to I$ is bijective, $(id + A_-)^{-1} : I_n \to E_n := E \cap H_n^-$ is continuous and $H = I \oplus F$ topologically with $\dim F = k < \infty$, since $ind H_n^- = \overline{H}_-$ by Remark 1.5.
Let Π be the continuous projection onto I with $\ker \Pi = F$. Let $R_- := {}^t(\mathcal{F}^{-1} \circ (id + A_-)^{-1} \circ \Pi \circ L_- \circ \mathcal{F})$ and $f \in C^\infty(-M, V)$.

$$\begin{aligned}\langle P(x,D)R_-(f), T\rangle &= \langle f, \mathcal{F}^{-1} \circ (id + A_-)^{-1} \circ \Pi \circ L_-(Q_k(-\cdot)\widehat{c_k T} + P_0(-\cdot)\hat{T})\rangle \\ &= \langle f, \mathcal{F}^{-1}(id + A_-)^{-1} \circ \Pi \circ (id + A_-)\mathcal{F}(T)\rangle = \langle f, T\rangle \text{ for } T \in \mathcal{F}^{-1}(E).\end{aligned}$$

R_- is *-tame, since $\mathcal{F}, \mathcal{F}^{-1}, L_-, \Pi$ and (by the *-tame equivalent change from $\tilde{H}_n^-$ to H_n^-, see Remark 1.5) $(id + A_-)^{-1}$ are *-tame w.r.t. $|\ |_{n,Kn}^-$. If

$$\sup_{k \le \ell} \|c_k\|_{n_0+N+1, Kn_0+C'} C_2 < 1/(2\ell),$$

then $id + A_{n_0}^- = id + A_- |_{H_{n_0}}$ is invertible, hence $0 = \ker(id + A_{n_0}^-) = G = F$ and $(id + A_-)$ is bijective. So R_- then is a *-tame right inverse for $P(x,D)$ on $C^\infty(-M, V)$.

b) We now consider H_+:

i) Let $p_n(f) := |\, f\, |_{n,Kn}^+$ with K defined as for *-tameness. Let $\tilde{H}_n^+ := \{f \in H(\mathbf{C}^N) \mid p_n(f) < \infty\}$ and let H_n^+ be the closure of H_+ in $\tilde{H}_n^+$. $A_+ : (H_+, p_{n-\epsilon}) \to H_n^+$ is continuous by (43). Let $\tilde{A}_n^+ : H_{n-\epsilon}^+ \to H_n^+$ be its (unique) continuous extension. Then

$$A_n^+ : H_n^+ \to H_n^+ \text{ is compact,} \tag{46}$$

since $id : \tilde{H}_n^+ \to \tilde{H}_{n-\epsilon}^+$ is compact.

ii) Let $G_n := \ker(id + A_n^+)$. Then $\dim G_n := k_n < \infty$ and G_n and k_n are decreasing, since $A_{n+1}^+ = A_n^+ |_{H_{n+1}^+}$. So $G_n = G_{n_0}$ for $n \ge n_0$, G_{n_0} is contained in $\cap H_n^+ = H_+$ and therefore $G_{n_0} = G = \ker(id + A_+)$ and $\dim G = k < \infty$. ${}^tA_+ : H_+' \to H_+'$ satisfies the estimates dual to (43) for the dual norms. H_+' being a (DFS)-space, we see as in a), that

$$\begin{aligned}\infty > k' := \dim\ker {}^t(id + A_+) &= \dim(im(id + A_+))^\circ = \\ &= codim(\overline{im(id + A_+)}) = codim(im(id + A_+)).\end{aligned}$$

The final equation is seen in

iii) $id + A_+$ has closed range

Proof: Let $f_k \to f$ in H_+ with $f_k \in I := im(id + A_+)$. Let E_n be a topological complement of G in H_n^+. Then $f_k \in I_n := im(id + A_n^+)$ and there are $f^n \in E_n$ such that $(id + A_n^+)f^n = f$ by (46).

$$(id + A_{n_0}^+)(f^{n_0} - f^n) = (id + A_{n_0}^+)f^{n_0} - (id + A_n^+)f^n = 0 \text{ for } n \geq n_0$$

and therefore $f^{n_0} - f^n \in G \cap E_{n_0} = 0$, $f^{n_0} = f^n \in H_n^+$ for $n \geq n_0$ and $f^{n_0} \in H_+$ and $(id + A_+)f^{n_0} = f$.

iv) $(id + A_+) : E \to I$ is bijective, where E is a topological complement of G in H_+, and $(id + A_+)^{-1} : I \to E$ is continuous w.r.t. p_n for any $n \geq n_0$, since it is the restriction of $(id + A_n^+)^{-1}$. Now let Π be a continuous projection onto I and let $R_+ = {}^t(\mathcal{F}^{-1} \circ (id + A_+)^{-1} \circ \Pi \circ L_+ \circ \mathcal{F})$ and $T \in C^\infty(M,V)'$. As in a) we see, that R_+ is a right inverse for $P(x,D)$ on $\mathcal{F}^{-1}(E)$. R_+ is *-tame, since any of its composition factors is *-tame w.r.t. $| \ |_{n,Kn}^+$.
If $\sup_{k \leq \ell} \|c_k\|_{n_0+N+1,Kn_0+C'} C_3 < 1/(2\ell)$ (see the proof of 4.3), then $\|A_+\| < 1/2$ w.r.t. p_{n_0} and therefore $\|A_{n_0}^+\| < 1/2$. Hence $G = 0$ and R_+ is a right inverse for $P(x,D)$ on H_+.

If $P(x,D)$ is surjective on $C^\infty(M,V)'$ (or $C^\infty(-M,V)$, resp.), then we get a right inverse for $P(x,D)$ on $C^\infty(M,V)'$ (and $C^\infty(-M,V)$; resp.) from Theorem 4.4 without the assumption, that the coefficients c_k should be small. Indeed, on the finitedimensional (hence complemented) exceptional space coming from the proof of Theorem 4.4 $P(x,D)$ is clearly *-tamely invertible to the right.

Since the restrictions on the decay of the coefficients c_k depend on the choice of V, we were interested to handle slowly growing functions V in this paper, e.g. $V(t) = \ell n\, t$. The result is especially simple in this case:

4.5. Corollary: Let $P(x,D) = P_0(D) + \sum_{|a:\tilde{d}|<1} c_a(x)D^a$, where P_0 is semielliptic with index $\tilde{d}$ and $c_a \in S(\mathbf{R}^N)$ for any a. Let $m_i(t) + k/t$ be increasing to ∞ for any $k \in \mathbf{N}$ and let

$$\exists \gamma : \ t^{\frac{1}{\gamma}} \leq m_i(t) \leq t^\gamma \text{ for large } t \text{ and } i \leq N.$$

Then $P(x,D)$ has a ℓ-tame right inverse on a subspace of finite codimension in $S(M)'$ (and in $S(-M)$).

Proof: $S(\pm M) = C^\infty(\pm M, \sum_{i \leq N} \ell n(1+ | t_i |))$ and $C^\infty(0, \sum_{i \leq N} \ell n(1+ | t_i |)) = S(\mathbf{R}^N)$ by definition. The assumptions for ℓ-tame estimates are satisfied by Example 1.10, the assumptions on the perturbation term follow from Remark 4.1.

To obtain almost 1-tame estimates in Corollary 4.5, we only have to assume, that also (13) holds.

REFERENCES

[1] L. Hörmander: *The analysis of linear partial differential operators I + II*, Grundlehren **256/257**, Springer 1983.

[2] M. Langenbruch: 'Kolmogorov diameters in solution spaces of systems of partial differential equations', *Manuscr. Math* **53** (1985), 35-64.

[3] M. Langenbruch: 'Bases in solution sheaves of systems of partial differential equations', *J. Reine Angew. Math.* **373** (1987), 1-36.

[4] M. Langenbruch: 'Partial differential equations without solution operators in weighted spaces of (generalized) functions', *Manuscr. Math.* **56** (1986), 353-374.

[5] M. Langenbruch: 'Power series spaces and weighted solution spaces of partial differential equations', *Math. Z.* **194** (1987), 71-88.

[6] M. Langenbruch: 'Complemented kernels of partial differential operators in weighted spaces of (generalized) functions', *Studia Math.*. **89** (1988), 37-63.

[7] M. Langenbruch: 'Solution operators for partial differential equations in weighted Gevrey spaces', preprint.

[8] S. Lojasiewicz, E. Zehnder: 'An inverse function theorem in Fréchet spaces, *J. Funct. Anal.* **33** (1979), 165-174.

[9] R. Meise, B.A. Taylor: 'Splitting of closed ideals in (DFN)-algebras of entire functions and the property (DN)' *Trans. Am. Math. Soc.* **302** (1987), 341-370.

[10] R. Meise, B.A. Taylor, D. Vogt: 'Characterisation des opérateurs linéaires aux dérivées partielles avec coefficients constants sur $E(\mathbf{R}^N)$ admettant un inverse á droite qui est linéaire et continu', *C.R. Acad. Sci. Paris Sér. I. Math.*. **307** (1988), 239-242.

[11] K. Nyberg: 'A tame splitting theorem for Köthe spaces', to appear in *Arch. Math.*

[12] B.A. Taylor: 'Linear extension operators for entire functions', *Mich. Math. J.*, **29** (1982), 185-197.

[13] D. Vogt: 'Power series space representations of nuclear Fréchet spaces, preprint.

[14] D. Vogt: 'Tame spaces and power series spaces', *Math. Z.*, **196** (1987), 523-536.

[15] D. Vogt: 'Operatoren in Fréchetträumen, Stetigkeitscharakteristik, Zahmheit und Bedingungen für die Existenz von Rechtsinversen', Felix-Klein-Kolloquium, Düsseldorf Mai 1985.

STEIN SPACES M FOR WHICH O(M) IS ISOMORPHIC TO A POWER SERIES SPACE

AYDIN AYTUNA
Department of Mathematics
Middle East Technical University
06531 Ankara-Turkey

Spaces of analytic functions form a stimulating example in the theory of nuclear Fréchet spaces. For example, the works of the Russian school on the problem of finding a common basis for the space of analytic functions on an open domain and a closed set contained in that domain [16],[39],[36] and some ideas of Kolmogorov concerning the spaces of analytic functions [59] has strongly motivated the introduction of linear topological invariants for Fréchet spaces like diametral dimension ([60],[22])(later generalized Mitiagin invariants [126]) and the investigation of quasi- equivalence of basis in nuclear Fréchet spaces [35].

On the other hand this influence has not been one sided indeed there are numerous examples where the results of the theory of nuclear Fréchet spaces has been successfully employed in some problems of complex analysis (see for example [41],[25]). The work of Mitiagin-Henkin ([75]) demonstrates the relevance of the linear topological structure of analytic function spaces in attempts to linearize certain theorems like the solution of Cousin problem I and the extension of analytic functions from subvarieties. For example it is not possible to find a continuous linear extension operator from the space of analytic functions on a closed subvariety of $\mathbf{C}^N$ that is biholomorphic to a polydisc into the space entire functions since any such operator must necessarily be compact [21],[110]. For the possible applications of the well developed theory of Köthe spaces (see [37]) to complex analysis it is necessary to obtain sequence space representations of spaces of analytic functions. Whether such a sequence space representation exists for the space of analytic functions on a given Stein manifold is still an unsolved problem even for domains in the complex plane .

Open Problem: ([48], [73]). Let X be a Stein manifold. Does the space $\mathcal{O}(X)$ of analytic functions on X, with the compact open topology, have a basis?

An answer to this question will also contribute to the efforts of finding "natural" nuclear

T. Terzioğlu (ed.), Advances in the Theory of Fréchet Spaces, 115–154.

Fréchet spaces without basis, that started after the solution of Grothendieck's basis problem in the negative by Mitiagin-Zobin [128] (see also [33]). On the other hand in order to apply abstract functional analytic results to complex analysis one **has** to consider the following question of "How much and, what sort of information does the space $\mathcal{O}(X)$ carry about the complex analytic structure of X"? If we take the algebraic structure of these spaces also into account the topological algebra $\mathcal{O}(X)$ is a complete invariant of the Stein manifold X in the sense that for two Stein manifolds X_1, and $X_2, \mathcal{O}(X_1)$ is isomorphic to $\mathcal{O}(X_2)$ as topological algebras if and only if X_1 and X_2 are biholomorphic to each other [41].

Although diametral dimension considerations show that $\mathcal{O}(X_1) \not\cong \mathcal{O}(X_2)$ as Fréchet spaces if $\dim X_1 \neq \dim X_2$ ([60]). For Stein manifolds X with the same dimension, the Fréchet space $\mathcal{O}(X)$ is far from being a complete invariant. For example Rolewicz showed that $\mathcal{O}(X)$, for a finitely connected domain X in the complex plane, is isomorphic to one and only one of the following three prototype Fréchet spaces, $\mathcal{O}(\Delta^1), \mathcal{O}(\Delta^1 - \{0\}), \mathcal{O}(\mathbf{C})$, depending upon the topological structure of the complement of X in $\mathbf{C}$. (Δ^1 is the unit disc in $\mathbf{C}$) [87]. The proof of this assertion uses results, like the Riemann mapping theorem, that are peculiar to the theory of functions of a single complex variable. In higher dimensions the situation is rather hazy. The invariant $\mathcal{O}(\cdot)$ does not, for example, distinguish between the unit polydisc and the unit ball in $\mathbf{C}^n$. Yet it is refined enough to differentiate a continuum of complete Reinhardt domains as was shown by Zaharyuta([126]) using the so called generalized Mitiagin invariants. An investigation about the classification of complex manifolds with respect to the linear topological type of the spaces of analytic functions on them was already suggested by A. Grothendieck [48]. He also pointed out some properties of the Fréchet space $\mathcal{O}(\mathbf{C})$ that are not shared by $\mathcal{O}(\Delta^1)$ (see also D. Vogt [114]).

The spaces $\mathcal{O}(\mathbf{C})$ and $\mathcal{O}(\Delta^1)$ are members of a well studied class of nuclear Fréchet spaces namely, power series spaces. Let $\alpha = (\alpha_n)$ be an increasing, unbounded sequence of real numbers. For $0 < R \leq +\infty$, set

$$\Lambda_R(\alpha) = \{\xi = (\xi_1, \xi_2, \cdots) \in \mathbf{C}^{\mathbf{N}};\ \ \|\xi\|_\rho \doteq \sum_n |\xi_n| \rho^{\alpha_n} < \infty,\ \ \forall 0 < \rho < R\}.$$

Equipped with the norms $\|\cdot\|_\rho$, $\Lambda_R(\alpha)$ is a Fréchet space. $\Lambda_R(\alpha)$ is called a *power series space of finite (infinite) type* in case $R < \infty (R = \infty)$. The spaces $\Lambda_R(\alpha)$, $0 < R < \infty$ are all isomorphic to $\Lambda_1(\alpha)$ [37]. With this notation, the space of analytic functions in the unit polydisc, Δ^d, of $\mathbf{C}^d$ is isomorphic to $\Lambda_1(\{n^{\frac{1}{d}}\})$ and the space of entire functions on $\mathbf{C}^d$ is isomorphic to $\Lambda_\infty(\{n^{\frac{1}{d}}\})$ [87]. In the structure theory of power series spaces, the following linear topological invariants introduced by Vogt, play an important role. A Fréchet space E with a fundamental system of seminorms $\{\| \ \|_k\}$ is said to have the property $(\underline{DN}), (DN), (\Omega), (\overline{\Omega})$ in case the corresponding statements below hold,

$$(\underline{DN}) \ : \ \exists n_0 \ \forall k \ \exists K, C, d: \ \| \ \|_k^{1+d} \leq C \| \ \|_{n_0} \| \ \|_K^d$$

$$(DN) \ : \ \exists n_0 \ \forall k \ \exists K, C: \ \| \ \|_k^2 \leq C \| \ \|_{n_0} \| \ \|_K$$

$$(\Omega) \; : \; \forall k \; \exists n \; \forall p \; \exists d, C : \; \| \; \|_n^{*1+d} \leq C \| \; \|_k^{*d} \| \; \|_p^*$$
$$(\overline{\Omega}) \; : \; \forall k \; \exists n \; \forall p \; \exists C : \; \| \; \|_n^{*2} \leq C \| \; \|_k^* \| \; \|_p^* .$$

Here $\| \; \|_k^*$ denotes the dual norm of $\| \; \|_k$, that is $\|L\|_k^* = \sup\{| \; L(x) \; | : \; \|x\|_k \leq 1\}$. The properties $(\underline{DN}), DN$ are inherited by subspaces and $\Omega, \overline{\Omega}$ by quotient spaces. These linear topological invariants, together with necessary nuclearity conditions, characterize the subspaces and quotients of certain nuclear power series spaces [113].

A large number of Fréchet spaces occuring in analysis turns out to be power series spaces [111], and for further investigations one needs to know the exponent sequence, $\alpha = \{\alpha_n\}$ occuring in the sequence space representation of such spaces. The identification of the exponent sequence can be accomplished if one knows the diametral dimension of the space in question ([85],[99]). But this affair is actually controlled by a weaker invariant. For a nuclear Fréchet space E satisfying the conditions $(\underline{DN})$ and Ω there is a unique (up to equivalence) sequence $\varepsilon(E)$, called the *associated exponent sequence* of E which has the property that if E is isomorphic to a power series space then it is isomorphic to $\Lambda_1(\varepsilon(E))$ or $\Lambda_\infty(\varepsilon(E))$ depending on the type of the power series space [15]. Some techniques for computing associated exponent sequences, especially suitable for function spaces, are discussed in [15] and [100] (see also section II).

In this expository article we restrict our attention to Stein spaces M for which $\mathcal{O}(M)$ is isomorphic to a power series space, consider some examples and touch upon the points raised above in this context. The main theme of this talk is the interplay between the plurisubharmonic functions on X and the linear topological structure of $\mathcal{O}(X)$. The problem of classification of Stein manifolds X for which $\mathcal{O}(X)$ is a power series space is equivalent to the characterization of Stein manifolds of dimension d for which $\mathcal{O}(X)$ is isomorphic to $\mathcal{O}(\Delta^d)$ and to $\mathcal{O}(\mathbf{C}^d)$. In the first section after discussing some general properties of analytic function spaces, we take up the question of intrinsic characterization of Stein manifolds X for which $\mathcal{O}(X)$ is isomorphic to a power series space of finite type. The main result of this subsection is that $\mathcal{O}(X)$ is isomorphic to a finite type power series space if and only if there exists a bounded plurisubharmonic exhaustion function on X. We will call such manifolds hyperconvex, following Stéhle who studied such spaces in a completely different context [97]. Hyperconvex manifolds can be considered as multidimensional analogs of completely regular Riemann surfaces [91, p. 349] with respect to the complex potential theory as developed by Bedford and Taylor([19],[20]). We sketch two proofs of this result the first one uses Hörmander L^2-estimates on $\overline{\partial}$-equation whereas the other is more operator theoretic in nature. In the second subsection we consider infinite type power series spaces. The main result of this subsection is that $\mathcal{O}(X)$ is isomorphic to an infinite type power series space if and only if every bounded plurisubharmonic function on X is constant. These manifolds can be considered as multidimensional analogs of parabolic open Riemann surfaces. In this section we also give some examples of Stein manifolds M for which $\mathcal{O}(M)$ is isomorphic to a power series space.

In the second section we consider some generalizations of the results given in section I. The proofs sketched in section one also applies to reduced Stein spaces and hence yields a characterization of Stein spaces X for which $\mathcal{O}(X)$ is isomorphic to a power series space. Also in this section we consider the space of global sections of coherent analytic sheaves and the properties of $\mathcal{O}(X)$, relevant to the structure theory, that are carried over to these spaces. In general (Ω) type, but not (DN) type, conditions of $\mathcal{O}(X)$ are carried over to these spaces. These considerations yield sequence space representations for global sections of certain coherent analytic sheaves over Stein spaces for which $\mathcal{O}(X)$ is a power series space.

In the third section, we consider a special case of a problem of Mitiagin- Henkin on the possibility of finding a linear continuous extension operator for analytic functions defined on a closed subvariety of a given Stein space M, namely the case when $\mathcal{O}(V)$ is isomorphic to a power series space.

More generally, we consider for a given, quadruple (M, V, W, T), M a Stein space, V a closed subvariety of M, W a Stein space for which $\mathcal{O}(W)$ is isomorphic to a power series space and T a continuous linear operator from $\mathcal{O}(W)$ into $\mathcal{O}(V)$ the possibility of finding a continuous linear operator $\tilde{T}$, such that the diagram

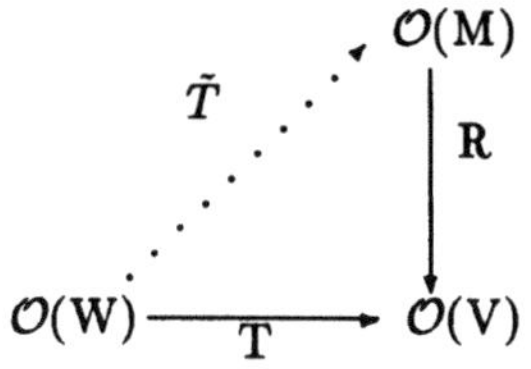

commutes, where R is the restriction operator. An answer to this question is given in terms of the "location" of a certain plurisubharmonic function, induced by the operator T, with respect to the restrictions of plurisubharmonic function on M to V.

We shall use the standard terminology and notation of the theory of locally convex spaces as in ([61], [85], [37]) and of the theory of several complex variables as in ([49], [52], [68]). Throughout Stein manifolds are assumed to be connected and second countable unless otherwise stated. The set of plurisubharmonic functions on M will be denoted by $PSH(M)$ and we will use the symbol $\| \ \|_D$ to denote the sup norm on $D \subset M$.

§I: Stein manifolds M for which $\mathcal{O}(M)$ is isomorphic to a power series space.

Let M be a Stein manifold. The space $\mathcal{O}(M)$ of analytic functions on M is a nuclear Fréchet space with the topology of uniform convergence on compact subsets of M. For a closed set $K \subset M$ we will denote the space of germs of analytic functions defined near K

with the inductive limit topology by $\mathcal{O}(K)$. We collect some general structural properties of these Fréchet spaces in the following:

I.1. Proposition: Let M be a Stein manifold of dimension d. Then;

(1) $\mathcal{O}(M)$ has the property (Ω).

(2) $\mathcal{O}(M)$ is isomorphic to a closed subalgebra of $\mathcal{O}(\Delta^d)$.

(3) $\mathcal{O}(M)$ contains a closed subspace isomorphic to $\mathcal{O}(\mathbf{C}^d)$.

(4) The associated exponent sequence of $\mathcal{O}(M)$ is $\{n^{\frac{1}{d}}\}_n$.

The first assertion follows from the imbedding theorem of Stein manifolds, [49, p.226], Oka-Cartan theorem [49, p.245] and the fact that for any $N, \mathcal{O}(\mathbf{C}^N)$, which is isomorphic to $\Lambda_\infty(n^{\frac{1}{N}})$, has the property (Ω) [117, 2.6 Satz]. The assertion (2) is Lemma 4 of [13]. Proofs of (3) and (4) can be found in [6] (see also [15, 2.2 Theorem] for a more general version) and ([15]). If $\mathcal{O}(M)$ is isomorphic to a power series space then in view of (4), it is isomorphic to either $\mathcal{O}(\mathbf{C}^d)$ or $\mathcal{O}(\Delta^d), d = \dim M$. It follows from (2) that $\mathcal{O}(M)$ has the property (DN). Under the additional assumption of existence of a basis, some more information on the structure of these spaces were obtained in [13] and [79].

A continuous linear operator T from a power series space X into $\mathcal{O}(M), M$ a complex manifold, naturally induces a plurisubharmonic function ρ_T on M. Indeed if $X = \Lambda(\alpha)$ and if we denote by $\{e_n\}$ the canonical basis of X, then $\{\frac{\ell n|T(e_n)|}{\alpha_n}\}_n$ is a locally bounded family of plurisubharmonic functions on M. In view of the continuity of T we set:

$$\rho_T(z) = \overline{\lim}_{\xi \to z} \ \overline{\lim}_n \ \frac{\ell n \mid T(e_n)(\xi) \mid}{\alpha_n}.$$

Then ρ_T is a plurisubharmonic function on M [103, p.74]. Statements about structural similarities of $\mathcal{O}(M)$ to power series spaces can be sometimes transformed via ρ_T's introduced above, to statements about the set of plurisubharmonic functions on M. For example([7])

I.2. Proposition: Let M be a Stein manifold. Then there exists a continuous non-compact operator from $\mathcal{O}(\Delta)$ into $\mathcal{O}(M)$ if and only if there exists a non-constant bounded plurisubharmonic function on M.

Proof: $(\Rightarrow)$ Suppose that there are no non-constant bounded plurisubharmonic functions on M.

Fix a non-zero continuous linear operator $T : \mathcal{O}(\Delta) \to \mathcal{O}(M)$. The plurisubharmonic function ρ_T is strictly negative on M, so it must be equal to a negative constant say λ. Choose a number β with $\lambda < \beta < 0$. For a given compact set K in M, in view of Hartog's lemma ([52, p.21], cf. [89]), there exists an m such that $\|T(e_n)\|_K \leq e^{n\beta}$, for $n \geq m$. Now choose an R_0 with $e^\beta < R_0 < 1$. It follows that for every compact set K of M there exists a $C(K) > 0$ such that;

$$\|T(e_n)\|_K \leq C(K)\ \|e_n\|_{R_0},\ \ n = 1, 2, \cdots$$

where $\|\ \ \|_{R_0}$ is the sup norm on the disc around zero with radius R_0. But this precisely means that T is compact.

($\Leftarrow$) Without loss of generality we can assume that there exists a negative nonconstant $\rho \in PSH(M)$ with $\sup_{z\in M}\rho(z) = 0$. We choose an exhaustion $\{G_n\}$ of M by pseudo-convex relatively compact domains and a sequence $\{\rho_n\}$ of C^∞ negative functions with $\rho_n \in PSH(G_n)$, $\forall n$, such that $\rho_n(z) \downarrow \rho(z)$ for every $z \in M$ ([96],[89]). We also choose an exhaustion $\{K_n\}$ of M by holomorphically convex compact sets K_n with $K_n \subset G_n$, $\forall n$. Let $M_\alpha^\beta = \sup_{z\in K_\beta}\rho_\alpha(z)$, $\alpha > \beta$. These numbers are strictly negative, since $\rho_n < 0$ forall n. We fix a sequence $\varepsilon_n \downarrow 0$ and set $\eta_n = 2\varepsilon_n, n = 1, 2, \cdots$. For a given n, there exists analytic functions $f_1, \cdots, f_s$ on G_n and positive integers $a_1, \cdots, a_s$ such that

$$\rho_n(z) - \varepsilon_n \leq \sup_{1\leq i\leq s}\{\frac{\ell n \mid f_i(z) \mid}{\alpha_i}\} \leq \rho_n(z),\quad \forall z \in K_n.$$

([69, p.55],[24]).

In view of Runge's theorem we can find for each $1 \leq k \leq n$ an analytic function $F_n^k \in \mathcal{O}(M)$ and a natural number b_n^k such that

$$\frac{\ell n\|F_n^k\|_\beta}{b_n^k} \leq (\frac{1}{2})\, M_n^\beta,\ \ \beta = 1, 2, \cdots n,\ \ k = 1, \cdots n. \tag{1}$$

$$\frac{\ell n\|F_n^k\|_k}{b_n^k} \geq M_n^k - \eta_n,\ \ k = 1, \cdots, n. \tag{2}$$

Continuing in this fashion we obtain functions $F_n^k \in \mathcal{O}(M)$ and natural numbers b_n^k, $k = 1, \cdots, n$, $n = 1, 2, \cdots$, such that (1) and (2) hold. By changing F_i^j with its suitable power if necessary, we can assume that the sequence $\{b_1^1, b_2^1, b_2^2, \cdots, b_n^1, \cdots, b_n^n, \cdots\}$ is strictly increasing. We will denote the above sequence by $I = \{a_n\}$ and the correspondingly ordered sequence of F_i^j, s by $\{g_n\}$. We define an operator T from the subspace of $\mathcal{O}(\Delta)$ spanned by $\{e_n\}, n \in I$, into $\mathcal{O}(M)$ by setting $T(e_{a_n}) = g_n$, $n = 1, 2, \cdots$,. Let n be given. For m large enough g_m will be equal to F_k^j for some k and j with $k > n$. Since $M_k^n \leq M_n^n$, in view of (1) and (2), we have that

$$\frac{\ell n\|g_m\|_n}{a_m} \leq \frac{1}{2}\, M_n^n.$$

Now let R_n be a positive number such that $e^{\frac{1}{2}M_n^n} < R_n < 1$. Then for all m sufficiently large we have

$$\|T(e_{a_m})\|_n = \|g_m\|_n \leq R_n^{a_m} = \|e_{a_m}\|_{R_n}.$$

It follows that T extends to a continuous linear operator from the step space $\mathcal{O}(\Delta)_I = \overline{span\{e_k\}_{k\in I}}$ into $\mathcal{O}(M)$. We now show that T is non-compact. Suppose the contrary. Then there exists an R with $O < R < 1$ such that $\forall n \ \exists\, C(n) > 0$ with

$$\|T(e_{a_m})\|_n \leq C(n)R^{a_m} \quad \text{for} \quad m = 1, 2, \cdots. \tag{3}$$

Since $\sup_n(\frac{1}{2}M_n^n) = O$ there exists an n_0 such that $e^{\frac{1}{2}M_{n_0}^{n_0}} > R$. We choose a positive number C such that $\frac{1}{2}M_{n_0}^{n_0} + C < 0$ and find an n satisfying $M_m^n \geq \frac{1}{2}M_{n_0}^{n_0} + C$ for $m \geq n$. In view of (1), (2), (3) and our choice of n_0, for all m sufficiently large we have

$$e^{\frac{C}{2}b_m^n} \leq e^{(M_m^n - \eta_m - \frac{1}{2}M_{n_0}^{n_0})b_m^n} \leq C(n).$$

But this gives a contradiction since $b_m^n \to \infty$ as $m \to \infty$. Hence T is a non-compact continuous operator. Composing the natural projection from $\mathcal{O}(\Delta)$ onto $\mathcal{O}(\Delta)_I$ with T, we obtain a non-compact continuous operator from $\mathcal{O}(\Delta)$ into $\mathcal{O}(M)$. This finishes the proof of I.2 Proposition.

Existence of a non-compact continuous operator between two nuclear Fréchet spaces, in most cases, imply that the structure of these spaces have some common features ([110], [79], [101]). Indeed in our case the existence of a non-compact continuous operator from $\mathcal{O}(\Delta)$ into $\mathcal{O}(M)$ yields the existence of a closed subspace of $\mathcal{O}(M)$ that is isomorphic to a step space of $\mathcal{O}(\Delta^d)$, $d = \dim M$. This can be seen by combining the results of [64] and Proposition 1 of [79].

We now take up the question of characterizing Stein manifolds M for which $\mathcal{O}(M)$ is isomorphic to a power series space. If $\mathcal{O}(M)$ is isomorphic to a power series space then it is isomorphic to either a finite type power series space or to an infinite type power series space. We will treat these cases separately.

(i) Finite case: Let M be a Stein manifold of dimension d. Now let us suppose that $\mathcal{O}(M)$ is isomorphic to a finite type power series space. In view of the above discussions $\mathcal{O}(M)$ must be isomorphic to $\mathcal{O}(\Delta^d)$. We fix an isomorphism T between $\mathcal{O}(\Delta^d)$ and $\mathcal{O}(M)$. Using multi-index notation we set $e_n = T(z^n)$, $n \in \mathbb{N}^d$. With this notation the plurisubharmonic function induced by T in the above discussion takes the form

$$\rho_T(z) = \overline{\lim}_{\xi \to z} \ \overline{\lim}_{|n|} \ \frac{\ell n \mid e_n(\xi) \mid}{\mid n \mid}.$$

In view of I.1 Proposition, ρ_T takes only negative values and is not constant. The fact that T is an isomorphism imposes some additional properties on ρ_T as was observed by Zaharyuta ([124, Theorem 4.3]). For an $\varepsilon > 0$ let $M_\varepsilon = \{z \in M;\ \rho_T(z) < -\varepsilon\}$. Then M_ε, if not empty, is an open subset of M. Employing Hartog's lemma to ρ_T on M_ε yields for each compact subset of K a constant $A(K) > 0$ such that; $\|e_m\|_K \leq A(K)e^{-\frac{|m|\varepsilon}{2}}$, $\forall m$. Taking into account that T^{-1} is also continuous and that $\{e_m\}$'s form a basis in $\mathcal{O}(M)$, the above expression implies the existence of a compact set $K_0 \subset M$ with the following property:

$\forall$ compact subset $K \subset M$, $\exists\, C = C(K) > 0$ such that

$$\|f\|_K \leq C\|f\|_{K_0}.$$

This in particular implies that M_ε is contained in the holomorphically convex hull of K_0 and hence is relatively compact since M is a Stein manifold.

Let us agree to call a function $\rho : M \to [-\infty, b)$, $b \in \mathbf{R}$, *proper* in case $\overline{\{z : \rho(z) < c\}}$ is compact in M for every $c < b$. Stein manifolds which admit a bounded plurisubharmonic proper function were introduced and studied by Stehlé in connection with a conjecture of Serre [97]. Following [97] we give the following definition.

I.3 Definition: A Stein manifold M is called *hyperconvex* in case there exists a negative plurisubharmonic function on M that is proper.

The discussion above exhibits the hyperconvexity of M as a necessary condition for the existence of an isomorphism between $\mathcal{O}(M)$ and $\mathcal{O}(\Delta^d)$, $d = \dim M$. It turns out that this condition is also sufficient. This was first stated in [127] and independently later in [9], (cf.[10]).

I.4 Theorem: Let M be a Stein manifold of dimension d. Then $\mathcal{O}(M)$ is isomorphic to $\mathcal{O}(\Delta^d)$ as Fréchet spaces if and only if M is hyperconvex.

We will sketch two proofs of this theorem. We start by describing the so called "Dead-end space method" of Mitiagin and Henkin [75] (cf. [74]). We follow the discussion given by Vogt in [108]. Let $(H_0, \langle\ \rangle_0)$ and $(H_1, \langle\ \rangle_1)$ be two Hilbert spaces and suppose that a continuous one to one compact operator $T : H_1 \to H_0$ with dense range is given. Since T is compact it can be represented as, $T(x) = \sum_{n=1}^{\infty} \lambda_n \langle x, f_n\rangle_1 e_n$ for some orthonormal bases $\{e_n\}$ and $\{f_n\}$ of H_0 and H_1 respectively and for some sequence $\{\lambda_n\}$ with $\lambda_n > 0$, $\lambda_n \downarrow 0$. Identifying H_1 with $T(H_1)$ in H_0, we set $|\ e_n\ |_1\ (= \frac{1}{\lambda_n}) = a_n$, $n \in \mathbf{N}$. Using the terminology of [80, p.131] for each $0 < t < 1$ we form the quadratic interpolation space H_t by;

$$H_t \doteq \{x \in H_0;\ |\ x\ |_t = \{\sum_{n=1}^{\infty} |\ \langle x, e_n\rangle_0\ |^2 \quad a_n^{2t}\}^{\frac{1}{2}} < \infty\}.$$

Then the set $Q(H_1 : H_0) = \bigcap_{t<1} H_t (\subset H_0)$ equipped with the topology coming from the family of norms $\{|\ \ |_t\}_{t<1}$ (finite centre of a Hilbert scale in the Russian terminology (see [74])) is a finite type power series space and all finite type power series spaces can be obtained in this way. For further details see [75], [74]. We will say that a nuclear Fréchet space is represented by a $Q(H_1 : H_0)$ in case there is a continuous inclusion from X into H_0 and X thought as a subset of H_0 is equal to $Q(H_1; H_0)$ both set theoretically and topologically. The dead-end space method investigates the posibility of representing a given nuclear Fréchet space X as $Q(H_\infty; H_0)$ for a suitable Hilbert space H_∞ (dead-end space) which can be continuously imbedded into X, and $H_0 = (\widehat{X, \langle,\rangle_0})$ for some suitable continuous inner product $\langle,\rangle_0$ on X with respect to the natural inclusion of H_∞ into H_0. Of course it is not always possible to find such a dead-end space and one needs some a priori knowledge about the norms of $\{e_n\}$ as well as the dual norms of $\{e_n^*\}$ to ensure the convergence in X of the orthogonal expansion of a given element of X in H_0 with respect to the orthonormal basis $\{e_n\}_n$. If for a given nuclear Fréchet space X and a continuos inner product $\langle,\rangle_0$ on it, there exists a Hilbert space $(H_\infty, \langle,\rangle_\infty)$ continuously imbedable into X such that the couple H_∞ and $H_0 \doteq \widehat{(X, \langle,\rangle_0)}$ satisfies,

(α) For every continuous seminorm $\|\ \|$ on X there exists a number $0 < \lambda < 1$ and a $C > 0$ such that

$$\|\cdot\| \leq C\|\cdot\|_0^\lambda \ \|\cdot\|_\infty^{1-\lambda}$$

(ω) For every μ with $O < \mu < 1$ there exists a continuous norm $\|\ \|$ such that

$$\|\ \|^* \leq \|\ \|_0^{*\mu} \|\ \|_\infty^{*1-\mu}$$

(where $\|L\|^* \doteq \sup\{| L(x) |: \|x\| \leq 1\}$, $\forall L \in X^*$, is the dual norm of $\|\ \|$) or equivalently $\forall \mu \in (0,1)$ there exists a neighborhood of zero W such that

$$W \subset r^{1-\mu} U_\infty + \frac{1}{r^\mu} U_0, \ \forall r > 0$$

where $U_\infty = \{x \in X : \|x\|_\infty \leq 1\}$ and $U_0 = \{x \in X : \|x\|_0 \leq 1\}$ [117].

then one has a control on the norms and the dual norms of the common orthonormal basis of H_∞ and H_0 and indeed under these assumptions it follows that X can be presented as $Q(H_\infty; H_0)$ [108]. This is the version of the dead-end space method that we will use. (For other versions of this method see [65], [37, 2.2.9 Theorem,]). We will employ the symbol $\hookrightarrow$ to denote continuous inclusions.

In the case when X has the property ($\underline{DN}$), for any Hilbert space $H_\infty \hookrightarrow X$, condition ($\alpha$) is satisfied for any continuous Hilbertian norm $\|\ \|_0$ dominating $\|\ \|_{n_0}$ of the condition ($\underline{DN}$), so in this case the coupling condition (ω) is the significant condition.

The motivation of the first proof comes from the observation that the equivalent form of the condition (ω), in the case $X = \mathcal{O}(M)$, can be considered as a condition on approximating analytic functions on compact subsets of M by analytic functions with specified growth. We now describe this proof, following [10]. Let a hyperconvex Stein manifold M be given and fix a negative proper plurisubharmonic ρ on M. One chooses a sequence of negative numbers $\alpha_n \uparrow 0$ in such a way that $D_{\alpha_n} \subset\subset D_{\alpha_{n+1}}$, $n = 1, \cdots$, where $D_{\alpha_k} = \{z \in M : \rho(z) < \alpha_k\}$, $k = 1, \cdots,$. For a suitable measure μ ([10, Lemma 1]) on M that is of the form $c\,dV$ for some positive continuous function on M and a volume element dV on M, one considers the norms,

$$\|f\|_k = \{\int_{D_{\alpha_k}} | f |^2 \, d\mu\}^{\frac{1}{2}} \quad k = 1, 2, \cdots.$$

Cleary the family $\{\|\ \|_k\}$ generates the topology of $\mathcal{O}(M)$. Let k be given and choose an m such that $\overline{D_{\frac{\alpha_k}{4}}} \subset D_{\alpha_m}$. For a given element f of $\mathcal{O}(\overline{D}_{\alpha_m})$ with $\int_{D_{\alpha_m}} | f |^2 \, d\mu \leq 1$, one gets an a family $\{G_L\}_L$ of analytic functions on M approximating f on D_{α_k}, by solving for each $L > 0$ a certain $\overline{\partial}$-equation with bounds by employing the weight factor $e^{-\rho_L}$ where $\rho_L = \tau_L \circ \rho$ and

$$\tau_L(t) = \begin{cases} -\frac{L}{\alpha_k}t + L & \text{if } \alpha_k < t < 0 \\ 0 & \text{if } t \leq \alpha_k \end{cases} \qquad L > 0.$$

If we let

$$U_s = \{f \in \mathcal{O}(M) : \|f\|_s \leq 1\}, s = 1, 2, \cdots, \text{ and } U_\infty = \{f \in \mathcal{O}(M) : \int_M |f|^2 \, d\mu \leq 1\}$$

then the above procedure yields:

$\forall\, k$ and m with $\overline{D}_{\frac{\alpha_k}{4}} \subset D_{\alpha_m}$ there exists a $C > 0$ such that;

$$(*) \qquad U_m \subset C\, r\, U_\infty + \frac{1}{r} U_k, \quad \forall r > 0.$$

Now let $H_\infty = \{f \in \mathcal{O}(M) : \|f\|_\infty \doteq \{\int_M |f|^2 \, d\mu\}^{\frac{1}{2}} < \infty\}$ and choose a k such that $\| \ \|_k$ dominates the n_0, norm appearing in the condition $(\underline{DN})$ for $\mathcal{O}(M)$ (I.1. Proposition). Dualizing (*) as in ([117, 2.2 Korollar];) and iterating the obtained expression one gets the condition (ω) for the pair H_∞ and $H_0 \doteq \overline{\mathcal{O}(M) \cap L^2(D_{\alpha_k}, d\mu)}$. Since this pair also satisfies (α) by our choice of k, we obtain a representation of $\mathcal{O}(M)$ as $Q(H_\infty : H_0)$. This concludes the description of the first proof of I.4 Theorem.

The second proof of I.4 Theorem that we will give, following Zaharyuta [126] ,[127] utilizes a special extremal plurisubharmonic function. Following the terminology of Sadullahev [89] we define

I.5 Definition: Let M be a Stein manifold and $K \subset M$ a compact set. The plurisubharmonic function,

$$\omega_K^M = \overline{\lim}_{\xi \to z} \sup\{u(\xi) : u \in PSH(M), \ u \leq 0 \text{ on } M \text{ and } u \leq -1 \text{ on } K\}$$

is called the ***plurisubharmonic measure*** (***$\mathcal{P}$-measure***) of K relative to M.

These functions are natural complex counterparts of the harmonic measures of the classical potential theory [50, p.165]. Plurisubharmonic measures were studied by various authors in different contexts, for example in the context of Bernstein-Walsh approximation theorem and related problems by Siciak [95], Zaharyuta [124] and in the framework of the complex potential theory based on the complex Monge-Ampère equation by Bedford [17], Bedford-Burns [18], Bedford-Taylor [19], Sadullahev [89]. A compact set K in a Stein manifold M is said to be *regular* in case $\omega_K^D \equiv -1$ on K for every pseudoconvex domain $D \subset K$ ([124]). In dimensions greater than one on $\mathcal{P}$-measures contrary to the one dimensional case, need not have nice regularity properties. For example in general they need not be continuous [17]. But nevertheless if K is regular, ω_K^M satisfies (in a generalized sense) the homogeneous complex Monge-Ampère equation $(dd^c u)^n = 0$ off K ([20] see also [89]). We shall need the following property of $\mathcal{P}$-measures: Let X be a Stein manifold and K a compact subset of X. For an nonconstant $f \in \mathcal{O}(X)$ satisfying $\|f\|_X \leq e^{c_1}$, $\|f\|_K \leq e^{c_0}$, the plurisubharmonic function

$$\frac{\ell n \mid f \mid - c_1}{c_1 - c_0}$$

is negative on X and is dominated by -1 on K. Therefore it is dominated by the $\mathcal{P}$-measure ω_K^X and hence we have,

$$\ell n \mid f(z) \mid \leq c_1(\omega_K^X(z) + 1) - c_0 \omega_K^X(z) : \quad z \in X.$$

This can be thought as the analog in several dimensions, of the classical "two-constant theorem" [78, p.212] cf. [89]. We will return to $\mathcal{P}$-measures and discuss some additional general properties of it in the next subsection but now we wish to restrict our attention to $\mathcal{P}$-measures on hyperconvex Stein manifolds.

If M is a hyperconvex Stein manifold and if K is a regular compact set then the $\mathcal{P}$-measure, ω_K^M is a negative proper plurisubharmonic function on M. Moreover, in this case, ω_K^D enjoys an additional regularity property, namely that it is *continuous* on M as was shown by Zaharyuta [124] (cf. [89, 104. Theorem]). One of the main ingredients of the second proof is the following powerful interpolation theorem of Zaharyuta([124, Theorem 4.1])

I.6 Theorem: Let M be a Stein manifold and let D be a domain in M of the form, $D = \{z \in D_1 : \rho(z) < 0\}$, where D_1 is a pseudoconvex domain of $M, D \subset\subset D_1$, and ρ is a continuous plurisubharmonic function on D_1. Then for a compact regular set K with $\overline{K^\circ} = K$, and a pair of Hilbert spaces H_0 and H_1 with continuous inclusions;

$$\mathcal{O}(\overline{D}) \hookrightarrow H_1 \hookrightarrow \mathcal{O}(D)$$

$$\mathcal{O}(K) \hookrightarrow H_0 \hookrightarrow (\mathcal{O}(K^\circ) \cap C(K))$$

the interpolation spaces of H_1 and H_0 with respect to the inclusion induced by the natural inclusion of $\mathcal{O}(D)$ into $\mathcal{O}(K)$ satisfy;

$$\mathcal{O}(F_t) \hookrightarrow H_t \hookrightarrow \mathcal{O}(D_t), \qquad 0 \leq t \leq 1$$

where

$$D_t = \{z \in D; \ \omega_K^D(z) < t - 1\}$$

and

$$F_t = \{z \in D; \ \omega_K^D(z) \leq t - 1\}.$$

If we take two polydiscs, around zero in $\mathbf{C}^d$, with polyradii $R(1) = (R_1(1), \cdots, R_d(1))$ and $R(0) = (R_1(0), \cdots, R_d(0))$, such that $R_i(0) < R_i(1)$, $i = 1, \cdots, d$, then the quadratic interpolation spaces of the Hardy spaces [88] $H^2(\Delta^d(R(1))$ and $H^2(\Delta^d(R(0))$ can naturally be identified with the Hardy space $\{H^2(\Delta^d(R(t)))\}_{0<t<1}$ where $R_i(t) = R_i(1)^t\, R_i(0)^{1-t}$, $i = 1,\cdots,d$ and $0 \le t \le 1$. Since

$$\omega_{\Delta(R(0))}^{\Delta(R(1))}(z_1,\cdots,z_d) = \max_{1\le i\le d}\{\frac{\ell n \mid z_i \mid - \ell n\, R_i(1)}{\ell n\, R_i(1) - \ell n\, R_i(0)}\}$$

it follows that there exists continuous inclusions $\mathcal{O}(\overline{\Delta}_t) \hookrightarrow H^2(\Delta^d(R(t)) \hookrightarrow \mathcal{O}(\Delta_t)$, where

$$\Delta_t = \{z \in \Delta^d(R(0)) : \ \omega_{\Delta(R(0))}^{\Delta(R(1))}(z) < t-1\}, \ \ 0 \le t \le 1\}.$$

The proof of I.6 Theorem, in the special case when $D = \Delta(R(1))$ and $K = \overline{\Delta(R(0))}$ can be obtained, in view of the above observation, from the well-known functorial property (in the terminology of [80, p.135] of the quadratic interpolation ([63] ,[70]) . The strategy of the proof of I.6 Theorem given in [124] is to reduce the general case to the special case above by using Oka-Weil domains ([49, p.211]).

An immediate consequence of the above theorem is that for domains D described in the statement of I.6, one can represent $\mathcal{O}(D)$ as $Q(H_1; H_0)$ with K, H_1, H_0 choosen so that the hypotheses of the theorem are satisfied [123, Theorem 3]. In particular for a hyperconvex M, by choosing a regular compact set K, one gets information about the structure of the spaces of analytic functions defined on domains like $\{z \in M;\ \omega_K^M(z) < \beta\}$, $\beta < 0$, which are the building blocks of M.

Now fix a hyperconvex Stein manifold M and a regular compact set K. Denoting the $\mathcal{P}$-measure of K relative to M by ω, set $M_\alpha = \{z : \ \omega(z) < \alpha\}$ for $-1 < \alpha < 0$. For a fixed $\alpha_0 \in (-1,0)$, and $0 > \alpha > \alpha_0$, applying the above theorem to $D = M_\alpha$, $K = \overline{M}_{\alpha_0}$ one gets a represention of $\mathcal{O}(M_\alpha)$ as $Q(H_1; H_0)$ for some suitably choosen Hilbert space H_0 and H_1. The dual space of $Q(H_1; H_0)$ has a relatively simple structure, namely it is a sequence space, and I.6 Theorem allows one to use standard sequence space methods in studying the dual space of $\mathcal{O}(M_\alpha)$. Indeed it is not difficult in this case, to show, that for a domain $\Omega \subset\subset M_\alpha$ with $\Omega \supset \{z : \ \omega_{M_{\alpha_0}}^{M_\alpha}(z) \le t\}$ for some $t > -\frac{1}{2}$, there exists $C > 0$ such that

$$\| \ \|_\Omega^* \le C \| \ \|_{M_\alpha}^{*\frac{1}{2}} \| \ \|_{M_{\alpha_0}}^{*\frac{1}{2}}$$

where the norms are sup norms on the respective domains. Now taking into account that

$$\omega_{M_{\alpha_0}}^{M_\alpha} = \frac{\omega - \alpha_0}{\alpha_0 - \alpha},$$

one has for $0 > \beta > \frac{\alpha_0}{4}$ the inclusion $M_\beta \supset \{z : \omega_{M_{\alpha_0}}^{M_\alpha} \le -\frac{1}{3}\}$ for α near zero and hence one obtains

$$\forall \alpha_0 \text{ and } \frac{\alpha_0}{4} < \beta < 0, \ \ \forall \alpha \ \exists C > 0 \ \text{ such that}$$

$$\| \ \|_{M_\beta}^* \le C \| \ \|_{M_\alpha}^{*\frac{1}{2}} \ \| \ \|_{M_{\alpha_0}}^{*\frac{1}{2}}.$$

This is precisely the condition $\overline{\Omega}$ of Vogt-Wagner. Although this condition is a condition on the "local" structure of the space it is powerful enough, when coupled with the condition $(\underline{DN})$, to allow one to *construct* a dead end space for which (α) and (ω) hold, as was shown by Vogt [108, 1.4 Lemma] . So I.4 Theorem follows from the following result of Vogt [108, 1.6 Satz]:

I.7 Theorem: . A nuclear space E is isomorphic to a finite type power series space if and only if E has the properties $(\underline{DN})$ and $(\overline{\Omega})$.

If we consider the two proofs sketched above each seems to have some advantages. The first one appears to be more direct and exhibits the dead end space in a rather explict way, which in turn allows one to control the rate of growth of the basis vectors constructed, whereas in the second one abstract Functional Analysis techniques play a dominant role and the proof in some sense has a "local" character. We will explore this aspect of the second proof in section 2. The considerations in the first proof yield([10](cf. [9])).

I.8 Corollary: Let D be a bounded hyperconvex domain in some $\mathbf{C}^N$. Then there exists an orthogonal basis for the Bergman space $L^2(D) \cap \mathcal{O}(D)$, that is at the same time a basis for the Fréchet space $\mathcal{O}(D)$. Moreover if D is strictly pseudoconvex with C^4 boundary then one can find a basis for $\mathcal{O}(D)$ consisting of functions that are analytic in a fixed neighborhood of $\overline{D}$.

For bounded pseudoconvex domains D in a Stein manifold, of dimension d, the existence of an isomorphism between $\mathcal{O}(D)$ and $\mathcal{O}(\Delta^d)$ is a local property of the boundary of D. This can be seen by employing the arguments given in [54, Proposition 1] (cf. [97]). In the next proposition we gather some equivalent conditions for a domain to be hyperconvex [10].

I.9 Proposition: For a bounded pseudoconvex domain, D, in a Stein manifold M of dimension d, the following are equivalent;

(1) $\mathcal{O}(D) \simeq \mathcal{O}(\Delta^d)$.

(2) For every $p \in \partial D$, there is a ball B_p centred at p (in some coordinate neighborhood) of p such that $\mathcal{O}(D \cap B_p) \simeq \mathcal{O}(\Delta^d)$.

(3) There exists a strictly pseudoconvex C^∞ function that is negative and proper on D.

(4) For every sequence $\{z_n\}$ of points in D without a limit point (in D), there exists a $u \in PSH(M)$, $u < 0$ with $\lim u(z_n) = 0$.

(5) There exists a negative proper plurisubharmonic function on M which satisfies the homogeneous complex Monge-Ampére equation $(dd^c u)^n = 0$ outside a compact subset of M.

In the condition (1) of the above proposition, we could have employed the unit ball, B^d, instead of Δ^d, since $\mathcal{O}(B^d)$ and $\mathcal{O}(\Delta^d)$ are isomorphic as Fréchet spaces. In this connection we would like to mention a result of Burns [26] and Stoll [98] which says that a necessary and sufficient condition for $\mathcal{O}(M)$ to be isomorphic to $\mathcal{O}(B^d)$ as Fréchet *algebras* is the existence of a strictly plurisubharmonic C^∞, proper $\rho : M \to [0,1)$ such that $\ell n\rho$ satisfies the homogeneous complex Monge-Ampère equation $(dd^c u)^n = 0$ on M.

Klimek [58] has introduced an extremal plurisubharmonic function u_Ω^w a domain $\Omega \subset \mathbf{C}^n$ and $\omega \in \Omega$ which can be viewed as a counterpart of the classical Green's function with a pole at a given point, as:

$$u_\Omega^w(z) \doteq \sup\{u(z)\}$$

where the supremum is taken over all negative funtions $u \in PSH(\Omega)$ (including-∞) such that $u(z) - \ell n \mid w - z \mid = O(1)$ as $z \to w$.

This function and functions related to this were also studied by Demailly [32] and Lelong [69]. Adopting the same definition for a Stein manifold M and arguing as in Corollary 1.3 of [58], it follows that $u_M^w \in PSH(M)$, $\forall w \in M$ and $u_M^w - \ell n \mid w - z \mid = O(1)$ as $z \to w$. By using Proposition 9.1 of [20] and adopting the argument given in p.179 of [50] one has, (cf.also [58]).

I.10 Proposition: Let M be a hyperconvex Stein manifold. Then for every $w \in M$ the function u_M^w is a negative proper plurisubharmonic function on M and satisfies $(dd^c u_M^w)^n = O$ in $M - \{w\}$.

Hence Stein manifolds M for which $\mathcal{O}(M) \simeq \mathcal{O}(\Delta^d)$, can be considered, from the point of view of complex potential theory, as generalizations of regular hyperbolic Riemann surfaces [91, p. 349] to higher dimensions.

We now turn to examples of Stein manifolds M of dimension greater than 1 for which $\mathcal{O}(M)$ is isomorphic to a finite type power series space. First concrete non-trival examples of such spaces discovered were bounded complete Rienhardt domains [3], and (p,q)-circular domains [1]. The spaces of analytic functions on the these spaces are in a natural way Köthe spaces. Later examples considered, were among Stein manifolds M for which $\mathcal{O}(M)^*$ admits a "good" description [2], [71], [72], and also among domains with a "rich" (semi) group of automorphisms [73, Proposition 23]. The possibility of representing elements $\mathcal{O}(M)^*$ as analytic functions allows one to use results like Hadamard's three circles theorem when dealing with (ω) type conditions introduced above (see [102]). An example belonging to the first group is a bounded convex domain in some $\mathbf{C}^n$ [75, Lemma 1.3]. This example and a spliting result for a certain exact sequence of Fréchet spaces were the main ingredients of the proof given in [75] of $\mathcal{O}(D) \simeq \mathcal{O}(\Delta^d)$ for a strictly pseudoconvex domain, $D \subset\subset \mathbf{C}^d$. Using I.4 Theorem we could add to this list smoothly bounded relatively compact pseudoconvex domains with Lipschitz boundary in Stein manifolds [32] (cf. [31]) and Stein manifolds that are complete with respect to the Carathéodory metric [10]. For domains of holomorphy $D \subset\subset \mathbf{C}^n$ with $\mathcal{O}(D) \simeq \mathcal{O}(\Delta^n)$, the plurisubharmonic function constructed on D in I.4 Theorem provides a subharmonic barrier [50, p.168] for each boundary point of D thought as a subset of $\mathbf{R}^{2n}$. In particular these domains are regular for the Dirichlet problem in the sense of potential theory [50, p.171]. But this property fails to characterize domains of holomorphy D for which $\mathcal{O}(D) \simeq \mathcal{O}(\Delta^n)$. For example (cf. [124]) the bounded domain of holomorphy $\Omega = \{(z,\omega) \in \mathbf{C}^2 : |\, z \,|< 1,\ |\, \omega \,|<|\, z \,|\}$ is a regular domain in $\mathbf{R}^4$ and $\mathcal{O}(\Omega)$ contains a closed complemented subspace isomorphic to $\mathcal{O}(V)$, $V = \Omega \cap \{\omega = 0\}$, in view of Oka-Cartan theory. Since $\mathcal{O}(V) \simeq \mathcal{O}(\Delta - \{0\}) \simeq \mathcal{O}(\Delta) \times \mathcal{O}(\mathbf{C})$ ([87]) it follows that $\mathcal{O}(\Omega)$ has a complemented subspace isomorphic to an infinite type power series space $\mathcal{O}(\mathbf{C})$ and therefore cannot be isomorphic to a finite type power series space .

We now take up an example, variants of which were considered by various authors [122] [34] that illustrates some of the pathologies concerning the isomorphism $\mathcal{O}(M) \simeq \mathcal{O}(\Delta^n)$ for Stein manifolds M. Let $K_n = \{z \in \mathbf{C} :|\, z - \frac{1}{e^n} \,|\leq \frac{1}{e^{n^3}}\}$ $n = 1, 2, \cdots$. Since the sequence $\{e^n(e^{-(n+1)^3} + e^{-n^3})\}$ converges to zero, there is an n_0 such that, $K_n \cap K_{n+1} = \phi$ for $n \geq n_0$. Let

$$\Omega = \Delta - (\bigcup_{n \geq n_0+1} K_n \cup \{0\}).$$

Since

$$\sum \frac{n}{\ell n[(\frac{1}{e^{n^3}})^{-1}]} < \infty,$$

it follows from a result of Wiener [50, p.222] that the point 0 is an irregular point of Ω and so $\mathcal{O}(\Omega) \not\simeq \mathcal{O}(\Delta)$ in view of the above remarks. For any $N \in \mathbf{N}$, set $\Omega_N = \Omega \times \Delta^{N-1}$. Then Ω_N is a domain holomorphy in $\mathbf{C}^N$ and $\mathcal{O}(\Omega_N)$ contains $\mathcal{O}(\Omega)$ as a complemented subspace, so it follows that $\mathcal{O}(\Omega_N) \not\simeq \mathcal{O}(\Delta^N)$, $N = 1, \cdots$.. The usual proof of the hyperconvexity of bounded convex domains in $\mathbf{C}^n$ and also the proof given by Range [86] for the existence of a bounded plurisubharmonic exhaustion functions on smoothly bounded domains of holomorphy raises the question as to whether a bounded domain of holomorphy which can be "approximated" from the outside by "nice" domains is necessarily hyperconvex. (cf. also Question 6.4 of [75]). Indeed it is proved in [54] that a bounded domain of holomorphy in $\mathbf{C}^n$ that has the *property* S (see ([54, p.173])) for the definition of the property S) is hyperconvex (cf. also [38]. Some aspects of this issue can be settled in the negative by considering the Ω_N's introduced above. Since for each N, Ω_N has no complementary envelope of holomorphy these spaces can be employed in providing answers in the negative to Question 6.4 of [75] and also to a question raised by Stéhle in [97] (see [10]). Observe that bounded analytic functions on the domain Ω introduced above is dense in $\mathcal{O}(\Omega)$ in view of Runge's theorem and so this example demonstrates that the presence of "natural" dead-end spaces for $\mathcal{O}(D)$ will not in general imply the isomorphism $\mathcal{O}(D) \simeq \mathcal{O}(\Delta^d)$.

An interesting class of hyperconvex domains can be obtained by considering domains D of the form $D = \{(z,\omega) \in \mathbf{C}^{n+1};\ z \in D_1,\ |\, \omega \,|< e^{-\phi(z)}\}$ where D_1 is a domain of holomorphy in $\mathbf{C}^n$ with the property that $\mathcal{O}(D_1) \simeq \mathcal{O}(\Delta^n)$ and ϕ is a continuous plurisubharmonic function on D. Domains of this type have been studied by various authors. In particular N. Sibony constructed a domain of this type to show that a bounded domain of holomorphy need not be the natural domain of definition of a bounded analytic function ([94]). Mitiagin-Henkin after showing the validity of $\mathcal{O}(V) \simeq \mathcal{O}(\Delta^d)$, $d = \dim V$, for a connected closed submanifold of Δ^n, asked whether a Stein manifold M with $\mathcal{O}(M) \simeq \mathcal{O}(\Delta^d)$ is necessarily biholomorphic to a closed submanifold of some Δ^n [75, Question 6.1]. It follows from I.4 theorem that the same conclusion is valid for Stein manifolds M which admit a proper holomorphic function into a polydisc (or more generally into a hyperconvex manifold). Hence one can relax the question of Mitiagin-Henkin little bit and ask whether there exists a proper holomorphic mapping from a Stein manifold M with $\mathcal{O}(M) \simeq \mathcal{O}(\Delta^d)$ into some Δ^n. A negative answer to this question is given in [10] by using the Sibony example mentioned above. We would like to refer the reader to [46] for some examples of hyperconvex domains appearing in complex differential geometry.

For a hyperconvex Stein manifold M, $\mathcal{P}$-measures were employed, especially in the second proof sketched above, in producing a basis for the Fréchet space $\mathcal{O}(M)$. We would like to close this subsection by describing how, in turn, $\mathcal{P}$-measures can be obtained from a basis of $\mathcal{O}(M)$ by functional analysis considerations. Following the ideas given in [121] and [102] we have,

I.11. Proposition: Let M be a hyperconvex Stein manifold and K a compact holomorphically convex, regular subset of M. Then there exists a basis $\{e_n\}$ of $\mathcal{O}(M)$ and positive numbers $d_n, n = 1, 2, \cdots$, such that

$$\overline{\lim}_{\xi \to z}\, \overline{\lim}_n \frac{\ell n \mid e_n(\xi) \mid}{d_n} = 1 + \omega_k(z) \text{ for } z \notin K.$$

Moreover if $M \subset\subset \mathbf{C}^n$ is a hyperconvex domain one can choose a basis $\{e_n\}$ of $\mathcal{O}(M)$ as above to satisfy the additional property that it forms a complete orthogonal system in the Bergman space $L^2(M) \cap \mathcal{O}(M)$.

Proof: Since we will be dealing with $1 + \omega_K$ throughout the proof, we will give it a name and set $\rho_K = 1 + \omega_K$. We fix a measure $d\varepsilon$ on M, with the properties mentioned in the first proof of I.4 Theorem and set $H_\infty = \{f \in \mathcal{O}(M) : \|f\|_\infty \doteq \{\int \mid f \mid^2 d\varepsilon\}^{\frac{1}{2}} < \infty\}$. Choose a Hilbert space H with continuous inclusions $\mathcal{O}(K) \hookrightarrow H \hookrightarrow AC(K)$ in the notation of the second proof of I.4 Theorem ([124]). Arguing as in the first proof, we see that the pair $\{H_\infty, H\}$ satisfies the condition (γ). On the other hand if we set $D_\alpha = \{z \in M : \rho_K(z) < \alpha\}$, $0 < \alpha < 1$, then as above (see the discussion given after I.5 Definition) for each $f \in \mathcal{O}(M)$ and $0 < \alpha < 1$ with $\|f\|_K = e^{C_0}$, $\|f\|_{D_\alpha} = e^{C_1}$, we have the relation

$$\ell n \mid f(z) \mid \leq C_1(\omega_K^{D_\alpha}(z) + 1) - C_0 \omega_K^{D_\alpha}(z); \quad z \in D_\alpha. \tag{1}$$

Since $\omega_K^{D_\alpha} = (\omega_K + 1 - \alpha)/\alpha$, the above relation immediately implies that the pair $\{H_\infty, H\}$ also satisfies the condition (ω). Hence $\mathcal{O}(M)$ can be represented as $Q\{H_\infty, H\}$. Now let $\{e_n\}$ be the common orthonormal basis of H_∞ and H with $\|e_n\|_H = 1$, $n = 1, \cdots$.. Set $a_n \doteq \|e_n\|_\infty$ and $d_n \doteq \ell n\, a_n, n = 1, \cdots$.. So $\{e_n\}$ is a basis of $\mathcal{O}(M)$. Let c_0 denote the norm of the continuous inclusion $H \hookrightarrow AC(K)$. From the continuous inclusion $H_\infty \hookrightarrow \mathcal{O}(X)$ we can find, for every $0 < \alpha < 1$, a $c(\alpha) > c_0$ such that $\| \quad \|_{D_\alpha} \leq c(\alpha) \| \quad \|_\infty$, where $D_\alpha = \{z \in M : \rho_K(z) < \alpha\}$. Now for a fixed $0 < \alpha < 1$, we have;

$$\frac{\ell n \mid e_n(z) \mid}{d_n} - \frac{\ell n\, c(\alpha)}{d_n} \leq 1; \quad z \in D_\alpha$$

$$\frac{\ell n \mid e_n(z) \mid}{d_n} - \frac{\ell n\, c(\alpha)}{d_n} \leq \frac{\ell n\, c_0}{d_n} - \frac{\ell n\, c(\alpha)}{d_n} \leq 0; \qquad z \in K, \quad n = 1, 2, \cdots.$$

Hence it follows that

$$\frac{\ell n \mid e_n(z) \mid}{d_n} - \frac{\ell n\, c(\alpha)}{d_n} \leq 1 + \omega_K^{D_\alpha}(z) = \frac{1}{\alpha} \rho_K(z) \qquad \text{for} \qquad z \in D_k, \quad n = 1, 2, \cdots.$$

Therefore $\{\ell n \mid e_n \mid / d_n\}$ is a locally bounded sequence of plurisubharmonic functions on M and so

$$\rho(z) \doteq \overline{\lim}_{\xi \to z} \overline{\lim}_n \frac{\ell n \mid e_n(\xi) \mid}{d_n}$$

is a plurisubharmonic function on M. Moreover from the above analysis it follows that $\rho \leq \rho_K$. We set $\Omega_\alpha = \{z \in M;\ \rho(z) < \alpha\}$, $0 < \alpha < 1$.

The fact that $\mathcal{O}(X)$ can be represented as $Q\{H_\infty, H\}$ implies the existence of functions $\sigma(t)$, $\tau(t)$ with $\sigma(t) \uparrow 1$, $\tau(t) \uparrow 1$ as $t \uparrow 1$ and continuous inclusions

$$\mathcal{O}(\overline{D}_{\sigma(t)}) \hookrightarrow H_t \hookrightarrow \mathcal{O}(D_{\tau(t)})$$

for $t < 1$ and near 1 where $H_t \doteq \{x \in H : \mid x \mid_t \doteq \{\sum \mid \langle x, e_n \rangle \mid^2 a_n^{2t}\}^{\frac{1}{2}} < \infty\}$, $0 < t < 1$. For a fixed $0 < t < 1$, near 1, applying Zaharyuta's interpolation theorem (I.6 Theorem) on $D_{\sigma(t)}$ to the data $\mathcal{O}(\overline{D}_{\sigma(t)}) \hookrightarrow H_t$, $\mathcal{O}(K) \hookrightarrow H$, one obtains a continuous inclusion $\mathcal{O}(\overline{D}_{s\sigma(t)}) \hookrightarrow H_{st}$, for each $0 < s < 1$. For a fixed $0 < \alpha < 1$ and $0 < \beta < 1$, choose an α_0 with $\beta < \alpha_0/\alpha < 1$ and a $0 < t < 1$ so that $\beta < \alpha_0 t/\alpha\sigma(t)$ and $\frac{\alpha_0}{\sigma(t)} < 1$. Then it follows from the above considerations that, $\mathcal{O}(\overline{D}_{\alpha_0}) \hookrightarrow H_{\alpha_0 t/\sigma(t)}$. Therefore for $f \in \mathcal{O}(M)$, we have:

$$\{\sum_n \mid \langle f, e_n \rangle \mid^2 a_n^{2\alpha\beta}\}^{\frac{1}{2}} \leq \sum_n \mid \langle f, e_n \rangle \mid^2 a_n^{\frac{2\alpha_0 t}{\sigma(t)}} \leq C \, \|f\|_{D_\alpha} \tag{2}$$

for some $C > 0$ which does not depend upon the given $f \in \mathcal{O}(M)$. Now fix an α, $0 < \alpha < 1$ and let $z_0 \in \Omega_\alpha$. Since ρ is not constant (see I.2 Proposition) we can assume that $\rho(z_0) < \alpha - \varepsilon$ for some $\varepsilon > 0$. Choose a closed ball B around z_0 with $B \subset\subset \Omega_{\alpha-\varepsilon}$. In view of Hartog's lemma, there exists a $C > 0$ such that

$$\|e_n\|_B \leq C\, e^{(\alpha-\varepsilon)d_n}, \ n = 1, 2, \cdots. \tag{3}$$

We choose a β with $\frac{\alpha-\varepsilon}{\alpha} < \beta < 1$. Now for an $f \in \mathcal{O}(M)$, the series $\sum \langle f, e_n \rangle e_n$ converges, uniformly on compact subsets of M, to f. In view of (2) we have an estimate of the form; $| \langle f, e_n \rangle | \leq C\, a_n^{-\beta\alpha} \|f\|_{D_\alpha}$, $n = 1, 2, \cdots$.. Combining (2) and (3) we obtain;

$$| f(z) | \leq \sum | \langle f, e_n \rangle | \; | e_n(z) | \leq C(\sum a_n^{-\beta\alpha+\alpha-\varepsilon}) \|f\|_{D_\alpha} \leq C\, \|f\|_{D_\alpha}; \quad z \in B \quad (4)$$

where C does not depend upon f. Since (4) is true for every $f \in \mathcal{O}(X)$, applying it to f^s, $s \in \mathbf{N}$ and then by first taking the sth root of both sides and then letting $s \to \infty$, we see that B is in the holomorphically convex hull of $\overline{D}_\alpha$ and hence is a subset of $\overline{D}_\alpha$ since $\overline{D}_\alpha$ is holomorphically convex. It follows that $\Omega_\alpha \subset \overline{D}_\alpha$ for every $0 < \alpha < 1$. Now for a fixed $z \in M$ with $\rho_K(z) > 0$, we have that $\rho(z) > 0$, say $\rho(z) = \alpha$. Then for every $\varepsilon > 0$ the inclusion $\Omega_{\alpha+\varepsilon} \subset \overline{D}_{\alpha+\varepsilon}$ implies that $\rho(z) \leq \alpha + \varepsilon = \rho(z) + \varepsilon$. Hence it follows that $\rho_K \leq \rho$ on the set $\{z : \rho_K(z) > 0\}$. But this set in view of (1) coincides with K^c, since K is holomorphically convex. Combining this with the first part of the proof we conclude that $\rho(z) = \rho_K(z)$ for $z \in K^c$. The second assertion of the proposition follows immediately from the fact that in the case when M is a relatively compact hyperconvex domain in some $\mathbf{C}^N$, one can take the dead-end space H_∞ in the above proof as the Bergman space $L^2(M) \cap \mathcal{O}(M)$ (see I.8 Corollary). This finishes the proof of I.11 Proposition.

(ii) **Infinite case:** We now investigate the structure of Stein manifolds M of dimension d, for which $\mathcal{O}(M)$ is isomorphic to a power series space of infinite type. A prime example of such a Stein manifold is the d-dimensional complex Euclidean space. In fact, in view of the above discussions the Stein manifolds M for which $\mathcal{O}(M)$ is isomorphic to an infinite type power series space is precisely the class of Stein manifolds M for which $\mathcal{O}(M) \simeq \mathcal{O}(\mathbf{C}^d)$, $d = \dim M$. If M is such a manifold then the Fréchet space $\mathcal{O}(M)$, has, in particular the property (DN). We will first discuss the restrictions that this linear topological property of $\mathcal{O}(M)$ imposes upon M. To this end we fix convex compact set with nonempty interior, $K_0 \subset M$, and with the property that $\| \;\|_{K_0}$ dominates the dominating norm $\| \;\|_0$ given in the condition (DN) (see §0). We choose a compact regular set K, $K^0 \supset K_0$, and an exhaustion $\{D_n\}_{n=1}^\infty$ of M with relatively compact hyperconvex domains such that $\overline{D}_n$ is holomorphically convex for every n, and

$$\forall n \;\; \exists C > 0 : \; \| \;\|_{D_n}^2 \leq C \| \;\|_K \; \| \;\|_{D_{n+1}}. \quad (1)$$

The fact that we can choose such an exhaustion follows immediately from the condition (DN). Upon iterating (1) we get;

$$\forall n, s \;\; \exists C > 0 : \; \| \;\|_{D_n} \leq C \| \;\|_{K_0}^{1-\frac{1}{2^s}} \; \| \;\|_{D_{n+s}}^{\frac{1}{2^s}}. \quad (2)$$

Therefore in view of Runge's theorem, for every n and s there exists a $C > 0$ such that

$$\ell n \; | f(z) | \leq \ell n \; C + (1 - \frac{1}{2^s}) \ell n \; \|f\|_{K_0} + \frac{1}{2^s} \, \ell n \; \|f\|_{D_{n+s}}; \; \forall z \in D_n, \; f \in \mathcal{O}(D_{n+s+1}) \quad (3)$$

Now fix an n_0 and consider for each s, the $\mathcal{P}$-measure $\omega_K^{D_{n_0+s+1}}$. Since each D_ℓ is hyperconvex and K is regular these plurisubharmonic functions are continuous. We choose a representation of $\omega_K^{D_{n_0+s+1}}$ on D_{n_0+s} as

$$\omega_K^{D_{n_0+s+1}}(z) = \overline{\lim}_{\xi \to z} \overline{\lim}_n \frac{\ell n \mid f_n(\xi) \mid}{c_n} \tag{4}$$

with $c_n > 0$, $n = 1, \cdots$, and $f_n \in \mathcal{O}(D_{n_0+s+1})$, $n = 1, 2, \cdots$. . In view of Hartog's lemma for a fixed $\varepsilon > 0$, there exists an N such that for $n > N$:

$$\frac{\ell n \|f_n\|_{K_0}}{c_n} \leq -1 + \varepsilon, \qquad \frac{\ell n \|f_n\|_{D_{n_0+s}}}{c_n} \leq \varepsilon. \tag{5}$$

Combining (5) and (3) we obtain

$$\overline{\lim}_n \frac{\ell n \mid f_n(z) \mid}{c_n} \leq (1 - \frac{1}{2^s})(-1 + \varepsilon) + \frac{\varepsilon}{2^s}; \quad z \in D_n.$$

Therefore in view of (4), since ε is at our disposal, we get an estimate of the form

$$\omega_K^{D_{n_0+s+1}}(z) \leq (1 - \frac{1}{2^s})(-1) \qquad \forall z \in D_{n_0},\ s \geq 1. \tag{6}$$

Now for any relatively compact domain Ω of M for m large the sequence of plurisubharmonic functions $\{\omega_K^{D_n}\}_{n=m}$ is a decreasing sequence on Ω and so $\lim_n \omega_K^{D_n}$ is a negative plurisubharmonic function on M. Moreover it is constantly equal to -1 on K since K is regular. Hence $\lim_m \omega_K^{D_m} \leq \omega_K^M$. On the other hand since $\omega_K^M \leq \omega_K^{D_m}$ for every m, it follows that $\lim_m \omega_K^{D_m} = \omega_K^M$.

Now (6) implies that $\omega_K^M \equiv -1$. In particular if ρ is a nonconstant negative plurisubharmonic function on M, then from the relation

$$\frac{\rho}{-\sup_{z \in K} \rho(z)} \leq \omega_K^M,$$

it follows that ρ must be a constant. We note that in the above analysis we can take K_0 to be a small closed ball in M. Indeed it follows from the Bishop's proof of the "three domains" theorem [77, p.130]. that we can take the norm corresponding to such a set as a weak dominating norm ($\| \ \|_0$ in the condition ($\underline{DN}$)). Connecting the dominating norm to this norm by the condition ($\underline{DN}$) one sees that upon replacing $\frac{1}{2}$ with a suitable λ_0, $0 < \lambda_0 < 1$, the above analysis is valid for K_0 equal to a relatively compact closed ball in M. Hence the task of finding a regular compact set containing K_0 in the above argument can easily be achieved. Therefore it follows that if $\mathcal{O}(X)$ has the property (DN) then every bounded plurisubharmonic function on M must be constant.

Now fix a Stein manifold M with the property that every bounded plurisubharmonic function on M reduces to a constant. Choose a regular compact set K with nonempty interior and an exhaustion of M with relatively compact hyperconvex domains $\{D_n\}$. As we have seen above the sequence of plurisubharmonic functions $\{\omega_K^{D_n}\}$ (or rather its tail) decreases pointwise to ω_K^M. But since K is regular, our assumption on M implies that $\omega_K^M \equiv -1$. Now for a fixed n, in view of Dini's theorem, there exists an m_0 such that for all $m \geq m_0$ we have

$$\omega_K^{D_m}(z) \leq -\frac{1}{2}, \quad z \in \overline{D}_n. \tag{7}$$

For a fixed n, and $m \geq m_0$ as above, (7) and the analog of two constants theorem (see subsection (i)) gives for each $f \in \mathcal{O}(M)$ with $\|f\|_K \leq 1$ an estimate of the form $\|f\|_{D_n} \leq \|f\|_{D_m}^{\frac{1}{2}}$. Applying this to $\frac{f}{\|f\|_K}$ for a given $f \in \mathcal{O}(M)$ we get

$$\forall n\ \exists m; \ \|f\|_{D_n} \leq \|f\|_{D_m}^{\frac{1}{2}} \|f\|_K^{\frac{1}{2}}; \quad \forall f \in \mathcal{O}(M). \tag{8}$$

But this is precisely the condition (DN) for $\mathcal{O}(M)$. To summarize, the above gives a sketchy proof of the equivalence between the conditions of; $\mathcal{O}(X)$ has the property (DN) and there are no nonconstant bounded plurisubharmonic functions on X. The later property is sometimes refered to as the "strong Liouville property". This relationship was observed independently by several authors [9], [114],[124]. In [9] it was obtained as a corollary of I.2 Proposition and a characterization of nuclear Fréchet spaces with the property (DN), due to Vogt [110], as those spaces E for which each continuous linear operator from a stable finite type power type power series space into E is compact. We note that a nuclear Fréchet space E has the property (DN) if and only if E is isomorphic to a closed subspace of s, the space of rapidly decreasing sequences [105].
We now state the main result of this subsection.

I.12 Theorem: Let M be a Stein manifold of dimension d. Then the following conditions are equivalent

(1) $\mathcal{O}(M)$ has the property (DN).

(2) For some compact regular set $K \subset M$ and an exhaustion of M with relatively compact hyperconvex domains $\{D_n\}$ we have $\omega_K^{D_n} \downarrow -1$.

(3) Every bounded plurisubharmonic function on M reduces to a constant.

(4) $\mathcal{O}(M) \simeq \mathcal{O}(\mathbf{C}^d)$.

The equivalence of the first three conditions follows from the discussion given above. In fact the equivalence of (2) and (3) is rather old, for example it appears implicitly in [67] . The proof of the equivalence of (4) and (1) appears in [14]. One of the main ingredients of the proof is Pelczynski's decomposition method [84] as adopted by Vogt [109] which says that if a Fréchet space E is isomorphic to a complemented subspace of a stable nuclear power series space $\Lambda_\infty(\alpha)$ and $\Lambda_\infty(\alpha)$ in turn is isomorphic to a complemented subspace of E, then E and $\Lambda_\infty(\alpha)$ are in fact isomorphic. From the structure theory of nuclear Fréchet spaces it follows that if $\mathcal{O}(M)$ has the property (DN) then it is isomorphic to a complemented subspace of $\mathcal{O}(\mathbf{C}^d)$, $d = \dim M$ [118]. On the other hand I.1 Proposition (3) says that there is an imbedding of $\mathcal{O}(\mathbf{C}^d)$ into $\mathcal{O}(M)$ as a closed subspace. Using this imbedding we were able to show that if $\mathcal{O}(M)$ has the property (DN) then it has a complemented subspace that is isomorphic to $\mathcal{O}(\mathbf{C}^d)$, $d = \dim M$. These considerations with analytic function spaces lead us to a solution of a special case of a problem of Pelczynski namely that a complemented subspace of s with stable diametral dimension is isomorphic to an infinite type power series space. [14, Theorem 2.2]. For more information on this and for diverse applications of this result we refer the reader to the reports of Terzioğlu [100] and Krone [66] for this conference.

We would like to draw the readers attention to the differences in our treatment of the finite and the infinite cases. In the first case interpolation methods were employed which allowed one to choose a basis from a predetermined Hilbert space whereas in the second case abstract functional analytic results like Pelczynski's decomposition method play a dominant role and so for example one does not have a control over the growth of the basis vectors constructed. Nevertheless in the presence of a "good" exhaustion function, interpolation techniques can be applied. Indeed harmonic exhaustion functions were used implicitly by Zaharyuta in his characterization of domains $D \subset \mathbf{C}$ for which $\mathcal{O}(D) \simeq \mathcal{O}(\mathbf{C})$ as the ones whose complement has logarithmic capacity zero. Evans-Selberg potential [91, Chapter 5, section 2] together with L^2-estimates for the $\bar{\partial}$-equation were used in [7] cf. [9] in showing the isomorphism $\mathcal{O}(M) \simeq \mathcal{O}(\mathbf{C})$ for a parabolic open Riemann surface. (For a different proof of this see [65]). In this case a basis for $\mathcal{O}(M)$ can be extracted from the Hilbert space $H = \{f \in \mathcal{O}(M) : \int |f|^2 e^{-\phi^2} dV < \infty\}$ where ϕ is the Evans- Selberg potential and dV is a suitable volume form on M. Stein manifolds M for which $\mathcal{O}(M)$ is isomorphic to $\mathcal{O}(\mathbf{C}^d)$ as *Fréchet algebras* can be characterized, by using a result of Stoll [98] and Burns [26], as the ones which admit a strictly plurisubharmonic, C^∞, proper function $\rho : M \to [0, \infty)$ such that $(dd^c(\ell n\, \rho))^n = 0$. The presence of a continuous plurisubharmonic exhaustion function on a Stein manifold M that satisfies the homogeneous complex Monge-Ampère equation $(dd^c)^n u = 0$ off a compact set not only gives the conditions (Ω) and (DN) for $\mathcal{O}(M)$, but also provides some information on how these conditions are connected. For such Stein manifolds interpolation considerations of Vogt on Fréchet spaces with (DN) and (Ω) type properties that are connected by a coupling condition, (see [116]) applies to $\mathcal{O}(M)$. Smooth algebraic subvarieties of $\mathbf{C}^N$ are among the Stein manifolds considered above ([47], [30]). The isomorphism $\mathcal{O}(V) \simeq \mathcal{O}(\mathbf{C}^d)$, $d = \dim V$, for an algebraic subvariety V was first given by Zaharyuta [125] in connection with a question raised by Mitiagin-Henkin [75]. Djakov and Mitiagin gave another proof of this result by considering the linear topological structure of polynomial ideals in the algebra of entire functions [33] . Using the decomposition method described above, yet another proof was given in [106],(cf. also[109]).

It is also possible to find intrinsic geometric conditions with respect to the geometry induced by a Kähler metric on a Stein manifold M which guaranties that $\mathcal{O}(M)$ has the property (DN). Any Stein manifold admits a complete Kähler metric [43]. For a given Kähler metric, G, if the associated real hermitian metric is given locally as $\sum g_{i_j} dx^i dx^j$ then the Laplace-Beltrami operator associated to G is given by the equation

$$\Delta f \doteq \frac{1}{\sqrt{g}} \sum_{i,j} \frac{\partial}{\partial x^j}(\sqrt{g} g^{ij} \frac{\partial f}{\partial x^i});$$

where $g = \det\{g_{ij}\}$. A differentiable function f is called *harmonic (subharmonic)* in case $\Delta f = 0$ $(\Delta f \geq 0)$. As in the Euclidean case subharmonic function can be characterized as those functions that satisfy a certain sub-mean value property [119] [91] and with the aid of this characterization one can extend the definition of subharmonicity to upper semicontinuous functions (cf. [27]). The link between the function theory on M and the potential theory on M given by a Kähler metric is the fact that for any $f \in \mathcal{O}(M)$, $\ln \mid f \mid$ is subharmonic in the above sense. This link allows one to obtain "Hadamard three circles" type theorems for analytic functions by employing the maximum principles of potential theory. For example let M be an open Riemann surface which admits a complete Kähler metric with non-negative Gaussian curvature. If we fix a point p_0 in M and denote by $r(p)$ the distance (geodesic) of p to p_0, then r is a continuous exhaustion function since the metric is complete. Moreover r satisfies the differential inequality $\Delta r \leq \frac{1}{r}$ in a weak sense [27, Theorem 3]. Hence the function $\rho(z) = \ell n(r(z))$ satisfies the weak differential inequality $\Delta \rho = 0$. Since ρ is an exhaustion, for each $n, D_n = \{z : \rho(z) < n\}$ is relatively compact in M. Now fix $f \in \mathcal{O}(M)$ and let $M_n = \sup_{z \in D_n} \ \ell n \mid f(z) \mid$. For a fixed n we consider the function

$$\psi = (\frac{M_{n+2} - M_n}{2})\rho - (\frac{M_{n+2} - M_n}{2})n + M_n.$$

In view of the subharmonicity of $\ell n \mid f \mid$ we have that $\ell n \mid f \mid -\psi$ satisfies the weak differential inequality $\Delta(\ell n \mid f \mid -\psi) \geq 0$ on the domain $U = D_{n+2} - \overline{D}_n$. So in view of the generalized Hopf maximum principle [27, Theorem 1] one has $\ell n \mid f \mid \leq \psi$ on U. In particular this inequality is valid on the set $\partial D_{n+1} = \{z : \rho(z) = n + 1\}$. Therefore we obtain

$$\|f\|_{D_{n+1}} \leq \|f\|_{D_n}^{\frac{1}{2}} \|f\|_{D_{n+2}}^{\frac{1}{2}}.$$

But this is precisely the condition (DN) for $\mathcal{O}(M)$ since $\{\| \ \|_{D_n}\}$ forms a fundamental system of norms for this Fréchet space. Hence we obtain, in view of I.11 Theorem, an old result of Blanc-Fiala-Huber which states that a complete open Riemannian surface with nonnegative Gaussian curvature is parabolic [51]. Following [91] (cf. also [92]) we will call a noncompact Riemannian manifold *parabolic* in case it admits no nonconstant bounded subharmonic function. Parabolic Riemannian manifolds were studied by several authors in the frame work of classification theory for Riemannian manifolds and various equivalent conditions for parabolicity were obtained ([92]). One such condition is the existence of an Evans potential on M [91, Theorem 1, p.400]. This function gives an exhaustion of M

and is harmonic off a point in M. Now if a Stein manifold M admits a complete Kähler metric that makes M parabolic then the argument given above with ψ replaced with an Evans potential implies that $\mathcal{O}(M)$ has the property (DN). So geometrical conditions that exists in literature which imply parabolicity for Riemannian manifolds, can be employed in producing Stein manifolds M for which $\mathcal{O}(M) \simeq \mathcal{O}(\mathbf{C}^d)$, $d = \dim M$. We would like to mention one such condition due to L.Karp [53] (cf. also [29]). Following [53] we say that a complete noncompact Riemannian manifold X has *moderate volume growth* if there exist a point x_0 in X and a positive nondecreasing function $F(r)$ such that $\int_a^\infty \frac{dr}{rF(r)} = \infty$ for some $a > 0$ and

$$\limsup_{r \to \infty} \frac{volume(B_r(x_0))}{r^2 F(r)} < \infty$$

where $B_r(x_0)$ is the geodesic ball of radius r around x_0. Now Theorem 3.5 of [53] combined with the arguments given in [91, Chapter V.7B], (cf. also [92, p.400]) gives that a noncompact Riemannian manifold X with moderate volume growth is parabolic in the above sense. Hence it follows that if a Stein manifold M with a complete Kähler metric has moderate volume growth then $\mathcal{O}(M) \cong \mathcal{O}(\mathbf{C}^d)$, $d = \dim M$. We note a Stein Kähler manifold M could very well be non-parabolic and satisfy $\mathcal{O}(M) \simeq \mathcal{O}(\mathbf{C}^d)$, $d = \dim M$, at the same time as the example of $\mathbf{C}^n, n > 1$ with the usual metric shows. For more information on the relationship between the existence of complete Kähler metrics with special properties on a given complex manifold M and some "non-existence theorems" for certain "analytic" objects on M we refer the reader to [120].

In search for other linear topological invariants of $\mathcal{O}(M)$ that implies the isomorphism $\mathcal{O}(M) \simeq \mathcal{O}(\mathbf{C}^d)$, one can ask as to whether such an isomorphism exists if the diametral dimensions of the spaces $\mathcal{O}(M)$ and $\mathcal{O}(\mathbf{C}^d)$ are equal. This point was taken up in [15]. For domains in the complex plane the answer is yes [15, 1.7 Corollary]; but for dimensions greater then one the diametral dimension considerations alone fails to give the required isomorphism. For example for the domain $D = \Delta \times \mathbf{C} \subsetneq \mathbf{C}^2$ one has that $\Delta(\mathcal{O}(D)) = \Delta(\mathcal{O}(\mathbf{C}))$ yet $\mathcal{O}(D) \simeq \mathcal{O}(\Delta) \hat{\otimes} \mathcal{O}(\mathbf{C}) \not\simeq \mathcal{O}(\mathbf{C}^2)$ (see [15]). However the condition $\Delta(\mathcal{O}(M)) = \Delta(\mathcal{O}(\mathbf{C}^d)$ for a Stein manifold M forces $\mathcal{O}(M)$ to contain a complemented copy of $\mathcal{O}(\mathbf{C}^d)$ [15, 1.3. Corollary].

Another class of examples of complex manifolds M for which $\mathcal{O}(M)$ has the property (DN) can be obtained by considering domains $\Omega \subsetneq \mathbf{C}^n$ that are of the form $\Omega = \mathbf{C}^n - F$ where F is a closed removable set with respect to bounded plurisubharmonic functions. Clearly such a domain admits no nonconstant bounded plurisubharmonic functions but in general it need not be a domain of holomorphy. Nevertheless if we consider the envelope of holomorphy $\hat{\Omega}$ of such an Ω, then $\hat{\Omega}$ is a Stein manifold and the Fréchet spaces $\mathcal{O}(\Omega)$ and $\mathcal{O}(\hat{\Omega})$ are isomorphic to each other [52, p.131]. Moreover every continuous linear operator from $\mathcal{O}(\Delta)$ into $\mathcal{O}(\hat{\Omega})$ must be compact since a noncompact operator from $\mathcal{O}(\Delta)$ into $\mathcal{O}(\hat{\Omega})$ will induce a nonconstant negative plurisubharmonic function on Ω in view of the argument given in the proof of I.2 Proposition. Hence it follows that $\mathcal{O}(\Omega)$ has the property (DN).

I.12 Theorem allows one to extend the classical Liouville's theorem to the class of Stein manifolds M for which $\mathcal{O}(M)$ has the property (DN)(cf. also[106]). But this class does not coincide with the class of Stein manifolds which admit no nonconstant bounded analytic function. For example one can consider a domain Ω of the form $\Omega = \mathbf{C} - F$ where F is a generalized Cantor set with analytic capacity zero, but with logarithmic capacity positive [4, p.252]. Then $\mathcal{O}(\Omega) \not\cong \mathcal{O}(\mathbf{C})$ since Ω admits a nonconstant negative subharmonic function yet every bounded holomorphic function on Ω is constant.

We would like to close this section by noting that with the solution of Problem 6.2 of [75] in the positive by Aleksandrov [5] all the questions in [75] are provided with answers.

§II. Some generalizations

In this section we consider generalizations of the results discussed in section one, to Stein spaces and to coherent analytic sheaves on such spaces. By a Stein space we mean a *reduced* Stein space in the sense of [45] which has a Hausdorff, separable topology. Analytic functions on a Stein space X, $\mathcal{O}(X)$, equipped with the topology of uniform convergence on compact subsets of X is a nuclear Fréchet space [49, p. 158]. These spaces, as in the manifold case, have the property (Ω). This was shown by Vogt [62], (cf. [11]) and Schottenloher ([93]) independently. On the other hand not every $\mathcal{O}(X)$, X a connected Stein space, enjoys the property $(\underline{DN})$ since such spaces need not even possess a continuous norm. For example take the closed variety of $\mathbf{C}^2$ that is made up of the union of an infinite number of vertical complex lines with a horizontal complex line. In this connection we have: [11].

II.1 Proposition: For a Stein space X the following conditions are equivalent:

(1) $\mathcal{O}(X)$ has a continuous norm.

(2) X has finite number of irreducible components.

(3) $\mathcal{O}(X)$ is isomorphic, as a topological algebra, to a closed subalgebra of $(\mathcal{O}(\Delta^d))^k$ for some d and k.

(4) $\mathcal{O}(X)$ has the property $(\underline{DN})$.

For a Stein space X with finite number of irreducible components one can take in the statement (3), $d = \dim X$ and k = number of irreducible components. We note that although the spaces $\mathcal{O}(\Delta^d)^k$, $k > 1$, $d \in \mathbf{N}$ and $\mathcal{O}(\Delta^d)$ are not isomorphic as topological algebras, they are certainly isomorphic as Fréchet spaces.

Following, [90] an upper semicontinuous function on a Stein space X will be called *plurisubharmonic* in case it is locally bounded from above near each singular point of X and its restriction to the manifold part of X is a plurisubharmonic function on this manifold.

For equivalent definitions of plurisubharmonicity we refer to [44], [40], [76]. Most of the properties of plurisubharmonic functions on a Stein manifold are still valid in the Stein space case [90], [40]. In particular the properties of $\mathcal{P}$-measures and the proof of Zaharyuta's interpolation theorem (I.2) carry over to Stein space situation with minor modifications. Adopting the notion of hyperconvexity to this set up we have

II.2 Theorem: Let X be an irreducible Stein space with dimension d. Then

(1) $\mathcal{O}(X)$ is isomorphic to a finite type power series space if and only if $\mathcal{O}(X) \simeq \mathcal{O}(\Delta^d)$ if and only if X is hyperconvex.

(2) $\mathcal{O}(X)$ is isomorphic to an infinite type power series space if and only if $\mathcal{O}(X) \simeq \mathcal{O}(\mathbf{C}^d)$, if and only if every bounded plurisubharmonic function on X is constant.

In the case when X has finite number of irreducible components one obtains similar necessary and sufficient conditions upon analysing the relation between $\mathcal{O}(X)$ and the spaces of analytic functions on its irreducible components as in [109], [106], [11],

The space of analytic functions on a Stein space X is in the family of spaces formed by taking the global sections of coherent analytic sheaves on X. For a Stein space X, the set of coherent analytic sheaves on X will be denoted by $Coh(X)$ and for $\mathcal{F} \in Coh(X)$, $D \subset X$, the space of sections of $\mathcal{F}$ over D will be denoted by $\mathcal{F}(D)$. Let X be a Stein space and $\mathcal{F} \in Coh(X)$. For a given holomorphically convex compact set K of X we can introduce a seminorm on $\mathcal{F}(X)$ by first choosing a finite number of sections $\sigma_1, \cdots, \sigma_p \in \mathcal{F}(X)$ that generate $\mathcal{F}_x$ as a $\mathcal{O}_x$ module for each x in a neighborhood Ω of K (such a choice is possible in view of Cartan theorem A) and then set for $\sigma \in \mathcal{F}(X)$;

$$\|\sigma\|_K \doteq \inf \max_i \|c_i\|_K$$

where the infimum is taken over all $(c_1 \cdots, c_p) \in \mathcal{O}(\Omega)^p$ such that $\sigma = \sum_{i=1}^p c_i \sigma_i$ [52, p.171]. This definition of course depends upon the choice of $\sigma_1, \cdots, \sigma_p$, but another choice of generators will produce an equivalent seminorm [52, p.171]. The space $\mathcal{F}(X)$ with the topology coming from these seminorms is a nuclear Fréchet space [49, p.240] ([41], [56]). We note that with respect to this topology, the maps induced from sheaf morphisms are automatically continuous. The linear topological properties of these Fréchet spaces were studied by several authors in connection with certain problems of complex analysis (eg. [41], [45], [75]). We will be mainly interested in obtaining sequence space representations of $\mathcal{F}(X)$, $\mathcal{F} \in Coh(X)$ for Stein spaces X appearing in II.2 Theorem. The support of an $\mathcal{F} \in Coh(X)$ will be denoted by $s(\mathcal{F})$. Recall that $s(\mathcal{F})$ is a analytic subvariety of X. [45, p.76]. We first consider some properties of these spaces relevant to the structure theory of Fréchet spaces. As we have already seen not every Fréchet space of the from $\mathcal{F}(X)$, has the property $(\underline{DN})$. We will say that a coherent analytic sheaf $\mathcal{F}$ is *of type* $(\underline{DN})$ in case $\mathcal{F}(X)$ has the property $(\underline{DN})$. We do not know of a general criterion that characterizes coherent analytic sheaves of type ($\underline{DN}$). Nevertheless there are some techniques that can be used in this connection. One such technique stems from Bishop's proof of the "Three Domains theorem" of several complex variables [77, p.130] (cf. [107]). Arguing along the same lines as [107, Section 5] one obtains that a locally free sheaf on an irreducible Stein space is of type ($\underline{DN}$). Another technique is to uses the results of Palamodov ([81], [82]). For example, following Mitiagin-Henkin [75, Proposition 6.1] one can show that if D is a relatively compact pseudoconvex domain in a Stein manifold M and if $\mathcal{F} \in Coh(M)$ then $\mathcal{F}$ restricted to D is a sheaf of type ($\underline{DN}$). Still another class of sheaves of type ($\underline{DN}$) for irreducible Stein spaces X can be obtained by taking the coherent analytic subsheaves of $\mathcal{O}_X^p$, $p \in \mathbf{N}$. This follows immediately from the "closure of modules" theorem of Cartan which implies, for analytic subsheaf $\mathcal{F}$ of $\mathcal{O}_X^p$, that $\mathcal{F}(X)$ is a closed subspace of $\mathcal{O}(X)^p$ [49, p.235]. In our next result we gather some general properties of these spaces [11].

II.3 Proposition: Let X be a Stein space and $\mathcal{F} \in Coh(X)$. We have:

(1) $\mathcal{F}(X)$ has the property (Ω).

(2) If $\mathcal{F}$ is of type ($\underline{DN}$) then the associated exponent sequence of $\mathcal{F}(X)$ is $\{n^{\frac{1}{d}}\}_n$ where $d = \dim s(\mathcal{F})$.

(3) If $\mathcal{F}$ is of type ($\underline{DN}$) then $\mathcal{F}(X)$ is isomorphic to a closed subspace of $\mathcal{O}(\Delta^d)$ and is a quotient space of $\mathcal{O}(\mathbf{C}^{d+1})$, $d = \dim s(\mathcal{F})$. Moreover there is a closed subspace of $\mathcal{F}(X)$ that is isomorphic to $\mathcal{O}(\mathbf{C}^d)$ where d is as above.

For Stein manifolds (1) of II.3 Proposition is due to Vogt [106]. The general case can be reduced to the special case when $X = \mathbf{C}^d$, with the use of Oka-Weil domains [11, Proposition 1.1], and (3) follows from some general consideration of the structure theory of nuclear Fréchet spaces [118, 3.4 Satz], [108, 3.2 Satz] and [15, 2.2 Theorem]. For (2) one employs a rather technical simple lemma which we hope will find further applications [11, 1.4 Lemma]. A nuclear Fréchet space E satisfying the conditions ($\underline{DN}$) and (Ω) with an increasing sequence $\{\|\ \|_k\}$ of Hilbertian norms is said to be in *standard form* in case.

$$\forall k \;\; \exists \lambda \in (0,1),\; C > 0 : \;\; \|\ \|_{k+1} \leq C \,\|\ \|_k^{\lambda}\, \|\ \|_{k+2}^{1-\lambda}$$

$$\forall k \;\; \exists j,\; C > 0 : \;\; U_{k+1} \subset C\, r^j U_p + \frac{1}{r} U_k, \qquad \forall r > 0$$

where U_k is the neighborhood of zero associated with $\|\ \|_k$, $k = 1, 2, \cdots$.

II.4 Lemma: Let $\{X, |\ |_k\}^{\infty}$ and $\{Y, \|\ \|_k\}$, $k = 0, 1, \cdots$, be two nuclear Fréchet spaces with properties ($\underline{DN}$), (Ω) and in standard form. Denoting by $\{U_k\}$ and $\{V_k\}$ the neighborhoods of zero associated with the norm systems $\{|\ |_k\}$ and $\{\|\ \|_k\|\}$ respectively we have;

(1) If there exists a linear (not necessarily continuous) mapping $T : X \to Y$, integers k_1, k_2, s, $k_1 < k_2$ and constants $c_1 > 0$, $c_2 > 0$ with

$$\overline{T(U_{k_2})} \supset c_1 V_s \quad (\text{ in } (Y, \widehat{\| \ \|_0}))$$

$$T(U_{k_1}) \subset c_2 V_0$$

then

$$\varepsilon(X) = O(\varepsilon(Y)).$$

(2) If there exists a linear (not necessary continuous) mapping $S : X \to Y$ integers s_1, s_2, k, $s_1 < s_2$ and constants $c_3 > 0$, $c_4 > 0$ with

$$S(X) \cap V_{s_1} \subset c_3 S(U_0)$$

$$S(U_k) \subset c_4 V_{s_2}$$

then

$$\varepsilon(Y) = O(\varepsilon(X)).$$

Now we turn to sequence space representations of $\mathcal{F}(X)$'s in the case of Stein spaces X for which $\mathcal{O}(X)$ is isomorphic to a power series space. We take up the finite case first

II.5 Theorem: Let X be a hyperconvex Stein space. Then for any $\mathcal{F} \in Coh(X)$ of type ($\underline{DN}$), the Fréchet spaces $\mathcal{F}(X)$ and $\mathcal{O}(\Delta^d)$ are isomorphic to each other where $d = \dim s(\mathcal{F})$.

The main point in the proof of II.5 Theorem is the observation that if $\mathcal{O}(X)$ has the property $(\overline{\Omega})$ then so does $\mathcal{F}(X)$ for any $\mathcal{F} \in Coh(X)$ [11, II.1 Proposition]. Since for a given closed submodule M of $\mathcal{O}(X)^p$, $p \in \mathbf{N}$ and X a Stein space, one can find a coherent analytic sheaf $\mathcal{M}$ on X such that $\mathcal{M}(X) = M$ [28] one obtains as a Corollary of II.5 Theorem([11, III.3 Corollary]).

II.6 Corollary: Let X be a hyperconvex Stein space of dimension d then for any closed submodule M of $\mathcal{O}(X)^p$, M and $\mathcal{O}(\Delta^d)$ are isomorphic as Fréchet spaces.

This corollary in particular applies to closed ideals of $\mathcal{O}(X)$.

For Stein spaces X for which $\mathcal{O}(X)$ is isomorphic to an infinite type power series space, we cannot hope to obtain a result analogous to II.5 Theorem since any coherent analytic sheaf on a given Stein manifold can be realized as a coherent analytic sheaf on some $\mathbf{C}^N$. We will restrict our attention to two particular classes of sheaves, namely subsheaves of the structure sheaf and locally free sheaves (e.g. sheaf of germs of holomorphic section of a holomorphic vector bundle [45, p.90]. In this case we have; [11, III.1 Theorem].

II.7 Theorem: Let X be a Stein space of dimension d with the property that $\mathcal{O}(X)$ is isomorphic to a power series space of infinite type. Let $\mathcal{F} \in Coh(X)$ be either a locally free sheaf or a subsheaf of $\mathcal{O}^p$ for some $p \in \mathbf{N}$. Then the Fréchet spaces $\mathcal{F}(X)$ and $\mathcal{O}(\mathbf{C}^d)$ are isomorphic to each other.

One proves this theorem by first showing that the property (DN) passes to $\mathcal{F}(X)$'s for $\mathcal{F}$'s as in the statement of the theorem, and then applies [14, 2.2 Theorem]. In view of II.2 Theorem the above result in particular applies to algebraic varieties.

II.8 Corollary: Let $X \subset \mathbf{C}^N$ be an algebraic variety. Then for any holomorphic vector bundle on X, the Fréchet space of global holomorphic sections of this bundle is isomorphic to $\mathcal{O}(\mathbf{C}^d)$, $d = \dim X$.

We also have the following analog of II.6 Corollary in this setting.

II.9 Corollary: Let X be a Stein space of dimension d, for which $\mathcal{O}(X)$ is isomorphic to an infinite type power series space. Then for any closed submodule M of $\mathcal{O}(X)^p$, $p \in \mathbf{N}$, M and $\mathcal{O}(\mathbf{C}^d)$ are isomorphic as Fréchet spaces.

In the case when X is an algebraic variety the above corollary is due to Djakov-Mitiagin [34].

§III. Linear Extension Operators.

Let M be a Stein space and V a closed subvariety of M. One of the consequences of the Oka-Cartan theory is that every analytic function on V can be extended to an analytic function on M. The question as to whether one can find a continuous linear extension operator from $\mathcal{O}(V)$ into $\mathcal{O}(M)$ was raised and studied by Mitiagin-Henkin in [75]. Such an operator, if it exists, will imbed $\mathcal{O}(V)$ into $\mathcal{O}(M)$ as a closed complemented subspace. This simple observation exhibits an immediate obstruction, in some cases, for the existence of a continuous linear extension operator . Indeed if one knows, a priori that the Fréchet space $\mathcal{O}(V)$ is not isomorphic to a closed subspace of $\mathcal{O}(M)$ then such an operator cannot exist. This situation occurs if $\mathcal{O}(V)$ has no continuous norm or if $\mathcal{O}(V)$ is isomorphic to a finite type power series space and $\mathcal{O}(M)$ is isomorphic to an infinite type power series space. In the later case it is known that every continuous linear mapping from $\mathcal{O}(V)$ into $\mathcal{O}(M)$ is compact [21]. On the other hand for a strictly pseudoconvex domain M in a Stein manifold and for a closed variety V of the form $V = D \cap \tilde{V}$ with $\tilde{V}$ a closed manifold in a neighborhood of $\overline{M}$, intersecting ∂M transversally the existence of a continuous extension operator was proved in [75] by using $\overline{\partial}$-methods. Another important situation where an extension operator exists is the case $M = \mathbf{C}^N$ and V an algebraic variety([34]).

In this section we take up a special case of the above mentioned question, namely the case when $\mathcal{O}(V)$ is isomorphic to a power series space. More generally we consider for a given data consisting of a Stein space M a subvariety V of M, a Stein space W for which $\mathcal{O}(W)$ is a isomorphic to a power series space and a continuous linear operator $T : \mathcal{O}(W) \to \mathcal{O}(V)$, the problem of finding a continuous linear operator $\tilde{T}$ such that the following diagram commutes:

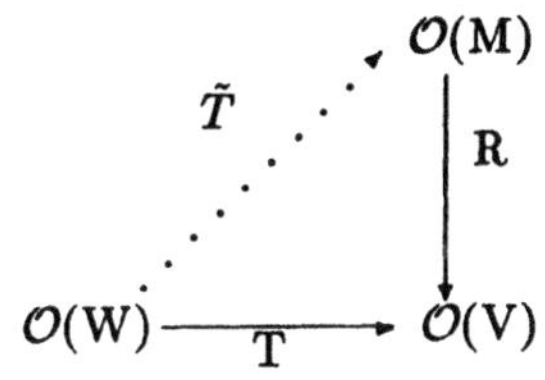

where R is the restriction operator.

The obstruction to finding $\tilde{T}$ for an arbitrary T in the above set up is due to the possible non vanishing of the first derived functor $Ext^1(\cdot,\cdot)$ of the functior Pro, in the terminology of the locally convex homological algebra as developed by Palamodov [83] (cf [115]). Indeed for a given M, V and W as above, denoting by $I(V)$ the ideal of analytic functions on M which vanish on V, the short exact sequence $O \to I(V) \to \mathcal{O}(M) \overset{R}{\to} \mathcal{O}(V) \to O$ gives rise to, a long exact sequence;

$$O \to L(\mathcal{O}(W), I(V)) \to L(\mathcal{O}(W), \mathcal{O}(M)) \to L(\mathcal{O}(W), \mathcal{O}(V)) \overset{\delta}{\to} Ext^1(\mathcal{O}(W), I(V)) \quad (1)$$
$$\to Ext^1(\mathcal{O}(W), \mathcal{O}(M)) \to Ext^1(\mathcal{O}(W), \mathcal{O}(V)) \to O.$$

as was explained in [115]. Here $L(E, F)$ denotes the space of all continuous linear operators from E into F. A connection between the structures of the Fréchet spaces E and F and the space $Ext^1(E, F)$ is given by the following powerful vanishing theorem of Vogt-Wagner[118, 1.9 Satz].

III.1 Theorem: . For two nuclear Fréchet spaces E and F, if E has the property (DN) and if F has the property (Ω) then $Ext^1(E, F) = 0$.

Since the ideal of a given variety V, $I(V)$, of a Stein space always have the property (Ω) (see §II), in view of III.1 Theorem, the above stated problem has a positive solution for *every* operator T in the case when $\mathcal{O}(W)$ is isomorphic to an infinite type power series space. In particular the condition that $\mathcal{O}(V)$ has the property (DN) characterizes among the closed varieties V of $\mathbf{C}^d$ the ones that admit a continuous linear extension operator . Combining this with the results of §I we have [14]

III.2 Corollary: Let V be a closed subvariety of some $\mathbf{C}^N$. Then a continuous linear extension operator $\mathcal{E} : \mathcal{O}(V) \to \mathcal{O}(\mathbf{C}^N)$ exists if and only if $\mathcal{O}(V)$ is isomorphic to $\mathcal{O}(\mathbf{C}^d)$ as Fréchet spaces, $d = \dim V$.

We note that sufficiency of the isomorphism $\mathcal{O}(V) \simeq \mathcal{O}(\mathbf{C}^d)$ in the above result was already stated in [75, Proposition 6.4].

Now we take up the case when $\mathcal{O}(W)$ is isomorphic to a finite type power series space. In this case $\mathcal{O}(W)$ is isomorphic to $\mathcal{O}(\Delta^d)$, $d = \dim W$ (§I), and due to the nature of the problem posed above we can, without loss of generality, take $W = \Delta^d$. Unlike the infinite type case, the functor Ext^1 will not vanish on pairs $(\mathcal{O}(\Delta^d)\ \ I(V))$ [115]. For a nuclear Fréchet space E and a subvariety V of a Stein manifold M, the space $Ext^1(E, I(V))$ can be identified with the first Cech cohomology group on M of a suitable E^*-valued coherent analytic sheaf (I^{E^*}) [8],[104],[112]. The non vanishing of Ext^1 in this case reflects the failure of the Cartan theorem (B) in the case of coherent analytic sheaves with values in a DF space [112] (cf. [8]). For Fréchet valued coherent analytic sheaves the validity of Cartan theorem (B) was established by E. Bishop [23] and Bungart [25].

Now let a Stein space M and a closed subvariaety V be given. We fix a continuous linear operator $T : \mathcal{O}(\Delta^d) \to \mathcal{O}(V)$. As in section I we will denote by ρ_T the plurisubharmonic function associated to T i.e.;

$$\rho_T(z) = \overline{\lim}_{\xi\to z}\ \overline{\lim}_{|n|} \frac{\ell n \mid T(z^n) \mid}{\mid n \mid}.$$

If there exists a $\tilde{T} : \mathcal{O}(\Delta^d) \to \mathcal{O}(M)$ such that $R \circ \tilde{T} = T$ then ρ_T is dominated by the restriction to V of a negative plurisubharmonic function on M, namely by $\rho_{\tilde{T}}$.

It turns out that this condition is also sufficient for the existence of such an operator $\tilde{T}$ [12].

III.3 Theorem: Let M be a Stein space and V a closed subvariety of M. For an continuous linear operator T from some $\mathcal{O}(\Delta^d)$ into $\mathcal{O}(V)$ the following conditions are equivalent

(i) There exists a continuous linear operator $\tilde{T} : \mathcal{O}(\Delta^d) \to \mathcal{O}(M)$ such that $R \circ \tilde{T} = T$, where R is the restriction operator from $\mathcal{O}(M)$ onto $\mathcal{O}(V)$.

(ii) There exists a negative plurisubharmonic function ρ on M such that $\rho_T \leq \rho\mid_V$ on V.

The proof of the above result uses only the structure of analytic functions on Hartog's domains and Cartan theorem (B). In particular it does not empty $\overline{\partial}$-techniques. This result in particular gives a description of the kernel of the operator δ appearing in the long exact sequence (2) which is constructed by abstract homological algebra techniques. Our next result gives a necessary and sufficient condition for this operator to be the identically zero operator [12].

III.4 Corollary: Let M be a Stein space and V a closed subvariety of M. Then the following conditions are equivalent

(i) For every Stein space W for which $\mathcal{O}(W)$ is a finite type power space and for *every* continuous linear operator $T : \mathcal{O}(W) \to \mathcal{O}(V)$ there exists a continuous linear operator $\tilde{T} : \mathcal{O}(W) \to \mathcal{O}(M)$ such that $R \circ \tilde{T} = T$.

(ii) For every negative plurisubharmonic function ρ on V there exists a negative plurisubharmonic function $\tilde{\rho}$ on M such that $\rho \le \tilde{\rho}\,|_V$.

To prove this result first one expresses a given negative function ρ on V as a Hartog's function, then using this representation one defines a continuous linear operator: $T : \mathcal{O}(\Delta) \to \mathcal{O}(V)$ and applies III.3 Theorem.

The above corollary can be used to characterize among the hyperconvex varieties V of a Stein space M the ones which admit a continuous linear extension operator $\mathcal{E} : \mathcal{O}(V) \to \mathcal{O}(M)$. Since any negative plurisubharmonic function on a Stein space is dominated by a constant multiple of a $\mathcal{P}$-measure one can reexpress the condition (ii) above using $\mathcal{P}$-measures (§1).

III.5 Corollary: Let M be a Stein space and V a hyperconvex subvariety of M. Then the following conditions are equivalent.

(i) There exists a continuous linear extension operator $\mathcal{E} : \mathcal{O}(V) \to \mathcal{O}(M)$.

(ii) There exists compact regular sets $K \subseteq V$, $S \subseteq M$ with non empty interiors and a constant $C > 0$ such that $\omega_K^V \le C\omega_S^M\,|_V$.

In the case when M itself is hyperconvex in the condition (ii) above one can employ generalized Green's functions introduced in section I, but then one must require the domination to take place off a compact subset of V.

Although we have chosen to treat the case when $\mathcal{O}(W)$ is isomorphic to an infinite type power series space by making use of some general considerations, we note that the line of reasoning sketched above can also be used in this case. Indeed the existence of an operator $\tilde{T} : \mathcal{O}(W) \to \mathcal{O}(M)$ with $R \circ \tilde{T} = T$ for any $T : \mathcal{O}(W) \to \mathcal{O}(V)$ can be deduced, in this case, from the fact that for any plurisubharmonic function ρ on V there exists a plurisubharmonic function $\tilde{\rho}$ on M such that $\rho \le \tilde{\rho}\,|_V$.

In the case when $\mathcal{O}(M)$ is isomorphic to an infinite type power series space and when W is hyperconvex, III.3 Theorem characterizes the operators T for which such a $\tilde{T}$ exists as the ones for which $\sup_{z \in V} \rho_T(z) < 0$. This family is precisely the family of all compact operators from $\mathcal{O}(W)$ into $\mathcal{O}(V)$. This can also be derived from the general extention properties of compact operators and the fact that every continuous operator from a finite type power series space into an infinite type power series space is compact .

We now look at some examples. For a smoothly bounded relatively compact domain D with C^2 boundary in a Stein manifold and a negative plurisubharmonic function ρ on D

one has a bound of the form

$$\rho(z) < C\{-d(z, \partial D)\}, \quad z \in D$$

for some $C > 0$, where $d(z, \partial D)$ is the distance of z from ∂D. [75, Lemma 3.2]. In the case when D is given by, $D = \{z : u(z) < 0\}$, for some C^2 plurisubharmonic function u defined in a neighborhood of $\overline{D}$ since $-d(\cdot, \partial D)$ is dominated by a positive constant multiple of u, we have that any negative plurisubharmonic function on D is dominated by a positive constant multiple of u. Moreover this property remains valid for submanifolds of D of the form $D \cap M'$ where M' is a closed (analytic) submanifold in a neighborhood of $\overline{D}$ which intersects ∂D transversally since in this case $D \cap M'$ can be represented as $\{z \in M' : u(z) < 0\}$ [75]. In view of the results of §I, we obtain, upon applying III.5 Corollary, a slight generalization of Theorem 4.2 of [75].

III.6 Corollary: Let M be a Stein manifold adn $D \subset\subset M$ a smoothly bounded domain in M of the form $D = \{z : u(z) < 0\}$ for some C^2, plurisubharmonic function u defined in a neighborhood of $\overline{D}$. For a complex submanifold M' of a neighborhood of $\overline{D}$ which intersects ∂D transversally there exists a continuous linear extension operator $\mathcal{E} : \mathcal{O}(D \cap M') \to \mathcal{O}(D)$.

Even if we drop the transversality condition in the above corollary we can still get some information about the class of continuous operators $T : \mathcal{O}(\Delta^d) \to \mathcal{O}(D \cap M')$ for which there exists a continuous linear $\tilde{T} : \mathcal{O}(\Delta^d) \to \mathcal{O}(D)$ such that $R \circ \tilde{T} = T$. These are precisely the operators for which $\rho_T \leq Cu$ on $D \cap M'$, for some $C > 0$. This observation can be used, for example, in constructing concrete operators for which no such $\tilde{T}$ exists. Following Mitiagin-Henkin, Example 5.3 [75] we let

$$D = \{(z, w) \in \mathbf{C}^2;\ | z |^2 + | w - 1 |^2 < 1\} \text{ and } M' = \{(z, w) \in \mathbf{C}^2;\ w = z^2\}.$$

Then the operator T defined by the assignment $T(f)(z, w) \doteq f(e^{z^3})$ for $f \in \mathcal{O}(\Delta)$ and $(z, w) \in D \cap M'$ is a continuous, linear (in fact algebraic) operator from $\mathcal{O}(\Delta)$ into $\mathcal{O}(D \cap M')$. Taking into account that in this case $\rho_T(z, w) = \ell n \mid e^{z^2} \mid$ and $u(z, w) = | z |^2 | w - 1 |^2 - 1$ an easy computation, for points near $(0, 0)$, shows the impossibility of finding a $C > 0$ such that $\rho_T \leq Cu$ on $D \cap M'$. Hence it follows that no such $\tilde{T}$, with the specifications above, exists for this operator T.

Finally we would like to briefly mention yet another instance where the obstruction to the linearization program of Mitiagin-Henkin can be expressed in terms of plurisubharmonic functions. For two domains of holomorphy D_1 and D_2 with $D_1 \cap D_2 \neq \emptyset$ and for an $f \in \mathcal{O}(D_1 \cap D_2)$, it is possible in view of Cartan theorem B to separate the "singularities" of f by writing $f = f_1 - f_2$ on $D_1 \cap D_2$ for some analytic functions $f_1 \in \mathcal{O}(D_1)$ and $f_2 \in \mathcal{O}(D)$. (Solvability of Cousin problem I [52, p.136]). The possibility of separating singularities by continuous linear operators was studied by Mitiagin-Henkin [75, Section 3]. Due to some obstructions of linear topological type, it is not always possible to separate singularities in a continuous linear fashion. For example no linear separation of singularities exists in the case $D_1 = \{z \in C^1;\ Im\,z < 1\}$, $D_2 = \{z :\ Im\,z > -1\}$ [75]. Following the line of reasoning given in §III of [75] one has:

III.7 Proposition: Let M be a Stein manifold and $D_1 \subset\subset M$ and $D_2 \subset\subset M$ be two hyperconvex domains with $D_1 \cap D_2 \not\equiv \phi$. Then the following are equivalent

(i) There exists continuous linear operators $L_1 : \mathcal{O}(D_1 \cap D_2) \to \mathcal{O}(D_1)$ and $L_2 : \mathcal{O}(D_1 \cap D_2) \to \mathcal{O}(D_2)$ such that for every $f \in \mathcal{O}(D_1 \cap D_2)$ one has $f = L_1(f)\,|_{D_1 \cap D_2} - L_2(f)\,|_{D_1 \cap D_2}$.

(ii) For every negative plurisubharmonic function ρ on $D_1 \cap D_2$, there exists negative plurisubharmonic functions ρ_1 on D_1 and ρ_2 on D_2 such that

$$\rho \leq \max\{\rho_1, \rho_2\} \text{ on } D_1 \cap D_2.$$

In view of the results in sections I and II one can consider, III.5 Corollary and III.7 Propositon as necessary and sufficient conditions for the splitting of certain short exact sequences of Fréchet spaces each of which is isomorphic to a finite type power series space. Short exact sequences of this type we also studied by Vogt [108] and Ketonen-Nyberg [55].

References

[1] L.A. Aizenberg: 'Spaces of functions analytic in (p,q)-circular domains'. *Soviet Math. Dokl.* **2** (1961), 79-82.

[2] L.A. Aizenberg: 'The general form of a continuous linear functional in spaces of functions holomorphic in convex domains of $\mathbf{C}^n$'. *Soviet Math. Dokl.* **7** (1966), 198-202.

[3] L.A. Aizenberg, B.S. Mitiagin: 'The space of functions analytic in multicircular domains'. *Sibirsk. Mat. Zh.* **1** (1960), 153-170 (Russian).

[4] L.V. Ahlfors, L. Sario: *Riemann Surfaces.* Princeton University Press 1960.

[5] A.B. Aleksandrov: 'Proper holomorphic mappings from the ball into a polydisc'. *Soviet Math. Dokl.* **33** (1986), 1-5.

[6] A. Aytuna: Analitik Fonksiyon Uzaylarinin Lineer Topolojik Yapilari Uzerine. Habilitation thesis, 1981 (Turkish).

[7] A. Aytuna: 'On Stein manifolds M for which $\mathcal{O}(M)$ has the property (DN)'. Preprint.

[8] A. Aytuna: 'On the vector valued Cousin Problem'. Preprint.

[9] A. Aytuna: 'On the linear topological structures of spaces of analytic functions'. *Doğa Turkish J. Math.* **10** (1986), 46-49.

[10] A. Aytuna: 'On Stein manifolds M for which $\mathcal{O}(M)$ is isomorphic to $\mathcal{O}(\Delta^n)$ as Fréchet spaces'. *Manuscripta Math.* **62** (1988), 297-315.

[11] A. Aytuna: 'The Fréchet space structure of global sections of certain coherent analytic sheaves'. To appear in *Math. Balkanica.*

[12] A. Aytuna: 'Extension operators for analytic functions defined on certain closed subvarieties of a Stein space'. Preprint.

[13] A. Aytuna, T. Terzioğlu: 'On certain subspaces of a nuclear power series space of finite type'. *Studia Math.* **69** (1980) 79-86.

[14] A. Aytuna, J. Krone, T. Terzioğlu: 'Complemented infinite type power series subspaces of nuclear Fréchet spaces'. *Math. Ann.* **283** (1989), 193-202.

[15] A. Aytuna, J. Krone, T. Terzioğlu: 'Imbedding of power series spaces and spaces of analytic functions'. Preprint.

[16] K.I. Babenko: 'On the entropy of a class of analytic functions'. *Nauchn. Dokl. Vyssh. Shkol. Ser. Fiz-Mat. Nauk* **2** (1958), 9-13 (Russian).

[17] E. Bedford: 'Extremal plurisubharmonic functions for certain domains in $\mathbf{C}^2$'. *Indiana University Mathematics Journal* **28** (1979), 613-626.

[18] E. Bedford, D. Burns: 'Holomorphic mapping of annuli in $\mathbf{C}^n$ and the associated extremal function'. *Ann. Scuola Norm. Sup. Pisa* S.IV, Vol.VI, No.3 (1979), 381-414.

[19] E. Bedford, B.A. Taylor: 'The Dirichlet problem for a complex Monge-Ampère equation'. *Invent. Math.* **37** (1976), 1-44.

[20] E. Bedford, B.A. Taylor: 'A new capacity for plurisubharmonic functions'. *Acta Mathematica,* Vol.149 (1982), 1-4.

[21] Cz. Bessaga: 'Some remarks on Dragilev's theorem'. *Studia Math.* **31** (1968), 307-318.

[22] Cz. Bessaga, A. Pelczynski, S. Rolewicz: 'On diametral approximative dimension and linear homogeneity of F-spaces'. *Bull. Acad. Polon. Sci. Sér. Math.* **9** (1961), 307-318.

[23] E. Bishop: 'Analytic functions with values in a Fréchet spaces', *Pacif. J. Math.* **12** (1962), 1177-1192.

[24] H. Bremermann: 'On a generalized Dirichlet problem for plurisubharmonic functions and pseudoconvex domains. Characterization of Silov boundaries'. *Trans. Amer. Math. Soc.* **91** (1959), 246-276.

[25] L. Bungart: 'Holomorphic functions with values in locally convex spaces and applications to integral formulas'. *Trans. Amer. Math. Soc.* **111** (1964), 317-344.

[26] D. Burns: 'Curvature of Monge-Ampére foliations and parabolic manifolds'. *Annals of Mathematics*, **115** (1982), 349-373.

[27] E. Calabi: 'An extension of E. Hopf's maximum principle with an application to Riemannian geometry'. *Duke Math. J.* **25** (1957), 45-56.

[28] H. Cartan: 'Idéaux et modules de fonctions analytiques de variables complexes'. *Bull. Soc. Math. France* **78** (1950), 29-64.

[29] S.Y. Cheng, S.T. Yau: 'Differential equations on Riemannian manifolds and their geometric applications'. *Communications on pure and applied Math.* **28** (1975), 333-354.

[30] M. Cornalba, P. Griffiths: 'Analytic cycles and vector bundles on non-compact algebraic varieties'. *Inv. Math.* **28** (1975), 1-106.

[31] K. Diederich, J.E. Fornaess: 'Pseudoconvex domains, bounded strictly plurisubharmonic exhaustion functions'. *Inv. Math.* **39** (1977), 129-141.

[32] J.P. Demailly: 'Measure de Monge-Ampére et measures pluriharmoniques'. *Math. Z.* **194** (1987), 519-564.

[33] P.B. Djakov, B.S. Mitiagin: 'Modified construction of a nuclear Fréchet space without basis', *J. Funct. Anal.* **23** (1976), 415-104.

[34] P.B. Djakov, B.S. Mitiagin: 'The structure of polynomial ideals in the algebra of entire functions'. *Studia Math.* **68** (1980), 84-104.

[35] M.M. Dragilev: 'The canonical form of a basis in a space of analytic functions'. *Uspehi Mat. Nauk.* **15** (1960), 181-188 (Russian).

[36] M.M. Dragilev: 'Extendable bases of analytic functions'. *Amer. Math. Soc. Translations* Ser.2, **43** (1964).

[37] E. Dubinsky: *The Structure of Nuclear Fréchet spaces.* Lecture Notes in Math. 720, 1979.

[38] J.L. Ermine: 'Conjecture de Serre et espaces hyperconvexes'. Séminaire Norguet in; Lecture Notes in Math. 670, 1978, 124-139.

[39] V.D. Erokhin: 'Best linear approximation of functions analytically continuable from a continuum to a given region'. *Russian Math. Surveys.* **23** (1968), 94-134.

[40] J.E. Fornaess, R. Narasimhan: 'The Levi problem on complex spaces with singularities'. *Math. Ann.* **248** (1980), 47-72.

[41] O. Forster: 'Zur Theorie der Steinschen Algebren und Moduln'. *Math. Zeitschr.* **97** (1967), 376-405.

[42] O. Forster, K. Knorr: 'Ein Beweis der Grauertschen Bildgarbensatzes nach Ideen von B. Malgrange'. *Manuscripta Math.* **5** (1971), 19-44.

[43] H. Grauert: 'Charakterisierung der Holomorphiegebiete durch die vollständige Kählersche Metrik'. *Math. Ann.* **131** (1956), 38-75.

[44] H. Grauert, R. Remmert: 'Plurisubharmonische Funktionen in komplexen Räumen'. *Math. Zeitschr.* **65** (1956), 175-194.

[45] H. Grauert, R. Remmert: *Coherent Analytic Sheaves.* Springer-Verlag, Berlin-Heidelberg-New York, Tokyo 1984.

[46] R.E. Greene, H. Wu: *Function Theory on Manifolds which Possess a Pole*, Lecture Notes in Math. Vol.699, 1979.

[47] P.A. Griffiths, J. King: 'Nevanlinna theory and holomorphic mappings between algebraic varieties', *Acta. Math.* **130** (1973), 145-220.

[48] A. Grothendieck: *Produits Tensoriels Topologiques et Espaces Nucléaires*, Mem. Amer. Math. Soc. 16, 1953.

[49] R. Gunning, H. Rossi: *Analytic Functions of Several Complex Variables.* Englewood Cliffs, 1965.

[50] L.L. Helms: *Introduction to Potential Theory.* John Wiley and Sons, 1969.

[51] A. Huber: 'On subharmonic functions and differential geometry in the large'. *Comment. Math. Helv.* **32** (1957), 13-72.

[52] L. Hörmander: *An Introduction to Complex Analysis in Several Variables.* North-Holland Publishing Company, 1973.

[53] K. Karp: 'Subharmonic functions on real and complex manifolds'. *Math. Z.* **179** (1982), 535-554.

[54] N. Kerzman, I.P. Rosay: 'Fonctions plurisubharmoniques d'exhaustions bornées et domaines taut'. *Math. Ann.* **257** (1981), 171-184.

[55] T. Ketonen, K. Nyberg: 'Twisted sums of nuclear Fréchet spaces'. *Ann. Acad. Sci. Fenn. Ser A. I. Math.* **7** (1982), 323-335.

[56] N.V. Khue: 'On the cohomology of sheaves $S \in L$'. *Studia Math.* **72** (1982), 183-197.

[57] M. Klimek: 'Extremal plurisubharmonic functions and L-regular sets in $\mathbf{C}^n$'. *Proc. Royal Irish Acad.* Vol.82A (1982), 217-230.

[58] M. Klimek: 'Extremal plurisubharmonic functions and invariant pseudodistances'. *Bull. Soc. Math. France.* **113** (1985), 123-142.

[59] A.N. Kolmogoroff: 'Über die beste Annäherung von Funktionen einer gegebenen Funktionenklasse'. *Ann. of Math.* (2) **37** (1936), 107-110.

[60] A.N. Kolmogoroff, V.M. Tikhomirov: 'ε-entropy and ε-capacity of sets in function spaces'. *Amer. Math. Soc. Transl.* (2), **17** (1961), 277-364.

[61] G. Köthe: *Topologische Lineare Räume.* Springer-Verlag, 1960.

[62] B. Kramm: 'A duality theorem for nuclear function algebras in Aspects of Mathematics and its applications'. (ed. J.A. Barraso) *Elsevier Science Publishers* B.V. (1986), 495-531.

[63] S.G. Krein; 'On an interpolation theorem in operator theory'. *Soviet Math. Dokl.* **1** (1960), 61-64.

[64] J. Krone: 'Existence of Bases and the Dual Splitting Relation for Fréchet spaces'. to appear in *Studia Math.*

[65] J. Krone: 'On projections in power series spaces and the existence of bases'. to appear in *Proc. Amer. Math. Soc.*

[66] J. Krone: 'On Pelczynski's problem'. This volume.

[67] Y. Kusunoki: 'On the generalized Dirichlet problem for plurisubharmonic functions'. *J. Math. Kyoto Univ.* **4-1** (1964), 123-147.

[68] P. Lelong: *Plurisubharmonic Functions and Positive Differential Forms.* New York, Gordon and Breach 1969.

[69] P. Lelong: 'Notions capacitaires et fonctions de Green pluricomplexes dans les espaces de Banach'. *C.R. Acad. Sci. Paries, t.305 Série I*(1987), 71-76.

[70] J.L. Lions: 'Espaces intermédiaires entre espaces Hilbertiens et applications'. *Bull. Math. Soc. Sci. Math. Phys. R.P. Romaine*, **50** (1958), 419-432.

[71] A. Martineu: 'Indicatries des fonctionelles analytiques et inversion de la transformée de Fourier-Borel par la transformation de Laplace'. *C.R. Acad. Sci. Paris t.255, Série I* (1962), 1845-1847, 2888-2890.

[72] A. Martineu: 'Sur la topologie des espaces de fonctions holomorphes'. *Math. Ann.* **163** (1966), 62-88.

[73] B.S. Mitiagin: 'Approximate dimension and bases in nuclear spaces'. *Russian Math. Surveys*, **16** (1961), 59-127.

[74] B.S. Mitiagin: 'The equivalence of bases in Hilbert scales'. *Studia Math.* **37** (1970), 111-137 (Russian).

[75] B.S. Mitiagin, G.M. Henkin: 'Linear problems of complex analysis'. *Russian Math. Surveys* **26** (1971), 99-164.

[76] R. Narasimhan: 'The Levi problem for complex spaces I'. *Math. Ann.* **142** (1961), 335-365.

[77] R. Narasimhan: *Several Complex Variables.* The University of Chicago Press, Chicago 1971.

[78] R. Nevanlinna, V. Paatero: *Introduction to Complex Analysis.* Addison-Wesley Publishing Company 1969.

[79] Z. Nurlu, T. Terzioğlu: 'Consequences of existence of a non-compact operator between Köthe spaces'. *Manuscripta Math.* **47** (1984), 1-12.

[80] R.S. Palais (ed.): *Seminar on the Atiyah-Singer Index Theorem.* Ann. of Math. Studies 57, Princeton Univ. Press, Princeton 1965.

[81] V.P. Palamodov: 'Coherent analytic sheaves', *Tzdat. "Nauka", Moscow* (1966), 246-247 (Russian).

[82] V.P. Palamodov: 'Differential operators in coherent analytic sheaves'. *Math USSR Sbornik* Vol.6, No.3 (1968), 365-391.

[83] V.P. Palamodov: 'Homological methods in the theory of locally convex spaces', *Russian Math. Surveys* **26:1** (1971), 1-64.

[84] A. Pelczynski: 'Projections in certain Banach spaces'. *Studia Math.* **19** (1960), 209-228.

[85] A. Pietsch: *Nuclear Locally Convex Spaces.* Springer- Verlag 1972.

[86] R.M. Range: 'A remark on bounded strictly plurisubharmonic exhaustion functions'. *P.A.M.S.* Vol.81, No.2 (1981), 220-222.

[87] S. Rolewicz: 'On spaces of holomorphic functions'. *Studia Math.* **21** (1961-62), 135-160.

[88] W. Rudin: *Function Theory in Polydiscs.* W.A. Benjamin, New York - Amsterdam 1969.

[89] A. Sadullaev: 'Plurisubharmonic measures and capacities on complex manifolds'. *Russian Math. Surveys* **36:4** (1981), 61-119.

[90] A. Sadullaev: 'An estimate for polynomials on analytic sets'. *Math. USSR Izvestiya* Vol.20 (1983), 493-502.

[91] L. Sario, M. Nakai: *Classification Theory of Riemann Surfaces.* Grundlehren 164 Springer-Verlag 1970.

[92] L. Sario, M. Nakai, C. Wang, L.O. Chung: *Classification Theory of Riemannian Manifolds.* Lecture Notes in Mathematics 605 Springer-Verlag 1977.

[93] M. Schottenloher: 'Embedding of Stein spaces into sequence spaces'. *Manuscripta Math.* **39** (1982), 15-29.

[94] N. Sibony: 'Prolongement des fonctions holomorphes bornées et métrique de Carathéodory'. *Inventiones Math.* **209** (1975), 205-230.

[95] J. Siciak: 'Extremal plurisubharmonic functions in $\mathbf{C}^n$'. *Ann. Polon. Math.* **319** (1981), 175-211.

[96] Y.T. Siu: 'Every Stein subvariety admits a Stein neighborhood'. *Invent. Math.* **38** (1976), 89-100.

[97] J.L. Stéhle: 'Fonctions plurisubharmoniques et convexité holomorphe de certains fibrés analytiques'. Séminaire Lelong 73-74, Lecture Notes in Mathematics 474 1975, 155-180.

[98] W. Stoll: 'The characterization of strictly parabolic manifolds'. *Ann. Sc. Nor. Sup. Pisa*, S.IV, Vol.VII, No:11, (1980), 87-154.

[99] T. Terzioğlu: 'Die diametrale dimension von lokalkonvexen Räumen'. *Collect. Math.* **20** (1969), 49-99.

[100] T. Terzioğlu: 'Some invariants of Fréchet spaces and imbeddings of smooth sequence spaces'. This volume.

[101] T. Terzioğlu, M. Yurdakul: 'Restrictions of unbounded continuous linear operators on Fréchet spaces'. *Arch. Math.* **46** (1986), 547-550.

[102] N.T. Van: 'Bases de Schauder dans certains espaces de fonctions holomorphes'. *Ann. Inst. Fourier, Grenoble* **22** (2) (1972), 169-253.

[103] V.S. Vladimirov: *Methods of the Theory of Functions of many Complex Variables.* M.I.T. Press, 1966.

[104] D. Vogt: 'Vektorwertige Distributionen als Randverteilungen holomorpher Funktionen'. *Manuscripta Math.* **17** (1975), 267-290.

[105] D. Vogt: 'Charakterisierung der Unterräume von s'. *Math. Z.* **155** (1977), 109-117.

[106] D. Vogt: Über die Existenz von Ausdehnungsoperatoren für holomorphe Funktionen auf analytischen Mengen in C^n. Preprint.

[107] D. Vogt: 'Charakterisierung der Unterräume eines nuklearen stabilen Potenzreihenraumes von endlichem Typ'. *Studia Math.* **71** (1982), 251-270.

[108] D. Vogt: 'Eine Charakterisierung der Potenzreihenräume von endlichem Typ und ihre Folgerungen', *Manuscripta Math.* **37** (1982), 269-301.

[109] D. Vogt: 'Ein Isomorphiesatz für Potenzreihenräume'. *Arch. Math.* **38** (1982), 540-548.

[110] D. Vogt: 'Frécheträume zwischen denen jede stetige lineare Abbildung beschränkt ist'. *J. Reine Angew. Math.* **345** (1983), 182-200.

[111] D. Vogt: 'Sequence Space Representations of Spaces of Test Functions and Distributions'. Advances in Functional Analysis, Holomorphy and Approximation Theory. (ed. G.I. Zapata). New York - Basel 1983, 405-443.

[112] D. Vogt: 'On the solvability of $P(D)f = g$ for vector functions'. *RIMS Kokyovoku* **508** (1983), 168-182.

[113] D. Vogt: 'Structure theory of nuclear stable power series spaces'. Proceedings of the Conference on Functional Analysis/Banach Space Geometry (ed. W. Ruess) 1983 Essen 319-329.

[114] D. Vogt: 'Some results on continuous linear maps between Fréchet spaces in Functional Analysis: Surveys and Recent Results III'. (ed. K.D. Bierstedt - B. Fuchssteiner). *North-Holland Math. Studies* **90** (1984), 349-381.

[115] D. Vogt: 'On the functors $Ext^1(E,F)$ for Fréchet spaces'. *Studia Math.* **85** (1987), 163-197.

[116] D. Vogt: 'Power series representation of nuclear Fréchet spaces'. Preprint.

[117] D. Vogt, M.J. Wagner: 'Charakterisierung der Quotientenräume von s und eine Vermutung von Martineau'. *Studia Math.* **67** (1980), 225-240.

[118] D. Vogt, M.J. Wagner: 'Charakterisierung der Unterräume und Quotientenräume der nuklearen stabilen Potenzreihenräume von unendlichem Typ'. *Studia Math.* **70** (1981), 65-80.

[119] H. Wu: 'An elementary method in the study of nonnegative curvature'. *Acta. Math.* **142** (1979), 57-78.

[120] H. Wu: 'Liouville theorems'. *Complex Analysis III.* (ed. C.A. Berenstein) Lecture Notes in Mathematics 1277, Springer-Verlag 1987, 331-349.

[121] V.P. Zaharyuta: 'Continuable bases in spaces of analytic functions of one and of several variables'. *Siberian Math. J.* **8** (1967), 204-216.

[122] V.P. Zaharyuta: 'Spaces of functions of one variable, analytic in open sets and on compacta'. *Math. USSR Sbornik* Vol.II (1970), 75-88.

[123] V.P. Zaharyuta: 'Isomorphism of spaces of holomorphic functions of several complex variables'. *Func. Anal. and Appl.* **5** (1972) 326-328.

[124] V.P. Zaharyuta: 'Extremal plurisubharmonic functions, Hilbert scales, and the isomorphism of spaces of analytic functions of several variables, I, II'. *Theor. Funkcii Funkcional Anal. i. Prilozen. Vyp.* **19** (1974), 133-157, **21** (1974), 65-83 (Russian).

[125] V.P. Zaharyuta: 'Spaces of analytic functions on algebraic varieties in $\mathbf{C}^n$'. *Izv. Severo-Kavkaz. Nauchn. Tsentia Vyssh. Shkoly Ser. Estestv. Nauk* **4** (1977), 52-55.

[126] V.P. Zaharyuta: 'Generalized Mitiagin invariants and a continuum of pairwise non-isomorphic spaces of anaytic functions'. *Func. Anal. and Appl.* **11** (1978), 182-188.

[127] V.P. Zaharyuta: 'Isomorphisms of spaces of analytic functions'. *Sov. Math. Dokl.* **22** (1980), 631-634.

[128] N.M. Zobin, B.S. Mitiagin: 'Examples of nuclear linear metric spaces without a basis'. *Funct. Anal. and Appl.* **8** (1975), 303-313.

MONOMIAL EXPANSIONS IN INFINITE DIMENSIONAL HOLOMORPHY

by

Seán Dineen

(University College Dublin)

Let E and F denote locally convex spaces over $\mathbf{C}$ and let U denote an open subset of E. A function $f : U \to F$ is called holomorphic if

(a) it is continuous,

(b) for each $a \in U$, $v \in E$ and $\phi \in F'$ the mapping $\lambda \in \mathbf{C} \to \phi \circ f(a + \lambda v)$ is holomorphic as a function of one complex variable on its domain of definition.

For the sake of clarity I will only consider $\mathbf{C}$–valued holomorphic functions in this lecture. We let $H(U)$ denote the set of all $\mathbf{C}$– valued holomorphic functions on U. Mappings which only satisfy condition (b) are called Gateâux holomorphic functions. The study of holomorphic mappings between locally convex spaces is known as infinite dimensional holomorphy. This area has grown rapidly in the last twenty years and it is not now possible to survey even briefly the different aspects of the subject. We use [5] as a general reference.

My aim in this lecture is to focus attention on one tool – the monomial expansion – and to explain some aspects of the topological theory using this expansion. Before developing the main theme of the lecture I will make some general remarks and give a number of elementary examples.

We begin with a few observations on the nature of holomorphy vis a vis linear functional analysis.

(i) In linear functional analysis one adds while in infinite dimensional holomorphy one adds and multiplies. Indeed the closure, in a suitable topology, of the algebra generated by the linear functionals is frequently the space of all holomorphic functions. This point of view emphasises elements of E' as functions on E rather than points or vectors in E'. (see application 1).

T. Terzioğlu (ed.), Advances in the Theory of Fréchet Spaces, 155–171.

(ii) The linear theory is local while holomorphy is a global theory. By this we mean that the behaviour of a linear function near a point determines in a transparent fashion the behaviour near any other point while in holomorphy the behaviour near a point, although globally determining (the identity principle), does not lead to an easily predictable set of global phenomena. This leads to results like Zorn's theorem (discussed below) and concepts like domain of holomorphy which do not appear in the linear theory.

(iii) Two of the more natural topologies in infinite dimensional holomorphy, τ_ω and τ_δ (see below), are rarely discussed in the linear theory while the topology of uniform convergence on bounded sets cannot be defined in many infinite dimensional settings.

(iv) In the context of Fréchet spaces we find that the Montel property (see application 3) is the critical property in moving from nuclear spaces to Banach spaces.

Let $\Delta : E \longrightarrow E^n := \underbrace{E \times \cdots \times E}_{n \text{ times}}$ denote the diagonal mapping i.e. $\Delta(x) = \underbrace{(x, \ldots, x)}_{n \text{ times}}$. An n-homogeneous polynomial on E, P, is a mapping of the form $L \circ \Delta$ where L is an n-linear mapping on E^n. Let $\mathcal{P}(^nE)$ denote the vector space of all continuous n-homogeneous polynomials on E. A polynomial is a finite sum of homogeneous polynomials.

Example 1. If $\phi \in E'$ (the continuous linear forms on E) then $\phi^n \in \mathcal{P}(^nE)$. The elements of the algebra generated by ϕ^n, $n \in N$ and $\phi \in E'$ are called polynomials of finite type.

Example 2. Let E denote a Banach space and let μ denote a regular countably additive Borel measure of bounded variation on U (the closed unit ball of E' endowed with the weak* topology). The mapping

$$x \in E \longrightarrow \int_U \phi(x)^m \, d\mu(\phi)$$

defines an element of $\mathcal{P}(^nE)$. Polynomials of this kind are called polynomials of integral type.

Example 3. Let E denote a locally convex space with basis $(e_n)_n$. If $m \in N^{(N)}$, i.e. $m = (m_1, m_2, m_3, \ldots, m_n, \ldots)$ and $m_i = 0$ for all i sufficiently large, we denote by z^m the mapping $\sum_{n=1}^{\infty} z_n e_n \longrightarrow z^m := z_1^{m_1} z_2^{m_2} \ldots z_n^{m_n} \ldots$ where we let $a^0 = 1$ for any real number a. Thus z^m is a product of coordinate evaluations relative to a given basis and defines an element of $\mathcal{P}(^{|m|}E)$ where $|m| = \sum_i |m_i|$. We call the functions z^m, $m \in N^{(N)}$, monomials (relative to the basis $(e_n)_n$).

Using polynomials we obtain the familiar Taylor series expansion of holomorphic functions. A mapping $f : U \subset E \to \mathbf{C}$ is holomorphic if and only if for each $\xi \in U$ there exists a sequence $(P_{n,\xi})_{n=0}^{\infty}$, $P_{n,\xi} \in \mathcal{P}(^n E)$ for all n, such that

$$f(\xi + z) = \sum_{n=0}^{\infty} P_{n,\xi}(z) \qquad (*)$$

for all z in some neighbourhood of the origin and the series converges uniformly to f on some neighbourhood of the origin. We call $(*)$ the Taylor series expansion of f at ξ. By the one dimensional Cauchy integral formula

$$P_{n,\xi}(z) = \frac{1}{2\pi i} \int_{|\lambda|=\delta} \frac{f(\xi + \lambda z)}{\lambda^{n+1}}\, d\lambda$$

for all $\xi \in U$, $n \in \mathbf{N}$, $z \in E$ and $\delta = \delta(\xi)$ sufficiently small. Hence the Taylor series expansion can be related to partial derivatives of f and we have

$$P_{n,\xi}(z) = \frac{1}{n!} \frac{\partial^n f}{\partial z^n}(\xi).$$

The natural domain of convergence of $(*)$ is a balanced subset of E and when U itself is balanced the Taylor series expansion at the origin converges to the function at all points of U.

Example 4. E a Fréchet space, $(\phi_n)_{n=1}^{\infty}$ a sequence in E' then $f := \sum_{n=1}^{\infty} \phi_n^n \in H(E)$ if and only if $\phi_n \to 0$ as $n \to \infty$ uniformly on compact subsets of E. In this case $\sum_{n=1}^{\infty} \phi_n^n$ is the Taylor series expansion of f at the origin.

If E is a locally convex space with basis $(e_n)_n$ and $f \in H(E)$ then for each $m \in N^{(N)}$ we define the monomial coefficients of f, $(a_m)_{m \in N^{(N)}}$, as in the finite dimensional case, by

$$a_m = \frac{1}{(2\pi i)^r} \int \overset{\dots}{{}_T} \int \frac{f(\eta_1, \dots, \eta_r, 0 \dots)}{\eta_1^{m_1+1} \eta_2^{m_2+2} \dots \eta_r^{m_r+1}}\, d\eta_1 \dots d\eta_r = \frac{\partial^{|m|} f}{\partial z^m}(0)$$

where r is chosen so that $m_i = 0$ for $i > r$,

$$T = \left\{ \sum_{i=1}^{r} \eta_i e_i;\ |\eta_i| = \alpha_i > 0 \text{ for } i = 1, \dots, r \right\}$$

and $d\eta_1 \dots d\eta_r$ is Lebesgue measure on T. The coefficients $(a_m)_{m \in N^{(N)}}$ do not depend on r or on the positive scalars α_i, $i = 1, \dots, r$. Thus we have a monomial expansion of f

$$\sum_{m \in N^{(N)}} a_m z^m \qquad (**).$$

The question now is how to intrepret, in terms of convergence, this monomial expansion. The finite dimensional theory implies uniform convergence on compact sets of the domain which are contained in the span of a finite number of basis elements. As it stands the sum in (**) is unordered and, consequently, we are obliged to consider, at least initially, either unconditional convergence or absolute convergence. The natural domain of convergence of a monomial expansion is a polydisc. Even with these restrictions the situation is still unclear as the following example shows.

Example 5. If H is a Hilbert space with orthonormal basis $(e_n)_n$ then the mapping

$$P\left(\sum_{n=1}^{\infty} z_n e_n\right) = \sum_{n=1}^{\infty} z_n^2$$

defines a continuous 2–homogeneous polynomial on H and hence a holomorphic function. The monomial expansion of P is $\sum_{n=1}^{\infty} z_n^2$ but this series does not converge uniformly to P on any neighbourhood of the origin.

Thus to consider monomial expansions we must consider other kinds of convergence and topologies on spaces of holomorphic functions.

We now look at the three main topologies that have been considered on $H(U)$ in recent years. First, we have the compact open topology τ_0. Next, a semi–norm p on $H(U)$ is said to be ported by a compact subset K of U if for each open subset V of U, $K \subset V$, there exists $C(V) > 0$ such that

$$p(f) \leq C(V)\,\|f\|_V$$

for all f in $H(U)$. The topology generated by all semi–norms ported by compact subsets of U is called the τ_w topology. A semi–norm p on $H(U)$ is called τ_δ continuous if for each increasing countable open cover of U, $(V_n)_{n=1}^{\infty}$, there exists a positive integer N and $C > 0$ such that $p(f) \leq C\|f\|_{V_N}$ for each f in $H(U)$. The τ_0, τ_w and τ_δ topologies were originally motivated by considerations in point set topology, several complex variables theory and functional analysis respectively. We always have $\tau_0 \leq \tau_w \leq \tau_\delta$ and a good deal of research on the topological (or functional analytic) aspects of infinite dimensional holomorphy has been devoted to finding conditions under which some or all of these topologies coincide. It is not difficult to show that the Taylor series of a holomorphic function on a balanced domain converges to the function in each of the above topologies. The same is not true for the monomial expansion (this follows for τ_w and τ_δ by considering example 5). So to consider monomial expansions we consider restricting the class of spaces and we consider the weakest of the above topologies – the τ_0 topology.

Let E denote a Fréchet nuclear space with (Schauder) basis. Then E is a Köthe sequence space i.e. $E \approx \Lambda(P)$ where P is a sequence of weights $(w_m)_m$ which, by the Grothendieck–Pietsch criterion, can be chosen so that $\frac{w_m}{w_{m+1}} \in \ell_1$ i.e. if $w_m = (w_{n,m})_{n=1}^{\infty}$ then $\sum_n \frac{w_{n,m}}{w_{n,m+1}} < \infty$. Then $E = \left\{(z_n)_n;\ \sum_n |z_n| w_{n,m} < \infty \text{ all } m\right\}$ and the topology of E is generated by the sequence of semi-norms

$$\left\|(z_n)_n\right\|_m = \sum_n |z_n| w_{n,m}.$$

Now let V denote an open polydisc in E i.e. there exists a sequence of non-negative numbers $(\beta_n)_n$ such that

$$V = \{(z_n)_n \in E;\ \sup_n |z_n \beta_n| < 1\}$$

and $(\beta_n)_n$ is a continuous weight on E. If K denotes a compact subset of V then there exists a sequence of positive numbers $(\alpha_n)_n$ such that

$$K \subset \widetilde{K} := \{(z_n)_n \in E;\ \sup_n |z_n \alpha_n| \leq 1\}$$

and, moreover, $\widetilde{K}$ is a compact polydisc in V. Using the Grothendieck–Pietsch criterion we can find a sequence of positive numbers, $(\delta_n)_n$, such that $\delta_n > 1$ all n, $\sum_{n=1}^{\infty} \frac{1}{\delta_n} < \infty$ and $\delta\widetilde{K} = \{(\delta_n z_n)_n;\ (z_n)_n \in \widetilde{K}\}$ is also a compact subset of V. Let $f \in H(V)$ have monomial expansion $\sum_{m \in N^{(N)}} a_m z^m$. By the Cauchy integral formula in several variables

$$\|a_m z^m\|_K \leq \|a_m z^m\|_{\widetilde{K}} = \frac{1}{\delta^m}\|a_m z^m\|_{\delta\widetilde{K}} \leq \frac{1}{\delta^m}\|f\|_{\delta\widetilde{K}}.$$

Hence

$$\begin{aligned} \sum_{m \in N^{(N)}} \|a_m z^m\|_K &\leq \left(\sum_{m \in N^{(N)}} \frac{1}{\delta^m}\right) \|f\|_{\delta\widetilde{K}} \\ &= \|f\|_{\delta\widetilde{K}} \prod_{n=1}^{\infty} \frac{1}{1 - \frac{1}{\delta_n}} := C\|f\|_{\delta\widetilde{K}}. \end{aligned} \qquad (\dagger)$$

By the same argument

$$\left\|f - \sum_{m \in J} a_m z^m\right\|_K \leq \|f\|_{\delta\widetilde{K}} \sum_{m \in N^{(N)} \setminus J} \left(\frac{1}{\delta}\right)^m$$

for any finite subset J of $N^{(N)}$ and hence we deduce that the monomial expansion of a holomorphic function on an open polydisc converges absolutely to the function in the τ_0 topology. Furthermore (†) shows that the monomials form an absolute basis for $(H(V), \tau_0)$. The same type of argument applies to the τ_w and τ_δ topologies and to other classes of nuclear spaces e.g. DFN spaces with basis.

The two expansions we have discussed – the Taylor series expansion and the monomial expansion – are complementary. The Taylor series applies to more general situations while the monomial expansion gives more precise estimates and a combination of both can often lead to stronger results.

Our first application compares topologies. In this case the Taylor series approach leads to the stronger result while the monomial approach gives us additional information.

Application 1. [3, 6] Let V denote a polydisc in the Fréchet nuclear space E with basis. If $T \in (H(V), \tau_0)'$ then we have

$$T\left(\sum_{m \in N^{(N)}} a_m z^m\right) = \sum_{m \in N^{(N)}} a_m b_m \qquad (***)$$

where $b_m = T(z^m)$ for all $m \in N^{(N)}$. Now E'_β is a DFN space with basis and so it is natural to consider the expansion $\sum_{m \in N^{(N)}} b_m w^m$ where $w = (w_n)_n \in E'_\beta$. We call $V^M = \{(w_n)_n \in E'_\beta;\ \sup_n |w_n z_n| \leq 1 \text{ all } (z_n)_n \in V\}$ the multiplicative polar of V. The set V^M is a compact polydisc in E'_β. We let $H(V^M)$ denote the set of all functions which are defined and holomorphic on some neighbourhood of V^M. This is the space of holomorphic germs on V^M and it is endowed with a locally convex space structure by taking the inductive limit of the bounded holomorphic functions on the different neighbourhoods of V^M. The correspondence given in $(***)$ is then a linear topological isomorphism between the locally convex spaces $(H(V), \tau_0)'_\beta$ and $H(V^M)$. An analysis of the same kind shows that $(H(V), \tau_w)'_\beta$ is linearly isomorphic to $H_{HY}(V^M)$ (the hypocontinuous germs on V^M). Now it is not difficult to show that if $(e_n)_n$ is an absolute basis for (E, τ_1) and (E, τ_2) then $\tau_1 = \tau_2$ if and only if $(E, \tau_1)' = (E, \tau_2)'$. Applying this result to the monomials as absolute basis for τ_0 and τ_w we get that $\tau_0 = \tau_w$ on $H(V)$, V an arbitrary polydisc in E, if and only if $H(W) = H_{HY}(W)$ for every polydisc W in E'_β. Since DFN spaces are k–spaces this latter condition is always satisfied and so $\tau_0 = \tau_w$ on $H(V)$ for any open polydisc V in E.

The basis requirement in the above result has been removed by Meise, Mujica has extended the result to Schwartz spaces with basis and for recent developments on Fréchet Montel spaces we refer to [1, 2] and to the article of Ansemil in these proceedings.

The same type of approach is used to show that $\tau_0 = \tau_\delta$ on $H(E)$, E Fréchet nuclear space, if and only if E is a DN space ([5, 6]).

Application 2. We have seen that nuclearity leads to an absolute monomial expansion and it is natural to ask if nuclearity is essential for this to happen. In view of the well known connections between absolute convergence and nuclearity we should not be surprised if this turns out to be the case. We now consider this problem [4].

If the monomials form an absolute basis for $\big(H(E), \tau_0\big)$, E a Fréchet space with basis, then E'_{τ_0} will have an absolute basis and so we consider one of the simplest non–nuclear cases in which this can happen i.e. we let E denote a Fréchet Montel space with basis $(e_n)_n$ and we suppose E'_β $(= E'_{\tau_0})$ has an absolute basis i.e. $E'_\beta = \Lambda(W)$. In this case E has a fundamental system of compact sets consisting of compact polydiscs. Let $K = \left\{\sum_n z_n e_n \in E;\ \sup_n |z_n| a_n \leq 1\right\}$ denote a compact polydisc in E. If $\big(H(E), \tau_0\big)$ has an absolute basis then there exists $C > 0$ and a compact polydisc

$$\widetilde{K} = \left\{\sum_n z_n e_n \in E;\ \sup_n |z_n| b_n \leq 1\right\}$$

such that

$$\sum_{m \in J} |\alpha_m| \, \|z^m\|_K \leq C \sup_{z \in \widetilde{K}} \left|\sum_J \alpha_m z^m\right| \tag{1}$$

for any set of scalars $(\alpha_m)_{m \in N^{(N)}}$ and any finite subset J of $N^{(N)}$.

By duality it suffices to show that E'_β is nuclear and, because of the polarity between compact subsets of E and neighbourhoods of 0 in E'_β, we can suppose that $(a_n)_n$ and $(b_n)_n$ belong to W. To show that E'_β is nuclear we may use the Grothendieck–Pietsch criterion if we can show that there exists $\delta > 0$ such that for every $(a_n)_n$ in W there exists a $(b_n)_n$ in W satisfying (1) with

$$\sum_n \left(\frac{a_n}{b_n}\right)^\delta < \infty.$$

A rephrasing of (1) says that

$$\sum_{m\in J} |a_m| \, \|z^m\|_{\{|z_n|\le r_n\}} \le C \sup\left\{\left|\sum_{m\in J} a_m z^m\right| ; \ |z_n|\le 1\right\} \tag{2}$$

where $r_n := \frac{a_n}{b_n}$, J a finite subset of $N^{(N)}$, and we require $\sum_n r_n^\delta < \infty$ for some positive δ.

An old inequality of H. Bohr says that if $f(z) = \sum_{n=0}^{\infty} a_n z^n$ is a holomorphic function on the unit disc $\mathbf{D}$ in $\mathbf{C}$ and $\|f\|_{\mathbf{D}} \le 1$ then $\sum_{n=0}^{\infty} |a_n| r^n \le 1$ whenever $r \le \frac{1}{3}$ and this result is best possible. This implies that $r_n \le \frac{1}{3}$ for all n but this is not sufficient.

We consider (1) again but now only look at a_m with $|m| = 2$. In this case (1) implies

$$\sum_{j,k=1}^{n} \frac{a_j}{b_j}\frac{a_k}{b_k} |\alpha_{j,k}| \le C.\sup\left\{\left|\sum_{j,k=1}^{n} \alpha_{j,k} z_j z_k\right| ; \ \max_j |z_j| \le 1\right\} \tag{3}$$

for any positive integer n and any symmetric $n \times n$ matrix $(\alpha_{j,k})_{j,k}$.

A result of Hardy–Littlewood [11] (based on earlier work of Toeplitz) says that if $\alpha_{j,k} = \exp(2\pi i jk/n)$, $i = \sqrt{-1}$ and $1 \le j, \ k \le n$, then

$$\sup\left\{\left|\sum_{j,k=1}^{n} \alpha_{j,k} z_j w_k\right| ; \ \max_j |z_j| \le 1, \ \max_k |w_k| \le 1\right\} \le n^{3/2}.$$

Let $(\beta_j)_{j=1}^{\infty}$ denote the non–increasing rearrangement of $\left(\frac{a_j}{b_j}\right)_j$. By the Hardy–Littlewood result and (3) there exists a positive number C_1 such that $\left(\sum_{j=1}^{n} \beta_j\right)^2 \le C_1 n^{3/2}$ and hence $(n\beta_n)^2 \le C_1 n^{3/2}$ i.e. $\beta_n \le \frac{C_1^{1/2}}{n^{1/4}}$ for all n. Thus

$$\sum_{n=1}^{\infty} \left(\frac{a_n}{b_n}\right)^8 = \sum_{n=1}^{\infty} \beta_n^8 \le C_1^4 . \sum_{n=1}^{\infty} \frac{1}{n^2} < \infty$$

and we have shown that E'_β and hence E is nuclear. The above also leads to a generalisation of the inequality of Bohr.

From the above we see that the problem was reduced to the case of monomials of degree 2 and thus to 2–homogeneous polynomials. A further analysis of the proof, using the well known correspondance between polynomials of finite type and symmetric tensors in $E' \otimes E'$, shows that the problem we have discussed is essentially a problem of deciding under what conditions the ϵ and π topologies coincide on $E' \otimes E'$. Using this reduction and absolutely summing operators Defant and de Morães [4] showed that the result is valid in a much more general setting. They proved the following;

if E is a locally convex space, (E', τ_0) has an absolute basis and the monomials corresponding to this basis form an absolute basis for $(H(E), \tau_0)$ then (E', τ_0) is nuclear.

Under certain circumstances they also show that the presence of an unconditional monomial expansion relative to the compact open topology implies nuclearity. A result of Ryan [12], given below, shows that nuclearity is not essential in this case.

The results we have just given suggest that monomial expansions might only be useful in the presence of nuclearity. Ryan [12], however, was able to proceed by initially restricting the class of holomorphic functions to be considered and by looking at norms of the ℓ_1 type. In contrast to the finite dimensional situation the Taylor series expansion at the origin, $\sum_{n=0}^{\infty} P_n$, of an entire holomorphic function on a Banach space E, f, need not converge uniformly to the function on every bounded set. We have for B the unit ball of the Banach space E

$$\sup\{r > 0,\ \|f\|_{rB} < \infty\} = \frac{1}{\limsup_{n\to\infty} \|P_n\|^{1/n}}$$

$$= \sup\left\{r > 0,\ \sum_{n=0}^{k} P_n \to f \text{ uniformly on } rB\right\}$$

$$:= r_f(0) \quad \text{(the radius of boundedness of } f \text{ at the origin).}$$

By choosing a sequence $(\phi_n)_n$ in example 4 which does not converge to 0 in norm we get an example of an entire function with finite radius of convergence at all points. Let $H_b(E)$ denote the set of all entire functions with infinite radius of convergence at one (and hence at every) point of E. For any infinite dimensional Banach space $H_b(E)$ is a proper subspace of $H(E)$.

Ryan showed that if $\sum_{m\in N^{(N)}} a_m z^m$ is the monomial expansion of $f \in H_b(\ell_1)$ then

$$\lim_{|m|\to\infty} \left(|a_m| m^m / |m|^{|m|}\right)^{1/|m|} = 0 \tag{4}$$

and that, conversely, any set of scalars $(a_m)_{m\in N^{(N)}}$ which satisfies (4) form the monomial coefficients of an element of $H_b(\ell_1)$. He also showed that the monomial expansion converged absolutely and uniformly on bounded subsets of ℓ_1 to the function. A key role in his investigation was the equality

$$\|z^m\|_B = \frac{m^m}{|m|^{|m|}} \qquad \text{for } m \in N^{(N)} \text{ and } B \text{ the unit ball of } \ell_1.$$

He proceeded as follows for arbitrary holomorphic functions. If K is an absolutely convex compact subset of E let $E_K = \bigcup_{n=1}^{\infty} nK$ and let $\|x\|_K = \inf\{\lambda > 0;\ x \in \lambda K\}$ for x in E_K. Then E_K is a Banach space and $E = \varinjlim_{K}(E_K, J_K)$ where J_K is the canonical injection of E_K into E. A mapping $f : E \to \mathbf{C}$ is holomorphic if and only if $f \circ J_K \in H_b(E_K)$ for every absolutely convex balanced compact subset K of E. If $\xi = (\xi_n)_{n=1}^{\infty} \in c_0^+$ let $K_\xi = \left\{(z_n)_n \in \ell_1;\ \sum_{n=1}^{\infty} \frac{|z_n|}{\xi_n} \leq 1\right\}$. K_ξ is a compact subset of ℓ_1 and $(K_\xi)_{\xi\in c_0^+}$ forms a fundamental system of absolutely convex compact subsets of ℓ_1. If $f : \ell_1 \to \mathbf{C}$ and $\xi \in c_0^+$ let $f_\xi(z) = f(\xi z)$. It follows from the above results that $f \in H(\ell_1)$ if and only if $f_\xi \in H_b(\ell_1)$ for all $\xi \in c_0^+$ and on applying (4) Ryan showed the following.

Proposition 1. *A collection $\{a_m,\ m \in N^{(N)}\}$ of scalars form the monomial coefficients of an element of $H(\ell_1)$ if and only if for all $\xi \in c_0^+$*

$$\sup_m \frac{|a_m| m^m}{|m|^{|m|}} \xi^m < \infty.$$

He also showed that the monomial expansion converged absolutely and uniformly on compact subsets to the function it defined. [**Warning**: absolute and uniform convergence on compact sets is not the same as saying that one has an absolute basis.] From this he obtained the following corollary.

Corollary. *The monomials form an equicontinuous unconditional Schauder basis for $(H(\ell_1), \tau_0)$.*

Ryan also obtained concrete τ_0 and τ_ω generating families of semi–norms in terms of the monomial coefficients.

Application 3. [7] In this application we consider a particular case of the following general problem;

In what circumstances can condition (a) in the definition of holomorphic function be replaced by an apparently weaker condition?

For example, if U is a connected domain in a Banach space E and f is a Gateâux holomorphic function on U then a theorem of Zorn says that f is holomorphic if all its derivatives at one point are continuous. In particular, continuity at one point implies continuity at all points.

Let $E = \lim\limits_{\overrightarrow{n}}(E_n, \tau_n)$ denote a countable inductive limit of Banach spaces and suppose E is a DFM space. We can suppose in this case that $(B_n)_n$ is a fundamental sequence of compact subsets of E where B_n is the closed unit ball of E_n, all n. If $f : E \to \mathbf{C}$ is holomorphic then $f \circ \tau_n \in H_b(E_n)$ for all n. Conversely, using a truncation of the Taylor series and the radius of convergence, it can be shown that if $f : E \to \mathbf{C}$ is Gateâux holomorphic and $f \circ \tau_n \in H_b(E_n)$ all n then $f \in H(E)$. Here we are interested in whether or not $f \circ \tau_n \in H(E_n)$ all n implies $f \in H(E)$. If E is a DFS space then the result of Zorn and the compact linking maps can be used to show that the result is true. If one replaces holomorphic mappings by continuous mappings and considers DFM spaces which are not DFS then the result is not true. Since $H_b(E) \neq H(E)$ for infinite dimensional Banach spaces the result is also not true in this case. Thus we have another example, as for instance in the results of Ansemil–Ponte and Ansemil–Taskinen, where in moving from the nuclear to the Banach space cases the critical stage appears when the Montel hypothesis is introduced. Here we use the results of Ryan to obtain a positive result for a certain class of DFM spaces.

Let A denote a Köthe matrix i.e. $A = (\alpha_{i,n})_{i,n}$ is a matrix of non–negative terms such that $\alpha_{i,n} \leq \alpha_{i,n+1}$ for all i and n. For the sake of simplicity we suppose $\alpha_{i,n} > 0$ for all i and n. Let

$$k_n(A) = \left\{ (z_i)_i ;\ \sum_{i=1}^{\infty} \frac{|z_i|}{\alpha_{i,n}} := \left\| (z_i)_i \right\|_n < \infty \right\}$$

for n a positive integer. The space $(k_n(A), \|\cdot\|_n)$ is a Banach space and $k_1(A) \subset k_2(A) \subset \cdots$. Let $k(A) = \lim\limits_{\overrightarrow{n}} k_n(A)$ and suppose $k(A)$ is a DFM space.

Proposition 2. [7] If $f : k(A) \to \mathbf{C}$ is Gateâux holomorphic and $f|_{k_n(A)} \in H(k_n(A))$ for all n then $f|_{k_n(A)} \in H_b(k_n(A))$ for all n.

Our proof proceeds by contradiction and follows the general principle – if a counterexample exists then a nice counterexample also exists.

Let f be a counterexample. Since each $k_n(A)$ is, modulo a diagonal transformation, isometric to ℓ_1 and $k(A) = \bigcup\limits_n k_n(A)$ it follows that f has a monomial expansion $\sum\limits_{m \in N^{(N)}} a_m z^m$. We may now suppose, without loss of generality, that $f|_{k_1(A)} \notin H_b(k_1(A))$. The mapping

$$T_n : \ell_1 \longrightarrow k_n(A)$$
$$(z_i)_i \longrightarrow (\alpha_{i,n} z_i)_i$$

is an isometry. Hence $f \circ T_1 \in H(\ell_1)$ and $f \circ T_1 \notin H_b(\ell_1)$. We have

$$f \circ T_1(z) = \sum_{m \in N^{(N)}} a_m (\alpha_{\cdot,1} z)^m$$

for all $z \in \ell_1$ and hence by Ryan (4)

$$\limsup_{|m| \to \infty} \left[|a_m| \frac{m^m}{|m|^{|m|}} (\alpha_{\cdot,1})^m \right]^{1/|m|} = \delta > 0.$$

On replacing $f(z)$ by $f(\lambda z)$ we may suppose $\delta > 1$. Hence we can choose a sequence $(m_j)_j$ in $N^{(N)}$ such that $|m_j|$ is strictly increasing and

$$|a_{m_j}| \frac{m_j^{m_j}}{|m_j|^{|m_j|}} (\alpha_{\cdot,1})^{m_j} \geq 1$$

for all j. Let

$$\beta_j = \frac{|m_j|^{|m_j|}}{m_j^{m_j} (\alpha_{\cdot,1})^{m_j}}$$

and let

$$g(z) = \sum_{j=1}^{\infty} \beta_j z^{m_j}.$$

If $z \in k(A)$ then

$$\sum_{j=1}^{\infty} |\beta_j| \, |z^{m_j}| \leq \sum_{m \in N^{(N)}} |a_m| \, |z^m| < \infty$$

and, since $|m_j|$ is strictly increasing, the result of Zorn shows that $g|_{k_n(A)} \in H(k_n(A))$ for all n. Since $\beta_j \cdot \frac{m_j^{m_j}}{|m_j|^{|m_j|}} (\alpha_{\cdot,1})^{m_j} = 1$ for all j the result of Ryan (4) shows that $g \notin H_b(k_1(A))$. We have thus shown that the function

$$g(z) = \sum_{j=1}^{\infty} \frac{|m_j|^{|m_j|}}{m_j^{m_j} (\alpha_{\cdot,1})^{m_j}} z^{m_j} \qquad (*)$$

will also be a counterexample. Note that $(*)$ is both the Taylor series expansion at the origin and the monomial expansion of g.

Let B_n denote the unit ball of $k_n(A)$. Since $g\big|_{k_n(A)}$ is holomorphic it has positive radius of convergence at the origin, i.e.

$$\limsup_{j\to\infty}\left[\frac{|m_j|^{|m_j|}}{m_j^{m_j}(\alpha_{\cdot,1})^{m_j}}\,\|z^{m_j}\|_{B_n}\right]^{1/|m_j|} < \infty.$$

Since

$$\|z^{m_j}\|_{B_n} = \frac{m_j^{m_j}}{|m_j|^{|m_j|}}(\alpha_{\cdot,n})^{m_j}$$

this shows that

$$\limsup_{j\to\infty}\left[\left(\frac{\alpha_{\cdot,n}}{\alpha_{\cdot,1}}\right)^{m_j}\right]^{1/|m_j|} := C(n) < \infty \qquad (A)$$

for all n. Now

$$g\circ T_1(z) = \sum_{j=1}^{\infty}\frac{|m_j|^{|m_j|}}{m_j^{m_j}}\,z^{m_j}$$

defines a holomorphic function on ℓ_1 and hence, by proposition 1, we have

$$\sup_j \xi^{m_j} < \infty \qquad \text{for all } \xi\in c_0^+ \qquad (B).$$

Using Zorn's result once more one can show quite easily that a counterexample exists if and only if (A) and (B) are satisfied by $(m_j)_j$ with $|m_j|$ strictly increasing. For each j let $m_j = (m_{1,j}, m_{2,j}, \cdots, m_{n,j}, \cdots)$. The finiteness criterion given in (B) ensures that a certain proportion of the "power" goes to infinite. Specifically, by using a diagonal process and taking subsequences if necessary, one can show that there exists $\delta > 0$ such that

$$\frac{\sum_{i>j} m_{i,j}}{|m_j|} = \frac{\sum_{i>j} m_{i,j}}{\sum_{i=1}^{\infty} m_{i,j}} \geq \delta$$

for all j.

Next, using δ, we find a strictly increasing sequence of positive integers $(k_j)_j$ such that if

$$m_j' = (\underbrace{0,\ldots,0}_{k_j}, m_{k_j+1}, \ldots, m_{k_{j+1}}, 0\ldots)$$

then (A) and (B) are still satisfied by $(m_j')_j$ and $\left(|m_j'|\right)_j$ is strictly increasing. In other words the monomials in a counterexample may be taken to have disjoint support. Now let $\dfrac{\alpha_{i,n}}{\alpha_{i,n-1}} = \exp(\beta_{i,n})$

for all i and n. We have $\beta_{i,n} \geq 0$ and $\frac{\alpha_{i,n}}{\alpha_{i,1}} = \exp\left(\sum_{\ell=2}^{n} \beta_{i,\ell}\right)$. Condition (A) then becomes

$$\sup_j \left(\sum_{i=k_j+1}^{k_{j+1}} \frac{m_{i,j}}{|m_j|} \beta_{i,n} \right) = C'(n) < \infty$$

for every n. For each j and n let

$$A(j,n) = \{i;\ \beta_{i,n} \geq 4^n C'(n),\ k_j + 1 \leq i < k_{j+1}\}.$$

Then

$$C'(n) \geq \sum_{i \in A(j,n)} \frac{m_{i,j}\beta_{i,n}}{|m_j|} \geq 4^n C'(n) \,.\, \sum_{i \in A(j,n)} \frac{m_{i,j}}{|m_j|}$$

and

$$\sum_{i \in A(j,n)} \frac{m_{i,j}}{|m_j|} \leq \frac{1}{4^n}$$

for all j and n. Hence, for each j,

$$\sum_n \sum_{i \in A(j,n)} \frac{m_{i,j}}{|m_j|} \leq \sum_{n=1}^{\infty} \frac{1}{4^n} = \frac{1}{3}.$$

Since $\sum_i \frac{m_{i,j}}{|m_j|} = 1$ it follows that for each j, $\exists \ell_j$, $k_j + 1 \leq \ell_j < k_{j+1}$ such that $\ell_j \notin \bigcup_n A(j,n)$. Then $\beta_{\ell_j,n} < 4^n C'(n)$ for all j and $\sup_j \beta_{\ell_j,n} < \infty$ for all n. Hence

$$\sup_j \frac{\alpha_{\ell_j,n}}{\alpha_{\ell_j,1}} < \infty$$

for all n and the closed complemented subspace of $k(A)'_\beta$ generated by $e_{\ell_j} := (0, \ldots, 1, 0 \ldots)$

$\hookrightarrow \ell_j$th positi

is isomorphic to c_0. This contradicts the fact that $k(A)'_\beta$ is a Fréchet Montel space and completes the proof.

Application 4. A function $v : E \to [-\infty, +\infty)$ which is not identically equal to $-\infty$ is called plurisubharmonic if it is upper semicontinuous and if, for each a and b in E,

$$v(a) \leq \frac{1}{2\pi} \int_0^{2\pi} v(a + be^{i\theta})\, d\theta.$$

A subset A of E is called polar if there exists a plurisubharmonic function v such that

$$A \subset \{x \in E;\ v(x) = -\infty\}.$$

Polar sets are the natural negligible sets of complex analysis and this is particularly so in infinite dimensional complex analysis where sets of first category or sets of Lebesgue measure zero may either be unsuitable or not available. In two papers [8, 9] the author, Meise and Vogt looked at the question of whether or not Fréchet nuclear spaces and DFN spaces contained compact sets (and, hence, small sets from the functional analyst's point of view) which were non–polar (and, hence, large from the complex analyst's point of view). Linear characterisations of exactly when this was the case were obtained and the following is typical of the results obtained

"A DFN space E contains a bounded non–polar set if and only if E'_β has $(\underline{DN})$."

Here I will give one simple example of a result of this type using monomial expansions. This proof was later rendered obsolete by finer techniques and so did not appear in print.

Let $E = \lambda(A)$ denote a nuclear Fréchet space with basis where $A := (a_{j,k})_{j,k}$ is a positive Köthe matrix. If $(b_j)_j \in E^+$ then $B := [(b_j)_j] = \{(z_j)_j;\ |z_j| \le b_j \text{ all } j\}$ is a compact subset of E. Let E_B denote the subspace of E generated by B. It is not difficult to show that B is polar if and only if E_B is polar and a result of Coeuré says that E_B is non–polar if and only if every holomorphic function f on an open set containing E_B extends to a holomorphic function on E. Let $f(z) = \sum_{m \in N^{(N)}} a_m z^m$ be the monomial expansion of f. For any positive number c we have, since f is holomorphic on E_B,

$$f(c) := \sup_m |a_m| c^{|m|} b^m < \infty.$$

Since f converges absolutely on some neighbourhood of 0 there exists a positive integer k such that

$$\sup_m \left[|a_m| \left(\frac{1}{\alpha_{\cdot,k}} \right)^m \right] = K < \infty.$$

To show that f extends we must show that

$$\sup_{m \in N^{(N)}} |a_m|\, |\alpha^m| < \infty$$

for every $(\alpha_j)_j$ in E.

Now suppose that for every positive integer k there exists a positive integer $n = n(k)$ such that for every $(\alpha_j)_j \in E$ there exists j_0 with

$$\left| \frac{\alpha_j^n}{b_j}\, a_{j,k}^{n-1} \right| \le 1 \quad \text{for all } j \ge j_0.$$

Let $m = (m_1, m_2, \ldots, m_s, 0, \ldots) \in N^{(N)}$. A calculation shows

$$(|a_m \alpha^m|)^n \leq |a_m| b^m . \left[|a_m| . \left(\frac{1}{a_{.,k}}\right)^m\right]^{n-1} \left|\left(\frac{\alpha_1^n a_{1,k}^{n-1}}{b_1}\right)^{m_1} \cdots \left(\frac{\alpha_{j_0}^n a_{j_0,k}}{b_{j_0}}\right)^{m_{j_0}}\right|.$$

Let

$$C = \left[\sup_{j \leq j_0} \left|\frac{\alpha_j^n a_{j,k}^{n-1}}{b_j}\right|\right] + 1.$$

Then

$$\begin{aligned}[|a_m \alpha^m|]^n &\leq |a_m| b^m . \left[|a_m| \left(\frac{1}{a_{.,k}}\right)^m\right]^{n-1} . (C^{j_0})^{|m|} \\ &\leq f(C^{j_0}) . K^{n-1}\end{aligned}$$

and, hence,

$$\sup_m |a_m| . |\alpha^m| \leq \left[f(C^{j_0}) . K^{n-1}\right]^{1/n} < \infty.$$

This implies that $\left[(b_j)_j\right]$ is non–polar in E. Now we take E to be $H(\mathbf{D})$, $\mathbf{D}$ the open unit disc in $\mathbf{C}$. $H(\mathbf{D})$ is a Fréchet nuclear space with Köthe matrix given by $a_{j,k} = \left(1 - \frac{1}{k}\right)^j$. Let $b_j = \frac{1}{2^j}$ for all j. For the positive integer k choose n such that

$$\left(1 - \frac{1}{k}\right)^{n-1} \leq \frac{1}{8}.$$

If $\sum_{j=0}^{\infty} \alpha_j z^j \in H(\mathbf{D})$ then $\limsup_{j \to \infty} |\alpha_j|^{1/j} \leq 1$ and hence we can choose j_0 such that $|\alpha_j|^n \leq 2^j$ for all $j \geq j_0$. Hence

$$\begin{aligned}\left|\frac{\alpha_j^n}{b_j} a_{j,k}^{n-1}\right| &= \left[2\left(1 - \frac{1}{k}\right)^{n-1}\right]^j . |\alpha_j|^n \\ &\leq \left(\frac{1}{4}\right)^j . 2^j = \frac{1}{2^j}\end{aligned}$$

for all $j \geq j_0$. This shows that

$$\left\{f \in H(\mathbf{D}) : f(z) = \sum_{n=0}^{\infty} a_n z^n,\ |a_n| \leq \frac{1}{2^n} \text{ all } n\right\}$$

is a non–polar compact subset of $H(\mathbf{D})$.

References

[1] **J. M. Ansemil, S. Ponte**, The compact open and the Nachbin ported topologies on spaces of holomorphic functions, Arch. Math., 51, 1988, 65–70.

[2] **J. M. Ansemil, J. Taskinen**, On a problem of topologies in infinite dimensional holomorphy, preprint 1988.

[3] **P. J. Boland, S. Dineen**, Holomorphic functions on fully nuclear spaces, Bull.S.M.Fr., 106, 1978, 311–336.

[4] **A. Defant, L. A. de Morães**, Absolute and unconditional bases in spaces of holomorphic functions in infinite dimensions, preprint 1987.

[5] **S. Dineen**, Complex analysis in locally convex spaces, North–Holland Math. Studies, 57, 1981.

[6] **S. Dineen**, Analytic functions on fully nuclear spaces, Studia Math., 73, 1982, 11–32.

[7] **S. Dineen, M. L. Lourenco**, Holomorphic functions on strong duals of Fréchet–Montel Spaces II, Arkiv der Math., (to appear).

[8] **S. Dineen, R. Meise, D. Vogt**, Characterisation of nuclear Fréchet spaces in which every bounded set is polar, Bull.S.M.Fr., 112, 1984, 41–68.

[9] **S. Dineen, R. Meise, D. Vogt**, Polar subsets of locally convex spaces, *Aspects of Mathematics and its Applications*, Ed. J. A. Barroso, Elsevier Science Publishers, B.V., 1986, 295–319.

[10] **S. Dineen, R. M. Timoney**, Absolute basis, tensor products and a theorem of Bohr, Studia Math. (to appear).

[11] **J. E. Littlewood**, On bounded bilinear forms in an infinite number of variables, Quart. J. Math., 1, 1930, 164–174.

[12] **R. A. Ryan**, Holomorphic mappings on ℓ_1, T.A.M.S., 302, 2, 1987, 797–811.

RELATIONS BETWEEN τ_0 AND τ_ω ON SPACES OF HOLOMORPHIC FUNCTIONS

J.M. ANSEMIL[1]
Departamento de Análisis Matemático
Facultad de Matemáticas
Universidad Complutense
28040 - Madrid
Spain

Abstract

In this paper we give a survey about the recent results concerning the coincidence or not of the compact open and the Nachbin ported topologies on the space of all holomorphic functions on an open subset of a complex Fréchet space.

Let us consider a complex locally convex space E, an open subset U of E, and let us denote by $\mathcal{H}(U)$ the space of all holomorphic functions on U.

τ_0 and τ_ω will denote the compact open and the Nachbin ported topologies on $\mathcal{H}(U)$ respectively. τ_0 is the locally convex topology generated in $\mathcal{H}(U)$ by the seminorms $|f|_K$; $f \in \mathcal{H}(U)$, where K ranges over the compact subsets of U. τ_ω is the locally convex topology generated in $\mathcal{H}(U)$ by the family of all seminorms on $\mathcal{H}(U)$ which are ported by some compact subset of U: A seminorm p on $\mathcal{H}(U)$ is ported by the compact subset K of U if for every open subset V, $K \subset V \subset U$, there exists a constant $C > 0$ such that

$$p(f) \leq C \mid f \mid_V \quad \text{for all} \quad f \in \mathcal{H}(U).$$

The τ_ω topology was introduced by L. Nachbin in 1967 motivated by certain properties of analytic functionals on $\mathbf{C}^n$ ([N1]), see also [N2] and [B]). This topology has been studied

[1] Supported by the D.G.I.C.Y.T. of Spain. Project PB87-1031.

T. Terzioğlu (ed.), Advances in the Theory of Fréchet Spaces, 173–180.

by several authors: Nachbin [N2], Barroso [B], Schottenloher [S], Aron [Ar], Dineen [D1, D2, D3, D4], Bierstedt-Meise [Bi-M] and Mujica [Mu1, Mu2] among others.

It is clear that $\tau_0 \leq \tau_\omega$. When E has finite dimension $\tau_0 = \tau_\omega$, and when E is a Banach space of infinite dimension $\tau_0 \neq \tau_\omega$ (see Proposition 1 bellow). For general locally convex spaces τ_0 may coincide with τ_ω (for example when E is the space $\mathcal{D}(\Omega)$ of all tests functions on an open subset Ω of $\mathbf{R}^n$), or may be different (for example when $E = \mathcal{D}'(\Omega)$, the space of distributions on Ω) (see [Bo-D-2]).

From now on we will assume that E is a (complex) Fréchet space. In this situation it may happen that $\tau_0 = \tau_\omega$ on $\mathcal{H}(U)$; but the following result is well-known (see [D4]):

Proposition 1. Let E be a Fréchet space. If for some open subset U of E, $\tau_0 = \tau_\omega$ on $\mathcal{H}(U)$, then E is a Montel space.

For several years some authors have been interested in the converse, i.e.: Let E be a Fréchet-Montel space and U an open subset of E. Does τ_0 coincide with τ_ω on $\mathcal{H}(U)$?

Several partial affirmative answers to the previous question have been obtained by different authors in the last twenty years; but very recently, Ansemil and Taskinen [A-T] have shown that $\tau_0 \neq \tau_\omega$ on $\mathcal{H}(U)$ for every open subset U of the Fréchet-Montel space F which Taskinen defined in [T] to solve (in the negative) the "Problème des topologies" of Grothendieck.

In this paper we present a survey about the results concerning this question. We begin with a short exposition of the partial affirmative answers:

When E is the space $\mathbf{C}^{\mathbf{N}}$ of all complex sequences, Barroso in 1971 showed in [B] that $\tau_0 = \tau_\omega$ on $\mathcal{H}(\mathbf{C}^{\mathbf{N}})$ and later, Barroso-Nachbin [B-N] and Schottenloher [S] showed independently that also $\tau_0 = \tau_\omega$ on $\mathcal{H}(U)$ for every open subset U of $\mathbf{C}^{\mathbf{N}}$. The crucial point in the proofs is the fact that every holomorphic function on $\mathbf{C}^{\mathbf{N}}$ depends only on a finite number of variables.

For nuclear Fréchet spaces E, the question was answered (in the affirmative) for every balanced open subset of E, first by Boland-Dineen [Bo-D-1] in 1978 assuming that E has a Schauder basis, and later in 1981 by Meise [M] for general nuclear Fréchet spaces. Recently (in 1986) Mujica [Mu2] has shown that $\tau_0 = \tau_\omega$ on $\mathcal{H}(U)$ for every polynomially convex open subset U of a nuclear Fréchet space. The result of [Bo-D-1] was obtained using the fact that every holomorphic function on a nuclear Fréchet space with a basis has a monomial expansion. (See the paper of Dineen in this Proceedings.)

The most general result in this direction was obtained in 1984 by Mujica [Mu1] who showed that when E is a Fréchet-Schwartz space, then $\tau_0 = \tau_\omega$ on $\mathcal{H}(U)$ for every balanced open subset U of E. Moreover he also showed in 1985 ([Mu3]) that when E is a Fréchet-Schwartz space with a Schauder basis, then $\tau_0 = \tau_\omega$ on $\mathcal{H}(U)$ for every open subset U of E. Mujica obtains his results using germs of holomorphic functions.

In a paper recently published, Ansemil and Ponte [A-P] have shown -using Taylor series expansions- the following

Theorem 2. When E is a Fréchet-Montel space and U is a balanced open subset of E, the following conditions are equivalent:

a) $\tau_0 = \tau_\omega$ on $\mathcal{H}(U)$.

b) $\tau_0 = \tau_\omega$ on $\mathcal{P}({}^nE)$ for all $n \in \mathbf{N}$.

($\mathcal{P}({}^nE)$ denotes the space of all n-homogeneous continuous polynomials on E.)

Remark 3. Such conditions are equivalent to $\tau_0 = \tau_\omega$ on the space $\mathcal{H}(K)$ of holomorphic germs on any balanced compact subset of E. We recall that $\mathcal{H}(K)$ is the space of all cosets defined in $\cup\{H(V) : K \subset V, V \text{ open }\}$ by the equivalence relation: f is equivalent to g if f coincides with g on an open neighbourhood of K. The τ_0 and τ_ω topologies are defined in $\mathcal{H}(K)$ by

$$(\mathcal{H}(K), \tau_0) = ind_{V \supset K}(H(V), \tau_0) \quad \text{and} \quad (\mathcal{H}(K), \tau_\omega) = ind_{V \supset K}(H(V), \tau_\omega).$$

Indeed, the proof of Theorem 2 is obtained through spaces of germs using a result of Mujica which states that $(\mathcal{H}(K), \tau_0)$ is an inductive limit in the category of all topological spaces of a sequence of compact spaces.

Concerning the equality $\tau_0 = \tau_\omega$ on $\mathcal{P}({}^nE)$ we have the following result ([A-P], Proposition 3):

Theorem 4. Let E be a Fréchet-Montel space. Then, for every $n \in \mathbf{N}$, the following conditions are equivalent:

a) $\tau_0 = \tau_\omega$ on $\mathcal{P}({}^nE)$

b) $(\mathcal{P}({}^nE), \beta)$ is bornological

c) $(\mathcal{P}({}^nE), \beta)$ is barrelled

d) $(\mathcal{P}({}^nE), \beta)$ is a Montel space

e) $(\mathcal{P}({}^nE), \beta)$ is a DFM space.

(β denotes the topology of uniform convergence on bounded sets.)

Using these two theorems it is possible to prove that for some Fréchet-Montel (non Schwartz) spaces E, it may happen that $\tau_0 = \tau_\omega$ on $\mathcal{H}(U)$ for every balanced open subset U of E, and that for the Taskinen space $F, \tau_0 \neq \tau_\omega$ on $\mathcal{H}(U)$ for every open subset U of F.

Let us begin with the affirmative answer to our question.

Let E be a Fréchet-Montel space. Then $(\mathcal{P}({}^1E), \beta)$ which coincides with the strong dual E'_b of E, is a Montel space, hence $\tau_0 = \tau_\omega$ on $\mathcal{P}({}^1E)$.

For arbitrary n, $(\mathcal{P}({}^nE), \beta)$ is a complemented subspace of the space $\mathcal{L}_b({}^nE)$ of all n-linear continuous functions on E^n endowed with the topology of the uniform convergence on the bounded sets of E^n. Then, if $\mathcal{L}_b({}^nE)$ is barrelled, $(\mathcal{P}({}^nE), \beta)$ also is.

Using Buchwalter's results on duality it is possible to obtain the following isomorphism:

$$\mathcal{L}_b({}^nE) \cong E'_c\varepsilon(\overbrace{E \tilde{\otimes}_\pi \cdots \tilde{\otimes}_\pi E}^{n-1})'_c \cong [E \tilde{\otimes}_\pi (\overbrace{E \tilde{\otimes}_\pi \cdots \tilde{\otimes}_\pi E}^{n-1})]'_c$$

(ε and c respectively denote, the Schwartz product and the topology of uniform convergence on compact subsets.)

If E and F are two Fréchet-Schwartz spaces, $E \tilde{\otimes}_\pi F$ is also a Fréchet-Schwartz space. Hence, by the previous isomorphism, $\mathcal{L}_b({}^nE)$ is a Montel space, and then $(\mathcal{P}({}^nE), \beta)$ is barrelled. This proves (using Theorems 2 and 4) that $\tau_0 = \tau_\omega$ on $\mathcal{H}(U)$ for every balanced open subset U of E, E being a Fréchet-Schwartz space. (This is a different proof of one of the Mujica's results mentioned before.)

For Köthe sequence spaces we have the following

Theorem 5. ([A-P], Proposition 4) Let E be a Fréchet-Montel-Köthe sequence space. Then $\tau_0 = \tau_\omega$ on $\mathcal{H}(U)$ for every balanced open subset U of E.

The proof of this Theorem is obtained from Theorems 2 and 4, either showing that the completion of the projective tensor product of two Fréchet-Montel-Köthe sequence spaces is also a Fréchet-Montel-Köthe sequence space, or using a result of Bonet-Defant [Bon-De].

Since there are Fréchet-Montel-Köthe sequence spaces which are not Schwartz spaces, we have the first examples of Fréchet-Montel (non Schwartz) spaces, such that $\tau_0 = \tau_\omega$ on $\mathcal{H}(U)$ for every balanced open subset U of them.

Other examples are the Fréchet-Montel, non Schwartz spaces, constructed with Floret's technique ([F]): Given a Fréchet-Montel space F, whose topology is defined by an increasing family of seminorms (p_n), with $\dim(F/\operatorname{Ker} p_1) = \infty$,

$$E = \{x = (x_n) \in F^{\mathbf{N}} : q_k(x) = \sum_{n=1}^{\infty} n^k p_{n,k}(x_n) < \infty \quad \text{for every} \quad k \in \mathbf{N}\}$$

where $p_{n,k} = p_n$ if $n \geq k$ and $p_{n,k} = p_k$ if $n < k$, is a Fréchet-Montel space (with the topology generated by the seminorms (q_k)) which is not a Schwartz space.

In this case we can directly check that the completion of the projective tensor product of spaces of this kind is also a space of this same kind, and is therefore a Montel space.

Now we will indicate that $\tau_0 \neq \tau_\omega$ on $\mathcal{H}(U)$ for every open subset U of the Taskinen space F. This is a very recent result due to Ansemil and Taskinen [A-T].

Let us observe the following: For every open subset U of a complex locally convex space E and for every $n \in \mathbf{N}$, $\mathcal{P}(^nE)$ is a complemented subspace of $(\mathcal{H}(U), \tau)$ where $\tau = \tau_0, \tau_\omega$. Then, if we show that $\tau_0 \neq \tau_\omega$ on $\mathcal{P}(^nE)$ for some n, we have that $\tau_0 \neq \tau_\omega$ on $\mathcal{H}(U)$ for every open subset U of E.

By Theorem 4, $\tau_0 \neq \tau_\omega$ on $\mathcal{P}(^nE)$ if and only if $(\mathcal{P}(^nE), \beta)$ is not a Montel space.

Let us consider Taskinen space F ([T]). It is a Fréchet-Montel (non Schwartz) space with the approximation property, such that $F \tilde{\otimes}_\pi F$ is not a Montel space because it "contains" a subspace isomorphic to ℓ_1.

If we take a look at Taskinen's construction we can see that ℓ_1 is also "contained" in the space $F_s \tilde{\otimes}_\pi F$, which is the projective completion of the space of all symmetric tensors. From this, using some results on duality for symmetric tensor products and that F has the approximation property, we obtain that $(\mathcal{P}(^2F), \beta)$, which is isomorphic to $(F_s \tilde{\otimes}_\pi F)'_c$ is not a Montel space. Then, we have the following

Theorem 6. ([A-T], Corollary 2) $\tau_0 \neq \tau_\omega$ on $\mathcal{H}(U)$ for every open subset U of Fréchet-Montel space F.

Remarks 7. 1) By Theorem 6 and Remark 3 we also have that $\tau_0 \neq \tau_\omega$ on $\mathcal{H}(K)$ for every compact subset K of F.

2) The property $\tau_0 = \tau_\omega$ on $\mathcal{H}(U)$ is not inherited by its quotients. Indeed, if we take a Fréchet-Montel-Köthe sequence space $\lambda(A)$ which is not a Schwartz space, then, by a result of Valdivia [V], $\lambda(A)$ has a quotient G which "contains" ℓ_1 and therefore, while $\tau_0 = \tau_\omega$ on $\mathcal{H}(U)$ for every balanced open subset U of $\lambda(A)$ (Theorem 5), $\tau_0 \neq \tau_\omega$ on $H(V)$ for every open subset of G (this happens because G is not a Montel space, (see Proposition 1)).

REFERENCES

[A-P] Ansemil, J.M., Ponte, S.: 'The Compact Open and the Nachbin Ported Topologies on Spaces of Holomorphic Functions', *Arch. Math.*, **51** (1988), 65-70.

[A-T] Ansemil J.M., Taskinen, J.: 'On a Problem of Topologies in Infinite Dimensional Holomorphy'. To appear in *Arch. Math.*

[Ar] Aron, R.: 'Holomorphic Functions on balanced subsets of a Banach space'. *Bull. A.M.S.*, **78** (1972), 624-627.

[B] Barroso, J.: 'Topologias nos Espaços de Aplicaçoes holomorfas entre espaços localmente convexos'. *Anais Acad. Bras. de Ciencias,* **43** (1971), 527-546.

[B-N] Barroso, J., Nachbin, L.: 'Some Topological Properties of Holomorphic Mappings in Infinitely many Variables'. *Advances in Holomorphy. Ed. J. Barroso, North Holland Math. Studies*, **34**, (1979), 67-91.

[Bi-M] Bierstedt, K.-D., Meise, R.: 'Aspects of Inductive Limits in Spaces of Germs of Holomorphic Functions on Locally Convex Spaces and Applications to a Study of $(\mathcal{H}(U), \tau_\omega)'$. *Advances in Holomorphy, Ed. J. Barroso. North Holland Math. Studies*, **34** (1979), 117-118.

[Bo-D-1] Boland, P., Dineen, S.: 'Holomorphic Functions on Fully Nuclear Spaces'. *Bol. Soc. Mat. France.* **106**, (1978), 311-336.

[Bo-D-2] Boland, P., Dineen, S. 'Holomorphy on Spaces of Distributions'. *Pac. Jour. Math.*, **92** No.1, (1981), 72-74.

[Bon-De] Bonet, J., Defant, A.: 'Projective Tensor Products of Distinguished Fréchet Spaces'. *Proc. R. Ir. Acad.,* **85A** (1985), 193-200.

[D1] Dineen, S.: 'Bounding Subsets of a Banach Space'. *Math. Ann.*, **192**, (1971), 61-70.

[D2] Dineen, S.: 'Holomorphic Functions on (c_0, X_b)-Modules'. *Math. Ann.*, **196**, (1972), 106-116.

[D3] Dineen, S.: 'Holomorphic Functions on Locally Convex Topological Vector Spaces: I. Locally Convex Topologies on $\mathcal{H}(U)$'. *Ann. Inst. Fourier, Grenoble,* **23** (1973), 19-54.

[D4] Dineen, S.: 'Complex Analysis in Locally Convex Spaces'. *North Holland Math. Studies.* **57** (1981).

[F] Floret, K.: 'Fréchet-Montel spaces which are not Schwartz spaces'. *Portugal. Math.*, **42**, No.1, (1983-1984), 1-4.

[M] Meise, R.: 'A Remark on the Ported and the Compact Open Topology for Spaces of Holomorphic Functions on Nuclear Fréchet spaces'. *Proc. R. Ir. Acad.*, **81A**, (1981), 217-223.

[Mu1] Mujica, J.: 'A Banach-Dieudonné Theorem for Germs of Holomorphic Functions'. *J. Func. Anal.*, **57** (1984), 31-48.

[Mu2] Mujica, J.: 'Polynomial Approximation in Nuclear Fréchet Spaces'. *Aspects of Mathematics and its Applications. Ed. J. Barroso. Elsevier Sc. Pub.*, (1986), 601-606.

[Mu3] Mujica, J.: 'Holomorphic Approximation in Infinite- Dimensional Riemann Domains'. *Studia Math.*, **LXXXII** (1985), 107-134.

[N1] Nachbin, L.: 'On the Topology of the Space of all Holomorphic Functions on a Given Open Subset'. *Indag. Math.*, **29** (1967), 366-368.

[N2] Nachbin, L.: 'Topology on Spaces of Holomorphic Mappings'. *Erg. der Math. Springer-Verlag*, **47** (1969).

[N3] Nachbin, L.: 'Some Aspects and Problems in Holomorphy'. *Extrac. Math.*, **1**, No.2, (1986), 57-72.

[S] Schottenloher, M.: '$\tau_\omega = \tau_0$ for Domains in $\mathbf{C}^{\mathbf{N}}$'. *Infinite Dimensional Holomorphy and Applications. Ed. M. Matos. North Holland Math. Studies.*, **12** (1977), 393-395.

[T] Taskinen, J.: 'The Projective Tensor Product of Fréchet-Montel Spaces'. To appear in *Studia Math.*

[V] Valdivia, M.: '(F)-Spaces and Strict (LF)-Spaces'. *Math. Z.*, **195**, (1987), 345-364.

SOME RECENT RESULTS ON $\mathcal{V}C(X)$

KLAUS D. BIERSTEDT
FB 17, Univ. – GH – Paderborn
Postfach 1621
D – 4790 Paderborn
Fed. Rep. of Germany

JOSÉ BONET
Matemáticas, E.T.S. Arquitectura
Universidad Politécnica de Valencia
E – 46071 Valencia
Spain

ABSTRACT. We report on the part of our article [4] which deals with weighted inductive limits of type $\mathcal{V}C(X)$ and their projective hulls $C\overline{V}(X)$, as well as on the recent contributions of F. Bastin [1] and D. Vogt [13]. Two general questions which had been open for some time have rather satisfactory answers now:

1. If $\mathcal{V} = (v_n)_n$ is a decreasing sequence of continuous weights on a $k_{\mathbb{R}}$-space X, $\mathcal{V}C(X)$ must always be complete.

2. Let $\mathcal{V}$ denote a decreasing sequence of continuous weights on a completely regular Hausdorff space X and $\overline{V} = \overline{V}(\mathcal{V})$. (Then $C\overline{V}(X)$ is a (DF)-space.)
$\mathcal{V}$ satisfies condition (D) (introduced in [5]) if and only if the bounded subsets of $C\overline{V}(X)$ are metrizable, and this implies the topological equality $\mathcal{V}C(X) = C\overline{V}(X)$. The converse of the last implication is not true (even for a normal space X), but it does hold if each $\overline{v} \in \overline{V}$ is dominated by some $\tilde{v} \in \overline{V} \cap C(X)$ (and hence e.g. if X is locally compact and σ-compact).

Introduction

In the Proceedings of the 1985 Silivri Conference on Functional Analysis, the first-named author and R. Meise [6] surveyed various aspects of weighted inductive limits and their projective descriptions. The survey was based mainly on the articles [7], [8] (together with W. H. Summers) and [5]. Since then, much work has been done on weighted inductive limits of vector-valued spaces of continuous functions (and on vector-valued sequence spaces), and we will report on this topic in another contribution.

In the present paper, however, we concentrate on the scalar case and on $\mathcal{O}$-growth conditions. In 1985, one (rather embarrassing) question and one (important) problem on $\mathcal{V}C(X)$ remained open: On the one hand, it was not known whether, for a decreasing sequence $\mathcal{V} = (v_n)_n$ of strictly positive continuous functions on a $k_{\mathbb{R}}$-space X, the regular (LB)-space $\mathcal{V}C(X)$ must always be complete. And on the other hand, there was no characterization of the topological equality $\mathcal{V}C(X) = C\overline{V}(X)$ in terms of the sequence $\mathcal{V}$. A "weak" sufficient condition (D) had been derived in [5], together with a counterexample showing that this condition was not necessary in general. For the case of sequence spaces, the desired charac-

T. Terzioğlu (ed.), Advances in the Theory of Fréchet Spaces, 181–194.

terization amounted to characterizing the class of the distinguished Köthe echelon spaces $\lambda_1(A)$ in terms of the Köthe matrices A of its elements, and even this special case had not been solved in 1985.

In the meantime, progress has been made mainly in our articles [2], [3] and [4], and together with the contributions [13] of D. Vogt and [1] of F. Bastin, we have quite satisfactory answers now which we are going to discuss here: The completeness question was finally solved affirmatively in [4]. In [2], we had already characterized the distinguished echelon spaces and, in fact, showed that in this case, condition (D) is necessary as well. A different characterization of the class of the distinguished spaces $\lambda_1(A)$, with a short direct proof, was then given by D. Vogt [13]. In connection with our investigations on S. Heinrich's density condition in the context of metrizable locally convex spaces and on its dual reformulations for (DF)-spaces, we have now been able to prove that condition (D) characterizes the two "dual density conditions" for ($\mathcal{V}C(X)$ and) $C\overline{V}(X)$ (see [4]). And F. Bastin [1] has then proved that condition (D) does nevertheless also characterize the topological equality $\mathcal{V}C(X) = C\overline{V}(X)$ if one assumes that each weight $\overline{v} \in \overline{V}$ is dominated by a continuous weight $\tilde{v} \in \overline{V}$. (The last assumption is always satisfied if $\mathcal{V} \subset C(X)$ and if the space X is locally compact and σ-compact.)

These results and sketches of their proofs, as well as of some of the ideas involved, are the highlights of the present survey which, at its end, also includes some open questions and suggestions for future research.

1. Definitions; formulation of the problems

In the sequel, we let X denote a completely regular Hausdorff space and $\mathcal{V} = (v_n)_{n\in\mathbb{N}}$ a decreasing sequence of strictly positive weights on X; i. e., upper semicontinuous (u.s.c.) functions.

We will sometimes require X to be a $k_{\mathbb{R}}$-space, which means that a real-valued function f on X whose restriction to each compact subset of X is continuous must already be continuous on X and which is equivalent to the completeness of the space $C(X)$ of all (real- or complex-valued) continuous functions on X under its (canonical) topology of uniform convergence on the compact subsets of X. On the side of the weights v_n, we will sometimes assume that

(*) for each $n \in \mathbb{N}$ and each compact set $K \subset X$, $\inf_{x\in K} v_n(x) > 0$.

This assumption certainly holds if all the weights v_n are continuous.

We are interested in the normed spaces

$$Cv_n(X) := \{f \in C(X);\ p_{v_n}(f) = \sup_{x\in X} v_n(x)|f(x)| < \infty\} \quad \text{and}$$

$$C(v_n)_0(X) := \{f \in C(X);\ v_n f \text{ vanishes at } \infty \text{ on } X \text{ (i. e., for each } \epsilon > 0, \text{ there is a compact set } K \subset X \text{ with } v_n(x)|f(x)| < \epsilon \text{ for all } x \in X \setminus K)\},$$

both endowed with the norm p_{v_n}, and in the weighted inductive limits

$$\mathcal{V}C(X) := \text{ind}_n\ Cv_n(X) \quad \text{and} \quad \mathcal{V}_0C(X) := \text{ind}_n\ C(v_n)_0(X).$$

If X is a $k_{\mathbb{R}}$-space and if (*) holds, all the spaces $Cv_n(X)$ and $C(v_n)_0(X) \subset Cv_n(X)$ are Banach spaces. Under these conditions, however, the (LB)-space $\mathcal{V}_0C(X)$ can be incomplete; indeed, there are even non-regular inductive limits $\mathcal{V}_0C(X) = \text{ind}_n\ C(v_n)_0(X)$ (i. e., there exist bounded subsets B of $\mathcal{V}_0C(X)$ which are not contained and bounded in one of the generating Banach spaces $C(v_n)_0(X)$). For a characterization of the complete or, equivalently, regular spaces $\mathcal{V}_0C(X)$ in the case of continuous weights v_n on a locally compact space X (by the "regularly decreasing" condition on $\mathcal{V}$), see [7]. — But it is easy to see that all the spaces $\mathcal{V}C(X)$ are regular, and hence we can ask:

Question. *If $\mathcal{V} = (v_n)_n$ is a decreasing sequence of strictly positive weights with (*) on a $k_{\mathbb{R}}$-space X, must then $\mathcal{V}C(X)$ necessarily be complete?*

We recall in this context that the only major problem which remains open from Grothendieck's work in functional analysis in the early 1950's asks whether each regular (LB)- (or even (LF)-) space must be complete. It is conjectured that the answer to this problem of Grothendieck should be negative, but despite several efforts by various people, it has not been possible so far to find examples of regular, but incomplete (LB)-spaces, and in particular, there are no such examples among the co-echelon spaces (of arbitrary order p, $1 \leq p \leq \infty$ or $p = 0$) or among the inductive limits "of Moscatelli type".

On the other hand, the obstruction to proving completeness of regular (LB)-spaces is that there exists no "good and useful" description of the continuous seminorms on general (LB)-spaces. Even in the case of the weighted inductive limits $\mathcal{V}C(X)$, we have still not been able to give a complete description of the continuous seminorms on a general space $\mathcal{V}C(X)$ in such a way that we could solve the question above directly by use of the seminorms.

In an effort to exhibit a system of weighted sup-seminorms on $\mathcal{V}C(X)$ which, at least in many important cases, generates the inductive limit topology, the associated system $\overline{V} = \overline{V}(\mathcal{V})$ of weights was introduced by

$$\begin{aligned} \overline{V} = \overline{V}(\mathcal{V}) \quad &:= \quad \{\overline{v} \geq 0 \text{ u.s.c. on } X;\ \text{for each } n \in I\!N,\ \sup_{x \in X}\ \overline{v}(x)/v_n(x) < \infty\} \\ &= \quad \{\overline{v} \geq 0 \text{ u.s.c.; there are } \alpha_n > 0,\ n = 1, 2, \ldots,\ \text{with } \overline{v} \leq \inf_n\ \alpha_n v_n \text{ on } X\}. \end{aligned}$$

The corresponding weighted spaces are

$$C\overline{V}(X) := \{f \in C(X);\ \text{for each } \overline{v} \in \overline{V},\ p_{\overline{v}}(f) = \sup_{x \in X}\ \overline{v}(x)|f(x)| < \infty\} \quad \text{and}$$

$$C\overline{V}_0(X) := \{f \in C(X);\ \text{for each } \overline{v} \in V,\ \overline{v}f \text{ vanishes at } \infty \text{ on } X\},$$

both endowed with the topology given by the system $(p_{\overline{v}})_{\overline{v} \in \overline{V}}$ of seminorms. One easily verifies that both $C\overline{V}(X)$ and its closed subspace $C\overline{V}_0(X)$ are complete under the hypotheses of the question above.

While it is possible to prove (see [7]) that for locally compact X, $\mathcal{V}_0C(X)$ is always a dense topological linear subspace of $C\overline{V}_0(X)$ and that hence for any sequence $\mathcal{V} \subset C(X)$, $\mathcal{V}_0C(X) = C\overline{V}_0(X)$ holds algebraically if and only if $\mathcal{V}_0C(X)$ is complete, the case of $\mathcal{O}$-growth conditions is completely different (again see [7]): We always have $\mathcal{V}C(X) = C\overline{V}(X)$ algebraically, and the two topologies even have the same bounded sets. Hence we obtain (in full generality)

$$\mathcal{V}C(X) = (C\overline{V}(X))_{\text{bor}} = \text{the bornological space associated with } C\overline{V}(X).$$

But there are examples which show that the inductive limit topology of $\mathcal{V}C(X)$ can be strictly stronger than the weighted topology of $C\overline{V}(X)$. $C\overline{V}(X)$ is some kind of "projective hull" of $\mathcal{V}C(X)$, and it is interesting to know exactly when the projective description "$\mathcal{V}C(X) = C\overline{V}(X)$ topologically" is true.

Problem. *Give necessary and sufficient conditions on $\mathcal{V}$ (and X) such that the topological equality $\mathcal{V}C(X) = C\overline{V}(X)$ holds.*

Of course, if one is willing to add a sufficient condition for $\mathcal{V}C(X) = C\overline{V}(X)$ topologically to the hypotheses of our question, then this question obviously has an affirmative answer. But we would not call that a "solution" of the question! For a satisfactory solution, only some additional assumption of a very general type, like $\mathcal{V} \subset C(X)$ and/or X locally compact, is allowed.

On the other hand, the problem asks for a condition on the sequence $\mathcal{V} = (v_n)_n$ which, at least if $\mathcal{V} \subset C(X)$ and if X is a "good" space (say, locally compact and σ-compact), is equivalent to the topological equality $\mathcal{V}C(X) = C\overline{V}(X)$. In this precise sense, both the question and the problem (dating back to [7]) have now been solved, and we are going to present the details of the solutions ([4], plus Bastin [1]) in the next sections.

2. Solution of the completeness question

Our approach in [4] to the completeness question was already suggested in [7] and demonstrates that, although the topologies of $\mathcal{V}C(X)$ and $C\overline{V}(X)$ may indeed be different, they are nonetheless always closely enough related such that we can deduce the (desired) completeness of $\mathcal{V}C(X)$ from the (easily established) completeness of $C\overline{V}(X)$.

The solution is based on the following abstract result:

The barrelled space associated with a complete locally convex space must itself be complete.

This result can be proved via transfinite induction by use of the canonical transfinite construction of the associated barrelled space: In each step, the barrels are taken as a basis of 0-neighborhoods for a stronger locally convex topology which satisfies the closed neighborhood condition with respect to the previous topology.

1. Proposition([4]). *If $\mathcal{V} = (v_n)_n$ is a decreasing sequence of continuous weights on a $k_{\mathbb{R}}$-space X, then $\mathcal{V}C(X)$ is the barrelled space associated with the complete space $C\overline{V}(X)$ and hence must itself be complete.*

Since we already have $\mathcal{V}C(X) = (C\overline{V}(X))_{\text{bor}}$, it only remains to be seen that the barrelled and bornological spaces associated with $C\overline{V}(X)$ coincide, and this follows from the next two lemmas. The first one exhibits an interesting property of the canonical fundamental sequence of bounded sets in $C\overline{V}(X)$.

2. Lemma. *If $\mathcal{V} = (v_n)_n \subset C(X)$, then the space $C\overline{V}(X), \overline{V} = \overline{V}(\mathcal{V})$, has a fundamental sequence $(B_n)_n$ of absolutely convex bounded sets with the property that for any finite* *collection* $\lambda_1,\ldots,\lambda_m$ $(m \in \mathbb{N})$ *of positive numbers, the set*

$$S_m = S_m(\lambda_1,\ldots,\lambda_m) := \sum_{j=1}^m \lambda_j B_j$$

is closed in $C\overline{V}(X)$.

(Sketch of) **Proof.** If

$$B_n := \{f \in C(X);\ p_{v_n}(f) \le 1\} = \{f \in C(X);\ |f| \le 1/v_n \text{ on } X\}$$

denotes the closed unit ball of $Cv_n(X)$, we know that $(nB_n)_n$ is a countable basis for the bounded subsets of $C\overline{V}(X)$ (since $\mathcal{V}C(X)$ and $C\overline{V}(X)$ have the same bounded sets and since the inductive limit $\mathcal{V}C(X) = \text{ind}_n\ Cv_n(X)$ is regular). Hence it suffices to show that for arbitrary $\lambda_j > 0,\ j = 1,\ldots,m\ \ (m \in \mathbb{N})$, $S_m = \sum_{j=1}^m \lambda_j B_j$ is closed in $C\overline{V}(X)$.
But clearly

$$S_m \subset D_m := \{f \in C\overline{V}(X);\ |f| \le \textstyle\sum_{j=1}^m \lambda_j/v_j \text{ on } X\},$$

and the last set is closed in $C\overline{V}(X)$ (since the topology of this space is stronger than pointwise convergence on X and all weights v_n are strictly positive and continuous). At this point, it is enough to prove the converse inclusion $D_m \subset S_m$, and this can be done with a trick of Ernst-Schnettler [10]:
For fixed $f \in D_m$, put

$$f_1(x) := \left\{ \begin{array}{ll} f(x) & ;\ |f(x)| \le \lambda_1/v_1(x) \\ \lambda_1 f(x)/v_1(x)|f(x)| & ;\ \text{otherwise} \end{array} \right\}.$$

One first checks that f_1 is continuous with f. Then $|f_1| \le \lambda_1/v_1$ implies $f_1 \in \lambda_1 B_1$. Finally, an easy computation shows that $h_1 := f - f_1 \in C\overline{V}(X)$ satisfies

$$|h_1| \le \textstyle\sum_{j=2}^m \lambda_j/v_j \text{ on } X.$$

The conclusion now follows by induction. □

Note that the continuity of the weights v_n was utilized in the proof of Lemma 2. — Incidentally, our (sketch of) proof also serves to show (after a moment's reflection) that the assertion of Lemma 1. remains true for $\mathcal{V}C(X)$ and if X is locally compact, for $C\overline{V}_0(X)$ and $\mathcal{V}_0C(X)$ instead of $C\overline{V}(X)$.

3. Lemma. *Let E denote a locally convex space which has a fundamental sequence $(B_n)_n$ of absolutely convex bounded sets such that for any finite collection $\lambda_1,\ldots,\lambda_m$ $(m \in \mathbb{N})$ of positive numbers, the set*

$$S_m = S_m(\lambda_1,\ldots,\lambda_m) = \sum_{j=1}^m \lambda_j B_j$$

is closed in E.

(a) If E is quasibarrelled, E must already be bornological.

(b) If E is complete, then the barrelled and bornological spaces associated with E coincide (and, in particular, E_{bor} must then be complete, too).

Hints at the proof. In [11], Lemme 4, A. Grothendieck showed that, under certain conditions, the closure of a union $\bigcup_{i\in\mathbb{N}} A_i$ in a (DF)-space coincides with the "algebraic closure". The proof of this fact uses a Hahn-Banach separation argument. (a) can be proved (by contradiction) with a very similar argument.

To deduce (b) from (a), we let τ denote the topology of E, τ^t and $\tau^\times$ the barrelled and the bornological topology associated with τ, respectively. Then the completeness of E implies $\tau \leq \tau^t \leq \tau^\times$, and τ and $\tau^\times$ have the same bounded sets. Now (E, τ^t) is a barrelled (DF)-space with the same fundamental sequence $(B_n)_n$ of absolutely convex bounded sets as (E, τ). An application of (a) yields (E, τ^t) bornological, whence $\tau^t = \tau^\times$. □

Proposition 1. answers the question of Section 1. affirmatively if $\mathcal{V} \subset C(X)$. (Thus, at least for such sequences $\mathcal{V}$, there is also no example of a regular, but incomplete (LB)-space $\mathcal{V}C(X)$.) The property of the canonical fundamental sequence of bounded subsets of $C\overline{V}(X)$ derived in Lemma 2. will be quite useful to us in the next section, too.

3. Projective description and the dual density conditions

The first attempt to solve the projective description problem of Section 1. dates back to the time when [5] was written. In that article, the first-named author and R. Meise modified a condition in Grothendieck's "thèse" ([12], II, p. 102) — note that Grothendieck's assertion that this condition on A implies the quasinormability of the echelon space $\lambda_1(A)$ was erroneous, see [8], 3.11 — by replacing the "Schwartz type" part of the condition by one of "Montel type". Thus, the following condition (D) was introduced:

4. Definition. The sequence $\mathcal{V} = (v_n)_n$ is said to satisfy *condition (D)* if there exists an increasing sequence $J = (X_m)_{m\in\mathbb{N}}$ of subsets of X such that

(N,J) for each $m \in \mathbb{N}$ there is $n_m \geq m$ with $\inf_{x\in X_m} \; v_k(x)/v_{n_m}(x) > 0$ for $k = n_m + 1, n_m + 2, \ldots$, while

(M,J) for each $n \in \mathbb{N}$ and each subset Y of X with $Y \cap (X \setminus X_m) \neq \emptyset$ for all $m \in \mathbb{N}$, there exists $n' = n'(n, Y) > n$ with $\inf_{y\in Y} \; v_{n'}(y)/v_n(y) = 0$.

Property (M, J) clearly implies $X = \bigcup_{m\in\mathbb{N}} X_m$. Hence condition (D) means exactly that X is the countable increasing union of subsets X_m with the following two properties: On one hand, all the weight conditions v_k from a certain index n_m on coincide on each X_m. On the other hand, for any sequence $Y = (y_m)_m$ with $y_m \in X \setminus X_m$ for all m, and for any n, there exists n' (depending on both Y and n) such that $v_{n'|Y}$ yields a weaker weight condition than $v_{n|Y}$ (in the sense that $\inf_{y\in Y} \; v_{n'}(y)/v_n(y) = 0$).

In [5], it was proved that condition (D), plus an additional technical assumption, implies $\mathcal{V}C(X) = C\overline{V}(X)$ topologically, but an example of a decreasing sequence $\mathcal{V} = (v_n)_n$ of strictly positive continuous functions v_n on a normal subspace of $\mathbb{R}^2$ also showed that, conversely, condition (D) is not necessary for the projective description in general.

The original proof of the sufficiency of condition (D) was quite interesting (and it also works, mutatis mutandis, for functions with values in a locally convex space with the countable neighborhood property, cf. [4]), but rather long. In the sequence space case, a shorter proof was given in [6]. The two lemmas of Section 2. and (an improvement of) the (DF)-result of [9] can now be utilized to carry this proof over to function spaces. (We follow [4], Section 5.)

5. Lemma. *Let E denote a locally convex space with a fundamental sequence $(B_n)_{n\in\mathbb{N}}$ of absolutely convex bounded sets and fix a basis $\mathcal{U}_0$ of absolutely convex 0-neighborhoods in E. Then E is a (DF)-space if and only if, for every sequence $(\lambda_n)_{n\in\mathbb{N}}$ of positive numbers and for every sequence $(W_n)_{n\in\mathbb{N}} \subset \mathcal{U}_0$, the set $\bigcap_{n\in\mathbb{N}}(W_n + \sum_{k=1}^n \lambda_k B_k)$ is a neighborhood of 0 in E.*

Proof. Necessity is easy. To show sufficiency, fix a sequence $(V_n)_{n\in\mathbb{N}}$ of closed absolutely convex neighborhoods of 0 in E such that $V := \bigcap_{n\in\mathbb{N}} V_n$ is bornivorous. For each $n \in \mathbb{N}$, we can find $\lambda_n > 0$ and $W_n \in \mathcal{U}_0$ with $\lambda_n B_n \subset 2^{-n-1}V$ and $W_n \subset 2^{-1}V_n$. Then we have

$$W_n + \textstyle\sum_{k=1}^n \lambda_k B_k \subset 2^{-1}V_n + \sum_{k=1}^n 2^{-(k+1)}V_n \subset V_n, \quad n = 1, 2, \ldots$$

By our hypothesis, $\bigcap_{n\in\mathbb{N}}(W_n + \sum_{k=1}^n \lambda_k B_k)$ is a 0-neighborhood in E, and it is contained in V. □

The preceding lemma permits to check the (DF)-property of a locally convex space E with a fundamental sequence of bounded sets by using only a *fixed* basis of 0-neighborhoods of E. (Note that the definition of a (DF)-space requires to consider arbitrary closed absolutely convex neighborhoods of 0.)

With Lemma 5. and the "Ernst-Schnettler trick" which already appeared in the proof of Lemma 2., we next improve the main result of [9] for sequences $\mathcal{V} \subset C(X)$.

6. Proposition. *Let $\mathcal{V} = (v_n)_n$ be a decreasing sequence of strictly positive continuous functions on a completely regular Hausdorff space X and $\overline{V} = \overline{V}(\mathcal{V})$. Then $C\overline{V}(X)$ is a (DF)-space.*

Proof. As in the proof of Lemma 2., we put $B_n := \{f \in C\overline{V}(X);\ v_n|f| \le 1 \text{ on } X\}$, $n = 1, 2, \ldots$ In view of Lemma 5., it suffices to fix a sequence $(\lambda_n)_n$ of positive numbers and a sequence $(\overline{v}_n)_n \subset \overline{V}$ and to check that $W := \bigcap_{n\in\mathbb{N}}(W_n + \sum_{k=1}^n \lambda_k B_k)$ is a 0-neighborhood in $C\overline{V}(X)$, where

$$W_n := \{f \in C\overline{V}(X);\ \overline{v}_n|f| \le 1 \text{ on } X\},\ n \in \mathbb{N}.$$

From [10], Lemma 1.2, we get

$$W_n + \sum_{k=1}^n \lambda_k B_k \;=\; \{f \in C\overline{V}(X);\ |f| \le \frac{1}{\overline{v}_n} + \sum_{k=1}^n \frac{\lambda_k}{v_k}\}$$

(taking $1/0 = \infty$ and, later on, $1/\infty = 0$; note that for an inductive application of [10], 1.2, one need not assume that the function $\overline{v}_n$ is continuous).

At this point , we let

$$a_n := \frac{1}{\overline{v}_n} + \sum_{k=1}^n \frac{\lambda_k}{v_k} \quad\text{and}\quad \overline{w}_n := \frac{1}{a_n}, \quad n = 1, 2, \ldots.$$

Clearly,

$$(*)\qquad \overline{w}_n \le \inf(v_1/\lambda_1, \ldots, v_n/\lambda_n, \overline{v}_n)$$

holds for each $n \in \mathbb{N}$, and hence $\overline{w}_n$ belongs to the Nachbin system $\overline{\overline{V}}$ associated with $\mathcal{V}$ on X with the discrete topology. Taking $\overline{w} := \sup_{n\in\mathbb{N}} \overline{w}_n$, the inequality (*) yields $\overline{w} \in \overline{\overline{V}}$, and we have

$$\{f \in C\overline{V}(X);\ \overline{w}|f| \le 1 \text{ on } X\} \subset W_n + \textstyle\sum_{k=1}^{n} \lambda_k B_k$$

for $n = 1, 2, \ldots$. But we can choose $\overline{v} \in \overline{V}$ with $\overline{w} \le \overline{v}$ (see [5], Subsection 4.2), and it follows that $\{f \in C\overline{V}(X);\ \overline{v}|f| \le 1 \text{ on } X\}$ is a 0-neighborhood in $C\overline{V}(X)$ which is contained in W. □

We are ready for the new proof of the sufficiency of condition (D). Since no additional technical assumption is needed here, the following proposition actually improves the second case of Theorem 6.9 in [5] slightly.

7. Proposition. *If $\mathcal{V} = (v_n)_n$ is a decreasing sequence of continuous weights on X which satisfies condition (D), then $\mathcal{V}C(X)$ equals $C\overline{V}(X)$ topologically.*

Proof. If $\mathcal{V} = (v_n)_n$ satisfies property (M,J), one can readily verify (see [5], Corollary 6.4.(3)) that on each bounded subset of $C\overline{V}(X)$, the induced topology is generated by the system $(q_{\overline{v},m})_{\overline{v}\in\overline{V}, m\in\mathbb{N}}$ of seminorms, where

$$q_{\overline{v},m}(f) := \sup_{x\in X_m} \overline{v}(x)|f(x)| \quad \text{for } f \in C\overline{V}(X).$$

On the other hand, it follows easily from property (N,J) (cf. [5], Remarks 3.9 and 3.10) that $C\overline{V}(X)_{|X_m}$ is normable for each $m \in \mathbb{N}$. Combining the two facts, we see that in the presence of condition (D), the bounded subsets of $C\overline{V}(X)$ must be metrizable.

By Proposition 7., $C\overline{V}(X)$ is a (DF)-space. The metrizability of the bounded subsets then implies $C\overline{V}(X)$ quasibarrelled, see Grothendieck [11], Théorème 5. Now Lemmas 2. and 3. of Section 2. yield $C\overline{V}(X)$ bornological, whence $\mathcal{V}C(X) = (C\overline{V}(X))_{\text{bor}} = C\overline{V}(X)$ holds topologically. □

The preceding proof shows a connection between the topological equality $\mathcal{V}C(X) = C\overline{V}(X)$ and the metrizability of the bounded subsets of the (DF)-space $C\overline{V}(X)$. The first results on (DF)-spaces whose bounded subsets are metrizable are due to Grothendieck [11]; a better understanding of the corresponding subclass of (DF)-spaces now follows from our work on the "dual density conditions" in [3].

8. Definition. Let E denote a locally convex space, $\mathcal{B}(E)$ the system of all closed absolutely convex bounded subsets of E (or any basis of bounded sets in E) and $\mathcal{U}(E)$ the system of all closed absolutely convex 0-neighborhoods of E (or any basis of neighborhoods of 0 in E). Then E is said to satisfy the *strong dual density condition (SDDC)* (respectively, the *dual density condition (DDC)*) if the following holds:

Given any function $\lambda : \mathcal{B}(E) \to \mathbb{R}_+ \setminus \{0\}$ and an arbitrary $A \in \mathcal{B}(E)$, there always exist a finite subset $\mathcal{B}$ of $\mathcal{B}(E)$ and $U \in \mathcal{U}(E)$ such that

$$A \cap U \subset \Gamma(\bigcup_{B \in \mathcal{B}} \lambda(B)B) \qquad \text{(respectively, } A \cap U \subset \overline{\Gamma}(\bigcup_{B \in \mathcal{B}} \lambda(B)B)\text{)},$$

where Γ (respectively, $\overline{\Gamma}$) denotes the absolutely convex hull (respectively, the closed absolutely convex hull).

The dual density conditions are dual reformulations of S. Heinrich's *density condition* (see [2]), and they are interesting mainly in the context of the class of (DF)-spaces.

It is obvious that (SDDC) $\Rightarrow$ (DDC), and one can find examples of (DF)-spaces with (DDC) which do not satisfy (SDDC). This, however, cannot happen under the hypothesis of Lemma 3.

9. Remark. *Let E denote a locally convex space with a fundamental sequence $(B_n)_n$ of closed absolutely convex bounded sets with the property that for any finite collection $\lambda_1, \dots, \lambda_m$ ($m \in I\!N$) of positive numbers, the set $\sum_{j=1}^{m} \lambda_j B_j$ is closed in E. Then (DDC) and (SDDC) are equivalent conditions for E.*

To prove that (DDC) implies (SDDC) for such a space E, one applies (DDC) with the sequence $(2^{-n}\lambda_n)_n$ instead of $(\lambda_n)_n$ and concludes (SDDC) from the hypothesis on E.

In particular, (DDC) and (SDDC) are equivalent for the spaces $\mathcal{V}C(X), C\overline{V}(X), \mathcal{V}_0C(X)$ and $C\overline{V}_0(X)$ if we assume $\mathcal{V} \subset C(X)$ and if, for the last two spaces, X is also assumed to be locally compact.

The following is the main characterication theorem of [3] for the dual density conditions in the setting of abstract (DF)-spaces.

10. Theorem. *Let E be a (DF)-space.*

(a) The following assertions are equivalent:

(1) E satisfies (DDC),

(2) the bounded subsets of E are metrizable,

(3) the space $l_\infty(E)$ of all bounded sequences in E is quasibarrelled.

(b) Similarly, the following assertions are equivalent:

(4) E satisfies (SDDC),

(5) $l_\infty(E)$ is bornological.

Since E is (topologically isomorphic to) a complemented subspace of $l_\infty(E)$, it clearly follows from 10. that any (DF)-space with (SDDC) (resp., (DDC)) must be bornological (resp., quasibarrelled), but the converse implications are false (even in the class of strong duals of Fréchet spaces). For more details on the dual density conditions and their application to spaces of type $\mathcal{L}_b(\lambda_1(A), E)$ where E is a locally complete (DF)-space, we refer to [3]. — We now return to $\mathcal{V}C(X)$ and $C\overline{V}(X)$ and immediately state the projective description theorem in its present (rather complete) form ([4], plus Bastin [1]).

11. Theorem. *Let $\mathcal{V} = (v_n)_n$ denote a decreasing sequence of continuous weights on a completely regular Hausdorff space and $\overline{V} = \overline{V}(\mathcal{V})$. We consider the following properties:*

(1) $\mathcal{V}$ satisfies condition (D) (for a sequence $J = (X_m)_m$ of closed *sets $X_m \subset X$),*

(2) the bounded subsets of $C\overline{V}(X)$ are metrizable,

(3) $C\overline{V}(X)$ satisfies (DDC) (or, equivalently, (SDDC)),
(4) $l_\infty(C\overline{V}(X))$ is quasibarrelled (or, equivalently, bornological),
(5) $l_\infty(C\overline{V}(X)) = \text{ind}_n\ l_\infty(Cv_n(X))$ holds algebraically and topologically,
(6) $\mathcal{V}C(X)$ or $\mathcal{V}_0C(X)$ or $C\overline{V}_0(X)$ satisfies (DDC) (or, equivalently, (SDDC)).
(I) $C\overline{V}(X)$ is quasibarrelled (or, equivalently, bornological),
(II) $\mathcal{V}C(X) = C\overline{V}(X)$ holds (algebraically and) topologically.

Then (1) to (5) are all equivalent, as are (I) and (II), and (1)–(5) implies (I)–(II). If X is locally compact, (6) is also equivalent to (1)–(5). — However, even if X is normal, (I)–(II) does not imply (1)–(5). On the other hand, this implication is true if each $\overline{v} \in \overline{V}$ is dominated by some $\tilde{v} \in \overline{V} \cap C(X)$, and in this case (1) to (5) and (I), (II) are equivalent properties.

We give a sketch of a large part of the **proof** of Theorem 11. ([4]).

From the proof of Proposition 7., from Remark 9. and Theorem 10., it is clear that (1) $\Rightarrow$ (2) $\Leftrightarrow$ (3) $\Leftrightarrow$ (4) $\Rightarrow$ (I). Invoking Lemmas 2. and 3.(a), we also see that the two properties in (I) are indeed equivalent, and we conclude (I) $\Leftrightarrow$ (II) from $\mathcal{V}C(X) = (C\overline{V}(X))_{\text{bor}}$. Next, (5) is equivalent to (4) (with bornological) since one can easily verify that

$$\text{ind}_n\ l_\infty(Cv_n(X)) = (l_\infty(C\overline{V}(X))_{\text{bor}}$$

holds algebraically and topologically. Also, in view of (3) $\Rightarrow$ (II), we have (3) $\Rightarrow$ (6) for $\mathcal{V}C(X)$. Moreover, if X is locally compact, we know that $\mathcal{V}_0C(X)$ always is a topological subspace of $C\overline{V}_0(X) \subset C\overline{V}(X)$. $\mathcal{V}_0C(X)$ and its completion $C\overline{V}_0(X)$ are (DF)-spaces, and thus (6) for $\mathcal{V}_0C(X)$ or $C\overline{V}_0(X)$ is also implied by (2).

Concerning (2) $\Rightarrow$ (1), we first note that 0 has a countable basis of neighborhoods in each bounded subset of $C\overline{V}(X)$ if condition (2) holds. And since $C\overline{V}(X)$ has a countable basis of bounded sets, it is clearly possible to generate the topology of the bounded subsets of $C\overline{V}(X)$ by only using the seminorms $p_{\overline{v}}$ for the weights $\overline{v}$ in a countable subset of $\overline{V}$. We can then replace this subset by an increasing sequence $(\overline{v}_m)_{m\in\mathbb{N}} \subset \overline{V}$. At this point, defining

$$X_m := \{x \in X;\ \overline{v}_m(x)/v_m(x) \geq 1\}, \quad m = 1,2,\ldots,$$

we have obtained an increasing sequence $J = (X_m)_m$ of closed subsets of X. It is a matter of routine to check that $\mathcal{V}$ must now satisfy both (N,J) and (M,J).

The implication (6) $\Rightarrow$ (2) for locally compact X follows in a similar way. But its complete proof involves the space $C_c(X)$ of continuous functions with compact support under the topology τ induced by $\mathcal{V}C(X), \mathcal{V}_0C(X)$ or $C\overline{V}_0(X)$ (and some of its properties), and we refer to [4], proof of Corollary 5.8 (and of Proposition 5.5.(b)) for the details.

From the example of [5] which we had already mentioned after Definition 4., we know that (II) does not imply (1) even if X is normal. But F. Bastin [1] has proved quite recently that the implication (I) $\Rightarrow$ (1) is indeed true under the additional *"continuous domination" assumption* (which may come as some kind of surprise at this point). □

We would now like to explain the relationship between (1)–(5) and (I)–(II) (which, as we have seen, is quite delicate), as well as Bastin's contribution to Theorem 11., in more detail.

Let us say that we are in the sequence space case if X is some index set I (not necessarily countable) with the discrete topology. Then $A := (a_n)_{n\in\mathbb{N}}$, $a_n := 1/v_n$, defines a *Köthe matrix* on I, and if I and A are fixed, we abbreviate the *Köthe echelon space* $\lambda_1(I,A)$ by λ_1.

In this case, $\mathcal{V}C(X)$ is the "inductive dual" $E_i' := \text{ind}_n\ E'_{U_n^\circ}$ of $E = \lambda_1$, where $(U_n)_{n\in\mathbb{N}}$ denotes a basis of neighborhoods of 0 in λ_1, and $C\overline{V}(X)$ equals the strong dual E_b'. Thus, the projective description problem in the sequence space case asks for a necessary and sufficient condition on A such that $\lambda_1 = \lambda_1(I,A)$ is distinguished; i.e., such that the strong dual $(\lambda_1)_b'$ is barrelled or, equivalently, bornological. In [5], it was proved that condition (D) is indeed sufficient (without any further technical assumption). In [2], we then showed that condition (D) is necessary, too, and thus solved the sequence space case of the projective description problem. Our main tool was *Heinrich's density condition* and the result that a Köthe echelon space λ_1 is distinguished if and only if it satisfies the density condition; i.e., if and only if the bounded subsets of $(\lambda_1)_b'$ are metrizable (which, in the present terminology, is equivalent to saying that $(\lambda_1)_b'$ satisfies (DDC) or, equivalently, (SDDC)).

Later on, D. Vogt [13] derived a different necessary and sufficient condition on $A = (a_n)_n$ for distinguishedness of $\lambda_1 = \lambda_1(I,A)$ and gave a short direct proof of the equivalence. In terms of $\mathcal{V} = (v_n)_n$, $v_n = 1/a_n$, Vogt's condition reads as follows:

(V_1) For every sequence $(\lambda_k)_{k\in\mathbb{N}}$ of positive numbers, there is a sequence $(C_k)_k$ of positive numbers such that for every $C > 0$ and every $n \in \mathbb{N}$, one can find $m \in \mathbb{N}$ with

$$\inf\{C/v_n,\ \sup_{k\in\mathbb{N}} C_k/v_k\} \leq \sup\{\lambda_1/v_1,\dots,\lambda_m/v_m\}.$$

Utilizing $\overline{V} = \overline{V}(\mathcal{V})$, this can also be expressed in the following way:

(V_2) For every sequence $(\lambda_k)_k$ of positive numbers, there is a weight $\overline{v} \in \overline{V}$ such that for every $C > 0$ and every $n \in \mathbb{N}$, one can find $m \in \mathbb{N}$ with

$$\inf\{C/v_n, 1/\overline{v}\} \leq \sup\{\lambda_1/v_1,\dots,\lambda_m/v_m\}.$$

Returning to the more general case of function spaces, here is Bastin [1], Theorem 1:

12. Theorem. *Let $\mathcal{V} = (v_n)_n$ denote a decreasing sequence of continuous weights on a completely regular Hausdorff space X such that it suffices to consider only the continuous weights in $\overline{V}$ (in the sense that each $\overline{v} \in \overline{V}$ is dominated by some $\tilde{v} \in \overline{V} \cap C(X)$; this is always true if X is locally compact and σ-compact, cf. [5], Corollary 6.7). Then the following assertions are equivalent:*

(a) $C\overline{V}(X)$ is bornological,

(b) for each sequence $(\lambda_k)_k$ of positive numbers, there exists $\overline{v} \in \overline{V}$ such that for every $C > 0$ and every $n \in \mathbb{N}$, one can find $m = m(C,n) \in \mathbb{N}$ with

$$\inf\{C/v_n, 1/\overline{v}\} \leq \textstyle\sum_{k=1}^{m} \lambda_k/v_k,$$

(c) $\mathcal{V}$ satisfies condition (D).

Clearly, condition (b) of Theorem 12. is a slight reformulation of condition (V_2) above (just replace $(\lambda_k)_k$ by $(2^{-k}\lambda_k)_k$ in one direction). — Bastin's **proof** of 12.(a) $\Leftrightarrow$ (b) is short and direct, only utilizing the "Ernst-Schnettler trick" (i.e., the equality $S_m = D_m$ in the proof of Lemma 2.) at one point. And this proof also applies when X is discrete. Thus, the equivalence (b) $\Leftrightarrow$ (c) simply follows from the known characterization of the distinguished Köthe echelon spaces. □

Moreover, Bastin remarks that her proof of the implication (b) $\Rightarrow$ (a) in Theorem 12. actually works without the assumption of the continuous domination property and that one can prove (b) $\Leftrightarrow$ (c) in this generality, too. Hence she has a different proof of our Proposition 7. above.

We close this section by noting that the continuous domination property also permits to derive another result which fails to hold in general.

In [5], the authors asked whether for two decreasing sequences $\mathcal{V}_i$ on completely regular Hausdorff spaces X_i and $\overline{V}_i := \overline{V}(\mathcal{V}_i)$, $i = 1, 2$,

($\times$) the topological equalities $\mathcal{V}_1C(X_1) = C\overline{V}_1(X_1)$ and $\mathcal{V}_2C(X_2) = C\overline{V}_2(X_2)$ imply $(\mathcal{V}_1 \otimes \mathcal{V}_2)C(X_1 \times X_2) = C\overline{V}(\mathcal{V}_1 \otimes \mathcal{V}_2)(X_1 \times X_2)$ topologically;

i.e., whether the projective description is inherited "by products".

As we observed in [4], the example at the end of [5] (mentioned after Definition 4. above) can be reinterpreted to yield a negative answer to this general question. However, Bastin [1], Theorem 2. has now pointed out that the inheritance property ($\times$) does hold under the assumptions that $\mathcal{V}_i \subset C(X_i)$ for $i = 1, 2$ and that it suffices to consider only the continuous weights in both $\overline{V}_1$ and $\overline{V}_2$.

4. Concluding remarks

The theory of projective descriptions of weighted (LB)-spaces of type $\mathcal{V}C(X)$ (and of the Köthe co-echelon spaces) has been clarified to a large extent during the last years. At the moment, one could still ask *questions* like:

1. Is the distinguishedness of the Köthe echelon space $\lambda_1 = \lambda_1(I, A)$ equivalent to some condition on possible "sectional subspaces" l_1 of λ_1 and on their "position" in λ_1?

2. If the inductive limit topology of $\mathcal{V}C(X)$ is strictly stronger than the weighted topology of $C\overline{V}(X)$, can one still describe (a basis of) the continuous seminorms for $\mathcal{V}C(X)$ in a "natural and useful" way (as "close" to the canonical seminorms of $C\overline{V}(X)$ as possible)?

1. is inspired by the well-known characterization of the Montel echelon spaces λ_1 and by a glance at condition (D) (especially, at part (N,J) of this condition) and seems to be an interesting question from the point of view of the structure theory of Köthe spaces.

Concerning 2., we note that under certain (reasonable) conditions, Ernst-Schnettler [10] show how to generate the (gDF)-topology associated with a weighted space $CV(X)$ (which has an absorbent sequence of bounded sets) by some system $(p_w)_{w \in W}$ of weighted sup-seminorms, where W is a concrete system of weights associated with V in a certain (rather technical) way. They are then also able to characterize the quasibarrelledness of $CV_0(X)$ with their methods. In view of our Proposition 1. above, one might wonder if the (quasi-) barrelled space associated with the weighted space $C\overline{V}(X)$ can be described "in a similar way". However, note that $\overline{V} = \overline{V}(\mathcal{V})$ clearly is the *maximal* Nachbin family associated with $\mathcal{V}$. If for each $x \in X$, there exists $f \in C\overline{V}(X)$ with $f(x) \neq 0$ (e.g., this obviously holds if X is locally compact), it follows from [10], Lemma 2.3 that there is even no stronger weighted topology on $C\overline{V}(X)$ which yields the same bounded sets. Hence the desired description of the topology of $\mathcal{V}C(X)$ in question 2. must necessarily involve seminorms of a different type; i.e., seminorms which are not weighted sup-seminorms.

But, in any case, it seems more interesting from this point on to switch from the setting of weighted *(LB)-spaces* to the corresponding weighted inductive limits $\mathcal{V}C(X)$ and $\mathcal{V}_0C(X)$, where $\mathcal{V} = (V_n)_n$ is a decreasing sequence of *systems* of weights (cf. [7]). More specifically, we propose:

3. Treat the (LF)-spaces of type $\mathcal{V}C(X)$ and $\mathcal{V}_0C(X)$ in relation to their projective hulls in more detail.

Actually, the very first results of [7] on $\mathcal{V}_0C(X)$ were already obtained in this more general setting. On the other hand, for finer theorems it clearly remains to find the "right" generalizations of, say, the "regularly decreasing" condition of [7], of condition (D), etc.

The most interesting problems on weighted inductive limits, however, require to go back "to the roots"; viz., to spaces of *holomorphic* functions, with many important applications, e.g. in Ehrenpreis's theory of "analytically uniform spaces". Our general proposal in this context reads as follows:

4. Consider the projective description problem in the setting of (LB)- and (LF)-spaces of type $\mathcal{V}H(X)$ and $\mathcal{V}_0H(X)$ (where "H" stands for holomorphic functions, just as "C" for continuous functions).

The open mapping lemma of A. Baernstein shows that the (DFS)-spaces (or, more generally, the [semi-] Montel (LB)-spaces) $\mathcal{V}H(X)$ are topological subspaces of $\mathcal{V}C(X)$, and this, together with the results on spaces of continuous functions, already yields plenty of interesting applications (see [7]). However, the following *question* seems to be open.

5. Is an (LB)-space of type $\mathcal{V}H(X)$ always a topological subspace of the corresponding $\mathcal{V}C(X)$?

On the other hand, the Baernstein lemma cannot be applied in the case of proper (LF)-spaces. But, under "mild" hypotheses, Retakh's theory (or Vogt's recent theory of projective and inductive limits) might help to find an affirmative solution for the topological subspace problem of $\mathcal{V}H(X)$ in $\mathcal{V}C(X)$ even for many (LF)-spaces.

In fact, there is some hope that the general method which was first developed in [7]; viz., to work with partitions of unity in the setting of $\mathcal{V}C(X)$ or $\mathcal{V}_0C(X)$ and to carry over the corresponding results to $\mathcal{V}H(X)$ and $\mathcal{V}_0H(X)$ via an affirmative solution of the "subspace problem", can still be used for (LF)-spaces, too.

References

[1] F. Bastin, 'On bornological spaces $C\overline{V}(X)$', to appear.

[2] K. D. Bierstedt, J. Bonet, 'Stefan Heinrich's density condition for Fréchet spaces and the characterization of the distinguished Köthe echelon spaces', *Math. Nachr.* **135**(1988), 149 - 180.

[3] K. D. Bierstedt, J. Bonet, 'Dual density conditions in (DF)-spaces, I.', *Resultate Math.* **14**(1988), 242 - 274.

[4] K. D. Bierstedt, J. Bonet, 'Dual density conditions in (DF)-spaces, II.', to appear in *Bull. Soc. Roy. Sci. Liège.*

[5] K. D. Bierstedt, R. Meise, 'Distinguished echelon spaces and the projective description of weighted inductive limits of type $\mathcal{V}_d C(X)$', pp. 169 - 226 in: *Aspects of Mathematics and its Applications*, Elsevier Sci. Publ., North-Holland Math. Library, 1986.

[6] K. D. Bierstedt, R. Meise, 'Weighted inductive limits and their projective descriptions', *Doğa Mat.* **10**, 1(1986) (Special Issue: Proceedings of the Silivri Conference 1985), 54 - 82.

[7] K. D. Bierstedt, R. Meise, W. H. Summers, 'A projective description of weighted inductive limits', *Trans. Amer. Math. Soc.* **272**(1982), 107 - 160.

[8] K. D. Bierstedt, R. G. Meise, W. H. Summers, 'Köthe sets and Köthe sequence spaces', pp. 27 - 91 in: *Functional Analysis, Holomorphy and Approximation Theory*, North-Holland Math. Studies **71**, 1982.

[9] J. Bonet, 'On weighted inductive limits of spaces of continuous functions', *Math. Z.* **192**(1986), 9 - 20.

[10] B. Ernst, P. Schnettler, 'On weighted spaces with a fundamental sequence of bounded sets', *Arch. Math. (Basel)* **47**(1986), 552 - 559.

[11] A. Grothendieck, 'Sur les espaces (F) et (DF)', *Summa Brasil. Math.* **3**(1954), 57 - 122.

[12] A. Grothendieck, 'Produits tensoriels topologiques et espaces nucléaires', *Mem. Amer. Math. Soc.* **16**, 1955 (reprint 1966).

[13] D. Vogt, 'Distinguished Köthe spaces', to appear.

PROJECTIVE DESCRIPTIONS OF WEIGHTED INDUCTIVE LIMITS: THE VECTOR-VALUED CASES

Klaus D. Bierstedt
FB 17, Univ.-GH-Paderborn
Postfach 1621
D-4790 Paderborn
F.R. of Germany

José Bonet
Matemáticas
E.T.S. Arquitectura
Univ. Politécnica
E-46071 Valencia
Spain

ABSTRACT. *This is a report on some recent research on the projective description for weighted inductive limits of spaces of vector-valued continuous functions and on topological properties of vector-valued echelon and co-echelon spaces. In the first part, it is pointed out that most of the theorems known in the scalar case remain valid if the functions take values in a (DF)-space and that, similarly, many properties of the classical sequence spaces carry over to echelon spaces with values in a Fréchet space resp. to (DF)-space-valued co-echelon spaces. In the second part, we turn to the case that the continuous functions take their values in Fréchet spaces. This is related to some properties in the structure theory of Fréchet spaces and work by D. Vogt; part of the results in this direction are due to A. Galbis.*

Introduction

At the Functional Analysis Conference at the Nazim Terzioglu Center in Silivri, Bierstedt and Meise [11] reported on weighted inductive limits of spaces of continuous and holomorphic functions and their projective descriptions. They included a brief report on the "state of art" in the vector-valued case at that time. The vector-valued case

T. Terzioğlu (ed.), Advances in the Theory of Fréchet Spaces, 195–221.

was originally treated by Bierstedt and Meise [9] due to their interest in the ε-tensor product representations for spaces of vector-valued functions and spaces of functions of several variables, with applications to the representation of vector-valued distributions and distributions of several variables as boundary values of holomorphic functions. Since then, some progress on projective descriptions of weighted inductive limits of spaces of (DF)- and Fréchet-valued continuous functions has been achieved, and our main purpose is to give a report on this progress here.

One of the original aims of the method of "projective description" of weighted inductive limits of spaces of continuous or holomorphic functions developed in [12] was to give a canonical, sufficiently general, and useful representation of the inductive limit topology and its continuous seminorms by associating a certain projective hull with it. For most applications to spaces of holomorphic functions, the original aim was already achieved in 1982,[12], in a result in which an open mapping lemma of A. Baernstein II and a result on spaces of continuous functions, in whose proof partitions of unity could be used, permitted to conclude. At that time, the corresponding inductive limits of spaces of continuous functions could be considered as a tool, but rather concentrating on them, it is natural to ask for characterizations (in terms of the decreasing sequence of weights) when the inductive limit can actually be identified algebraically and topologically with its associated projective hull. This is the projective description problem. It turned out that the ideas and methods developed for function spaces had interesting consequences for echelon and co-echelon spaces of arbitrary order p, $1 \le p \le \infty$, or $p=0$ (cf. [13]), for example in the characterization of the quasinormable spaces $\lambda_p(A)$ and in the characterization of the distinguished Köthe echelon spaces of order 1, $\lambda_1(A)$, in terms of the Köthe matrix A of its elements. We refer to our other contribution in these Proceedings for a report on the recent progress on projective descriptions of weighted inductive limits in the scalar case.

The treatment of weighted inductive limits of spaces of continuous functions with values in (DF)-spaces turned out to be helpful in our original characterization of the distinguished $\lambda_1(A)$-spaces (see [5]). Moreover it provided partial positive answers to the following problem of Grothendieck, recently solved in the negative by Taskinen [35,36]: If F is a Fréchet space and E a (DF)-space, must then $L_b(F,E)$ and $F'_b \otimes_\varepsilon E$ be (DF)-spaces?. On the other hand the case in which the functions take values in a Fréchet space is related to the theory of tensor products of Fréchet and (DF)-spaces ([29], [32], [39]), to the structure theory of Fréchet spaces and to the work of Vogt [38] on the identity L(E,F)=LB(E,F) for pairs of Fréchet spaces. The problem of a projective description of weighted inductive limits of spaces of vector-valued continuous functions has also connections with the question of the interchangeability of countable inductive limits and tensor products (cf. [30]).

After an introductory section with the necessary notations, this article is divided in two main sections. In the first one we shall concentrate on weighted inductive limits of spaces of continuous functions with values in (DF)-spaces. In this (DF)-valued setting the same results as in the scalar case hold, provided that some minor technical assumptions are satisfied. The Fréchet-valued case is completely different. Here some hypotheses on the sequence of weights and on the Fréchet space are needed to obtain positive results, and in many cases the corresponding conditions are necessary and sufficient. These conditions appeared in the work of Vogt on the characterization of pairs (E,F) of Fréchet spaces such that L(E,F)=LB(E,F) and on the study of the functor $\mathrm{Ext}^1(E,F)$ for pairs (E,F) of Fréchet spaces (cf. [38,37]).

1. Notation. Formulation of the problem

In the sequel, we let X denote a completely regular Hausdorff space, $\mathcal{V}=(v_n)_{n\in\mathbb{N}}$ a decreasing sequence of strictly positive continuous functions on X, $E\neq\{0\}$ a locally convex space and cs(E) the family of all continuous seminorms on E.

We introduce the weighted spaces of continuous functions

$Cv_n(X,E):=\{f\in C(X,E);\ q_{v_n,p}(f):=\sup_{x\in X} v_n(x)p(f(x))<\infty\ \forall p\in cs(E)\}$,

$C(v_n)_o(X,E):=\{f\in Cv_n(X,E);\ v_n(p\circ f)$ vanishes at ∞ on X $\forall p\in cs(E)\}$,

both endowed with the locally convex topology generated by the family $\{q_{v_n,p};\ p\in cs(E)\}$ of seminorms, and the weighted inductive limits

$\mathcal{V}C(X,E):=\operatorname{ind}_n Cv_n(X,E)$, $\mathcal{V}_oC(X,E):=\operatorname{ind}_n C(v_n)_o(X,E)$.

The maximal Nachbin system of weights (i.e. non-negative upper semicontinuous functions) associated with $\mathcal{V}$ is

$\bar{V}=\bar{V}(\mathcal{V}):=\{\bar{v}\geq 0$ upper semicontinuous on X; for each $n\in\mathbb{N}$ $\sup_{x\in X}\frac{\bar{v}(x)}{v_n(x)}<\infty\}$.

The corresponding weighted spaces which are the projective hulls of the weighted inductive limits are defined by:

$C\bar{V}(X,E):=\{f\in C(X,E);\ q_{\bar{v},p}(f)<\infty\ \forall\bar{v}\in\bar{V}\ \forall p\in cs(E)\}$,

$C\bar{V}_o(X,E):=\{f\in C\bar{V}(X,E);\ \bar{v}(p\circ f)$ vanishes at ∞ on X $\forall\bar{v}\in\bar{V}\ \forall p\in cs(E)\}$,

both endowed with the locally convex topology given by $\{q_{\bar{v},p};\ \bar{v}\in\bar{V},\ p\in cs(E)\}$. Clearly $\mathcal{V}C(X,E)$ and $\mathcal{V}_oC(X,E)$ are continuously injected in $C\bar{V}(X,E)$ and $C\bar{V}_o(X,E)$, respectively. If E is the field of real or complex numbers, we drop it from the notation and we write $\mathcal{V}C(X)$, $\mathcal{V}_oC(X)$, $C\bar{V}(X)$, $C\bar{V}_o(X)$.

If X is a discrete space (i.e. an index set I), the usual notation for co-echelon spaces is as follows:

$k_\infty(E)=\operatorname{ind}_n l_\infty(v_n,E)$ instead of $\mathcal{V}C(I,E)$,
$k_o(E)=\operatorname{ind}_n c_o(v_n,E)$ instead of $\mathcal{V}_oC(I,E)$,
$K_\infty(E)=K_\infty(I,\bar{V},E)$ instead of $C\bar{V}(I,E)$,
$K_o(E)=K_o(I,\bar{V},E)$ instead of $C\bar{V}_o(I,E)$.

We recall that in this case we associate with $\mathcal{V}$ a Köthe matrix $A=(a_n)_{n\in\mathbb{N}}$ on I by taking $a_n(i):=v_n(i)^{-1}$ for all $n\in\mathbb{N}$ and $i\in I$. In the scalar case, K_∞ is the strong dual of the Köthe echelon space $\lambda_1(A)=\lambda_1(I,A)$ and k_∞ is the bornological space associated with K_∞ (cf. [13]).

The general problem of a projective description of weighted inductive limits of spaces of vector valued continuous functions can now be stated as follows.

Problem: *Under which conditions on* X, $\mathcal{V}$ *and* E *do the identities*

$$\text{(a)}\quad \mathcal{V}_oC(X,E)=C\overline{V}_o(X,E)$$

$$\text{(b)}\quad \mathcal{V}C(X,E)=C\overline{V}(X,E)$$

hold algebraically and topologically?

Here is the solution of the projective description problem in the scalar case:

1. Theorem. (a) Let X be a locally compact space.

1. $\mathcal{V}_oC(X)$ is a topological subspace of $C\overline{V}_o(X)$.

2. The following conditions are equivalent (abbreviated in the sequel by "tfae"):

(i) $\mathcal{V}_oC(X)=C\overline{V}_o(X)$ algebraically,

(ii) $\mathcal{V}_oC(X)$ is regular,

(iii) $\mathcal{V}_oC(X)$ is complete,

(iv) $\mathcal{V}$ is regularly decreasing, i.e. $\forall n\ \exists m,\ m>n,$ $\forall\varepsilon>0\ \forall k\geq m\ \exists\delta>0$ with $v_k(x)\geq\delta v_n(x)$ if $v_m(x)\geq\varepsilon v_n(x)$,

(v) $\forall n\ \exists m\ \forall\varepsilon>0\ \exists\bar{v}\in\overline{V} :\ v_m\leq\max(\varepsilon v_n,\bar{v})$.

(b) Let X be a completely regular Hausdorff space.

1. $\mathcal{V}C(X)=C\overline{V}(X)$ holds algebraically and the two spaces have the same bounded sets.

2. Assume that every $\bar{v}\in\overline{V}$ is dominated by some $w\in\overline{V}\cap C(X)$ (which is true, e.g., if X is discrete or if X is locally compact and σ-compact). Then tfae:

(i) $\mathcal{V}C(X)=C\overline{V}(X)$ topologically,

(ii) $\mathcal{V}$ has condition (D), i.e. there is an increasing sequence $J=(X_m)_{m\in\mathbb{N}}$ of subsets of X such that

$$(N,J)\quad \forall m\ \exists n(m)\geq m :\quad \inf_{x\in X_m}\frac{v_k(x)}{v_{n(m)}(x)}>0\quad \forall k>n(m)$$

(M,J) $\forall n\ \forall Y\subset X$ with $Y\cap(X\setminus X_m)\neq\emptyset$ for all $m\in\mathbb{N}$ $\exists n'=n'(n,Y)>n$:

$$\inf_{y\in Y} \frac{v_{n'}(y)}{v_n(y)} = 0.$$

Parts (a) and (b)1 in theorem 1 were already established by Bierstedt, Meise and Summers in [12]. The regularly decreasing condition on $\mathcal{V}$ characterizes the quasinormable Köthe echelon spaces $\lambda_1(A)$. Condition (D) was introduced by the first author and Meise in [10], where it was proved that (D) and some technical assumptions imply $\mathcal{V}C(X)=C\bar{V}(X)$ topologically. An example showed that the converse was false for normal spaces X. For discrete spaces X, the converse was established in [5] by the authors, characterizing in this way the distinguished Köthe echelon spaces $\lambda_1(A)$. In fact condition (D) implies the topological identity $\mathcal{V}C(X)=C\bar{V}(X)$ without any further assumption (see our other article in these Proceedings). The necessity of condition (D) under the continuous domination property was recently proved by Bastin [1], using an idea of Vogt [41].

2. Weighted inductive limits of spaces of (DF)-valued continuous functions

In this section we restrict our attention to (DF)-spaces E. The class of the (DF)-spaces was introduced by Grothendieck [28] as an extension of the class of the strong duals of Fréchet spaces. It also contains all the countable inductive limits of normed spaces. Every (DF)-space, and even every (gDF)-space (cf. [31], [34]) satisfies the countable neighbourhood property (which we shall abbreviate by c.n.p.), i.e. $\forall (p_n)_{n\in\mathbb{N}} \subset cs(E)$ $\exists (\lambda_n)_{n\in\mathbb{N}}$, $\lambda_n>0$ $(n\in\mathbb{N})$ $\exists p\in cs(E)$ such that $p_n\le\lambda_n p$ for all $n\in\mathbb{N}$ or, equivalently, for every sequence $(U_n)_{n\in\mathbb{N}}$ of 0-neighbourhoods in E there are $\lambda_n>0$, $n\in\mathbb{N}$, such that $\bigcap_{n\in\mathbb{N}} \lambda_n U_n$ is also a 0-neighbourhood in E. We refer to [24] and [15] for a study of the class of the l.c.s. satisfying the c.n.p. Let us remark that there are even l.c.s. with the c.n.p., but without a fundamental sequence of bounded sets.

The countable neighbourhood property, shared by all (DF)-spaces, is the condition on E which is needed to establish our main projective description result, a vector-valued extension of theorem 1.1.

1. Theorem. Let E be a l.c.s. satisfying the countable neighbourhood property.

(a) Let X be a locally compact space.

1. $V_oC(X,E)$ is a topological subspace of $C\overline{V}_o(X,E)$.

2. $V_oC(X,E)=C\overline{V}_o(X,E)$ holds algebraically if and only if V is regularly decreasing. In this case, if E is complete, $VC(X,E)$ is complete ,too, and we have

$$V_oC(X,E) = V_oC(X)\hat{\otimes}_\varepsilon E = C\overline{V}_o(X)\hat{\otimes}_\varepsilon E = C\overline{V}_o(X,E).$$

(b) Let X be a completely regular Hausdorff space.

1. $VC(X,E)=C\overline{V}(X,E)$ algebraically and the two spaces have the same bounded sets.

2. If every weight $\bar{v}\in\overline{V}$ is dominated by some weight $w\in\overline{V}\cap C(X)$, then the topological equality $VC(X,E)=C\overline{V}(X,E)$ is equivalent to condition (D).

We will present now some indications of the proof of this main theorem. Part (a)1 was obtained by the second author in [14], using a direct partition of unity argument which was an adaptation of the one given in the normed case in [12]. The original proof of the sufficiency of condition (D) for the topological identity $VC(X)=C\overline{V}(X)$ given in [10], with some technical assumptions, also works to obtain (b)2, and the details can be seen in [7]. The necessity of (D) follows directly from Bastin's result, [1], mentioned in connection with theorem 1.1(b)2.

The parts of theorem 1 above which concern the algebraic coincidence will now be deduced from several results in a more general setting. These results will also be applied in section 3 to treat the algebraic coincidence in the case of Fréchet valued functions.

We first recall the following lemma from [6,2.2].

2. Lemma. Let E be a locally complete l.c.s. (i.e. every bounded disc in E generates a Banach space). Then $K_\infty(E)$ and $L_b(\lambda_1(A),E)$ coincide

algebraically and topologically.

Proof. Given $f\in L_b(\lambda_1(A),E)$, we define $\psi(f):=(f(e_i))_{i\in I}$, where the e_i's are the canonical unit vectors in $\lambda_1(A)$. Conversely, given $x\in K_\infty(E)$, we define $\Phi(x):\lambda_1(A)\longrightarrow E$ by $\Phi(x)(\alpha):=\sum_{i\in I}\alpha_i x_i$. (Here the local completeness of E is needed.) Now the mappings ψ and Φ are linear and continuous and each of them is the inverse of the other. ∎

In what follows we will denote by L(G,H) (resp. LB(G,H)) the linear space of all the continuous linear (resp. bounded linear) maps from a l.c.s. G into a l.c.s. H. ($f\in L(G,H)$ is bounded if it maps a certain 0-neighbourhood in G into a bounded subset of H.) A pair (G,H) is said to satisfy the localization property if every equicontinuous subset A of L(G,H) is equibounded, i.e., there is a 0-neighbourhood U in G such that $A(U):=\bigcup\{f(U);f\in A\}$ is bounded in H. With this notation we have the following consequence of lemma 2.

3. Proposition. Let E be a locally complete l.c.s.

(a) $K_\infty(E)=k_\infty(E)$ algebraically if and only if $L(\lambda_1(A),E)=LB(\lambda_1(A),E)$.

(b) Every bounded subset of $K_\infty(E)$ is bounded in some $l_\infty(v_n,E)$ if and only if $(\lambda_1(A),E)$ has the localization property.

To apply this proposition in the proof of our main result, we need the following remark, essentially due to Grothendieck.

4. Remark. Let G be a metrizable l.c.s.. If H has a fundamental sequence of bounded sets or satisfies the c.n.p., then (G,H) has the localization property.

Proof. If H has a fundamental sequence of bounded sets, the result follows from [21]. Assume that H has the c.n.p. and suppose that there exists an equicontinuous subset A of L(G,H) which is not equibounded. If $(U_n)_{n\in\mathbb{N}}$ is a basis of absolutely convex 0-neighbourhoods in G, for each $n\in\mathbb{N}$ we can find a 0-neighbourhood V_n in H such that $A(U_n)$ is not absorbed by V_n. By assumption there are $\lambda_n>0$, $n\in\mathbb{N}$, such that $V:=\bigcap_{n\in\mathbb{N}}\lambda_n V_n$ is a 0-neighbourhood in G. Since A is equicontinuous, there is $m\in\mathbb{N}$ with $U_m\subset\bigcap\{f^{-1}(V);\ f\in A\}$. This implies

$A(U_m) \subset \lambda_m V_m$, which is a contradiction.■

Here is the result which contains theorem 1(b)1.

5. Corollary. Let E be a l.c.s. with the c.n.p. or a l.c.s. such that its completion $\hat{E}$ admits a fundamental sequence of bounded sets. Then $\mathcal{V}C(X,E)=C\overline{V}(X,E)$ algebraically and both spaces have the same bounded sets. Moreover $\mathcal{V}C(X,E)$ is a regular inductive limit.

Proof. Let B be a bounded subset of $C\overline{V}(X,E)$. We denote by $K_\infty(\hat{E})$ the corresponding co-echelon space, when X is endowed with its discrete topology. Then B is a bounded subset of $K_\infty(\hat{E})$. Since the completion of a l.c.s. with the c.n.p. also satisfies the c.n.p., we can apply remark 4 and proposition 3 to obtain that B is bounded in some $l_\infty(v_n,\hat{E})$. Each element of B is a continuous function with values in E. This implies that B is a bounded subset of $Cv_n(X,E)$.■

Part (a)2 in theorem 1 extends a result of [14], and it is a consequence of corollary 5 and our next two results.

6. Proposition. Let E be a l.c.s. If $\mathcal{V}$ is regularly decreasing and $\mathcal{V}C(X,E)=C\overline{V}(X,E)$ algebraically, then $\mathcal{V}_oC(X.E)=C\overline{V}_o(X,E)$.

Proof. We fix $f\in C\overline{V}_o(X,E) \subset C\overline{V}(X,E)$. By assumption $f\in Cv_n(X,E)$ for some $n\in\mathbb{N}$. We select $m\in\mathbb{N}$ as in the regularly decreasing condition (1.1 (a) 2(v)) and we suppose that $f\notin C(v_m)_o(X,E)$. Then we can find $\varepsilon>0$, $p\in cs(E)$ such that for each compact subset K of X there is $x(K)\notin K$ with $p(f(x(K)))v_m(x(K))>\varepsilon$. On the other hand there is $M_p>0$ such that $\sup_{x\in X} v_n(x)p(f(x))\le M_p$. For $\varepsilon/M_p>0$ we find $\bar{v}\in\overline{V}$ with $v_m\le\max((\varepsilon/M_p)v_n,\bar{v})$. Given $\bar{v}\in\overline{V}$ we can choose a compact subset L of X with $\bar{v}(x)p(f(x))<\varepsilon$ if $x\in X\setminus L$. For any $x\notin L$ we have

$$v_m(x)p(f(x))\le\max((\varepsilon/M_p)v_n(x),\bar{v}(x))p(f(x))\le\varepsilon.$$

This is a contradiction.■

Observe that, since $\mathcal{V}C(X)=C\overline{V}(X)$, the proof above yields also a new proof of the fact that the regularly decreasing condition implies $\mathcal{V}_oC(X)=C\overline{V}_o(X)$.

7. Proposition. Let E be a l.c.s. and X a locally compact space. Tfae:
(i) every bounded subset of $C\overline{V}_o(X,E)$ is bounded in some $C(v_n)_o(X,E)$,
(ii)(1) $\mathcal{V}$ is regularly decreasing and (2) every bounded subset of $C\overline{V}(X,E)$ is bounded in some $Cv_n(X,E)$.

Proof. (ii) implies (i). Given a bounded subset B of $C\overline{V}_o(X,E)$, B is also bounded in $C\overline{V}(X,E)$. By (ii), there is $n\in\mathbb{N}$ such that B is bounded in $Cv_n(X,E)$. If $m\in\mathbb{N}$ is selected according to the regularly decreasing condition, we proceed as in the proof of proposition 6 to show $B \subset C(v_m)_o(X,E)$. Moreover B is bounded there because it is bounded in $Cv_m(X,E)$.

(i) implies (ii). Clearly (i) implies $\mathcal{V}_oC(X)=C\overline{V}_o(X)$ algebraically, hence $\mathcal{V}$ is regularly decreasing according to the scalar theorem 1.1. We prove (ii)(2). For each compact subset K of X we select $\varphi_K\in C(X)$ with compact support such that $\varphi_K=1$ on K and $0\le\varphi_K\le1$. Let $\mathcal{K}$ be the family of all compact subsets of X directed by inclusion. If B is a bounded subset of $C\overline{V}(X,E)$, then $B_o:=\{f\varphi_K; K\in\mathcal{K}, f\in B\}$ is a bounded subset of $C\overline{V}_o(X,E)$. By (i), B_o is bounded in some $C(v_n)_o(X,E)$. This implies that B is bounded in $Cv_n(X,E)$.■

Theorem 1 has a certain number of consequences for the topological properties of spaces of vector-valued sequences and continuous functions, ε-tensor products and spaces of continuous linear mappings. Our first lemma shows that the topological projective description in problem (a) is exactly equivalent to a result on the interchangeability of inductive limits and ε-tensor products. The proof is standard (cf. [16]).

8. Lemma. Let X be a locally compact space and E a l.c.s. $\mathcal{V}_oC(X,E)$ is a topological subspace of $C\overline{V}_o(X,E)$ if and only if
$\mathcal{V}_oC(X)\otimes_\varepsilon E=(\text{ind } C(v_n)_o(X))\otimes_\varepsilon E=\text{ind}(C(v_n)_o(X)\otimes_\varepsilon E)$ holds topologically.

In [7,4.7] we also present results about the interchangeability of ε-tensor products and ε-products of a l.c.s. E with weighted inductive limits of type $\mathcal{V}C(X)$ if the sequence $\mathcal{V}$ satisfies condition (D). We refer to that article for the details.

9. Proposition. Let X be a locally compact space.
(i) If E is (DF) (resp. (gDF), quasibarrelled (DF), bornological (DF)), then $\mathcal{V}_oC(X)\otimes_\varepsilon E$ is of the same type.
(ii) If E is (DF) (resp. (gDF), quasibarrelled (DF)), then $\mathcal{V}_oC(X,E)$ and $C\overline{V}_o(X,E)$ are of the same type.

Proof. In all the cases E has the c.n.p., hence $\mathcal{V}_oC(X,E)$ is a topological subspace of $C\overline{V}_o(X,E)$ by theorem 1. According to lemma 8, $\mathcal{V}_oC(X)\otimes_\varepsilon E = \text{ind}(C(v_n)_o(X)\otimes_\varepsilon E)$ holds topologically. Since $C(v_n)_o(X)$ is an ε-space (cf.[30]), it is enough to apply results of Defant, Govaerts [22] (also see [34, 11.5.10, 11.5.11]) to conclude (i). Part (ii) follows easily by density.■

Our next result will appear in [8].

10. Theorem. Assume that every weight $\bar{v}\in\overline{V}$ is dominated by a weight $w\in\overline{V}\cap C(X)$. If E is a (DF)-space (resp. (gDF)-space), then $C\overline{V}(X,E)$ and $C\overline{V}(X)\otimes_\varepsilon E$ are also (DF) (resp. (gDF)).

As a consequence we obtain partial positive answers to the problem of Grothendieck mentioned in the introduction. The following special case of theorem 10 (cf. lemma 2) was already proved in [6].

11. Corollary. (i) If E is a locally complete (DF)-space, then $L_b(\lambda_1(A),E)$ is a (DF)-space.
(ii) If E is a (DF)-space, then $\lambda_1(A)'_b\otimes_\varepsilon E$ is a (DF)-space.

It is important to recall at this point that, by the work of Taskinen [36], there are Fréchet-Montel spaces F with the bounded approximation property such that $L_b(F,F'_b)=F'_b\hat{\otimes}_\varepsilon F'_b$ is not a (DF)-space. Modifying Taskinen's constructions in [35] it is proved in [18] that for Fréchet spaces G,H, neither $L_b(G,H'_b)$ nor $G'_b\hat{\otimes}_\varepsilon H'_b$ needs to be a (gDF)-space. Nevertheless, Galbis [26] observed that the result corresponding to corollary 11 also holds for (gDF)-spaces instead of (DF)-spaces.

The characterization of the quasibarrelled spaces $L_b(\lambda_1(A),E)$ for a (DF)-space E is more delicate and was only obtained after our study of the dual density conditions for (DF)-spaces. Here is our main result [6,2.5,3.4].

12. Theorem. Let E be a (DF)-space.
(i) If E is locally complete, then $L_b(\lambda_1(A),E)$ is (quasi)barrelled if and only if (1) $\lambda_1(A)$ is Montel and E is (quasi)barrelled, or (2) $\lambda_1(A)$ is distinguished and E has metrizable bounded sets (or, equivalently, $l_\infty(E)$ is (quasi)barrelled).
(ii) $\lambda_1(A)'_b \otimes_\varepsilon E$ is quasibarrelled (resp. bornological) if and only if $\lambda_1(A)$ is distinguished and E is quasibarrelled (resp. bornological).

Observe that the characterization of the quasibarrelled locally complete (DF)-spaces of type $L_b(\lambda_1(A),E)$ gives rise to an interesting dichotomy which does not appear when we study the ε-tensor product $\lambda_1(A)'_b \otimes_\varepsilon E$. We also remark that in a certain sense theorem 12(i) reduces the study of the quasibarrelledness of the space $L_b(\lambda_1(A),E)=K_\infty(E)$ to the case when we only consider a single weight, viz., to the quasibarrelledness of $l_\infty(E)$.

It seems harder to characterize the locally complete bornological (DF)-spaces of type $L_b(\lambda_1(A),E)$ in a similar way and there is still a gap; for partial results, necessary resp. sufficient conditions, we refer to [6].

Given a decreasing sequence $\mathcal{V}=(v_n)_{n\in\mathbb{N}}$ of strictly positive weights on an index set I, and the corresponding Köthe matrix $A=(a_n)_{n\in\mathbb{N}}$, it is possible to define, in a canonical way, the vector-valued echelon and co-echelon spaces of order p, $1\le p<\infty$:

$\lambda_p(E)=\lambda_p(I,A,E)=$ proj $l_p(I,a_n,E)$,

$k_p(E)=k_p(I,\mathcal{V},E)=$ ind $l_p(I,v_n,E)$,

$$K_p(E)=K_p(I,V,E):=\{x=(x(i))\in E;\ r^p_{\bar{V},q}(x):=\left[\sum_{i\in I}(\bar{v}(i)q(x(i)))^p\right]^{1/p}<\infty$$

$$\forall\bar{v}\in\bar{V}\ \forall q\in cs(E)\}.$$

There is a full duality between echelon spaces with values in Fréchet spaces and co-echelon spaces with values in (DF)-spaces, which extends the scalar results given in [13].

13. Proposition. (i) If E is a Fréchet space, then $\lambda_p(E)'_b \cong K_q(E'_b)$, where $p^{-1}+q^{-1}=1$ ($q=\infty$ if $p=1$ and $q=1$ if $p=0$).
(ii) If E is a (DF)-space, then $K_p(E)'_b \cong \lambda_q(E'_b)$, where $p^{-1}+q^{-1}=1$, $1<p<\infty$, and $k_o(E)'_b = K_o(E)'_b \cong \lambda_1(E'_b)$.

A direct proof of proposition 13(i) for p=1 was established in [5]. The other results contained in proposition 13 were obtained by Galbis [25]. The projective description result in the setting of co-echelon spaces of order p, $1 \le p < \infty$, with values in a (DF)-space follows next (cf. [6]).

14. Proposition. Let E be a l.c.s. satisfying the c.n.p. Then, for $1 \le p < \infty$, we have $k_p(E) = K_p(E)$ algebraically and topologically.

The duality established above and the characterization theorem 12 yield the following individual characterization of the distinguished Fréchet spaces $\lambda_1(A)\hat{\otimes}_\pi E$ (cf. [5]).

15. Corollary. Let E be a Fréchet space. Then $\lambda_1(A)\hat{\otimes}_\pi E = \lambda_1(E)$ is distinguished if and only if $\lambda_1(A)$ is Montel and E is distinguished or $\lambda_1(A)$ is distinguished and E'_b has metrizable bounded sets (or, equivalently, $l_1(E) = l_1\hat{\otimes}_\pi E$ is distinguished).

In contrast to this we note the following result (cf. [29],[34],[5]).

16. Proposition. Let E be a Fréchet space. Then $\lambda_1(A)\hat{\otimes}_\pi E = \lambda_1(E)$ is nuclear (resp. Schwartz, Montel, quasinormable, reflexive, resp. a space with the density condition) if and only if both $\lambda_1(A)$ and E are.

3. Weighted inductive limits of spaces of Fréchet-valued continuous functions

In this section E always stands for a Fréchet space with an increasing fundamental sequence $(\|.\|_k)_{k\in\mathbb{N}}$ of continuous seminorms; the sets $U_k:=\{x\in E; \|x\|_k\le 1\}$, $k\in\mathbb{N}$, form a basis of 0-neighbourhoods in E. The dual norms will be denoted by $\|.\|_k^*$, $\|u\|_k^* := \sup\{|\langle u,x\rangle| ; \|x\|_k\le 1\}$, $u\in E'$. We set E_k' for the subspace of E' of all u such that $\|u\|_k^*$ is finite.

We first deal with the algebraic identity in the projective description problem, and we can give a complete solution. But before we state the final results 3.1 and 3.2, let us first look at one particular case, viz. that of sequence spaces, to get some idea. Thus, we consider a Köthe matrix A on the index set $\mathbb{N}$ and a Fréchet space E. According to proposition 2.3, $\mathcal{k}_\infty(E)=K_\infty(E)$ holds algebraically if and only if $L(\lambda_1(A),E)=LB(\lambda_1(A),E)$. By [38, 1.3], this is equivalent to the following condition:

$$\forall (n(l)) \subset \mathbb{N}\ \exists n\ \forall k\ \exists l_o=l_o(k),\ C_k>0\ \forall i\in\mathbb{N}\ \forall e\in E :$$
$$\|e\|_k v_n(i) \le C_k \max(\|e\|_l v_{n(l)}(i) : 1\le l\le l_o).$$

This condition is very well understood from the work of Vogt [38,39]. An adequate extension to function spaces yields the complete characterization for spaces of continuous functions (cf. [19, section 3]).

1. Proposition. Let X be a locally compact space and E a Fréchet space. Tfae:

(i) Every bounded subset of $C\overline{V}(X,E)$ is bounded in some $Cv_n(X,E)$.

(ii) $\mathcal{V}C(X,E)=C\overline{V}(X,E)$ holds algebraically.

(iii) $\forall (n(l)) \subset \mathbb{N}\ \exists n\ \forall k\ \exists l_o=l_o(k),\ C_k>0\ \forall x\in X\ \ \forall e\in E$:
$$\|e\|_k v_n(x) \le C_k \max(\|e\|_l v_{n(l)}(x) : 1\le l\le l_o).$$

In this case $\mathcal{V}C(X,E)$ is a regular (LF)-space.

Now from an application of proposition 2.7 and Grothendieck's factorization theorem we obtain the following consequence.

2. Proposition. Let X be a locally compact space and E a Fréchet space. Tfae:

(i) Every bounded subset of $C\overline{V}_o(X,E)$ is bounded in some $C(v_n)_o(X,E)$.

(ii) $\mathcal{V}_oC(X,E)=C\overline{V}_o(X,E)$ holds algebraically.

(iii) (1) $\mathcal{V}$ is regularly decreasing and (2) $\mathcal{V}C(X,E)=C\overline{V}(X,E)$ holds

algebraically (which is equivalent to condition 1(iii)).
In this case $\mathcal{V}_o C(X,E)$ is a boundedly retractive, hence quasi-complete, (LF)-space.

It is also possible to prove the corresponding characterization for co-echelon spaces of order p, $1 \le p < \infty$, on an index set I (which explains the meaning of Vogt's condition in this context).

3. Proposition. Let E be a Fréchet space and $1 \le p < \infty$. Tfae:
(i) Every bounded subset of $K_p(E)$ is bounded in some $l_p(v_n, E)$.
(ii) $k_p(E) = K_p(E)$ holds algebraically.
(iii) $\forall (n(1)) \subset \mathbb{N}\ \exists n\ \forall k\ \exists l_o = l_o(k),\ C_k > 0\ \forall i \in \mathbb{N}\ \forall e \in E$:

$$\|e\|_k \, v_n(i) \le C_k \max(\|e\|_l \, v_{n(l)}(i) : 1 \le l \le l_o).$$

From the work of Vogt [38,2.2,3.2] we obtain precise conditions on the Fréchet space E if we restrict our attention to sequences $\mathcal{V}$ such that $\lambda_1(A)$ is a power series space.

4. Theorem. Let E be a Fréchet space. Let $\beta = (\beta_i)_{i \in \mathbb{N}}$ be an exponent sequence such that $\sup(\beta_{i+1}/\beta_i) < \infty$.
(i) Let $\mathcal{V}$ be a decreasing sequence of weights on $\mathbb{N}$ such that $\lambda_1(A)$ is a power series space of finite type $\Lambda_1(\beta)$. For $1 \le p \le \infty$, tfae:
(1) $k_p(E) = K_p(E)$ holds algebraically,
(2) E satisfies the property (DN) of Vogt (cf. [42]).
(ii) Let $\mathcal{V}$ be a decreasing sequence of weights on $\mathbb{N}$ such that $\lambda_1(A)$ is a power series space of infinite type $\Lambda_\infty(\beta)$. For $1 \le p \le \infty$, tfae:
(1) $k_p(E) = K_p(E)$ holds algebraically,
(2) E satisfies the property (LB_∞) (cf. [38,3.1]).

5. Observation. For a Fréchet space E, $k_\infty(E) = K_\infty(E)$ holds algebraically for every decreasing sequence $\mathcal{V}$ of strictly positive weights on $\mathbb{N}$ if and only if E is normable.

If E is normable, the result was already given in [12]. Let E be a non-normable Fréchet space. If E does not have a continuous norm, then E contains ω complemented, therefore for every non-normable Köthe echelon space $\lambda_1(A)$ we have $L(\lambda_1(A),E) \ne LB(\lambda_1(A),E)$. This implies

$\mathcal{k}_\infty(E)\neq K_\infty(E)$ by proposition 2.3. If E has a continuous norm, we can apply a result of Bessaga, Pełczińsky, Rolewicz, [4], to obtain a closed subspace of E isomorphic to a Köthe echelon space $\lambda_1(A)$ with a continuous norm. For the corresponding sequence $\mathcal{V}$ we have $\mathcal{k}_\infty(E)\neq K_\infty(E)$.

To treat the problem of a topological projective description with values in a Fréchet space, we will from now on restrict our attention to the case when the space X is nothing but the set of natural numbers $\mathbb{N}$. In this case, by [13],there is $\bar{v}\in\bar{V}$ such that $\bar{v}>0$.

Therefore in the rest of the section, $\mathcal{V}=(v_n)_{n\in\mathbb{N}}$ is a decreasing sequence of strictly positive weights on $\mathbb{N}$ and $A=(a_n)_{n\in\mathbb{N}}$ is the corresponding Köthe matrix.

We first provide a complete individual characterization of the pairs $(\mathcal{V},E)$ of a sequence of weights $\mathcal{V}$ and a Fréchet space E such that $\mathcal{k}_p(E)$ is a topological subspace of $K_p(E)$, $1\le p<\infty$ or $p=0$, in terms of a condition of Vogt [38,1.4] equivalent to the identity $L(E,\lambda_\infty(A)) = LB(E,\lambda_\infty(A))$. The result for $p=0$ was obtained by the second author in [17]. Galbis [26] proved the corresponding result for $1\le p<\infty$, and he also showed in [27] that an adequate extension of Vogt's condition to function spaces characterized the pairs $(\mathcal{V},E)$ ($\mathcal{V}$ a decreasing sequence of strictly positive continuous weights on a locally compact σ-compact space X and E a Fréchet space) such that $\mathcal{V}_oC(X,E)$ is a topological subspace of $C\bar{V}_o(X,E)$.

6. <u>Theorem</u>. Let E be a Fréchet space and $1\le p<\infty$ or $p=0$. Tfae:

(i) $\mathcal{k}_p(E)$ is a topological subspace of $K_p(E)$.

(ii) $\forall (k(l))\subset\mathbb{N}\ \exists k\ \forall n\ \exists l_o=l_o(n),\ C_n>0\ \ \forall i\in I\ \ \forall u\in E'$:

$$\frac{\|u\|^*_k}{v_n(i)} \le C_n \max\left[\frac{\|u\|^*_{k(l)}}{v_l(i)} : 1\le l\le l_o\right].$$

Moreover, if (ii) is satisfied, then $(\mathrm{ind}\ c_o(v_n))\otimes_\varepsilon E$ is bornological, $K_o\otimes_\varepsilon E$ is quasibarrelled and $K_p(E)$ is barrelled for $1\le p<\infty$ or $p=0$.

For the proof of (i) implies (ii) one uses Grothendieck's factorization theorem. For the converse implication we only consider $p\neq 1$ here, and we denote by (G,t) the space $\mathrm{ind}\ l_p(v_n,E)$ and by s the coarser topology on G induced by $K_p(E)$. The dual of (G,t) can be canonically identified with the vector-valued echelon space $\lambda_q(E'_b)$, $p^{-1}+q^{-1}=1$ ($q=1$ if $p=0$). With this identification a subset of $\lambda_q(E'_b)$ is bounded if and only if it is equicontinuous in $(G,t)'$. Moreover a subset H of $\lambda_q(E'_b)$ is equicontinuous in $(G,s)'$ if and only if there are $\bar{v}\in\bar{V}$, $\bar{v}>0$, and $k\in\mathbb{N}$ with $\sum_{i=1}^{\infty}\left[\frac{\|u(i)\|_k^*}{\bar{v}(i)}\right]^q\leq 1$ for all $u\in H$ (observe that this also characterizes the dual of (G,s) as a subspace of $\lambda_q(E'_b)$). Now condition (ii) permits to conclude that every equicontinuous subset in $(G,t)'$ is also equicontinuous in $(G,s)'$, and the proof is complete.■

From Vogt's results in [38,4.2 and 5.2] we can then conclude the following.

7. Corollary. Let E be a Fréchet space. Let $\beta=(\beta_i)_{i\in\mathbb{N}}$ be an exponent sequence such that $\sup(\beta_{i+1}/\beta_i)<\infty$. Let $1\leq p<\infty$ or $p=0$.
(i) Let $\mathcal{V}$ be a decreasing sequence of weights on $\mathbb{N}$ such that $\lambda_1(A)$ is a power series space of finite type $\Lambda_1^{\infty}(\beta)$, then $k_p(E)$ is a topological subspace of $K_p(E)$ if and only if E has property $(\bar{\bar{\Omega}})$ (cf. [38,4.1]).
(ii) Let $\mathcal{V}$ be a decreasing sequence of weights on $\mathbb{N}$ such that $\lambda_1(A)$ is a power series space of infinite type $\Lambda_{\infty}^{\infty}(\beta)$, then $k_p(E)$ is a topological subspace of $K_p(E)$ if and only if E has property (LB^{∞}) (cf. [38,5.1]).

Next it is possible to derive a universal characterization. We recall that a Fréchet space E is called a quojection, [3], if every quotient of E with a continuous norm is a Banach space. Moscatelli [33] gave examples of quojections which are not isomorphic to a countable product of Banach spaces. Several authors proved recently (see the contribution [20] in this volume) that every quojection is the quotient of a countable product of Banach spaces.

A related class of Fréchet spaces was introduced by Bellenot and Dubinsky [3], viz. the Fréchet spaces which do not satisfy condition (*). For our purposes it suffices to recall that a separable Fréchet space has condition (*) if and only if it has a nuclear Köthe quotient with a continuous norm. Every quojection does not satisfy condition (*), and the converse is false due to examples in [2]. Vogt [40,0.3] proved that a Fréchet space does not satisfy condition (*) if and only if E" is a quojection. Now we can state the following result of Galbis [26].

8. Proposition. Let E be a Fréchet space and $1 \le p < \infty$ or $p=0$. Then $\mathcal{k}_p(E)$ is a topological subspace of $K_p(E)$ for every sequence of weights $\mathcal{V}$ on $\mathbb{N}$ if and only if E" is a quojection.

It remains to treat the most complicated case, viz. the following question: When is $\mathcal{k}_\infty(E)$ a topological subspace of $K_\infty(E)$? This is equivalent to the following question on the interchangeability of inductive and projective limits with spaces of continuous linear mappings: When is the inductive limit ind $L_b(l_1(a_n),E)$ a topological subspace of the space $L_b(\text{proj } l_1(a_n),E)$? Here we have only preliminary results.

First we observe that if $\mathcal{k}_\infty(E)$ is a topological subspace of $K_\infty(E)$ for some Fréchet space $E \ne \{0\}$, then $\mathcal{k}_\infty = K_\infty$ holds topologically and thus the sequence $\mathcal{V}$ must satisfy condition (D). The following observation is less trivial.

9. Proposition. Let E be a l.c.s. If $\mathcal{k}_\infty(E)$ is a topological subspace of $K_\infty(E)$, then $\mathcal{k}_o(E)$ must also be a topological subspace of $K_o(E)$. Hence in this case, if E is a Fréchet space, the condition 6(ii) must be satisfied.

Proof. Let W be a closed absolutely convex 0-neighbourhood in $\mathcal{k}_o(E) =$ ind $c_o(v_n,E)$. If $x \in E^{\mathbb{N}}$ and $m \in \mathbb{N}$, we write x_m for the m-th section of x. We put $U := \{x \in \mathcal{k}_\infty(E);\ x_m \in W \text{ for all } m \in \mathbb{N}\}$ and prove that U is a 0-neighbourhood in $\mathcal{k}_\infty(E)$. For each $n \in \mathbb{N}$ there is $p_n \in cs(E)$ with

$$\{x \in c_o(v_n,E);\ \sup_{i \in \mathbb{N}} v_n(i)p_n(x(i)) \le 1\} \subset W \cap c_o(v_n,E).$$

If $x \in l_\infty(v_n,E)$ and $\sup_{i \in \mathbb{N}} v_n(i)p_n(x(i)) \le 1$, then $x_m \in W$ for all $m \in \mathbb{N}$, hence $x \in U$ and $U \cap l_\infty(v_n,E)$ is a 0-neighbourhood in $l_\infty(v_n,E)$. Now by hypothesis, we can find $\bar{v} \in \bar{V}$, $p \in cs(E)$ such that

$$\{x\in k_\infty(E);\ \sup_{i\in\mathbb{N}} \bar{v}(i)p(x(i))\le 1\} \subset U.$$

If $z\in k_o(E)$ satisfies $\sup_{i\in\mathbb{N}} \bar{v}(i)p(z(i))\le 1$, there is $l\in\mathbb{N}$ with $z\in c_o(v_l,E)$. Then the sequence $(z_m)_{m\in\mathbb{N}}$ tends to z in $c_o(v_l,E)$. On the other hand, $z_m\in W$ for all $m\in\mathbb{N}$ and W is closed in $k_o(E)$. This yields $z\in W$. Consequently W is a 0-neighbourhood in $k_o(E)$ for the topology induced by $K_o(E)$.■

Our next theorem incorporates results of Vogt [39] and Krone, Vogt [32].

10. Theorem. Let E be a Fréchet space and V a decreasing sequence of weights on $\mathbb{N}$. We consider the following properties:

(i) $k_\infty(E)$ is a topological subspace of $K_\infty(E)$.

(ii) $K_\infty(E)$ is barrelled.

(iii) $K_o(E)$ is barrelled.

(iv) (V,E) satisfies condition (S) (cf. [37]):

$$\forall\mu\ \exists n_o,k\ \forall L,m\ \exists n,S\ \forall i\in\mathbb{N}\ \forall u\in E' :$$

$$\frac{\|u\|_k^*}{v_m(i)} \le S \max\left(\frac{\|u\|_L^*}{v_n(i)},\ \frac{\|u\|_\mu^*}{v_{n_o}(i)}\right).$$

(1) Then (i) implies (ii), (ii) implies (iii) and (iii) implies (iv).

(2) If $\lambda_1(A)$ is a Schwartz space (i.e. $\forall n\ \exists m>n$ such that v_m/v_n tends to 0) and if E is a Köthe echelon space of order one, then (iv) implies (iii).

(3) If $\lambda_1(A)$ is not normable and (iv) is satisfied, then E must be quasinormable.

(4) In general (ii) does not imply (i).

Proof. We prove (1). To establish that (i) implies (ii) we prove that if V satisfies condition (D), then $k_\infty(E)$ is dense in $K_\infty(E)$. Given $x\in K_\infty(E)$, if $J=(I_m)_{m\in\mathbb{N}}$ is the increasing sequence of subsets of $\mathbb{N}$ given by condition (D), we define $x_m=(x_m(i))_{i\in\mathbb{N}}$ by $x_m(i)=x(i)$ if $i\in I_m$ and 0 otherwise. First we show that every x_m belongs to $k_\infty(E)$. Indeed, (N,J) implies $\bar{v}_m:=v_{n_m}\chi_{I_m} \in \bar{V}$, and this yields $x_m\in l_\infty(v_{n_m},E)$. Now, condition (M,J) and [10,3.7] permit to conclude that $(x_m)_{m\in\mathbb{N}}$ tends to x in

$K_\infty(E)$.

To prove that (ii) implies (iii), we fix a barrel T_o in $K_o(E)$ and put $T:=\{x\in K_\infty(E);$ every section of x belongs to $T_o\}$. Clearly T is absolutely convex and closed. Moreover T is absorbent: For any $z\in K_\infty(E)$ the set B of all the sections of z is bounded in $K_o(E)$. Since this last space is complete, T_o is bornivorous, and hence one can find $\lambda>0$ with $\lambda B\subset T_o$, whereby $\lambda z\in T$. Now by (ii) there are $\bar{v}\in\bar{V}$, $k\in\mathbb{N}$ such that

$$\{x\in K_\infty(E);\ \sup_{i\in\mathbb{N}} \bar{v}(i)\|x(i)\|_k\le 1\}\subset T.$$

From this it follows that

$$\{x\in K_o(E);\ \sup_{i\in\mathbb{N}} \bar{v}(i)\|x(i)\|_k\le 1\}\subset T_o,$$

and T_o is a 0-neighbourhood in $K_o(E)$.

(iii) implies (iv) by an adaptation of the proof of [39,4.4], but in the present case the proof is easier and can be obtained as in [32,2.3]. The idea is as follows. If condition (S) is not satisfied, proceeding as in [32,2.3], for $\mu=1$, we can find a subset $\{i(k,n);k,n\in\mathbb{N}\}$ of $\mathbb{N}$ and $\{u(k,n);k,n\in\mathbb{N}\}\subset E_1'$ such that

$$\frac{\|u(k,n)\|_k^*}{v_k(i(k,n))} > 2^n \max\left[\frac{\|u(k,n)\|_{k+1}^*}{v_n(i(k,n))}, \frac{\|u(k,n)\|_1^*}{v_{k-1}(i(k,n))}\right]$$

for all $k,n\in\mathbb{N}$ and such that $i(k,n_1)\ne i(k,n_2)$ or $u(k,n_1)\ne u(k,n_2)$ for all k, $n_1\ne n_2$. For all $n\ge k$, we put $Y_{k,n}(i)=0$ if $i\ne i(k,n)$ and $Y_{k,n}(i(k,n)):= n\dfrac{v_k(i(k,n))}{\|u(k,n)\|_k^*}u(k,n)$. Clearly $Y_{k,n}:=(Y_{k,n}(i))_{i\in\mathbb{N}}$ can be considered as an element of $K_o(E)'$. We claim that the set $B:=\{Y_{k,n};\ n\ge k\}$ is weakly bounded in $K_o(E)'$. To see this, fix $x\in K_o(E)$; for every $k\in\mathbb{N}$ there are $m(k)\in\mathbb{N}$ and $C_k>0$ with $v_{m(k)}(i)\|x(i)\|_k\le C_k$ for all $i\in\mathbb{N}$. Put $l=m(1)$, $m=\max(m(l+1),l)$. Then for $C:=\{(k,n);n\ge k\}$, $A_1:=\{(k,n)\in C;n\in\mathbb{N},k=l+1,..,n\}$, $A_2:=\{(k,n)\in C;k=1,\dots,l,n=m,m+1,\dots\}$, we get that $C\setminus(A_1\cup A_2)$ is finite. Moreover, for $(k,n)\in A_1$ we have $|\langle x,Y_{k,n}\rangle|\le nC_1 2^{-n}$, and if $(k,n)\in A_2$, then we obtain $|\langle x,Y_{k,n}\rangle|\le nC_{l+1}2^{-n}$. Now by (iii), B is equicontinuous. According to the characterization of the equicontinuous subsets of $K_o(E)'$ obtained in the proof of theorem 6, there are $\bar{v}\in\bar{V}$, $\bar{v}>0$, and $l\in\mathbb{N}$ such that

$\sum_{i=1}^{\infty} \|\bar{v}(i)^{-1}Y_{k,n}(i)\|_1^* \le 1$ for all $n \ge k$ and $k \in \mathbb{N}$. This implies $nv_1(i(1,n)) = \|Y_{k,n}(i(1,n))\|_1^* \le \bar{v}(i(1,n))$ for all $n \ge 1$ which is a contradiction since $\bar{v} \in \bar{V}$ implies that $\bar{v}/v_1$ is bounded on $\mathbb{N}$.

Part (2) is exactly [32,2.1], and part (3) is a consequence of [39,5.9].

To show that (ii) does not imply (i), we first observe that if $\lambda_1(A)$ is Schwartz, then for every Fréchet space E, $k_o(E)=k_\infty(E)$ and $K_o(E)=K_\infty(E)$ hold topologically. In this case (i) is equivalent to $k_o(E)$ being a topological subspace of $K_o(E)$. If $\lambda_1(A)$ is even nuclear, this is in turn equivalent to $L(E,\lambda_1(A))=LB(E,\lambda_1(A))$. We take, for instance $v_n(i)=i^{-n}$, $i,n \in \mathbb{N}$. Then $\lambda_1(A)=s$. Taking also E=s, we know from [29,II, 4,no. 3,Cor.2] that $K_\infty(E)=L_b(s,s)$ is barrelled, and hence (ii) is satisfied. But (i) cannot hold because we certainly have $L(s,s) \neq LB(s,s)$.∎

From Vogt's work [39] we can also quote the following result.

11. Theorem. Let E be a nuclear Fréchet space and $\mathcal{V}$ a decreasing sequence of weights on $\mathbb{N}$.

(i) If $\lambda_1(A)$ is a nuclear power series space $\Lambda_\infty(\alpha)$ of infinite type with $\sup_{i \in \mathbb{N}} (\alpha_{i+1}/\alpha_i) < \infty$, then $K_\infty(E)$ is barrelled if and only if E satisfies property (Ω) (cf. [42]).

(ii) If $\lambda_1(A)$ is a nuclear power series space $\Lambda_1(\alpha)$ of finite type with $\lim_i (\alpha_{i+1}/\alpha_i)=1$, then $K_\infty(E)$ is barrelled if and only if E satisfies property $(\bar{\bar{\Omega}})$ (cf. [38,4.1]).

We now show that $k_\infty(E)$ need not be a topological subspace of $K_\infty(E)$ even if $\mathcal{V}$ satisfies condition (D) and $k_o(E)$ is a topological subspace of $K_o(E)$. In fact, it was quite a surprise to us that the fact that $k_\infty(\omega)$ is a topological subspace of $K_\infty(\omega)$ already implies that $\mathcal{V}$ must satisfy a certain condition which need not hold even if the corresponding space $\lambda_1(A)$ is quasinormable.

12. Proposition. If $k_\infty(\omega)$ is a topological subspace of $K_\infty(\omega)$, then we

have:

(*) $$\forall \bar{v}\in\bar{V}\ \exists l\in\mathbb{N}\ \forall n\in\mathbb{N}, D>0\ \exists C>0\ \forall i\in\mathbb{N}:$$
$$\frac{D}{v_n(i)} \le \max\left[\frac{C}{v_l(i)}, \frac{1}{\bar{v}(i)}\right].$$

Proof. The hypothesis is equivalent to the fact that $\text{ind}\ l_\infty(v_n)^{\mathbb{N}}$ is a topological subspace of $K_\infty^{\mathbb{N}}$. We fix $\bar{v}\in\bar{V}$ and set $B_{\bar{v}} := \{\alpha\in K_\infty;\ \sup_{i\in\mathbb{N}} \bar{v}(i)|\alpha(i)|\le 1\}$. It is enough to find $l\in\mathbb{N}$ with $K_\infty = l_\infty(v_l)+B_{\bar{v}}$ because this is clearly equivalent to the condition (*). First we choose $\lambda_j>0$, $j\in\mathbb{N}$, such that, for $V_j := \{\alpha\in l_\infty(v_j);\ \sup_{i\in\mathbb{N}} v_j(i)|\alpha(i)|\le\lambda_j\}$, we have $\sum_{j=1}^{\infty} V_j := \bigcup_{j=1}^{\infty} \sum_{s=1}^{j} V_s \subset B_{\bar{v}}$.
Now $W_n := \prod_{k\le n} V_n \times \prod_{k>n} l_\infty(v_n)$, $n\in\mathbb{N}$, defines a 0-neighbourhood in $l_\infty(v_n)^{\mathbb{N}}$ and $W := \sum_{j=1}^{\infty} W_j$ is a 0-neighbourhood in $\text{ind}\ l_\infty(v_n)^{\mathbb{N}}$. By assumption, there are $w\in\bar{V}$, $l\in\mathbb{N}$ such that

$$\left[\prod_{k<l} B_w \times \prod_{k\ge l} K_\infty\right] \cap \left[\text{ind}\ l_\infty(v_n)^{\mathbb{N}}\right] \subset W.$$

If $x\in K_\infty$, then $z := (0,\dots,0,\overset{(l+1)}{x},0,\dots)$ belongs to the intersection in the left hand side of the inclusion above. Therefore, we can find $m\ge l+1$ such that $z\in \sum_{j=1}^{m} W_j$. This implies

$$x \in l_\infty(v_l) + \sum_{j=l+1}^{m} V_j \subset l_\infty(v_l) + B_{\bar{v}}.$$

Since x is arbitrary, the proof is complete.■

13. Example. We consider the decreasing sequence of weights $\mathcal{V}=(v_n)_{n\in\mathbb{N}}$ on $\mathbb{N}\times\mathbb{N}$ given by $v_n(i,j)=j^{-i}$, $i\le n$, and $v_n(i,j)=j^{-n}$, $i\ge n+1$ (cf. [29, II p.99]). $\mathcal{V}$ is regularly decreasing (see [13, 4.11]), hence $\mathcal{V}$ satisfies condition (D). We put $\bar{v}(i,j) := 2\inf_{n\in\mathbb{N}} v_n(i,j)$, $(i,j)\in\mathbb{N}\times\mathbb{N}$. Clearly $\bar{v}(n,j)=2j^{-n}$ for all $n,j\in\mathbb{N}$. For all $p\in\mathbb{N}$ and all $j\in\mathbb{N}$ we have

$$j^{p+1} = \frac{1}{v_{p+1}(p+1,j)} > \frac{j^{p+1}}{2} = \max\left[\frac{j2^{-1}}{v_p(p+1,j)}, \frac{1}{\bar{v}(p+1,j)}\right].$$

This means that $\mathcal{V}$ does not satisfy the condition (*) in proposition

12. Consequently $k_\infty(\omega)$ is not a topological subspace of $K_\infty(\omega)$.

It is not surprising that a complete solution can be obtained if we restrict our attention to the class of the quojections.

14. Proposition. Let E be a non-normable quojection and V a decreasing sequence of weights on $\mathbb{N}$ satisfying condition (D). Then
(i) $K_\infty(E)$ is ultrabornological.
(ii) $k_\infty(E)$ is a topological subspace of $K_\infty(E)$ if and only if the condition (*) of proposition 12 is satisfied.

Proof. (i) is trivial for spaces E which are countable products of Banach spaces. Given an arbitrary quojection E, we apply [20] to find a countable product X of Banach spaces such that there is a quotient map q from X onto E. Moreover, an inspection of the proof of [20] shows that we can choose fundamental sequences of seminorms $(\|.\|_k)_{k\in\mathbb{N}}$ on X resp. E and a sequence of numbers $M_k>0$ such that for each $y\in E$ there is $x\in X$ with $q(x)=y$ and $\|x\|_k\le M_k\|y\|_k$ for all $k\in\mathbb{N}$. This implies that the canonical map Q from $K_\infty(X)$ into $K_\infty(E)$ defined by $Q((x(i))_{i\in\mathbb{N}}) := (q(x(i)))_{i\in\mathbb{N}}$ is linear, continuous, open and surjective. Consequently $K_\infty(E)$ is also ultrabornological.

Now we indicate how to obtain (ii). Every non-normable quojection contains ω complemented, so that the necessity of condition (*) in V follows directly from proposition 12. The fact that under the hypothesis of condition (*) on V, $k_\infty(E)$ must be a topological subspace of $K_\infty(E)$ if E is a countable product of Banach spaces is obtained directly by checking that every 0-neighbourhood of $k_\infty(E)$ contains a 0-neighbourhood of the topology induced by $K_\infty(E)$. For arbitrary quojections one proceeds in a similar way as we did in the proof of (i).■

It is implicitly included in [43,5.7] that the following condition is sufficient for $K_\infty(E)$ to be barrelled or ultrabornological if E is Fréchet and if V satisfies condition (D):

$$\exists n_o \ \forall \mu \ \exists k \ \forall L, m \ \exists n, S \ \forall i\in\mathbb{N} \ \forall u\in E' :$$

$$\frac{\|u\|_k^*}{v_m(i)} \le S \max\left(\frac{\|u\|_L^*}{v_n(i)}, \frac{\|u\|_\mu^*}{v_{n_o}(i)}\right).$$

For more information about this condition, called (S_1^*), we refer to [37] and [32].

If E is a Fréchet space such that E" is a quojection and if $\mathcal{V}$ satisfies condition (D), then $(\mathcal{V},E)$ satisfies (S_1^*) by [39,5.1]. Thus, in this case, $K_\infty(E)$ is again ultrabornological, which improves proposition 14(i).

4. Final remarks

The theory of the projective descriptions of weighted inductive limits of spaces of continuous functions with values in (DF)-spaces has been clarified to a fairly satisfactory extent as we have pointed out in section 2. In case the functions take their values in a Fréchet space, we have a rather good understanding of the algebraic identity in the projective description problem, and also of the topological description of the weighted inductive limits $\mathcal{V}_oC(X,E)$ with o-growth conditions. In any case it would be good to use the individual characterizations to obtain results of a universal type, fixing certain classes of sequences $\mathcal{V}$ or classes of Fréchet spaces E, and in particular to derive conditions which only depend on the space E if $\mathcal{V}$ satisfies certain properties. It is perhaps also possible to extend some of our results in the last part of section 3 to spaces of continuous functions.

On the other hand, it is quite obvious that more efforts are necessary to give a satisfactory picture of the problem of the projective description of the weighted inductive limits $\mathcal{V}C(X,E)$ of spaces of Fréchet-valued continuous functions with O-growth conditions. More specifically, one could ask the following questions:

(1) Characterize the bornological (DF)-spaces of type $L_b(\lambda_1(A),E)$.

(2) Are the regular (LF)-spaces in propositions 3.1 and 3.2 always complete?

(3) Find a complete characterization of the pairs $(\mathcal{V},E)$, $\mathcal{V}$ a decreasing sequence of strictly positive weights on $\mathbb{N}$ and E a Fréchet space, such that $k_\infty(E)$ is a topological subspace of $K_\infty(E)$.

(4) Let E be a Fréchet space and $\mathcal{V}$ a decreasing sequence of strictly positive weights on $\mathbb{N}$ satisfying condition (D). Suppose that the pair $(\mathcal{V},E)$ has condition (S) of Vogt (see 3.10). Is then $K_\infty(E)=L_b(\lambda_1(A),E)$ barrelled ?

REFERENCES

[1] F. Bastin, 'On bornological spaces $C\overline{V}(X)$', to appear in *Arch. Math.*

[2] E. Behrends, S. Dierolf, P. Harmand, 'On a problem of Bellenot and Dubinsky', *Math. Ann.* **275** (1986) 337-339.

[3] S.F. Bellenot, E. Dubinsky, 'Fréchet spaces with nuclear Köthe quotients', *Trans. Amer. Math. Soc.* **273** (1982) 579-594.

[4] C. Bessaga, A. Pełczyński, S. Rolewicz,'On diametral approximative dimension and linear homogeneity of F-spaces', *Bull. Acad. Polon. Sci.* **9** (1961) 677-683.

[5] K.D. Bierstedt, J. Bonet, 'Stefan Heinrich's density condition for Fréchet spaces and the characterization of the distinguished Köthe echelon spaces', *Math. Nachr.* **135** (1988) 149-180.

[6] K.D. Bierstedt, J. Bonet, 'Dual density conditions in (DF)-spaces I', *Results Math.* **14** (1988) 242-274.

[7] K.D. Bierstedt, J. Bonet, 'Dual density conditions in (DF)-spaces II', to appear in *Bull. Soc. Roy. Sci. Liège* (1988).

[8] K.D. Bierstedt, J. Bonet, J. Schmets, '(DF)-spaces of type CB(X,E) and $C\overline{V}(X,E)$', preprint 1988.

[9] K.D. Bierstedt, R. Meise,'Induktive Limites gewichteter Räume stetiger und holomorpher Funktionen', *J. reine angew. Math.* **282** (1976) 186-220.

[10] K.D. Bierstedt, R. Meise, 'Distinguished echelon spaces and projective descriptions of weighted inductive limits of type $\mathcal{V}C(X)$', pp.169-226 in : *Aspects of Mathematics and its Applications,* Elsevier Sci. Publ., North-Holland Math. Library, 1986.

[11] K.D. Bierstedt, R. Meise, 'Weighted inductive limits and their projective descriptions', *Doğa Tr. J. Math.* **10,** 1 (1986) (Special issue: Proceedings of the Silviri Conference 1985) 54-82.

[12] K.D. Bierstedt, R. Meise, W.H. Summers, 'A projective description of weighted inductive limits', *Trans. Amer. Math. Soc.* **272** (1980) 107-160.

[13] K.D. Bierstedt, R. Meise, W.H. Summers, 'Köthe sets and Köthe

sequence spaces', pp. 27-91 in : *Functional Analyis, Holomorphy and Approximation Theory,* North-Holland Math. Studies **71**, 1982.
[14] J. Bonet, 'A projective description of weighted inductive limits of spaces of vector valued continuous functions', *Collectanea Math.* **34** (1983) 115-125.
[15] J. Bonet, 'The countable neighbourhood property and tensor products', *Proc. Edinburgh Math. Soc.* **28** (1985) 207-215.
[16] J. Bonet, 'On weighted inductive limits of spaces of continuous functions', *Math. Z.* **192** (1986) 9-20.
[17] J. Bonet, 'Projective descriptions of inductive limits of Fréchet sequence spaces', *Arch. Math.* **48** (1987) 331-336.
[18] J. Bonet, A. Galbis, 'A note on Taskinen's counterexample to the problem of topologies of Grothendieck', to appear in *Proc. Edinburgh Math. Soc.*
[19] J. Bonet, A. Galbis, 'The identity L(E,F)=LB(E,F), tensor products and inductive limits', to appear in *Note di Mat.*
[20] J. Bonet, M. Maestre, G. Metafune, V.B. Moscatelli, D. Vogt, 'Every quojection is the quotient of a product of Banach spaces', preprint 1988.
[21] A. Defant, K. Floret, 'The precompactness-lemma for sets of operators', pp. 39-55 in : *Functional Analysis, Holomorphy and Approximation Theory II*, North-Holland Math. Studies **86**, 1984.
[22] A. Defant, W. Govaerts, 'Tensor products and spaces of vector-valued continuous functions', *Manuscr. Math.* **55** (1986) 433-449.
[23] S. Dierolf, 'On spaces of continuous linear mappings between locally convex spaces', *Note di Mat.* **5** (1985) 147-255.
[24] K. Floret, 'Some aspects of the theory of locally convex inductive limits', pp. 205-237 in: *Functional Analysis, Surveys and Recent Results II*, North-Holland Math. Studies **38**, 1980.
[25] A. Galbis, '*Commutatividad entre límites inductivos y productos tensoriales*', Thesis, University of Valencia, 1988.
[26] A. Galbis, 'Köthe sequence spaces with values in Fréchet or (DF)-spaces', *Bull. Soc. Roy. Sci. Liège* **57** (1988) 157-172.
[27] A. Galbis, 'Projective descriptions of weighted inductive limits of spaces of continuous functions with values in a Fréchet space', preprint 1988.
[28] A. Grothendieck, 'Sur les espaces (F) et (DF)', *Summa Brasil. Math.* **3** (1954) 57-112.
[29] A. Grothendieck, 'Produits tensoriels topologiques et espaces nucléaires', *Mem. Amer. Math. Soc.* **16**, 1955.
[30] R. Hollstein, 'Inductive limits and ε-tensor products', *J. reine angew. Math.* **319** (1980) 38-62.
[31] H. Jarchow, *'Locally Convex Spaces'*, B.G. Teubner, Stuttgart, 1981.
[32] J. Krone, D. Vogt, 'The splitting relation for Köthe spaces', *Math. Z.* **190** (1985) 387-400.
[33] V.B. Moscatelli: 'Fréchet spaces without continuous norms and without bases', *Bull. London Math. Soc.* **12** (1980) 63-66.
[34] P. Pérez Carreras, J. Bonet, '*Barrelled Locally Convex Spaces*', North-Holland Math. Studies **131**, 1987.
[35] J. Taskinen, 'Counterexamples to "problème des topologies" of

Grothendieck', *Ann. Acad. Sci. Fenn.*, Serie A, no **63**, 1986.
[36] J. Taskinen, The projective tensor product of Fréchet-Montel spaces', *Studia Math.* **91** (1988) 17-30.
[37] D. Vogt: 'On the functor $Ext^1(E,F)$ for Fréchet spaces', *Studia Math.* **85** (1987) 163-197.
[38] D. Vogt, Frécheträume, zwischen denen jede stetige lineare Abbildung beschränkt ist', *J. reine angew. Math.* **345** (1983) 182-200.
[39] D. Vogt, 'Some results on continuous linear maps between Fréchet spaces'. pp. 349-381 in *Functional Analysis: Surveys and Recent Results III* , North-Holland Math. Studies **90**, 1984.
[40] D. Vogt, 'On two problems of Mityagin', preprint, Wuppertal 1987.
[41] D. Vogt, 'Distinguished Köthe spaces", to appear in *Math. Z.*
[42] D. Vogt, 'On two classes of (F)-spaces', *Arch. Math.* **45** (1985) 225-266.
[43] D. Vogt, 'Lectures on projective spectra of (DF)-spaces', Wuppertal 1987.

ON TENSOR PRODUCT α-ALGEBRA BUNDLES

Athanasios Kyriazis
Mathematical Institute
University of Athens
57, Solonos Street
106 79 Athens
Greece

ABSTRACT. This note introduces a class of bundles, called topological α-algebra bundles, where the "coefficients" of the bundles belong to a topological algebra bundle α. Furthermore, given two such bundles ξ, η, we show the existence of the tensor product algebra bundle $\xi \hat{\otimes}_{\alpha} \eta$ and prove, under appropriate conditions, that the algebra of sections $\Gamma(\xi \hat{\otimes}_{\alpha} \eta)$ is isomorphic to $\Gamma(\xi) \hat{\otimes}_{\Gamma(\alpha)} \Gamma(\eta)$. On the other hand, given a topological algebra bundle ξ, we consider the (locally trivial fibre) "spectrum bundle" of ξ, $\mathfrak{M}(\xi)$, such that the bundles ξ, $\mathfrak{M}(\xi)$ are, in a sense, dual to each other. So, $\mathfrak{M}(\xi \hat{\otimes}_{\alpha} \eta)$ describes, under suitable conditions, the fibre product bundle of $\mathfrak{M}(\xi)$, $\mathfrak{M}(\eta)$ over $\mathfrak{M}(\alpha)$.

It is known [2,4] that for various classes of topological algebras the set $\mathfrak{M}(E \hat{\otimes}_{\mathbb{A}} F)$ of continuous non-zero characters (spectrum) of the complete tensor product $\mathbb{A}$-algebra is the pull-back of $\mathfrak{M}(E)$, $\mathfrak{M}(F)$ over $\mathfrak{M}(\mathbb{A})$.

In this paper, which continues the study begun in [5], we generalize the above situation in case of topological α-algebra bundles:

Thus, in Section 1 we introduce a more general notion of topological algebra bundles than that in [5]. Namely, given a topological algebra bundle ξ [5], we replace the structure group of ξ by a topological group acting continuously and effectively on the fibre of ξ. Now, if α is a topological algebra bundle, in the previous generalized sense, we define a topological α-algebra bundle, as one having "coefficients" from α instead of a topological algebra $\mathbb{A}$ (cf. [9]).

In Section 2 we consider the topological tensor product of two α-algebra bundles, as another such bundle, whose fibre is the (topological) tensor product $\mathbb{A}$-algebra of the fibres of the factors. This construction is achieved by considering the corresponding G-cocycles of the bundles concerned, where G is now a "locally equicontinuous", in an appropriate sense, topological group.

In Section 3 the section algebra $\Gamma(\xi \hat{\otimes}_{\alpha} \eta)$ of a tensor product

T. Terzioğlu (ed.), Advances in the Theory of Fréchet Spaces, 223–233.

α-algebra bundle is proved, under suitable conditions, to be the complete projective $\Gamma(\alpha)$-tensor product of the section algebras of the factors (Theorem 3.1).

In Section 4 we consider the "spectrum bundle" $\mathfrak{M}(\xi)$ of a topological algebra bundle ξ over a space X, realized as the spectrum of $\Gamma(\xi)$ (Theorem 4.1). Here X is not necessarily compact, getting thus a more general situation than that in [5].

As an application of the above, we get in the last section the expression of $\mathfrak{M}(\xi \hat{\otimes}_\alpha \eta)$ as $\mathfrak{M}(\xi) \underset{\mathfrak{M}(\alpha)}{\times} \mathfrak{M}(\eta)$ (Theorem 5.1). This constitutes the "bundle analogon" of the well-known decomposition of the spectrum of a topological tensor product algebra via the spectra of the factor algebras.

1. In [5] we have given the notion of a topological algebra bundle over a topological space X. This is a triplet $\xi = (\mathbf{E},\pi,X)$ with $\pi : \mathbf{E} \longrightarrow X$ continuous, whose fibre $\xi_x \equiv \pi^{-1}(x)$ is a topological ($\mathbb{C}$-) algebra, whereas the standard "local triviality" conditions are satisfied [5: p.400, (1.2),(1.3)]. More generally, one can replace the (structure) group $Aut_{\mathbb{C}}(M) \equiv Aut(M)$ of $\mathbb{C}$-automorphisms of the fibre $M \equiv \pi^{-1}(x)$, $x \in X$, of ξ with a topological group G acting continuously and effectively on M. We denote such a bundle by

$$\xi \equiv (\mathbf{E},\pi,X,M,G). \tag{1.1}$$

Locally convex, unital, commutative, nuclear Fréchet algebra bundles with locally equicontinuous spectra etc. will have the obvious meaning, referring to the pertinent class of topological algebras for the fibres.

Now, by a *system of transition functions* of a topological space X we mean a set

$$\Lambda \equiv \{(U_i),(\lambda_{ij}),M,G\} \tag{1.2}$$

consisted of an open covering $(U_i)_i$ of X, a topological algebra M and a G-cocycle (λ_{ij}) associated to $(U_i)_i$, with G a topological group as above.

Thus, a characterization of topological algebra bundles in terms of a system of transition functions can be given, as was also the case in [5 : Theorem 1.1]. Namely, let $\Phi(X)$ be the set of isomorphism classes of topological algebra bundles over X(cf. (1.1)) having locally equicontinuous structure groups and $H^1(X,G)$ the set of equivalence classes of systems of transition functions on X (cf. (1.2)). Then, one has

$$\Phi(X) = H^1(X,G) \tag{1.3}$$

within a bijection.

On the other hand, let M be a locally convex $\mathbb{A}$-algebra, with $\mathbb{A}$ a unital commutative locally convex algebra (cf. [4] for definitions) and $Aut(M), Aut(\mathbb{A})$ (resp. $Aut_{\mathbb{A}}(M)$) the topological groups of $\mathbb{C}$-automorphisms of M, $\mathbb{A}$(resp. $\mathbb{A}$-automorphisms of M), equipped with the simple convergence topology. Then, $Aut(\mathbb{A})$ "acts continuously" on $Aut(M)$ in the sense that the correspondence

(1.4) $$\mathrm{Aut}(\mathbb{A}) \times \mathrm{Aut}(M) \longrightarrow \mathrm{Aut}_{/\mathbb{A}}(M) : (\chi, f) \longmapsto \chi\cdot f$$

is continuous.

Now, if $\xi = (\mathbf{E},\pi,X,M,G)$ and $\alpha = (\mathbf{A},\rho,X,\mathbb{A},R)$ are two topological algebra bundles, we shall say that ξ is a *topological α-algebra bundle*, in case M is a topological $\mathbb{A}$-algebra, $G \lhd \mathrm{Aut}_{/\mathbb{A}}(M)$ and R acts continuously on G.

Furthermore, an analogous result to that of (1.3) for the above class of bundles is also true.

2. In this section we consider tensor product algebra bundles over a given bundle α. We have first to comment on some useful preliminary results.

So let G, H be two topological groups acting continuously on the topological $\mathbb{A}$-algebras M, N respectively. If $l_G : G\times M \longrightarrow M$, $l_H : H\times N \longrightarrow N$ are the corresponding (continuous) actions, the map

(2.1) $$l_{G\times H} : (G\times H) \times (M \otimes_{/\mathbb{A}} N) \longrightarrow M \otimes_{/\mathbb{A}} N : ((g,h), \sum_i x_i\otimes y_i) \longmapsto \sum_i gx_i \otimes hy_i$$

is an algebra representation of the group $G\times H$ on the ($\mathbb{A}$-)tensor product algebra $M \otimes_{/\mathbb{A}} N$. That is, one has

2.1. Lemma. *Let G, H be two topological groups acting continuously and effectively on the unital locally convex $\mathbb{A}$-algebras M, N respectively* (hence, one has $G \subsetneq \mathrm{Aut}_{/\mathbb{A}}(M)$; analogously for H). *Then, the topological group $G\times H$ acts continuously and effectively on $M \otimes^{\pi}_{/\mathbb{A}} N$.*

Proof. If $g\in G$ with $g \neq e_G$ (: neutral element of G), then there exists $x\in M$ with $gx \neq x$, such that $mx \otimes e_H \neq x \otimes e_H$, for any $m \in G\times H$, with $m \neq (e_G, e_H)$. On the other hand, if $\Phi : M\times N \longrightarrow M \otimes^{\pi}_{/\mathbb{A}} N$ is the canonical continuous map of tensor product, then the continuity of $l_{G\times H}$: $(G\times H) \times (M \otimes^{\pi}_{/\mathbb{A}} N) \longrightarrow M \otimes^{\pi}_{/\mathbb{A}} N$ (cf. (2.1)) follows from the relation $\Phi \circ (l_G \times l_H) = l_{G\times H} \circ (id_{G\times H} \times \Phi)$ and the relative topologies on $G\times H$ and $M \otimes^{\pi}_{/\mathbb{A}} N$ (cf. [4 : p.58, Appendix]).

2.2 Lemma. *Let M,N,M',N' be locally convex $\mathbb{A}$-algebras and K, L equicontinuous subsets of $\mathrm{Hom}_{/\mathbb{A}}(M,N)$, $\mathrm{Hom}_{/\mathbb{A}}(M',N')$, respectively. Then, $K\otimes L$ is an equicontinuous subset of $\mathrm{Hom}_{/\mathbb{A}}(M \otimes^{\pi}_{/\mathbb{A}} M', N \otimes^{\pi}_{/\mathbb{A}} N')$.*

Proof. For any W from a local basis of $N \otimes^{\pi}_{/\mathbb{A}} N'$, the set $\bigcap_{u\in K\otimes L} u^{-1}(W)$ is a neighborhood of zero in $M \otimes^{\pi}_{/\mathbb{A}} M'$: Indeed, from [8 : Corollary 2.1, p. 368], [4 : Appendix] it is enough to show that $\bigcap_{u\in K\otimes L} u^{-1}(\Gamma(U \otimes_{/\mathbb{A}} V))$ is a neighborhood of zero in $M \otimes^{\pi}_{/\mathbb{A}} M'$, with U, V running over local bases of N, N' respectively. Thus, the equicontinuity of K, L shows that

$\bigcap\limits_{u_1\in K} u_1^{-1}(U)$, $\bigcap\limits_{u_2\in L} u_2^{-1}(V)$ are neighborhoods of zero in M, M' respectively, such that $\Gamma((\bigcap\limits_{u_1} u_1^{-1}(U)) \otimes_{\mathbb{A}} (\bigcap\limits_{u_2} u_2^{-1}(V))) \subseteq \bigcap\limits_{u} u^{-1}(\Gamma(U \otimes_{\mathbb{A}} V))$ (resp. [8,4]).

Now, let G be a topological group acting continuously and effectively on a topological $\mathbb{A}$-algebra M with continuous multiplication, while $\mathbb{A}$ is a complete topological algebra. If $Aut_{\mathbb{A}}(\hat{M})$ ($\subseteq Hom_{\mathbb{A}}(\hat{M},\hat{M})_s$, cf. [5]) is the group of $\mathbb{A}$-automoprhisms of $\hat{M}$ (completion of M), then, under suitable conditions (e.g., if G is locally compact), G acts continuously on $\hat{M}$. Moreover, if G is locally equicontinuous and $\hat{G}$ is the subgroup of $Aut_{\mathbb{A}}(\hat{M})$, consisting of $g\in G$ to the completion of M, then $\hat{G}$ is also locally equicontinuous such that $\hat{G}$ acts continuously on $\hat{M}$ (cf. [8 : Chapter V; Theorem 8.1]).

Remarks. 1) The above Lemma 2.2 has also an inverse (cf., for instance, [4], [8]).

2) Let $M \hat{\otimes}_{\mathbb{A}} N$ denote the completion of the projective $\mathbb{A}$-tensor product. So, if any $\mathbb{A}$-automorphism of $M \hat{\otimes}_{\mathbb{A}} N$ leaves either one of the factor algebras M or N invariant, then one has the decomposition

(2.2) $$Aut_{\mathbb{A}}(M \hat{\otimes}_{\mathbb{A}} N) = Aut_{\mathbb{A}}(M) \times Aut_{\mathbb{A}}(N),$$

(see, for instance, [8 : Chapter XIII, Scholium 8.2]). As a result one remarks that if the topological groups G, H act continuously and effectively on M and N, respectively, then $G\hat{\times}H$ is locally equicontinuous iff this is the case for G, H.

Now, let $\xi = (\mathbf{E},\pi,X,M,G)$ and $\eta = (\mathbf{F},\sigma,X,N,H)$ be unital locally convex $\alpha = (\mathbf{A},\rho,X,\mathbb{A},R)$-algebra bundles with continuous multiplications and locally equicontinuous structure groups, with α being a unital commutative complete locally convex algebra bundle having continuous multiplication and locally equicontinuous structure group R. Moreover, let $\Lambda_\xi = \{(U_i),(\lambda_{ij}),M,G\}$, $\Lambda_\eta = \{(U_i),(\mu_{ij}),N,H\}$ be the transition functions for the bundles respectively. Then,

(2.3) $$(\lambda_{ij} \hat{\otimes} \mu_{ij}) ,$$

is a $G\hat{\times}H$-cocycle for a topological α-algebra bundle, denoted by $\xi \hat{\otimes}_\alpha \eta$; i.e., we have

(2.4) $$\xi \hat{\otimes}_\alpha \eta := (\mathbf{E} \hat{\otimes}_{\mathbf{A}} \mathbf{F}, \pi \hat{\otimes} \rho, X, M \hat{\otimes}_{\mathbb{A}} N, G\hat{\times}H)$$

where $\mathbf{E} \hat{\otimes}_{\mathbf{A}} \mathbf{F} \equiv \sum\limits_{x\in X} (\pi^{-1}(x) \hat{\otimes}_{\rho^{-1}(x)} \sigma^{-1}(x))$, and $\pi \hat{\otimes} \rho : \mathbf{E} \hat{\otimes}_{\mathbf{A}} \mathbf{F} \longrightarrow X$ the respective (canonical) projection. This is the so-called *tensor product α-algebra bundle* (or simply, *tensor product algebra bundle*).

3. In this section we examine the relation between the algebra of sections of $\xi \hat{\otimes}_\alpha \eta$ and those of ξ,η. First, we recall that given a topological algebra bundle ξ over X of fibre type M, the corresponding set of continuous sections of ξ, $\Gamma(\xi)$, is a topological algebra, such that

$$\Gamma(\xi) = B_\xi \tag{3.1}$$

within an isomorphism of topological algebras [5]. Here

$$B_\xi := \{(\tau_i)_i : \tau_i(x) = \lambda_{ij}(x)\tau_j(x),\ x \in U_i \cap U_j\} \subseteq \prod_{i\in J} C_c(U_i,M) \tag{3.2}$$

is a topological (sub)algebra of the cartesian product of the algebras of M-valued continuous functions on U_i, with (λ_{ij}) a G-cocycle defining ξ.

Thus, we have.

3.1. Theorem. *Let ξ,η be topological α-algebra bundles over a locally compact space X with a denumerable open covering of σ-compact subsets. Moreover, suppose that ξ,η,α are unital commutative nuclear Fréchet locally convex algebra bundles whose structure groups are locally equicontinuous. Then,*

$$\Gamma(\xi \hat{\otimes}_\alpha \eta) = \Gamma(\xi) \underset{\Gamma(\alpha)}{\hat{\otimes}} \Gamma(\eta) \tag{3.3}$$

within an isomorphism of $\Gamma(\alpha)$-locally convex algebras.

Proof. In view of (3.1), it is enough to show the next isomorphism of locally convex B_α-algebras

$$B_{\xi \hat{\otimes}_\alpha \eta} = B_\xi \underset{B_\alpha}{\hat{\otimes}} B_\eta . \tag{3.4}$$

Indeed, if (U_i) is a denumerable open covering of X consisting of σ-compact subspaces of X, then we have the next isomorphisms of locally convex algebras :

$$\begin{aligned}
\prod_i C(U_i, M \hat{\otimes}_A N) &= \text{(cf. [8 : Chapter XI, Theorem 1.1])}\ \prod_i (C(U_i)\hat{\otimes}(M \hat{\otimes}_A N)) \\
&= \text{(cf. [4 : p. 63, (5.15)])}\ \prod_i ((C(U_i) \underset{C(U_i)}{\hat{\otimes}} C(U_i))\hat{\otimes}(M \hat{\otimes}_A N)) \\
&= \text{(cf. [4 : Proposition 5.1])}\ \prod_i ((C(U_i)\hat{\otimes}M) \underset{C(U_i)\hat{\otimes}A}{\hat{\otimes}} (C(U_i)\hat{\otimes}N)) \\
&= \text{(cf. [8 : p. 391, (1.16)])}\ \prod_i (C(U_i,M) \underset{C(U_i,A)}{\hat{\otimes}} C(U_i,N)) \\
&= \text{(cf. [4 : p. 60, (5.6)])}\ \prod_i C(U_i,M) \underset{\prod_i C(U_i,A)}{\hat{\otimes}} \prod_i C(U_i,N).
\end{aligned}$$

So, the assertion follows from (3.1),(3.2),(2.3).

If α is complex line bundle, then (3.3) implies

$$\Gamma(\xi \hat{\otimes} \eta) = \Gamma(\xi) \underset{\Gamma(\alpha)}{\hat{\otimes}} \Gamma(\eta) , \tag{3.5}$$

within an isomorphism of locally convex $\Gamma(\alpha)$-algebras. In particular, if ξ,η,α are trivial bundles over X of fibre type M,N,A, one gets

$$C(X,M\hat{\otimes}_A N) = C(X,M) \underset{C(X,A)}{\hat{\otimes}} C(X,N) . \tag{3.6}$$

By considering C^∞-(complex)algebra bundles (resp. complex analytic algebra bundles) over a 2nd countable compact C^∞-manifold (resp. Stein manifold), one has an analogous decomposition to (3.3) for the corresponding algebras of sections. See also [6 : Scholium], [5 : Scholium 4.1] and [8].

4. In this section we define the *spectrum bundle* of a topological algebra bundle ξ over a not necessarily compact space X (cf., however, [5 : §4 Appendix]), and we identify it as the spectrum of the algebra $\Gamma(\xi)$.

Given a topological group G acting continuously and effectively on a topological algebra M, we get a "dual" action of G on the spectrum of M as follows

$$^tG := \{^tg : g\in G\} \subseteq Aut(\mathfrak{M}(M)) , \tag{4.1}$$

such that $^tg(\chi) := \chi\circ g$, $\chi\in\mathfrak{M}(M)$. Henceforth, *we assume that ^tG is a topological group acting continuously on $\mathfrak{M}(M)$*. (Take, for instance, M semi-simple and $\mathfrak{M}(M)$ locally equicontinuous). Thus, given a system of transition functions on X, say $\Lambda = \{(U_i),(\lambda_{ij}),M,G\}$, one gets a (dual) system of transition functions $\Lambda^* = \{(U_i),(\lambda^*_{ji}),\mathfrak{M}(M),{}^tG\}$ on X, with $\lambda^*_{ji}(x) := {}^t(\lambda_{ij}(x))$, $x\in U_i\cap U_j$.

So, through the "spectrum operator" $\mathfrak{M}$, the topological algebra bundle $\xi = (\mathbf{E},\pi,X,M,G)$ defines a (locally trivial) fibre bundle

$$\mathfrak{M}(\xi) = (S,p,X,\mathfrak{M}(M),{}^tG) \tag{4.2}$$

the so called *spectrum bundle* (cf. also [5]). Here the bundle space S is of the form

$$S \equiv (\biguplus_{i\in J}(U_i \times \mathfrak{M}(M))/\sim \; , \tag{4.3}$$

where the equivalence relation "$\sim$" on the disjoint union involved is given by

$$(x_i,\chi_i) \sim (x_j,\chi_j) \quad \text{iff} \begin{cases} x_i = x_j \equiv x\in U_i\cap U_j \\ \chi_j = \lambda^*_{ji}(x)(\chi_i) \; . \end{cases} \tag{4.4}$$

On the other hand, if $(U_i)_{i\in J}$ is an open covering of X, then for any $\alpha\in F(J)$ (set of finite subsets of J) we consider the topological algebra bundles

$$\xi_\alpha = (\mathbf{E}_\alpha,\pi_\alpha,X_\alpha,M,G) \tag{4.5}$$

with $X_\alpha = \bigcup_{i\in\alpha} U_i$, $\mathbf{E}_\alpha = \pi^{-1}(X_\alpha) \subseteq \mathbf{E}$, and π_α the canonical projection. Thus, the pair

$$(4.6) \qquad (\mathfrak{M}(\xi_\alpha),\ f_{\beta\alpha})_{\alpha\in F(J)}$$

defines a direct system of topological spaces, where $\mathfrak{M}(\xi_\alpha)$ is the spectrum bundle of ξ_α (cf. (4.2)) and, for any $\alpha \subseteq \beta$ in $F(J)$, $f_{\beta\alpha}$: $\biguplus_{i\in\alpha}(U_i \times \mathfrak{M}(M))/\sim_\alpha \longrightarrow \biguplus_{i\in\beta}(U_i \times \mathfrak{M}(M))/\sim_\beta$ are the canonical inclusion maps passing to the quotient.

The above in connection with [7 : Lemma 3.1] prove the next.

4.1. Lemma. *Let* $\xi = (\mathbf{E},\pi,X,M,G)$ *be a topological algebra bundle and* $(U_i)_{i\in J}$ *an open covering of* X. *Then*

$$(4.7) \qquad \mathfrak{M}(\xi) = \varinjlim_{\alpha\in F(J)} \mathfrak{M}(\xi)$$

within an homeomorphism.

Now, let $\xi = (\mathbf{E},\pi,X,M,G)$ be a topological algebra bundle and $\Lambda = \{(U_i),(\lambda_{ij}),M,G\}_{i\in J}$ a system of transition functions of ξ. For any $\alpha \in F(J)$ consider the next topological algebra, subalgebra of $\prod_{i\in\alpha} C(U_i,M)$;

$$(4.8) \qquad B_\alpha := \{\tau_\alpha \equiv (\tau_i)_{i\in\alpha} : \tau_{i,\alpha}(x) = \lambda_{ij}(x)\tau_{j,\alpha}(x)\ ,\ x \in U_i \cap U_j\}$$

(cf. (3.2)). The family

$$(4.9) \qquad (B_\alpha,\ g_{\alpha\beta}),\ \alpha \subseteq \beta \text{ in } F(J)$$

defines an inverse system of topological algebras, whose inverse limit is given by (3.2); i.e.

$$(4.10) \qquad B_\xi = \varprojlim_\alpha B_\alpha\ ,$$

within an isomorphism of topological algebras. Now, by (3.1),(4.5), (4.10) one has the next isomorphism of topological algebras of sections of the corresponding bundles

$$(4.11) \qquad \Gamma(\xi) = \varprojlim_\alpha \Gamma(\xi_\alpha)$$

So, by [8 : Chapter V, Lemma 7.1] and the above discussion we have the next.

4.1. Proposition. *Let* $\xi = (\mathbf{E},\pi,X,M,G)$ *be a topological algebra bundle with continuous multiplication and* $(U_i)_{i\in J}$ *an open covering of* X. *Moreover, let* $(\Gamma(\xi_\alpha))_{\alpha\in F(J)}$ *be the inverse system of the topological algebras of sections, as above* (cf. (4.9),(4.11)). *Then,* $(\mathfrak{M}(\Gamma(\xi_\alpha)))_\alpha$ *is a direct system of topological spaces, such that one gets the following continuous bijection*

(4.12) $$\underset{\overrightarrow{\alpha}}{lim}\, \mathfrak{M}(\Gamma(\xi_\alpha)) \xrightarrow{\sim} \mathfrak{M}(\Gamma(\xi)).$$

Now, by a *Gel'fand - Mazur* (topological) *algebra* one means a topological algebra M such that for every (2-sided) maximal regular closed ideal $I \subseteq M$ one has $M/I = \mathbb{C}$, within a topological algebraic isomorphism [8 : Chapter VIII, §9.(5)]. This definition is equivalent with saying that every (2-sided) closed regular maximal ideal of M is the kernel of a continuous character of M (ibid.). Moreover, by a *Waelbroeck algebra* we mean a unital (topological) Q-algebra (its group of the invertible elements is an open set) with continuous inversion (ibid., Chapter I, Definition 5.2).

So, we are in the position to prove the following result.

4.1. Theorem. *Let* $\xi = (\mathbf{E},\pi,X,M,G)$ *be a commutative complete Gel'fand - Mazur Waelbroeck algebra bundle, whose base is locally compact, having an open relatively compact covering* $(U_i)_{i\in J}$. *Then, one has* (see also (4.2))

(4.13) $$\mathfrak{M}(\xi) \equiv \biguplus_{i\in J} (U_i \times \mathfrak{M}(M))/\sim \; = \mathfrak{M}(\Gamma(\xi)) ,$$

within a homeomorphism.

Proof. By hypothesis we may assume that $X = \underset{\overrightarrow{\alpha\in F(J)}}{lim}\, \bar{U}_\alpha$, with $\bar{U}_\alpha = \overline{\bigcup_{i\in\alpha} U_i}$ compact subspaces of X. Thus, the topological algebra bundles $\xi_\alpha = (\mathbf{E}_\alpha,\pi_\alpha,X_\alpha,M,G)$, $\alpha \in F(J)$ (cf. (4.5)) satisfy the conditions of Theorem 4.1 in [5], such that one gets the next homeomorphism

(4.14) $$\mathfrak{M}(\Gamma(\xi_\alpha)) = \biguplus_{i\in\alpha} (V_{\alpha,i} \times \mathfrak{M}(M))/\sim_\alpha \quad , \; \alpha \in F(J),$$

where $V_{\alpha,i}$ denotes an open covering of the compact $\bar{U}_\alpha$. Thus, Proposition 4.1 and Lemma 4.1 yield a continuous bijection

(4.15) $$\omega : \biguplus_{i\in J} (U_i \times \mathfrak{M}(M))/\sim \;\longrightarrow \mathfrak{M}(\Gamma(\xi)) \quad ,$$

such that we next prove the continuity of ω^{-1}.

First if $p : \biguplus_{i\in J} (U_i \times \mathfrak{M}(M))/\sim \;\longrightarrow X$ is the canonical projection map one defines $p' : \mathfrak{M}(\Gamma(\xi)) \longrightarrow X$ with $p' := p\circ\omega^{-1}$, such that for any $\chi = g\circ\vartheta_{i,x} \in \mathfrak{M}(\Gamma(\xi))$, one has $p'(\chi) := x$, where $g \in \mathfrak{M}(M), x\in U_i$ and

(4.16) $$\vartheta_{i,x} : \Gamma(\xi) \longrightarrow M : \gamma \longmapsto \vartheta_{i,x}(\gamma) := \gamma(x)$$

is a continuous algebra morphism [5 : p. 415, (4.2)]. The map p' is clearly surjective and moreover continuous. Indeed, if

$\chi_\delta \xrightarrow[\delta]{} \chi \,|_{\mathfrak{M}(\Gamma(\xi))}$ we prove that $p'(\chi_\delta) = x_\delta \xrightarrow[\delta]{} x = p'(\chi)$. For, otherwise, let $x \in U$ and W, V open subsets of X such that $x \in W \subset \bar{W} \subset V \subset U$ and $x_\delta \notin V$ "frequently". Now, choose $f \in C(X)$ so that $0 \leqslant f(x) \leqslant 1$, $f = 1$ on $\bar{W}$, $f = 0$ off V and let $\gamma = f \cdot \mathbb{1}_M \in \Gamma(\xi)$ [5]. Then, if $x_{\delta'} \notin V$ one has $\chi_{\delta'}(\gamma) \longrightarrow \chi(\gamma)$ whereas $\chi_{\delta'}(\gamma) = g_{\delta'}(\vartheta_{j,x_{\delta'}}(\gamma)) = 0$ and $\chi(\gamma) = g(\vartheta_{i,x}(\gamma)) = \mathbb{1}_M$, a contradiction. Thus, p' is continuous.

Now we finally prove the *continuity of* ω^{-1} : Indeed, if $\chi_\delta \xrightarrow[\delta]{} \chi$ in $\mathfrak{M}(\Gamma(\xi))$, the continuity of p' implies $p'(\chi_\delta) = x_\delta \xrightarrow[\delta]{} x = p'(\chi)$, where $\chi_\delta = g_\delta \circ \vartheta_{i,x_\delta}$ and $\chi = g \circ \vartheta_{j,x}$. It suffices to show $g_\delta \xrightarrow[\delta]{} g$. If $g_\delta \not\longrightarrow g \,|_{\mathfrak{M}(M)}$ there exists a neighborhood of g, say $N(g) := \{g' : |g(\alpha_\lambda) - g'(\alpha_\lambda)| < \varepsilon,\ \lambda = 1,\dots,n\}$ with g_δ frequently not in $N(g)$. Moreover, we may assume that for open W and V one obtains $x \in W \subset \bar{W} \subset V \subset U_j$ and $x_{\delta'} \in W$ for all $\delta' \geqslant \delta(W)$. Thus, there exists $f_\lambda \in C(X,M)$ such that $f_\delta(x) = \alpha_\lambda$ on $\bar{W}$, $f_\lambda(x) = 0$ off V. If $\gamma_\lambda(x) = f_\lambda(x)$ (cf. (4.16)), let $N(\chi) := \{\chi' : |\chi'(\gamma_\lambda) - \chi(\gamma_\lambda)| < \varepsilon,\ \lambda = 1,\dots,n\}$. Then $|\chi_{\delta'}(\gamma_\lambda) - \chi(\gamma_\lambda)| = |g_{\delta'}(\alpha_\lambda) - g(\alpha_\lambda)| \geqslant \varepsilon$, i.e. a contradiction to $\chi_{\delta'} \longrightarrow \chi$ in $\mathfrak{M}(\Gamma(\xi))$.

The following is straightforward from Theorem 4.1.

4.1. Corollary. *The spectrum bundle of a complex line bundle* ζ *over a locally compact space* X, *is given by* $\mathfrak{M}(\zeta) = (X, id_X, X)$.

4.1. Remark. In case of a Fréchet locally convex algebra bundle ξ over a locally compact space X having a denumerable open covering of σ-compact subsets, the algebra of sections $\Gamma(\xi)$ is a Fréchet locally convex algebra (cf. (3.1),(3.2)). In this case the map (4.12) is a homeomorphism as follows from [8 : Chapter V, p. 176]. Hence, Theorem 4.1 is now an immediate consequence of Proposition 4.1, Lemma 4.1.

5. In this section we give an application of the above, pertaining to the canonical analysis of $\mathfrak{M}(\xi \hat{\otimes}_\alpha \eta)$ as Theorem 5.1 shows.

Thus, let ξ, η be topological α-algebra bundles and $\bar{\mu} : \mathfrak{M}(\xi) \longrightarrow \mathfrak{M}(\alpha)$, $\bar{\nu} : \mathfrak{M}(\eta) \longrightarrow \mathfrak{M}(\alpha)$ the "dual" bundle morphisms, defined by the canonical algebra bundle morphisms $\mu : \alpha \longrightarrow \xi$, $\nu : \alpha \to \eta$ (cf. [4 : p. 51, (2.5)]. If $(\lambda^*_{ji}), (\mu^*_{ji}), (\nu^*_{ji})$ are ${}^tG$, ${}^tH$, ${}^tR$ -cocycles respectively, defining the corresponding bundles $\mathfrak{M}(\xi)$, $\mathfrak{M}(\eta)$, $\mathfrak{M}(\alpha)$ (cf. (4.1) and remarks after this) the map

(5.1) $\lambda^{\star}_{ji} \underset{\nu^{\star}}{\times} \mu^{\star}_{ji} : U_i \cap U_j \to {}^tG \underset{R}{\times} {}^tH : x \mapsto (\lambda^{\star}_{ji} \underset{\nu^{\star}}{\times} \mu^{\star}_{ji})(x) := \lambda^{\star}_{ji}(x) \underset{\nu^{\star}_{ji}(x)}{\times} \mu^{\star}_{ji}(x)$

defines a ${}^tG \times_R {}^tH$ - cocycle, where ${}^tG \times_R {}^tH$ is a subgroup of $Aut(\mathfrak{M}(M) \underset{\mathfrak{M}(A)}{\times} \mathfrak{M}(N))$ (homeomorphisms of pull-back of $\mathfrak{M}(M)$, $\mathfrak{M}(N)$ over $\mathfrak{M}(A)$). So, if ${}^tG$, ${}^tH$ are locally equicontinuous topological groups (cf. (4.1)), the ${}^tG \times_R {}^tH$ - cocycle (5.1) defines a (locally trivial) fibre bundle, the *fibre product bundle of* $\mathfrak{M}(\xi)$, $\mathfrak{M}(\eta)$ *over* $\mathfrak{M}(\alpha)$, denoted by $\mathfrak{M}(\xi) \underset{\mathfrak{M}(\alpha)}{\times} \mathfrak{M}(\eta)$.

Thus, we have the next

5.1. Theorem. *Let* ξ, η *be two Waelbroeck topological* α*-algebra bundles over X, such that the conditions of Theorem 3.1 are valid. Furthermore, let the spectra of* $\Gamma(\xi)$, $\Gamma(\eta)$ *be locally equicontinuous. Then, one has*

$$\mathfrak{M}(\xi \hat{\otimes}_{\alpha} \eta) = \mathfrak{M}(\xi) \underset{\mathfrak{M}(\alpha)}{\times} \mathfrak{M}(\eta) \quad , \tag{5.2}$$

within an isomorphism of the fibre bundles concerned.

Proof. By Theorem 3.1 and [4 : Theorem 2.1] we have the following homeomorphisms

$$\mathfrak{M}(\Gamma(\xi \hat{\otimes}_{\alpha} \eta)) = \mathfrak{M}(\Gamma(\xi) \underset{\Gamma(\alpha)}{\hat{\otimes}} \Gamma(\eta)) = \mathfrak{M}(\Gamma(\xi)) \underset{\mathfrak{M}(\Gamma(\alpha))}{\times} \mathfrak{M}(\Gamma(\eta)) \ . \tag{5.3}$$

Now, the assertion follows by Theorem 4.1.

Remarks. 1) If α is a complex line bundle, (5.2), Corollary 4.1 give

$$\mathfrak{M}(\xi \hat{\otimes} \eta) = \mathfrak{M}(\xi) \oplus \mathfrak{M}(\eta) \ ; \tag{5.4}$$

(the second member of (5.4) means "Whitney sum").

2) One has a similar result to (5.2) by considering spectrum bundles of C^{∞}-(complex) algebra bundles and / or complex analytic ones (see also §3 and [5 : Scholium 4.1]).

ACKNOWLEDGEMENT. It is a pleasure to express my sincere thanks to Professor Anastasios Mallios for generous help and advice, as well as lively interest, during the preparation of this work.

REFERENCES

[1] N. Bourbaki: *General Topology II*. Hermann Paris, 1966.

[2] B.R. Gelbaum: 'Tensor products over Banach algebras' . *Trans. Amer. Math. Soc.* 118(1965), 131-149.

[3] A. Grothendieck : 'Produits tensoriels topologiques et espaces nucléaires' . *Mem. Amer. Math. Soc.* 16(1955).

[4] A. Kyriazis: 'On the spectra of topological A-tensor product A-algebras' . *Yokohama Math. J.* 31(1983), 47-65.

[5] A. Kyriazis: 'On topological algebra bundles'. *J. Austr. Math. Soc.*

43(1987), 398-419.
[6] A. Kyriazis: 'Tensor products of function algebras' . *Bull. Austr. Math. Soc.* 36(1987), 417-423.
[7] A. Kyriazis: 'An associative law for infinite topological $\mathcal{A}$-tensor product $\mathcal{A}$-algebras' . *Portugaliae Math. J.* (in press).
[8] A. Mallios: *Topological Algebras. Selected Topics.* North-Holland, Amsterdam, 1986.
[9] A. Mallios: 'Vector bundles and *K*-Theory over topological algebras'. *J. Math. Anal. Appl.* 92(1983), 452-506.

QUOJECTION AND PREQUOJECTIONS

G. METAFUNE
and
V.B. MOSCATELLI
Dipartimento di Matematica
Universita - C.P. 193
73100 Lecce - Italy

Abstract

We survey the progress made since 1985 concerning quojections and the recently discovered class of prequojections. Furthermore, in the Appendix we report on some related results regarding nuclear spaces, Fréchet and (LF)-spaces of Moscatelli type and primary Fréchet spaces.

Introduction.

We feel that it is time to report again on the progress made in the field of quojections since the second author's survey [40] in 1985. In so doing, we take the opportunity to look also at the rapid development in the study of the newly discovered and closely related class of prequojections. Indeed, the last three years have seen a considerable progress in the investigation on the structure and properties of these special classes of Fréchet spaces, progress which shows the necessity of an up-to-date survey in this field which may rightly be called "structure theory of Fréchet spaces in the absence of compactness". Because the various results have by now channelled into a well-defined research area, we felt that the chronological order in the exposition of the results adopted in [40] could no longer be maintained here. Thus, we have placed at the beginning of the various sections some results which, although discovered later, are relevant and useful to make the exposition clearer.

The paper is divided into two sections and an Appendix. The first section is devoted to quojections and is, in turn, divided into two subsections. In subsection 1.1 we highlight the

T. Terzioğlu (ed.), Advances in the Theory of Fréchet Spaces, 235–254.

role of quojections in the structure theory of Fréchet spaces, in the study of the functor Ext^1, in establishing the identity $L(E,F) = LB(E,F)$ and in the study of topological properties of tensor products. We do not treat here applications to infinite-dimensional holomorphy (for which we refer to [18] and [19]), to the problem of the interchangeability of inductive limits and tensor products, and to the projective description of weighted inductive limits (cf. [9]). Subsection 1.2 is devoted to twisted quojections and, precisely, to those originally constructed by the second author in 1980 [39] and here called "standard", which are still the most useful and interesting ones not only for the construction of counterexamples, but also for the structural properties they possess.

Prequojections are treated in the second section. Here we start with a brief historical survey since the paper of Bellenot and Dubinsky [3] in 1982, feeling justified by the fact that this is the first report on prequojections. Then we get to the main point of the section, represented by the general construction of a non-trivial prequojection out of any non-quasi-reflexive Banach space. We then conclude with some results on the structure of prequojections and on the role played by prequojections in the identity $L(E,F) = LB(E,F)$ and in the study of properties of projective tensor products.

Finally, in the Appendix we present some results which, although not always pertaining to the field of quojections and prequojections, nevertheless are sufficiently close in what concerns problems as well as methods of solution. We begin with some more constructions of twisted quojections. These are, at the moment, less usable than standard quojections; nevertheless, they are noteworthy and could prove themselves relevant to problems such as the complementation of twisted quojections in products, for which the solution for standard quojections is already known to be negative. Then, we adapt the construction of standard quojections to obtain nuclear spaces with "bad" topological properties. Next, we consider a different generalization of the standard quojection setting which yields further classes of spaces, studied in [7] and [8]. The latter turn out to be useful in many questions concerning Fréchet and (LB)-spaces such as, e.g., Taskinen's solution to Grothendieck's "problem of topologies" (cf. [45], [46]). We also mention the solutions to two problems of Grothendieck obtained by a more general construction leading to (LF)-spaces. Finally, the Appendix concludes with some highly non-trivial examples of primary non-normable Fréchet spaces, which are the only ones known up to date (except for ω, of course).

The notation we use is standard (cf., e.g., [25] and [26]), the only exception being that we denote by F' and F'' the strong dual and the strong bidual, respectively, of a Fréchet space F, and that by "subspace" we always mean a closed subspace.

1. QUOJECTIONS

A Fréchet space F is called a *quojection* if, for every continuous seminorm p on F, the space $F/\ker p$ is Banach when endowed with the quotient topology. It is easily seen that F is a quojection if and only if $F = proj_n(F_n, R_n)$, where each F_n is a Banach space and each map $R_n : F_{n+1} \to F_n$ is surjective. In such a case we write $F = quoj_n(F_n, R_n)$.

Quojections are characterized by the fact that every quotient with a continuous norm is a Banach space (cf. [3]) and this implies that a quotient of a quojection is again a quojection.

We recall that quojections were first considered by Grothendieck in his thesis [23]. Examples of quojections are furnished by finite or countable products of Banach spaces: these are *trivial* quojections. The first examples of quojections that are not isomorphic to products were given by the second author in [39]: such quojections are by now known as twisted quojections. These are dealt with in subsection 1.1 we present the main results on general quojections.

1.1 General Quojections

Our first result is a recently discovered theorem establishing the existence of quotient-universal quojections.

Theorem 1.1.1. ([10], [37]). *For any quojection F there exist a cardinal d and a continuous surjection $T : (\ell^1_d)^{\mathbb{N}} \to F$ such that* $\ker T$ *is a quojection.*

The cardinal d may be chosen equal to the density character of F and, consequently, every separable quojection is a quotient of $(\ell^1)^{\mathbb{N}}$ by a quojection subspace. Thus, the situation is similar to what happens in the case of Banach spaces, with one substantial difference. It is well-known (cf. [27, I, Theorem 2.f. 8, p.108]) that every separable Banach space $X \not\simeq \ell^1$ is the range of a continuous surjection from ℓ^1 whose kernel is unique up to an isomorphism (indeed up to an automorphism of ℓ^1). However, this is not the case for quojections and in [37] we show that there is a twisted quojection which may be realized as a quotient of $(\ell^1)^{\mathbb{N}}$ by either a Banach subspace or by a subspace which is a countable product of Banach spaces.

Proposition 1.1.2. [28]. *Let*

$$O \to F \to G \to E \to O$$

be an exact sequence of Fréchet spaces with E and F quojections. Then G is a quojection.

In other words, quojections enjoy the three-space property.

Since every quotient of a quojection is a quojection, if $T : F \to E$ is a continuous surjection between quojections, $\ker T$ need not be a quojection in general (indeed, it may even be nuclear). However, the situation is different if $E \simeq \omega$, i.e. if T is generated by an Eidelheit sequence in F'.

Theorem 1.1.3. [54]. *Let F be a non-normable Fréchet space. Then F is a quojection if and only if, for every continuous surjection $T : F \to \omega$, $\ker T$ is complemented in F. In such a case, $\ker T$ is either Banach or isomorphic to F.*

A similar result holds if E is Banach:

Proposition 1.1.4. [37]. *Let F be a quojection and let T be a continuous surjection onto a Banach space X. Then $\ker T$ is a quojection and a product if so is F.*

Clearly, in this case one cannot expect $\ker T$ to be complemented in F.

Now we come to the relationship between quojections and the first derived functor Ext^1. For the definition and properties of Ext^1 we refer to [51] and [53]; here we just recall that, for Fréchet spaces E and F, $Ext^1(E,F) = 0$ if and only if every exact sequence

$$O \to F \xrightarrow{i} G \xrightarrow{q} E \to O$$

splits, i.e. there exists $r : E \to G$ such that $qr = id_E$.

Theorem 1.1.5. ([51], [53]). *For a Fréchet space F, the following assertions are equivalent:*

(i) *F is a quojection;*

(ii) *$Ext^1(N,F) = O$ for every nuclear Fréchet space N;*

(iii) *$Ext^1(\omega,F) = O$;*

(iv) *$Ext^1((\ell^1)^{\mathrm{N}},F) = O$;*

(v) *$Ext^1((\ell^1_d)^{\mathrm{N}},F) = O$ for every cardinal d.*

Theorem 1.1.6. [37]. (a) *Let E be a separable quojection. Then $Ext^1(E,F) = O$ for every separable quojection F if and only if $E \simeq \ell^1, \omega, \ell^1 \times \omega$ or $(\ell^1)^{\mathrm{N}}$.*

(b) *Let F be a Fréchet space. Then $Ext^1(E,F) = O$ for every quojection E if and only if F is injective. In such a case, either $F \simeq X, \omega, X \times \omega$ with X an injective Banach space, or F contains a complemented copy of $(\ell^\infty)^{\mathrm{N}}$.*

We note that, in (a), $Ext^1(E,F) = 0$ for every separable quojection F if and only if E is complemented in $(\ell^1)^{\mathbf{N}}$: this follows from Theorem 1.1.1. In (b) the first part is obvious, since F is always a subspace of $(\ell_d^\infty)^{\mathbf{N}}$ for a suitable cardinal d. The non-trivial part of the theorem consists in establishing the indicated isomorphisms and this is done in [32] or [20].

If E and F are Fréchet spaces, together with the space $L(E,F)$ of all continuous linear maps from E to F we also consider the space $LB(E,F)$ of all such maps that are bounded, i.e. that map some 0-neighbourhood in E onto a bounded subset of F. Pairs of Fréchet spaces E, F for which $L(E,F) = LB(E,F)$ have been characterized in [51]. Concerning quojections we have

Theorem 1.1.7. ([5], [47]). *Let E, F be Fréchet spaces.*

(a) *If $L(E,F) = LB(E,F)$ for every F with a continuous norm, then E is a quojection.*

(b) *If E is a quojection, then $L(E,F) = LB(E,F)$ if and only if F has a continuous norm.*

We conclude this subsection with a further characterization of quojections involving topological properties of tensor products.

In his thesis [23] Grothendieck proved that if F is a quojection, then $F\tilde{\otimes}_\pi\varphi$ is barrelled. Actually, the following equivalence is true:

Theorem 1.1.8. [4]. *A Fréchet space F is a quojection if and only if $F\tilde{\otimes}_\pi G$ is barrelled for every barrelled locally convex space G.*

Note that the "only if" part can also be established using Theorem 1.1.1.

1.2. Twisted Quojections

If QUO denotes the class of quojections and PRO the subclass of trivial members of QUO, we put

$$QUOt = QUO \setminus PRO.$$

The members of QUOt are called *non-trivial*, or *twisted, quojections.*

Now, following [39], we introduce a special subclass of QUO. In this, we use the notation of [7] and [8]. Precisely, let L be a normal Banach sequence space. If (X_n) is a sequence of Banach spaces, we denote by $L(X_n)$ the Banach space

$$\{(x_n) \in \Pi_n X_n : (\|x_n\|)_n \in L\}$$

under the norm

$$\|(x_n)\| = \|(\|x_n\|)\|_L.$$

If Y_n is a subspace of X_n for all n, we put for every k,

$$F_k = L((X_n)_{n<k},\ (X_n/Y_n)_{n\geq k}). \tag{1}$$

It is clear that the identities of the spaces X_n and X_n/Y_n, together with the quotient maps $q_n : X_n \to X_n/Y_n$, induce surjective maps $R_k : F_{k+1} \to F_k$, so that we may define the quojection $quoj_k(F_k, R_k)$, which we denote by $Q[(X_n),(X_n/Y_n),(q_n);L]$, or $Q[(X_n),(X_n/Y_n);L]$ if there is no ambiguity about the maps q_n. In case all the spaces X_n (resp., Y_n) are equal to one fixed space X (resp, Y), we simply write $Q(X, X/Y; L)$. We shall agree to call *standard* any quojection isomorphic to one of the form $Q[(X_n),(X_n/Y_n);L]$.

The following considerations are in order. First, in case $Y_n = X_n$ for all n, we have $Q[(X_n),(O);L] = \Pi_n X_n$ for any L, showing that PRO is contained in the class of standard quojections. Second, it follows essentially from [39] (cf. also [35]) that $Q[(X_n),(X_n/Y_n);L]$ is twisted if and only if Y_n is not complemented in X_n for infinitely many n. Actually, more is true, since it is easy to see that the following generalization of Theorem 1.5 (b) in [35] holds:

Theorem 1.2.1. *A standard twisted quojection is never complemented in a countable product of Fréchet spaces with continuous norms.*

It is interesting to note that, if in a standard quojection $Q[(X_n),(X_n/Y_n);L]$ we choose as a fundamental sequence of seminorms (p_k) the norms of the steps F_k defined in (1), we obtain

$$\ker p_k = \Pi_{n\geq k} Y_n \quad \text{and} \quad Q[(X_n),(X_n/Y_n);L]/\ker p_k = F_k,$$

that is, for each k we obtain an exact sequence

$$O \to \Pi_{n\geq k} Y_n \to Q[(X_n),(X_n/Y_n);L] \to F_k \to O, \tag{2}$$

which obviously does not split. It follows that PRO does not have the three-space property.

In Theorem 1.1.1 we saw that every quojection is a quotient of some $(\ell^1_d)^{\mathbb{N}}$ by a quojection subspace. It is rather surprising to discover that every standard twisted quojection can be realized as a quotient of a $P \in PRO$ by a subspace which is a standard twisted quojection or even a Banach space. Before stating the theorem, we wish to give a useful representation of standard quojections as vector-valued sequence spaces, since this clarifies the structure of the former.

Proposition 1.2.2. [8]. *$Q[(X_n),(X_n/Y_n);L]$ coincides with the space*

$$\{(x_n \in \Pi_n X_n : (q_n(x_n)) \in L(X_n/Y_n)\}$$

topologized by the seminorms

$$p_1(x_n) = \|(\|q_n(x_n)\|)\|_L,$$

$$p_{k+1}(x_n) = p_1(x_n) + \sum_{n=1}^{k} \|x_n\|.$$

Now, using the above proposition, one can prove the following basic theorem, which we just state in terms of quojections $Q(X, X/Y; L)$ for simplicity of exposition.

Theorem 1.2.3. [37]. *Let the quojection $Q(X, X/Y; L)$ be given. For every pair (Z, s) of Banach space Z and continuous surjection $s : Z \to Y$ there is a quojection F and an exact sequence*

$$O \to F \to L(X) \times Z^{\mathbf{N}} \to Q(X, X/Y; L) \to O, \tag{3}$$

where $F = Q(Z, Z/\ker s; L)$. Consequently:

(a) *F is Banach if and only if $\ker s = \{0\}$;*

(b) *$F = \omega\times$ Banach if and only if $0 < \dim(\ker s) < \infty$;*

(c) *F is twisted if and only if $\ker s$ is not complemented in Z.*

Thus, combining (2) with Theorem 1.2.3 (a) we obtain

Theorem 1.2.4. *Given $Q(X, X/Y; L) \in QUOt$, there always exist a Banach space B and a $P \in PRO$ for which we have an exact sequence*

$$O \to P \to Q(X, X/Y; L) \to B \to O,$$

respectively,

$$O \to B \to P \to Q(X, X/Y; L) \to O.$$

We note that Theorem 1.2.3 is not exactly a refinement of Theorem 1.1.1, since the product involved depends on the given quojection. Such a product, however, may indeed be $(\ell^1)^{\mathbf{N}}$ in particular cases: take $X = Z = L = \ell^1$ and for Y a subspace of ℓ^1 not isomorphic to ℓ^1, to obtain from (3) the exact sequence

$$O \to Q(\ell^1, \ell^1/\ker s; \ell^1) \to (\ell^1)^{\mathbf{N}} \to Q(\ell^1, \ell^1/Y; \ell^1) \to O,$$

which tells us, in particular, that also $QUOt$ does not have the three-space property.

Note also that, by Theorem 1.2.1, the exact sequence (3) never splits for a twisted $Q(X, X/Y; L)$, respectively, F. Therefore, we have

Theorem 1.2.5. [37]. *Let $Q = Q(X, X/Y; L) \in QUOt$.*

(a) *There exists $F \in QUO$ such that $Ext^1(Q,F) \neq O$ and F may be chosen to be Banach, $\omega \times$ Banach , a member of PRO or a standard member of $QUOt$ if $Y \neq \ell^1_d$ for all cardinal numbers d.*

(b) *There exists $E \in QUO$ such that $Ext^1(E,Q) \neq O$ and E may be chosen to be a member of PRO or a standard member of $QUOt$ if X/Y is not injective.*

The first part of the theorem follows directly from Theorem 1.2.3, the assumption $Y \neq \ell^1_d$ for all d being necessary to ensure the existence of a pair (Z, s) with $s : Z \to Y$ surjective and $\ker s$ not complemented in Z (take, e.g., $Z = \ell^1_d$ with $d = dens(Y)$). The second part is also a consequence of Theorem 1.2.3 once Q is written as $Q(Z, Z/\ker s; L)$, the assumption of non-injectivity of X/Y guaranteeing the existence of a Banach space W containing X/Y as a non-complemented subspace (e.g., $W = \ell^\infty_d$ with $d = dens(X/Y)$). Observe, moreover, that here we cannot have E Banach or $\omega \times$ Banach, by Theorem 1.1.3 and Proposition 1.1.4.

We conclude with some results on the existence or not of bases in quojections. Basic for this topic is the following

Theorem 1.2.6. [22]. *Let F be a Fréchet space with an unconditional basis and without a continuous norm. Then F is isomorphic to a countable product of Fréchet spaces each of which has an unconditional basis and a continuous norm.*

It follows that no member of $QUOt$ can have an unconditional basis. It follows also, by Theorem 1.2.1, that no standard member of $QUOt$ can be complemented in a Fréchet spaces with an unconditional basis. But observe that the assumption that the basis be unconditional cannot be dropped in Theorem 1.2.6 since we have, regarding ℓ^p as a quotient of $L^q = L^q(0,1)$ for $2 < p < q < \infty$ [27, II, Corollary 2.f.5, p.212]).

Theorem 1.2.7. [30]. *The reflexive twisted quojection $Q(L^q, \ell^p; \ell^2)$ $(2 < p < q < \infty)$ has a basis.*

Finally, we mention that Theorem 1.2.6 has been generalized recently to Fréchet lattices, as follows.

Theorem 1.2.8. [14]. *Let F be an order complete Fréchet lattice with the weak-Fatou property and without a continuous norm. Then F is isomorphic to a countable product of Fréchet spaces each of which has a continuous norm and is a band in F.*

2. PREQUOJECTIONS

Prequojections are intimately related to the problem of characterizing those Fréchet spaces that have a nuclear quotient with a continuous norm and a basis, briefly, a nuclear Köthe quotient. In 1982 Bellenot and Dubinsky [3] proved that a separable Fréchet space F has a nuclear Köthe quotient if and only if it satisfies the following condition: there exists k such that for every j there is an ℓ with

$$\sup\{\|x'\|_1 : x' \in F'_k \text{ and } \|x'\|_j \leq 1\} = \infty, \tag{4}$$

where $\| \|_n$ is the dual norm of a fixed seminorm p_n on F and $F'_n = \{x' \in F' : \|x'\|_n < \infty\}$. Condition (4) means precisely that on F' the topology $\beta(F', F'')$ is not the topology arising from a strict inductive limit of a sequence of Banach spaces. Denoting by F'^x the strong dual of $(F', \beta(F', F''))$, it is easy to see that F does *not* satisfy (4) if and only if F'^x is a quojection. Actually, more is true, as shown by the following

Proposition 2.1. [54]. *If F does not satisfy* (4), *then it is quasinormable. Hence, F does not satisfy* (4) *if and only if F'' is a quojection.*

It is then natural to call a Fréchet space F a *prequojection* if (and only if) its bidual F'' is a quojection. Obviously, quojections are prequojections: they are the *trivial* ones. The existence of *non-trivial* prequojections (even with a continuous norm) has only been established recently. Before giving this result, let us take a step back and look at the attempts made to solve the problem.

It is easily seen (cf. [7]) that if there is a non-trivial prequojection, then there is also one such with a continuous norm, since the quotient of a prequojection is again a prequojection. Then the following question (asked by Vogt) arises naturally:

(Q) If F is a Fréchet space with a continuous norm, does F'' also have a continuous norm?

Clearly, an affirmative answer to (Q) would imply the non-existence of non-trivial prequojections, as it follows from the remarks made above. However, the answer to (Q) is negative.

Theorem 2.2. (a) *There exists a Fréchet space of Moscatelli type (cf. the Appendix for the definition) with a continuous norm whose bidual does not have a continuous norm.*

(b) *There exists a Köthe space with a continuous norm whose bidual does not have a continuous norm.*

Part (a) was first established in [16] and, in general, in [8]. Part (b) is proved in [48], where those Köthe matrices are characterized for which the situation described can occur.

Thus, question (Q) is of no help towards a solution of the problem concerning the existence of (non-trivial) prequojections. The right approach was taken once it was recognized that this problem is equivalent to a Banach-space problem (a fact which is rather surprising in itself). First, let us fix the notation.

Let X be a Banach space with topological dual X' and let M be a w^*-dense subspace of X' (usually, but not necessarily, assumed to be closed). Let also $B_{X'}$ be the unit ball if X' and put $B_M = M \cap B_{X'}$. Then, clearly,

$$\|x\|_M = \sup\{| \langle x, x' \rangle |: x' \in B_M\} \quad (x \in X)$$

is a new norm on X and M is said to be *non-norming* (over X) if $\| \ \|_M$ is not equivalent to the norm of X or, equivalently, if $\overline{B}_M^{w^*} \not\supset \varepsilon B_{X'}$ for all $\varepsilon > 0$. Now, given M we define $M^1 = \operatorname{span} \overline{B}_M^{w^*}$ and, inductively, $M^{n+1} = (M^n)^1$. Observe that M^1 is the collection of all limits of w^*-convergent and bounded nets in M and is called the *derived set* of M. In case X is separable, M^1 is just Banach's *weak derivation* of M (cf. [1, Annexe, §1, p.208]).

It is clear that M is non-norming if and only if $M^1 \neq X'$. Following [41] we shall then say that M is *strongly non-norming* if and only if $M^n \neq X'$ for all n. In such a case, necessarily $M^n \neq M^{n+1}$ for all n.

The basic link between Banach spaces and prequojections is then constituted by the following

Theorem 2.3. [17]. *A non-trivial prequojection exists if and only if there is a Banach space X such that X' contains a strongly non-norming subspace M.*

Actually, the proof of the above theorem shows that, if such a pair (X, M) exists, a prequojection F with continuous norm may be obtained by forming the projective limit of the Banach spaces $(X, \| \ \|_{M^n})$. Note that F is *countably normed* and the continuous injection $X \to F$ has a dense range.

Theorem 2.3, therefore, reduces the problem to that of finding (and, possibly, characterizing) Banach spaces with duals containing strongly non-norming subspaces. A glance at Banach's book [1, Annexe, §1, p.213] shows that there are uncountably many quotients of c_0 with this property. It follows

Theorem 2.4. [2]. *There is a prequojection with a continuous norm.*

However, the problem now is to characterize those Banach spaces whose duals contain strongly non-norming subspaces and to exhibit concrete examples of prequojections, this being necessary for any serious attempt to study these spaces. Unfortunately, we must say right away that prequojections are not "visible", the reason lying in Proposition 2.7 below. Nevertheless, one of the proofs of the following theorem (which provides the definitive solution) allows to observe the duals of some prequojections (cf. Theorem 2.6 below).

Theorem 2.5. *Let X be a Banach space. Then X' contains a strongly non-norming subspace M if and only if X is not quasi-reflexive* (i.e., $\dim X''/X = \infty$).

Theorem 2.5 is central to the study of prequojections and, therefore, it deserves some comments. At present, there are three proofs of it. The first, due to the second author [41], uses Davis and Lindenstrauss [12] elegant proof of the existence of a non-norming subspace in the dual of any non-quasi-reflexive Banach space. The second, due to Ostrovskij [43], proves more but only for separable X. Both proofs use induction and hence only serve to establish the existence of the required subspace M of X'. The third proof, also due to the second author [42], uses a result of Davis and Johnson [11] to actually *construct* the strongly non-norming subspace M. Without entering into details, we confine ourselves to saying that if X has a basis (x_n) of a special kind (i.e., as in Theorem 2 of [11]), then M is constructed from the biorthogonal functionals associated to the basis (x_n). Since, by [11], every non-quasi-reflexive Banach space Y has a basic sequence of such a special kind, one obtains the result for Y by means of a simple lifting lemma. Here we do not give the definition of such a special basis, but only say that a typical example is the standard basis of c_0. In this particular case, namely $X = c_0$, Theorem 2.5 enables us to prove the following

Theorem 2.6. [42]. *There is a prequojection F with a continuous norm such that $F' \simeq (\ell^1)^{\mathbb{N}}$ and hence $F'' \simeq (\ell^\infty)^{\mathbb{N}}$.*

Furthermore, the power of the constructive method is such that many variants and generalizations are possible yielding a number of remarkable results which show the close interplay between prequojections and the structure of Banach spaces. For all this, the interested reader is referred to [42] and to our forthcoming paper [38].

Now we come to some properties of prequojections. First, we observe that prequojections, like quojections, have the three-space property. This may be deduced from the corresponding result for quojections, i.e. Proposition 1.1.2, upon noting that, if

$$0 \to F \to G \xrightarrow{q} E \to O$$

is exact with E and F prequojections, then the map q "lifts" the bounded subsets since

$\ker q = F$ is quasinormable. The result may also be obtained from [54]. In fact, a Fréchet space is a prequojection if and only if it has property (Ω_1) [54, Lemma 0.2] and it is easy to see that the class of spaces having property (Ω_1) enjoys the three-space property.

More interesting is the following proposition, which explains why prequojections are hard to find.

Proposition 2.7. [9]. *A non-trivial prequojection F cannot have the bounded approximation property* (BAP).

Note, however, that there are prequojections which are countably normed, as remarked after Theorem 2.3.

The above proposition may be deduced, in case F has a continuous norm, from the following theorem, which shows the importance of prequojections in the study of the identity $L = LB$.

Theorem 2.8. ([33], [47]). *Let F be a prequojection and let G be a Fréchet space with a continuous norm and BAP. Then $L(F,G) = LB(F,G)$.*

Theorem 2.8 has some noteworthy implications. Recall that a Fréchet space is a quojection if and only if every quotient with a continuous norm is Banach. The analogue of this definition may now be obtained by combining Theorem 2.8 with the main result in [3]:

Theorem 2.9. [33]. *A separable Fréchet space F is a prequojection if and only if every quotient of F with BAP and a continuous norm is Banach.*

Now we apply Theorem 2.8 to the space F in Theorem 2.6 to obtain (with AP = approximation property)

Theorem 2.10. ([42], [33]). *The space F in Theorem 2.6 is countably normed and has AP but not BAP, even though F' has BAP (it has an absolute basis).*

The fact that F has AP follows from [13, Proposition 4.1 (b)], since F' has AP. The above Theorem thus tells us that the implication "F' has $BAP \Rightarrow F$ has BAP" is false and also that Proposition 2.7 cannot be improved to AP instead of BAP.

Now we come to the analogues of Theorem 1.1.3 and Proposition 1.1.4.

Proposition 2.11. (a) *Let*

$$O \to F \to E \to \omega \to O$$

be an exact sequence. Then E is a prequojection if and only if so is F.

(b) *Let*

$$O \to F \to E \to X \to O$$

be an exact sequence, with X Banach. Then E is a prequojection if and only if so is F.

Part (a) is proved in [54], while part (b) can be deduced from Proposition 1.1.4 by duality.

Concerning the identity $L = LB$, we have the following supplement to Theorem 2.8.

Theorem 2.12. ([5], [47], [50]). *A Fréchet space F is a prequojection if and only if $L(F,K) = LB(F,K)$ for every nuclear Köthe space K.*

As for the role of prequojections in the study of topological properties of projective tensor products we have

Theorem 2.13. ([4], [5]). *If F is a Fréchet space, the following assertions are equivalent:*

(i) *F is a prequojection;*

(ii) *$F \tilde{\otimes}_\pi K$ is barrelled for every nuclear Köthe space K;*

(iii) *$F \tilde{\otimes}_\pi G$ is barrelled for every quasi-Baire locally convex space G.*

Prequojections turn up also in the study of Eidelheit sequences (cf. [52] and [54]), in the interchangeability of inductive limits and tensor products and in the projective description of weighted inductive limits, for which we refer to [9] (where also further results on the identity $L = LB$ can be found). Here, we end this section with a result characterizing those Fréchet spaces which contain a dense subspace with a continuous norm. Apparently, this has little to do with prequojections. However, in the first place it is interesting to see how the same difference existing between a prequojection with continuous norm and its bidual can occur between a metrizable space and its completion. In the second place, a dense subspace with continuous norm of a quojection is a generalized prequojection in the sense of [33] (cf. also [9] and [47]) and for the latter spaces many of the results established for prequojections continue to hold.

Theorem 2.14. [36]. *Let F be a Fréchet space without continuous norm. Then F contains a dense subspace M with continuous norm if and only if there is a continuous seminorm p on F such that* $\dim F/\ker p = \infty$ *and dens* $(F/\ker p) =$ *dens* (F). *In such a case M does not have BAP.*

Note that the above theorem implies that a separable Fréchet space F has a dense subspace with continuous norm if and only if $F \not\simeq \omega$.

3. APPENDIX

We collect here some more results on quojections which, for organizational reasons, did not find a place in §1, as well as results on nuclear spaces and on Fréchet spaces of a special type which may be obtained by appropriate generalizations of the methods in §1. Then we conclude this appendix (and our survey) by exhibiting some primary Fréchet spaces.

Let us see first some more examples of twisted quojections. Because the construction is made in the dual space, we shall define a strict (LB)-space to be *twisted* if it is not isomorphic to a countable direct sum of Banach spaces. Note that the dual of a reflexive, twisted quojection is a twisted, strict (LB)-space and, conversely, the dual of a reflexive, twisted, strict (LB)-space is a twisted quojection. However, there exist twisted quojections (resp., twisted, strict (LB)-spaces) whose duals are not twisted (cf. [35]).

Now recall that a subspace Y of a Banach space X is called ***strongly pseudo-complemented*** in X if it is contained in a complemented subspace Z of X with $\dim X/Z = \infty$. Y is said to be ***pseudo-complemented*** in X if there is a subspace V of X such that $\dim V = \infty$, $Y \cap V = \{0\}$ and $Y + V$ is closed in X. Clearly, every strongly pseudo-complemented subspace is pseudo-complemented.

Theorem 3.1. [29]. *Let $E = s - ind_n E_n$ be the strict inductive limit of an increasing sequence (E_n) of Banach spaces. If, for some k, E_k is not strongly pseudo-complemented in E_n for all $n > k$, then E is twisted.*

Examples of such a chain (E_n) may be obtained by taking a Banach space X containing a non-pseudo-complemented subspace Y and then choosing any increasing sequence (E_n) of subspaces of X such that $E_1 \supset Y$ and $\dim E_{n+1}/E_n = \infty$. As a space X as above we may take, e.g., ℓ^1, ℓ^∞ or $L^p(0,1)$ $(2 < p \leq \infty)$ (cf. [44]). If, in Theorem 3.1, the spaces E_n are chosen to be reflexive, we immediately obtain that the strong dual E' is a twisted quojection. It must be noted, however, that the above construction, although simpler than that leading to standard quojections, yields spaces which are less manageable and whose structure, therefore, is more difficult to analyse.

Let us now give a rather singular example of a twisted quojection. Let (p_n) be an increasing sequence of real numbers, with $p_1 \geq 2$. As well-known, for every n there is a continuous surjection $q_n : L^{p_{n+1}}(0,1) \to L^{p_n}(0,1)$ (cf. [27, II, Corollary 2.f.5, p.212]); hence we may form the quojection $F = quoj_n(L^{p_n}(0,1), q_n)$ for which we obtain

Theorem 3.2. [35]. *F is twisted. If X is a Banach subspace of F, then $X \simeq \ell^2$. Consequently, if G is a quojection subspace of F, then either $G \simeq \ell^2$, ω, $\ell^2 \times \omega$, $(\ell^2)^{\mathbb{N}}$ or G is twisted.*

Note that at present no examples are known of twisted quojections containing no Banach subspace.

We now come to a generalization of the construction of standard quojections illustrated in subsection 1.2. This generalization replaces Banach spaces with Fréchet spaces with continuous norm and was already made in [39]. For simplicity, we confine ourselves to considering the case in which only two spaces are involved (plus the necessary sequence space, of course). Also, we restrict ourselves to the nuclear case.

Let then E be a nuclear Fréchet space with a continuous norm and let G be a non-complemented subspace of E such that also the quotient E/G has a continuous norm (cf., e.g., [49] or [55]). If L is a nuclear, Fréchet sequence space with continuous norm, we put for every n,

$$F_n = (\underbrace{E/G \oplus \cdots \oplus E/G}_{\text{n times}} \oplus E \oplus \cdots)_L.$$

Clearly, there are natural surjections $F_{n+1} \to F_n$ and we may form the projective limit of the nuclear Fréchet spaces F_n. Denoting by $F(E, E/G; L)$ such a space, we have

Theorem 3.3. [39]. *$F(E, E/G; L)$ is a nuclear Fréchet space which is twisted (i.e., not isomorphic to a countable product of Fréchet spaces with continuous norms). In particular, $F(E, E/G; L)$ has no basis (by Theorem 1.2.6).*

Note that, as in the case of quojections, G^{N} is a subspace of $F(E, E/G; L)$ and $F(E, E/G; L)/G^{\mathrm{N}} = L(E/G)$. Thus, choosing G so that both G and E/G have a basis (cf. [49] and [31]), we obtain

Theorem 3.4. [31]. *$F = F(E, E/G; L)$ has no basis but it contains a subspace H such that both H and F/H have a basis.*

Observe that this result does not depend on the existence or not of a basis in E (indeed, E may have a basis!)

It is not difficult to see that Theorems 1.2.3 and 1.2.4 can be extended to this context, with some obvious changes. In particular, it follows that all the situations concerning the existence or not of bases in subspaces and quotients of products of Fréchet spaces may occur (cf. [37]).

Now we shall analyse a class of Fréchet spaces whose definition is inspired by the construction of standard twisted quojections. As we will see, such a class is a source of many counterexamples in Fréchet-space theory.

Let $(X_n), (Y_n)$ be two sequences of Banach spaces and let $f_n : Y_n \to X_n$ be continuous linear maps with dense ranges. If L is a normal Banach sequence space we put, for each k,

$$F_k = L(Y_n)_{n<k}, \ (X_n)_{n\geq k}). \tag{5}$$

The maps f_n, together with the identities of the spaces X_n and Y_n induce continuous linear maps $S_k : F_{k+1} \to F_k$ with dense ranges and we may define the (reduced) projective limit $F = proj_k(F_k, S_k)$. Spaces such as F are called Fréchet spaces of *Moscatelli type* and were first studied systematically in [8] (see also [7]), although particular cases (outside the quojection context) had already been considered in [15], [16], [45] and [46].

As a sample of the results available for these spaces we give the following two theorems.

Theorem 3.5. [8]. *If $L = \ell^1$, then F is distinguished if and only if it is a quojection (i.e. there is n_0 such that all maps f_n are surjective for $n \geq n_0$).*

It is interesting to note that the non-distinguished space of Köthe-Grothendieck ([24], [26, §31.7]) is of Moscatelli type: for all n take $X_n = Y_n = \ell^1 = L$ and $f_n = f$, with $f(x_j) = (j^{-1}x_j)$ for $(x_j) \in \ell^1$.

Theorem 3.6. [8]. *The following assertions are equivalent:*

(i) F *is quasinormable;*

(ii) $\ell^1 \tilde{\otimes}_\pi \ F$ *is distinguished;*

(iii) F *has Vogt's property* (Ω);

(iv) F *is a quotient of some* $\ell^1_d \tilde{\otimes} s$;

(v) F'' *is a quojection;*

(vi) F *is a quojection.*

Moreover, F has Vogt's property (DN) if and only if it is Banach.

Theorems 3.5 and 3.6 show that, to require a minimum of regularity for a Fréchet space of Moscatelli type forces the space to be a quojection (or even Banach). In other words, such a space either is a quojection (and then is of the kind treated in §1), or else it has bad topological properties.

Another construction which sprang from the classical (LB)-construction in [39] was performed in [6] to answer two questions of Grothendieck (cf. [24]). Precisely, let F be a

Fréchet space and let G be a subspace of F. For each n we put

$$E_n = \underbrace{F \times \cdots \times F}_{\text{n-1 times}} \times G \times G \times \cdots.$$

This is equivalent to taking $X_n = G$, $Y_n = F$ and $L = \omega$ in the previous construction (5). Clearly, each E_n is a Fréchet space and a closed subspace of E_{n+1}, so that $E = ind_n E_n$ is a strict (LF)-space. Choosing F and G so that, if $j : G \to F$ is the canonical inclusion, the dual map $j' : F' \to G'$ does not have "good" properties, one obtains

Theorem 3.7. [6]. (a) *There is a strict (LF)-space whose bidual is not (LF), even though it is complete.*

(b) *There is a strict (LF)-space whose bidual is not complete.*

For another application to (LF)-spaces of the above construction, but with ω replaced by s, see [35].

Finally, we conclude this section (and our survey) with the only non-trivial examples of primary Fréchet spaces known up to date. Recall that a Fréchet space F is *primary* if $F = E \times G$ implies $E \simeq F$ or $G \simeq F$. Trivial examples of primary Fréchet spaces are ω and all primary Banach spaces. No other example was known prior to our paper [32]. A little later, the same result was rediscovered independently in [20]. The theorem consists essentially of the fact that $(c_0)^{\mathbf{N}}$ and $(\ell^p)^{\mathbf{N}}$ $(1 \leq p \leq \infty)$ are primary. Actually, it shows exactly what the complemented subspaces of such spaces are, from which the desired conclusion follows. We take the opportunity to recall that it is unknown whether a complemented subspace of a member of PRO is again in PRO (our conjecture being that this is false, in general). Thus, the theorem is interesting also in this direction.

Theorem 3.8. ([32], [20]). *Let $X = c_0$ or ℓ^p $(1 \leq p \leq \infty)$. If F is a complemented subspace of $X^{\mathbf{N}}$, then $F \simeq X$, $\omega, X \times \omega$ or $X^{\mathbf{N}}$. Consequently, $X^{\mathbf{N}}$ is primary.*

It is perhaps worth pointing out that ω is the only nuclear Fréchet space known to be primary (and even prime, i.e. every infinite-dimensional complemented subspace is isomorphic to ω). In particular, no power series space is primary (this fact can be verified directly, with some patience). Since the latter spaces have a continuous norm, it is, therefore, significant to have examples of primary Fréchet spaces with continuous norm. Again, these are provided by spaces built from the classical Banach spaces. Precisely, for $1 \leq p < \infty$ let $\ell^{p+} = \bigcap_{q>p} \ell^q$. Then we have

Theorem 3.9. [34]. *ℓ^{p+} is primary. Moreover, every nuclear power series space of finite*

type is a complemented subspace of ℓ^{p+}.

Thus, while $(\ell^p)^{\mathbb{N}}$ (and $(c_0)^{\mathbb{N}}$) are "almost" prime, in the sense that they are prime modulo trivial subspaces, their "continuous norm" counter parts ℓ^{p+} are highly non-prime, having nuclear complemented subspaces $\not\simeq \omega$.

REMARKS.

It is almost superfluous to point out that for most of the results presented in this survey (except those in §2, of course) there are dual analogues in the context of (LB)-spaces. In particular, we mention that dual constructions to those leading up to standard quojections and the spaces of Theorem 3.3 respectively, were first performed in [39]. Also, Fréchet spaces of Moscatelli type have their dual analogues in the (LB)-spaces of Moscatelli type, which are studied in detail in [7]. Theorem 3.2 has an obvious (LB)-space analogue, since the space F considered is reflexive. Moreover, Theorem 1.1.1 allows to prove that *every strict (LB)-space can be embedded as a subspace of $(\ell^\infty_d)^{\mathbb{N}}$ for a suitable cardinal number d* (cf. [37]). Finally, the following theorems have (LB)-space analogues which are explicitly or implicitly proved in the references indicated: 1.2.1 [39], [35], 1.2.3 [37], 1.2.5 [37], 1.2.6 [21], 1.2.7 [30], 3.4 [28], 3.8 [32], 3.9 [34].

Concerning prequojections, the assumption of separability in the result of Bellenot and Dubinsky quoted at the beginning of §2 has been removed recently by Önal and Terzioğlu ("Unbounded linear operators and nuclear Köthe quotients", preprint).

REFERENCES

[1] S. Banach, *Théorie des opérations lineaires*, Chelsea, New York 1955 (2nd ed.).

[2] E. Behrends, S. Dierolf and P. Harmand, 'On a problem of Bellenot and Dubinsky', *Math. Ann.* **275** (1986) 337-339.

[3] S. Bellenot and E. Dubinsky, 'Fréchet spaces with nuclear Köthe quotients', *Trans. Amer. Math. Soc.* **273** (1982), 579-594.

[4] J. Bonet, 'Quojections and projective tensor products', *Arch. Math.* **45** (1985) 169-173.

[5] J. Bonet, 'On the identity $L(E,F) = LB(E,F)$ for pairs of locally convex spaces E and F', *Proc. Amer. Math. Soc.* **99** (1987) 249-255.

[6] J. Bonet and S. Dierolf, 'A note on biduals of strict (LF)-spaces', *Resultate Math.* **13** (1988) 23-32.

[7] J. Bonet and S. Dierolf, 'On (LB)-space of Moscatelli type', *Doğa Mat.* to appear.

[8] J. Bonet and S. Dierolf, 'Fréchet spaces of Moscatelli type', preprint, 1988.

[9] J. Bonet and A. Galbis, 'The identity $L(E,F) = LB(E,F)$, tensor products and inductive limits', *Note Mat., Lecce* (to appear).

[10] J. Bonet, M. Maestre, G. Metafune, V.B. Moscatelli and D. Vogt, 'Every quojection is the quotient of a countable product of Banach spaces', These Proceedings.

[11] W. Davis and W. Johnson, 'Basic sequences and norming subspaces in non-quasi-reflexive Banach spaces', *Israel J. Math.* **14** (1973) 353-367.

[12] W. Davis and J. Lindenstrauss, 'On total nonnorming subspaces', *Proc. Amer. Math. Soc.* **31** (1972) 109-111.

[13] A. Defant, 'A duality theorem for locally convex tensor products', *Math. Z.* **190** (1985) 45-53.

[14] J.C. Diaz, 'Continuous norms on Fréchet lattices', *Arch. Math.* (to appear).

[15] S. Dierolf, 'On spaces of continuous linear mappings between locally convex spaces', *Note Mat., Lecce,* **5** (1985) 147-255.

[16] S. Dierolf and V.B. Moscatelli, 'A Fréchet space which has a continuous norm but whose bidual does not', *Math. Z.* **191** (1986) 17-21.

[17] S. Dierolf and V.B. Moscatelli, 'A note on quojections', *Funct. Approx. Comment. Math.* **17** (1987) 131-138.

[18] S. Dineen, 'Surjective limits of locally convex spaces and their application to infinite-dimensional holomorphy', *Bull. Soc. Math. France* **103** (1975) 441-509.

[19] S. Dineen, *Complex analysis in locally convex spaces,* North-Holland Math. Stud. **57** 1981.

[20] P. Domański and A. Ortyński, 'Complemented subspaces of product of Banach spaces', *Trans. Amer. Math. Soc.* to appear.

[21] K. Floret and V.B. Moscatelli, 'On basis in strict inductive and projective limits of locally convex spaces', *Pacific J. Math.* **119** (1985) 103-113.

[22] K. Floret and V.B. Moscatelli, 'Unconditional bases in Fréchet spaces', *Arch. Math.* **47** (1986) 129-130.

[23] A. Grothendieck, *Produits tensoriels topologiques et espaces nucléaires,* Mem. Amer. Math. Soc. **16**, 1955.

[24] A. Grothendieck, 'Sur les espaces F et DF' *Summa Brasil. Math.* **3** (1954) 57-123.

[25] H. Jarchow, *Locally convex spaces,* Teubner, Stuttgart, 1981.

[26] G. Köthe, *Topological vector spaces I,* Springer, Berlin, 1969.

[27] J. Lindenstrauss and L. Tzafriri, *Classical Banach spaces I, II,* Springer, Berlin, 1977, 1979.

[28] G. Metafune and V.B. Moscatelli, 'On the three-space problem for locally convex spaces', *Collect. Math.* **37** (1986) 287-296.

[29] G. Metafune and V.B. Moscatelli, 'Another construction of twisted spaces', *Proc. Roy. Irish Acad. Sect. A* **87** (1987) 163-168.

[30] G. Metafune and V.B. Moscatelli, 'A twisted Fréchet space with basis', *Monatsh. Math.* **105** (1988) 127-129.

[31] G. Metafune and V.B. Moscatelli, 'Nuclear Fréchet spaces with basis do not have the three-space property', *Arch. Math.* **50** (1988) 369-370.

[32] G. Metafune and V.B. Moscatelli, 'Complemented subspaces of sums and products of Banach spaces', *Ann. Mat. Pura Appl.* **153** (1989) 1-16.

[33] G. Metafune and V.B. Moscatelli, 'Generalized prequojections and bounded maps', *Resultate Math.* (to appear).

[34] G. Metafune and V.B. Moscatelli, 'On the space $\ell^{p+} = \bigcap_{q>p} \ell^{q}$', *Math. Nachr.* (to appear).

[35] G. Metafune and V.B. Moscatelli, 'On twisted Fréchet and (LB)-spaces', preprint, 1988.

[36] G. Metafune and V.B. Moscatelli, 'Dense subspaces with continuous norm in Fréchet spaces', preprint, 1988.

[37] G. Metafune and V.B. Moscatelli, 'On exact sequences of quojections', preprint, 1988.

[38] G. Metafune and V.B. Moscatelli, 'Banach spaces and prequojections', in preparation.

[39] V.B. Moscatelli, 'Fréchet spaces without continuous norms and without bases', *Bull. London Math. Soc.* **12** (1980), 63-66.

[40] V.B. Moscatelli, 'Strict inductive and projective limits, twisted spaces and quojections', *Rend. Circ. Mat. Palermo Suppl.* **10** (1985) 119-131.

[41] V.B. Moscatelli, 'On strongly non-norming subspaces', *Note Mat., Lecce,* **7** (1987) 311-314.

[42] V.B. Moscatelli, 'Strongly non-norming subspaces and prequojections', preprint, 1988.

[43] M.I. Ostrovskij, 'w^*-derived subspaces of transfinite order in dual Banach spaces', *Dokl. Akad. Nauk UKR, SSR,* Ser. A, **10** (1987) 9-12 (Russian).

[44] I. Singer, 'On pseudo-complemented subspaces of Banach spaces', *J. Funct. Anal.* **13** (1973) 223-232.

[45] J. Taskinen, 'Counterexamples to "Probléme des Topologies" of Grothendieck', *Ann. Acad. Sci. Fenn. Ser. A I Math. Dissertationes* **63** (1986).

[46] J. Taskinen, 'Examples of non-distinguished Fréchet spaces', preprint 1988.

[47] T. Terzioğlu, 'A note on unbounded operators and quotient spaces', *Doğa Mat.* **10** (1986) 338-344.

[48] T. Terzioğlu and D. Vogt, 'A Köthe space which has a continuous norm but whose bidual does not', *Arch. Math* (to appear).

[49] D. Vogt, 'Eine Charakterisierung der Potenzreihenräume von endlichem Typ und ihre Folgerungen', *Manuscripta Math.* **37** (1982) 269-301.

[50] D. Vogt, 'Fréchetträume zwischen denen jede stetige lineare Abbildung beschränkt ist', *J. Reine Angew. Math.* **345** (1983) 182-200.

[51] D. Vogt, 'Some results on continuous linear maps between Fréchet spaces', *Functional Analysis: Surveys and Recent Results* (eds. K.-D. Bierstedt and B. Fuchssteiner), North-Holland Math. Stud. **90**, 1984, 349-381.

[52] D. Vogt, 'Kernels of Eidelheit matrices and related topics', *Doğa Mat.* **10** (1986) 232-256.

[53] D. Vogt, 'On the functors $Ext^1(E, F)$ for Fréchet spaces', *Studia Math.* **85** (1987) 163-197.

[54] D. Vogt, 'On two problems of Mitiagin', *Math. Nachr.* (to appear).

[55] D. Vogt and M.J. Wagner, 'Charakterisierung der Unterräume und Quotientenräume der nuklearen stabilen Potenzreihenräume von unendlichem Typ', *Studia Math.* **70** (1981) 63-80.

NUCLEAR KÖTHE QUOTIENTS OF FRECHET SPACES

S. ÖNAL
Department of Mathematics
Middle East Technical University
06531 Ankara - Turkey

Throughout we let $\mathcal{U}(F)$ denote a base of neighborhoods of a locally convex space (lcs) F which consists of absolutely convex and closed sets. In this work every Köthe space will be assumed to admit a continuous norm.

We say a lcs F satisfies (y) if there is a neighborhood $V_1 \epsilon \mathcal{U}(F)$ such that

$$F' = \bigcup_{V \epsilon \mathcal{U}(F)} \overline{F'[V_1^0] \cap V^0}$$

where the closure is taken in the $\sigma(F', F)$-topology. We recall that the seminorm p_{v_1} defined by V_1 is in fact a norm on F if and only if $F'[V_1^0] =$ span of V_1^0 is $\sigma(F', F)$-dense in F'. Hence a lcs which satisfies (y) admits a continuous norm.

We refer to [4] for a more detailed discussion of the condition (y) and now state our main result.

Theorem 1: Assume E is Fréchet space and F is a lcs which satisfies (y). If there is a continuous, unbounded linear operator $T : E \to F$, then there is a nuclear Köthe Fréchet space $\lambda(A)$ and a quotient map $Q : F \to \lambda(A)$ such that $QT(E) = \lambda(A)$.

For the proof of this theorem we use the following lemma which is somewhat dual to the lemma in [6].

Lemma 2: Let $T : E \to F$ be an unbounded continuous linear map, where E is a Fréchet space and F a lcs which satisfies (y). Then there are sequences of zero neighborhoods U_n and V_n in E and F respectively such that $T(U_n) \subset V_n$, $T'(F'[V_1^0] \cap V_{n+1}^0)$ is not absorbed by U_n^0 and U_n forms a decreasing neighborhood base for the Fréchet space E.

T. Terzioğlu (ed.), Advances in the Theory of Fréchet Spaces, 255–258.

We have a weak imbedding from a lcs F which satisfies (y) into a Köthe space $\lambda^\infty(A)$. Here $\lambda^\infty(A)$ is a projective limit of $\ell_\infty(\Gamma)$ with diagonal maps. In case F is a Fréchet space we can imbed F into Fréchet space of the type $\lambda^\infty(A)$. Hence the condition (y) implies countable normability in the case F is a Fréchet space. In fact for Fréchet spaces the condition (y) is equivalent to the following stronger condition: For each $V \epsilon \mathcal{U}(F)$, there is $W \epsilon \mathcal{U}(F)$ such that

$$V^0 \subset \overline{F'[V_1^0] \cap W^0}.$$

In the case of Schwartz spaces, the condition (y) is equivalent to some other well-known conditions.

Proposition 3: Following are equivalent for a Schwartz space F.

(i) F is asymptotically normable.

(ii) F satisfies (y).

(iii) F has a neighborhood base $\mathcal{U}(F)$ such that completions of the corresponding linking maps are one to one.

The concept of asymptotical normability was introduced and examined by Terzioğlu and Vogt in [7].

Following [4], we say E satisfies the *boundedness condition* **(b)** if for each $U \epsilon \mathcal{U}(E)$ there is a $V \epsilon \mathcal{U}(E)$ such that for every $W \epsilon \mathcal{U}(E)$ we have $\rho > 0$, so that

$$W^0 \cap E'[U^0] \subset \rho V^0$$

holds. Next result was motivated by Terzioğlu [5].

Proposition 4: If E satisfies **(b)** and F satisfies (y) then $L(E,F) = LB(E,F)$.

Behrends, Dierolf and Harmand [1] constructed a proper Fréchet space which satisfies **(b)** and admits a continuous norm. It can be shown that this space is countably normed. By our proposition this space does not satisfy (y). Hence existence of a continuous norm or countable normability is not sufficient for the condition (y). However if F has the bounded approximation property and admits a continuous norm then F satisfies (y).

Proposition 5: Let F be lcs which has the bounded approximation property and a continuous norm. Then F satisfies (y).

Proposition 6: It F has the bounded approximation property and satisfies **(b)** then F satisfies the openness condition.

Bellenot and Dubinsky proved in [2] a theorem about the existence of nuclear Köthe quotients of Fréchet spaces which can be stated with our terminology as follows: a separable Fréchet space which does not satisfy condition **(b)** has a nuclear Köthe quotient. From Proposition 4 we know that condition **(b)** is an obstacle for the existence of a nuclear Köthe quotients which cannot be over come. However the assumption of separability in the theorem of Bellenot and Dubinsky can be removed.

Corollary 7: The following conditions are equivalent for a Fréchet space E

(i) E has a nuclear Köthe quotient.

(ii) E does not satisfy **(b)**.

(iii) The bidual E'' is not quojection.

Proof: The equivalence of (ii) and (iii) was proved by Vogt [8]. If E satisfies **(b)**, then we have $L(E, \lambda(A)) = LB(E, \lambda(A))$ for every nuclear Köthe space $\lambda(A)$ by Proposition 4. On the other hand, if E does not satisfy **(b)**, by a fairly standard method one can find a nuclear Köthe space $\lambda(B)$ and an unbounded, continuous, linear operator $T : E \to \lambda(B)$ ([5], Corollary 4). Since $\lambda(B)$ satisfies (y) by Proposition 5, E has a nuclear Köthe quotient by our theorem.

Another interesting characterization of those Fréchet spaces which are quojections is given in [3].

Corollary 8: Let E be a Fréchet subspace of a lcs F which satisfies (y). Then either E is a Banach space or there is a nuclear Köthe Fréchet space $\lambda(A)$ and a quotient map $Q : F \to \lambda(A)$ such that $Q(E) = \lambda(A)$.

Some of the results of this note and their proofs will appear in [4] and the others will appear elsewhere.

REFERENCES

[1] E. Behrends, S. Dierolf, P. Harmand., 'On a Problem of Bellenot and Dubinsky', *Math. Ann.* **275**, 337-339, (1986).

[2] S.F. Bellenot., Ed Dubinsky., 'Fréchet spaces with nuclear Köthe quotients', *Trans. A.M.S.* **273**, 579-594, (1982).

[3] J. Bonet, M. Maestre, G. Metafune, V.B. Moscatelli, D. Vogt., 'Every quojection is the quotient of a countable product of Banach Spaces', These Proceedings.

[4] S. Önal., T. Terzioğlu., 'Unbounded linear operators and nuclear Köthe quotients', To appear in *Arch. Math.*

[5] T. Terzioğlu., 'A note on unbounded linear operators and quotients spaces', *Doğa* **10**, 338-344, (1986).

[6] T. Terzioğlu, M. Yurdakul., 'Restrictions of unbounded continuous linear operators', *Arch. Math.* **46** 547-550, (1986).

[7] T. Terzioğlu, D. Vogt., 'On asymptotically normable Fréchet spaces', Preprint.

[8] D. Vogt., 'On two problems of Mitiagin', To appear in *Math. Nachr.*.

A NOTE ON STRICT LF-SPACES

JOSÉ BONET
Departamento de Matemáticas
Universidad Politécnica
E-46071 Valencia
(Spain)

SUSANNE DIEROLF
FB-IV - Mathematik
Universität Trier
D-5500 Trier
F.R. Germany

This note is an abstract of the talk delivered by the second author at this Workshop on Fréchet Spaces. Publication of a more detailed version containing all the proofs is intended.

1. The *problem* to be discussed in this note deals with the question, whether for a strict LF-space $F = ind\ F_n$ the continuous identity map $ind\ \ell^\infty(F_n) \to \ell^\infty(ind\ F_n)$ is a topological isomorphism.
Here we use the following notation: Given a locally convex space E and an index set I, we denote by $\ell_I^\infty(E)$ the space of all bounded families $(x_i)_{i\epsilon I}$ in E provided with the topology of uniform convergence on I; we write $\ell^\infty(E) := \ell_N^\infty(E)$. If E is locally complete, then $\ell_I^\infty(E)$ is topologically isomorphic to the space $\mathcal{L}_b(\ell_I^1, E)$ of all continuous linear maps from the Banach space ℓ_I^1 into E.

It is not hard to prove that the above problem has an affirmative answer if $F = ind\ F_n$ is a strict LB-space. Even more is true: if $F = ind\ F_n$ is a retractive LB-space, then $ind\ \ell^\infty(F_n)$ and $\ell^\infty(F)$ are topologically isomorphic by [4; (4.12) and (5.12)]. If $F = ind\ F_n$ is a regular LB-space, Bierstedt and Bonet [1; 1.5. Thm.] have proved that $ind\ \ell^\infty(F_n)$ and $\ell^\infty(F)$ are topologically isomorphic if and only if F satisfies the strong dual density condition (SDDC). Furthermore, by [4; (4.7)], there exists a (regular) LB-space $F = ind\ F_n$ where all the Banach spaces F_n are reflexive such that $ind\ \ell^\infty(F_n)$ and $\ell^\infty(F)$ are not topologically isomorphic.

In order to give an affirmative answer to the original problem $ind\ \ell^\infty(F_n) \overset{?}{\underset{top}{=}} \ell^\infty(F)$ for strict LF-spaces $F = ind\ F_n$, the following condition should be satisfied: For every sequence $(U_n)_{n\epsilon N}$ of 0-neighborhood U_n in F_n there exists another sequence $(W_n)_{n\epsilon N}$ of

T. Terzioğlu (ed.), Advances in the Theory of Fréchet Spaces, 259–264.

0-neighborhood W_n in F_n such that

$$\ell^\infty(F) \cap (\sum_{n \in N} W_n)^N \subset \sum_{n \in N} (\ell^\infty(F_n) \cap U_n^N).$$

(In fact, the left hand side is a typical 0-neighborhood in $\ell^\infty(F)$ and the right hand side is a typical 0-neighborhood in *ind* $\ell^\infty(F_n)$).

Since the crucial point of this decomposition lies already in the decomposability of finite sums, the (very) general problem behind is the following:Given a locally convex space X and subsets $M, N \subset X$, do there exist "nontrivial" subsets $\tilde{M} \subset M$, $\tilde{N} \subset N$ such that for every bounded subset $B \subset \tilde{M} + \tilde{N}$ there are bounded subsets $B_M \subset M$ and $B_N \subset N$ such that $B \subset B_M + B_N$ - or, equivalently, such that for every index set I one has

$$\ell_I^\infty(X) \cap (\tilde{M} + \tilde{N})^I \subset \ell_I^\infty(X) \cap M^I + \ell_I^\infty(X) \cap N^I \ ?$$

In the case that M or N are bounded subsets of X, this question trivially has a positive answer; if M and N are both unbounded, not much seems to be known about the problem. In the sequel we are going to treat the following special case.

2. Definition: Let X be a locally convex space, let $Y \subset X$ be a linear subspace, and let I be a set. We say that the "pair" $X \supset Y$ has the bounded decomposition property w.r. to I (I-BDP) if for every 0-neighborhood U in X there exists another 0-neighborhood W in X such that

$$\ell_I^\infty(X) \cap (W + Y)^I \subset \ell_I^\infty(X) \cap U^I + \ell_I^\infty(Y).$$

- or, equivalently, if the natural continuous map $Q_I : \ell_I^\infty \to \ell_I^\infty(X/Y)$, $(x_i)_{i \in I} \to (x_i + Y)_{i \in I}$ is open onto its range.We will abbreviate BDP := N-BDP.

Remarks:

a) If $X \supset Y$ has the I-BDP, then it follows easily that

$$\forall\, U\ \exists\, W\ \forall\, V \text{ 0-neighborhood in Y such that } U \cap Y \subset \alpha V$$

we have

$$\ell_I^\infty(X) \cap (W + V)^I \subset \ell_I^\infty(X) \cap U^I + (1 + 2\alpha) V^I.$$

This statement allows to deal also with a 0-neighborhood U in the "big" space X inconnection with a 0-neighborhood V in the "small" space Y such that V is "big" w.r. to $U \cap Y$.

b) The following stability properties are valid. Let X be a locally convex space and let $Y, Z \subset X$ be linear subspaces such that $X \supset Z \supset Y$.

If $X \supset Y$ has the I-BDP, then also $Z \supset Y$ has the I-BDP. If both $X \supset Z$ and $Z \supset Y$ have the I-BDP, then also $X \supset Y$ has the I-BDP.

The first statement is obvious, whereas for the proof of the second one needs a little computation.

3. Proposition: Let X be a locally convex space and $Y \subset X$ a linear subspace. Assume that one of the following conditions is satisfied

i) Y is complemented in X;

ii) Every bounded subset of X is precompact;

iii) Y is quasinormable.

Then the pair $X \supset Y$ has the I-BDP for all sets I.

We will present a proof of iii) since it is both short and typical for the methods used in our context: Given an absolutely convex 0-neighborhood U in X, choose an absolutely convex 0-neighborhood $W \subset \frac{1}{2}U$ in X such that for every $\epsilon > 0$ there is a bounded subset $B \subset Y$ satisfying $W \cap Y \subset \epsilon U + B$. Let $A \subset W + Y$ be bounded. Then $A \subset \lambda W$ for some $\lambda > 0$. We find $B \subset Y$ bounded such that $(1+\lambda)W \cap Y \subset \frac{1}{2}U + B$, whence $A \subset W + Y \cap (1+\lambda)W \subset U + B$. Consequently, $A \subset U \cap (A - B) + B$.

Condition iii) from above, which is a condition only on Y, even characterizes the I-BDP if one introduces a suitable quantifier.

4. Proposition: For a locally convex (resp. Fréchet) space Y the following are equivalent

i) Y is quasinormable;

ii) For every locally convex (resp. Fréchet) space X containing Y as a topological subspace and every set I, the natural map $Q_I : \ell_I^\infty(X) \to \ell_I^\infty(X/Y)$ is open onto its range.

If, in addition, Y is separable, then ii) is equivalent to

ii)' For every locally convex (resp. Fréchet) space X containing Y the map $Q : \ell^\infty(X) \to \ell^\infty(X/Y)$ is open onto its range.

For the proof of ii) $\Rightarrow$ i) one makes -of course- use of a natural "superspace" $X = \prod_{U \in \mathcal{U}} Y_U$, where $\mathcal{U}$ is a O-basis in Y and Y_U is the Banach space corresponding to a 0-neighborhood $U \in \mathcal{U}$.

Proposition 4, which is a characterization of the quasinormability of general locally convex spaces, should be compared with the following classical result of Palamodov [6], Merzon [5], De Wilde [3], and Cholodovskii [2].

5. Classical Theorem: For a Fréchet space Y the following are equivalent

i) Y is quasinormable;

ii) For every Fréchet space X containing Y as a topological subspace and every set I the natural map $Q_I : \ell_I^\infty(X) \to \ell_I^\infty(X/Y)$ is surjective (-or, equivalently, open).

It should be mentioned that this result cannot be extended to general locally convex spaces Y as can immediately be derived from (3.6) Example in [7; p.19].

Remark: From 4. and 5. we have that for a Fréchet space Y one has the equivalence

$$(\forall\ X, I \ \ Q_I \text{ is open onto its range}) \Leftrightarrow (\forall\ X, I \ \ Q_I \text{ is surjective}).$$

a) Without the quantifier $\forall X$ the implication " $\Rightarrow$ " fails. In fact, let X be the Grothendieck-Köthe Fréchet-Montel space which contains a linear subspace Y such that X/Y is topologically isomorphic to ℓ^1. Then, according to 3., the maps Q_I are all open onto their range but not surjective for infinite I.

b) Without the quantifier $\forall X$ and if Y is not metrizable, then also " $\Leftarrow$" fails. In fact, let $Z = ind\ Z_n$ be a regular LB-space, and let X be the locally convex direct sum $X := \bigoplus_{n \in N} Z_n$. Then $Y := \{(z_n)_{n \in N} \in X : \sum z_n = 0\}$ is a linear subspace of X such that $X/Y \underset{\text{top}}{\cong} Z$. By the regularity of $ind\ Z_n$, the map Q_I is surjective. If we choose $Z = ind\ Z_n$ in such a way that $\ell^\infty(Z)$ is not quasibarrelled [4; (4.7)], then Q_I cannot be open for any infinite I.

Our next aim is to apply the concept of the BDP to our original problem 1.

6. Theorem: Let $(F_n)_{n \in N}$ be a sequence of locally convex spaces such that, for every $n \in N$, F_n is a topological linear subspace of F_{n+1}, let $F := ind\ F_n$ be the corresponding (strict) inductive limit, and let I be a set.If for every $n \in N$ the pair $F_{n+1} \supset F_n$ has the I-BDP, then the natural continuous linear injection

$$J : ind\ \ell_I^\infty(F_n) \to \ell^\infty(F)$$

is a topological isomorphism onto its range.Consequently, if in addition the inductive limit $ind\ F_n$ is regular (e.g. if F_n is closed in F_{n+1} for all n) then J is a topological isomorphism.

(Note that J is surjective if and only if $ind\ F_n$ is regular.)

For a proof: Given a 0-neighborhood U in $ind\ \ell_I^\infty(F_n)$, one constructs inductively - making use of the BDP-hypothesis, of course- 0-neighborhoods W_n in $F_n (n \in N)$ such that

$$\left(\sum_j W_j\right)^I \cap \bigcup_n \ell_I^\infty(F_n) \subset U$$

7 Corollary: Let $F = ind\ F_n$ be a strict LF-space such that every F_n is quasinormable or a Montel space. Then $ind\ \ell_I^\infty(F_n)$ and $\ell_I^\infty(F)$ are topologically isomorphic for every set I.

For a certain special type of strict inductive limit the implication in 6.Theorem can even be reversed. In fact, we have

8 Proposition: Let $(X_k)_{k \epsilon N}$ be a sequence of locally convex spaces and for every $k \epsilon N$ let $Y_k \subset X_k$ be a linear subspace. For every $n \epsilon N$ let $F_n := \prod_{k<n} X_k \times \prod_{k \geq n} Y_k$ be the topological product, and let $F := ind\ F_n$ be the corresponding strict inductive limit (of so-called "generalized Moscatelli type"). Then for every set I the following are equivalent

i) $J : ind\ \ell_I^\infty(F_n) \rightarrow \ell_I^\infty(F)$ is a topological isomorphism onto its range;

ii) There is $k_0 \epsilon N$ such that the pair $X_k \supset Y_k$ has the I-BDP for all $k \geq k_0$.

Let us choose a non-quasinormable Fréchet space Y which is the projective limit proj Y_n of separable and reflexive Banach spaces Y_n. (Such spaces Y exist in the class of Köthe echelon spaces and in the class of Moscatelli-type-Fréchet spaces.) Then $X := \prod_{n \epsilon N} Y_n$ is a separable and reflexive Fréchet space and via 4. Proposition the pair $X \supset Y$ does not have the BDP. Thus we obtain with the help of 8.Proposition the following.

9 Example: There exist strict LF-spaces $F = ind\ F_n$, where all F_n are separable and reflexive Fréchet spaces, such that $ind\ \ell^\infty(F_n)$ and $\ell^\infty(F)$ are not topologically isomorphic (whence $\ell^\infty(F)$ is, for instance, not quasibarrelled).

REFERENCES

[1] Bierstedt, K.D., Bonet, J.: 'Dual density conditions in (DF)-spaces'. To appear.

[2] Cholodovskii, V.E.: 'On quasinormability of semimetrizable topological vector spaces', *Funk. Analiz* **7** (1976), 157-160.

[3] De Wilde, M.: 'Sur le relèvement des parties bornées d'un quotient d'espaces vectoriels topologiques', *Bull. de la Soc. Royale des Sciences de Liège* **43** (1974), 299-301.

[4] Dierolf, S.: 'On spaces of continuous linear mappings between locally convex spaces', *Note di Matematica* **5** (1985), 147-255.

[5] Merzon, A.E.: 'On the covering of bounded sets in locally convex spaces', *Funk. Analiz* **6** (1972), 91-92.

[6] Palamadov, V.P.: 'Homological methods in the theory of locally convex spaces', *Russian Math. Surveys* **26** (1971), 1-64.

[7] Roelcke, W., Dierolf, S.: 'On the three-space-problem for topological vector spaces', *Collectanea Math.* **32** (1981), 13-35.

AUTOMATIC CONTINUITY IN FRECHET ALGEBRAS

M. FRAGOULOPOULOU
Mathematical Institute
University of Athens
57, Solonos Street
10679 Athens
Greece

Abstract

This note is a survey of results on the automatic continuity of positive functionals and of *-representations of locally m-convex *-algebras. As a byproduct one obtains uniqueness of the topology of a Fréchet locally m-convex C^*-algebra. In particular, attempt has been made to avoid continuity of the involution.

Proofs of the results presented here together with further considerations are to be published elsewhere under the title "Automatic continuity of *-morphisms between non-normed topological *-algebras".

Now a *locally m-convex* (lmc) **-algebra* is an involutive topological algebra E (see also [8]), whose topology is defined by a directed family of *-preserving submulticative seminorms. In case the topology of E is defined by a directed family of C^*-seminorms, E is called an *lmc C^*-algebra*. A *positive functional* of E is a linear form f on E such that $f(x^*x) \geq 0$, for all $x \in E$. A **-representation* of E is an involution preserving morphism (*-morphism) φ of E in the C^*-algebra $L(H_\varphi)$ of all bounded linear operators on some Hilbert space H_φ.

T. Terzioğlu (ed.), Advances in the Theory of Fréchet Spaces, 265–268.

1. Continuity of Positive Functionals

It is known that every positive functional of an involutive Banach algebra with a bounded approximate identity (*bai*) is continuous (N. Th. Varopoulos [10]). Concerning the non-normed topological algebras we have the following:

1.1. Do-Sin Sya (1959) [9]. *Every positive functional of a unital Fréchet lmc *-algebra is continuous.*

1.2. T. Husain - R. Rigelhof (1969) [6; Theorem 4.2] (a variation of 1.1). *Every positive functional of a unital sequentially complete Q lmc *-algebra is continuous.* (A *Q-algebra* is a topological algebra the set of quasi-invertible elements of which is open [8; Definition I, 6.2]. The algebra $C^{\infty}[0,1]$ is a Fréchet Q lmc *-algebra).

1.3. P.G. Dixon (1981) [4; Theorem 4.3] (a generalization of 1.1). *Every positive functional of a Fréchet locally convex *-algebra with a left bai is continuous.*

In 1978 H.G. Dales [3; Theorem 11.1], using instructions of P.G. Dixon proved that the technique of 1.1 is valid for any "unital Fréchet topological *-algebra". He also remarks that the continuity of the involution "may not be necessary".

On the other hand, we shall see that 1.3 is still true by removing the continuity of the involution, at the cost however of the property Q and of the (local) "multiplicative convexity" of the topological algebra involved. Thus, one gets a generalization of Varopoulos' result in our case (see 2.7, (iii)).

2. Continuity of *-representations.

It is also known that every *-representation of an involutive Banach algebra is continuous [1; Theorem 37.3]. Regarding non-normed topological algebras, things appear as follows:

2.1. T. Husain - R. Rigelhof (1969) [6; Theorem 5.3]. *Every *-representation of a sequentially complete Q lmc *-algebra is continuous.*

The next general result shows, at least, that sequential completeness is redundant in 2.1.

2.2. M. Fragoulopoulou (1987) [5; Theorem 3.1]. *Every *-morphism of a Q lmc *-algebra in an lmc C*-algebra is continuous.*

2.3. R.M. Brooks (1972) [2; Lemma 3.1]. *Every *-representation of a unital Fréchet lmc *-algebra is continuous.*

2.4. G. Lassner (1972) [7; Theorem 4.1] (a generalization of 2.3). *If E is a unital barrelled locally convex *-algebra and φ a *-representation of E on H_φ, such that for every $\xi \in H_\varphi$ the positive functional*

$$f(x) = \langle \varphi(x)(\xi), \xi \rangle, \quad x \in E,$$

is continuous, then φ is continuous.

It is easily seen that 2.4 is also valid by removing unit and continuity of the involution and replacing barrelledness by the weaker concept of m-barrelledness (for the relevant terms see [8]). Thus, one gets the following.

2.5. Corollary. *Every *-representation of a Fréchet locally convex *-algebra is continuous.*

As a result we now obtain the next.

2.6. M. Fragoulopoulou (1987) [5; Theorem 3.4, Corollary 3.6]. *Every *-morphism of a Fréchet locally convex *-algebra in an lmc C^*-algebra is continuous. Consequently, the topology of a Fréchet lmc C^*-algebra is uniquely determined.*

Considering now the case of topological *-algebras, where no continuity of the involution is assumed we can prove the following.

2.7. M. Fragoulopoulou (1987) [5; Theorem 5.2, Proposition 5.6, Theorem 5.8]. *Let E be an involutive Fréchet Q lmc algebra. Then,*

(i) *Every *-representation of E is continuous.*

(ii) *Given $y \in E$ and a positive functional f of E, the induced positive functional*

$$f_y(x) := f(y^* x y), \quad x \in E,$$

of E, is continuous.

Based on (i), (ii) we now have Varopoulos' result in our case; i.e., one gets

(iii) *If E has moreover a bai, every positive functional of E is continuous.*

In this respect, one can give an example of a Fréchet lmc algebra with a discontinuous involution (cf. [5; Section 4]). Furthermore, one has criteria for the continuity of the involution of a locally convex barrelled Pták (take e.g. Fréchet) algebra (ibid.).

REFERENCES

[1] F.F. Bonsall - J. Duncan: *Complete Normed Algebras.* Springer-Verlag, Berlin Heidelberg New York, 1973.

[2] R.M. Brooks: 'On representing F^*-algebras'. *Pacific J. Math.* **39** (1971), 51-69.

[3] H.G. Dales: 'Automatic continuity: A survey'. *Bull. London Math. Soc.* **10** (1978), 129-183.

[4] P.G. Dixon: 'Automatic continuity of positive functionals on topological involution algebras'. *Bull. Austr. Math. Soc.* **23** (1981), 265-281.

[5] M. Fragoulopoulou: 'Automatic continuity of *-morphisms between non-normed topological *-algebras' (to appear).

[6] T. Husain - R. Rigelhof: 'Representations of MQ^*-algebras'. *Math. Ann.* **180** (1969), 297-306.

[7] G. Lassner: 'Topological algebras of operators'. *Reports on Math. Phys.* **3** (1972), 279-293.

[8] A. Mallios: *Topological Algebras. Selected Topics.* North-Holland, Amsterdam, 1986.

[9] D.-S. Sya (Shah Dao-Shing): 'On seminormed rings with involution'. *Izvestiya Akad. Nauk SSSR Ser. Mat.* **23** (1959), 509-528 (Russian).

[10] N. Th. Varopoulos: 'Sur les formes positive d'une algebre de Banach'. *C.R. Acad. Sci. Paris Sér. A-B*, **258** (1964), 2465-2467.

SOME SPECIAL KÖTHE SPACES

M. KOCATEPE
Bilkent University
Department of Mathematics
Ankara
Turkey

Z. NURLU
Middle East Technical University
Department of Mathematics
Ankara
Turkey

ABSTRACT. The aim of this survey is to gather the results concerning the structure of some special Köthe spaces, namely power series spaces, Dragilev spaces, Köthe spaces defined by transition functions and $S_g(a,r)$-spaces. The results that we summarize here are about subspaces, quotient spaces, compact operators, the functor Ext, the quasi-equivalence property and Bessaga's conjecture by means of which one can acquire quite extensive knowledge about the structure of these spaces. We also list some classical and recently established examples of function spaces which are isomorphic to one of these spaces.

1. Introduction

The spaces considered in this survey are some concrete examples of Köthe spaces which exhibit certain particular properties. Some of them emerged naturally in the course of history among function spaces while others were defined to answer questions on structure of Köthe spaces, mainly to serve as counter examples. However they have been and still are of utmost importance in the theory of nuclear Fréchet spaces since recent research concerning function spaces in various related fields such as weighted algebras of analytic functions, convolution operators, solution spaces of partial differential operators, etc. has shown that some of the newly emerging function spaces are in fact one of these spaces. Some illustrative examples depicting this relation are mentioned in the forthcoming sections. It would only reflect the truth if we say that the bulk of the research done in the theory of Köthe spaces is related to these spaces one way or other.

T. Terzioğlu (ed.), Advances in the Theory of Fréchet Spaces, 269–296.

2. Generalities

2.1. NOTATION AND TERMINOLOGY

In the sequel $E, F, \ldots$ will denote nuclear Fréchet spaces (NFS) unless otherwise stated. $U, V, \ldots$ will denote absolutely convex closed zero neighborhoods and $(\|.\|_k)$ will be used for a sequence of non-decreasing seminorms defining the topology of the space concerned. For $k \in \mathbf{N}$, $\|.\|_k^* : E' \mapsto [0, \infty]$ is defined by $\|y\|_k^* = \sup\{|y(x)| : \|x\|_k \leq 1\}$ and $E'_k = \{y \in E' : \|y\|_k^* < \infty\}$. A matrix (a_{kn}) of non-negative scalars satisfying
i) $\forall n, k \; a_{kn} \leq a_{k+1,n}$,
ii) $\forall n \; \sup_k a_{kn} > 0$
is called a **Köthe matrix** and the sequence space

$$K(a_{kn}) = \{(\xi_n) : \|(\xi_n)\|_k = \sum_n |\xi_n| a_{kn} < \infty \; \forall k\}$$

topologized by the seminorms $(\|.\|_k)$ is called a **Köthe space**. For a Köthe matrix we define

$$K^\infty(a_{kn}) = \{(\xi_n) : \|(\xi_n)\|_k = \sup_n |\xi_n| a_{kn} < \infty \; \forall k\}.$$

A Fréchet space E with an absolute basis (x_n) (that is, for every zero neighborhood U there exists a zero neighborhood V such that $\sum_n |\xi_n| p_U(x_n) \leq p_V(x)$ where $x = \sum_n \xi_n x_n$ and p_U is the Minkowski functional) is isomorphic to the Köthe space $K(a_{kn})$ where $a_{kn} = \|x_n\|_k$.

A Köthe matrix (a_{kn}) and also the Köthe space $K(a_{kn})$ are called of type (d_i) if the corresponding condition below is satisfied:

(d_0) $\forall k, n \;\; a_{k+1,n}/a_{kn} \leq a_{k+1,n+1}/a_{k,n+1}$ (also called **regular**)

(d_1) $\exists k \; \forall j \; \exists \ell \;\; \sup_n (a_{jn})^2/(a_{kn} a_{\ell n}) < \infty$

(d_2) $\forall k \; \exists j \; \forall \ell \;\; \sup_n a_{kn} a_{\ell n}/(a_{jn})^2 < \infty$

(d_3) $\forall k, n \;\; a_{k+1,n}/a_{kn} \leq a_{k+2,n}/a_{k+1,n}$

(d_4) $\forall k, n \;\; a_{k+1,n}/a_{kn} \geq a_{k+2,n}/a_{k+1,n}$

(d_5) $\exists M \geq 1 \; \forall k, n \;\; a_{k+1,n}/a_{kn} \leq (a_{k+2,n}/a_{k+1,n})^M$.

The conditions (d_0), (d_1) and (d_2) were introduced by M.M. Dragilev in [15], and (d_3), (d_4) and (d_5) by E. Dubinsky and W. Robinson in [19], [23] and [21]. A basis (x_n) of a NFS E is said to be of type (d_i) if there exists a fundamental system of norms $(\|.\|_k)$ such that $a_{kn} = \|x_n\|_k$ is of type (d_i). In this case we say that the space E is (d_i).

Basis-free formulation of properties (d_i) are as follows: A Fréchet space E is said to have the property (DN) (resp. $(\underline{DN}), (\Omega), (\overline{\Omega})$) if the corresponding condition below is satisfied:

(DN) $\exists k \; \forall j \; \exists \ell, C > 0 \;\; \|.\|_j^2 \leq C \|.\|_k \|.\|_\ell$

$(\underline{DN})$ $\exists k\, \forall j\, \exists \ell, \varepsilon, C > 0$ $\|\cdot\|_j^{1+\varepsilon} \leq C \|\cdot\|_k^{\varepsilon} \|\cdot\|_\ell$

(Ω) $\forall p\, \exists q\, \forall k\, \exists d > 0, C > 0$ $(\|\cdot\|_q^*)^{1+d} \leq C \|\cdot\|_k^* (\|\cdot\|_p^*)^d$

$(\overline{\Omega})$ $\forall p\, \exists q\, \forall k\, \exists C > 0$ $(\|\cdot\|_q^*)^2 \leq C \|\cdot\|_k^* \|\cdot\|_p^*$

The properties (DN) and $(\underline{DN})$ were introduced by D. Vogt in [53] and [54], (Ω) by D. Vogt and M.J. Wagner in [60] and $(\overline{\Omega})$ by M.J. Wagner in [62].

If a NFS E has a basis, then (d_1), (d_3) and (DN) are all equivalent, (d_2) is equivalent to $(\overline{\Omega})$, (d_4) is equivalent to (Ω), and (d_5) is equivalent to $(\underline{DN})$.

A Köthe space $K(a_{kn})$ is called

stable if $K(a_{kn}) \times K(a_{kn}) \cong K(a_{kn})$

weakly stable if $K(a_{kn}) \times \mathbf{K} \cong K(a_{kn})$ (here $\mathbf{K}$ denotes the scalar field)

unstable if $\exists s\, \forall p\, \exists q\, \forall r$ $\lim_n d_{n+1}(U_q, U_p)/d_n(U_r, U_s) = 0$ where $d_n(U,V)$ denotes the n-th Kolmogorov diameter of U with respect to V and $U_q = \{(\xi_n) : \sum_n |\xi_n| a_{qn} \leq 1\}$.

Unstable spaces are regular and Schwartz and hence unstability is equivalent to:

$$\exists s\, \forall p\, \exists q\, \forall r \quad \lim_n \left(\frac{a_{p,n+1}}{a_{q,n+1}} \right) \left(\frac{a_{rn}}{a_{sn}} \right) = 0.$$

2.2. SOME EXAMPLES

The classical Dynin-Mitiagin [24] theorem states that in a NFS with basis all bases are absolute and therefore such spaces have Köthe space representations. Some well-known classical examples are:

(a) $(s) = \{(\xi_n) : \lim_n \xi_n n^k = 0\, \forall k\} = K(\exp(k \log n))$= the space of all rapidly decreasing sequences.

(b) $A(\mathbf{C}^d) = K(\exp(kn^{1/d}))$= the space of all analytic functions on $\mathbf{C}^d$.

(c) $A(\Delta^d) = K(\exp(-n^{1/d}/k))$= the space of all analytic functions on Δ^d where $\Delta = \{z \in \mathbf{C} : |z| < 1\}$.

(d) $A(\mathbf{C}^r \times \Delta^{d-r}) = K(\exp(k(n_1 + \cdots + n_r) - (1/k)(n_{r+1} + \cdots + n_d)))$

(e) $C^\infty(\mathbf{R})$= the space of all functions $f : \mathbf{R} \mapsto \mathbf{C}$ having continuous derivatives of all orders, topologized by the seminorms $\|f\|_p = \sup_{|t| \leq p,\, k \leq p} |f^{(k)}(t)|$. We have that $C^\infty(\mathbf{R}) \cong (s)^{\mathbf{N}}$.

(f) $C^\infty[-1,1]$= the space of all infinitely differentiable functions f on [-1,1] with

$$\|f\|_p = \sup_{|t| \leq 1,\, k \leq p} |f^{(k)}(t)| < \infty.$$

We have that $C^\infty[-1,1] \cong (s)$.

(g) $C_0^\infty[-1,1]$= the space of all functions f in $C^\infty[-1,1]$ for which $f(\mp 1) = 0$ topologized with the seminorms in (f). We have that $C_0^\infty[-1,1] \cong (s)$.

(h) $S = \{f \in C^\infty(\mathbf{R}) : \|f\|_{p\ell} = \sup_{t\in\mathbf{R}, k\leq p} |t^\ell f^{(k)}(t)| < \infty \ \forall \ell \ \forall p\}$= space of all rapidly decreasing functions. We have that $S \cong (s)$.

(i) $C^\infty(\mathbf{R})^* = \{f \in C^\infty(\mathbf{R}) : f \text{ is } 2\pi\text{-periodic}\} \cong (s)$.

Of these examples (b),(c) and (d) are due to S. Rolewicz [52] and (e),(f),(g),(h),(i) are due to B.S. Mitiagin [45].

The Köthe space $K(\exp(r_k a_n))$, where r_k increases to $r \in (-\infty, \infty]$ and (a_n) is an increasing sequence of positive numbers with $\lim a_n = \infty$, is called a **power series space of type r** and denoted by $\Lambda_r(a)$ where $a = (a_n)$. It can be checked that $\Lambda_r(a) \cong \Lambda_1(a)$ if $r < \infty$ and that $\Lambda_r(a)$ is

stable if and only if $\sup_n a_{2n}/a_n < \infty$,

weakly stable if and only if $\sup_n a_{n+1}/a_n < \infty$,

unstable if and only if $\lim_n a_{n+1}/a_n = \infty$.

In this notation the above examples become:

(a) $(s) = \Lambda_\infty(a)$, $a_n = \log(n+1)$,

(b) $A(\mathbf{C}^d) = \Lambda_\infty(a)$, $a_n = n^{1/d}$,

(c) $A(\Delta^d) = \Lambda_1(a)$, $a_n = n^{1/d}$,

(d) $A(\mathbf{C}^r \times \Delta^{d-r}) \cong A(\mathbf{C}^r)\tilde{\otimes}_\varepsilon A(\Delta^{d-r}) \cong \Lambda_\infty(a)\tilde{\otimes}_\varepsilon \Lambda_1(b)$, $a_n = n^{1/r}$, $b_n = n^{1/(d-r)}$.

Some more recent examples of function spaces which can be represented as Köthe spaces are given below:

(j) Every infinite dimensional closed linear translation invariant subspace of $A(\mathbf{C})$ is isomorphic to some $\Lambda_\infty(a)$ (see [41]).

(k) For every surjective convolution operator on $\mathcal{E}(\mathbf{R})$, $\mathcal{E}_\omega(\mathbf{R})$ or $\mathcal{E}^{M^p}(\mathbf{R})$ with $\dim\ker T_\mu = \infty$, the kernel of T_μ is isomorphic to some $\Lambda_\infty(a)$. Here $\mathcal{E}_\omega(\mathbf{R})$ is the space of ultradifferentiable functions in the sense of Beurling, and $\mathcal{E}^{M^p}(\mathbf{R})$ is the space of ultradifferentiable fuctions in the sense of Komatsu (see [42]).

(l) The weighted algebras of entire functions $A_p^0(\mathbf{C}^n)$ are isomorphic to infinite type power series spaces for various radial weight functions p (see [40], [44], [43] and [8]).

2.3. BASIC TOOLS

Structural questions concerning power series spaces have been extensively studied by many authors among whom we can cite B.S. Mitiagin, E. Dubinsky, M. Alpseymen, V.V. Kashirin, Z. Nurlu, K. Nyberg, W.B. Robinson, D. Vogt and M.J. Wagner.

B.S. Mitiagin discovered as early as 1971 (see [48] and also D. Vogt [55]) that the so called dead-end-space method works in power series spaces of finite type. The method considers a NFS E satisfying certain conditions, namely $(\underline{DN})$ and $(\overline{\Omega})$, and uses a certain Hilbert space E_∞, the dead-end space, embedded in each of the Banach spaces $E_k =$

$(E,\widehat{\|.\|_k})$ and finds some complete orthogonal system in E_∞ by means of which one can show that E is isomorphic to a power series space of finite type. This consequence has found various applications in function spaces, namely one merely needs to check the two conditions and then one knows that the space considered has a basis and quite extensive knowledge is obtained for free on the structure of the space (subspaces, quotient spaces, compact operators, extension or lifting of operators, etc.)

In characterizing subspaces, quotient spaces and complemented subspaces, in general there are two conditions: one on the diametral dimension to ensure appropriate nuclearity and the other on the type of the fundamental norm system or the basis if the space has one. There are two basic approaches to deal with these structural problems: E. Dubinsky exploits the fact that the ℓ_1 and the ℓ_∞ norm systems on a nuclear Köthe space generate the same topology (thereby using the Grothendieck-Pietsch criterion for nuclearity) and constructs explicitly the embedding or the quotient map. On the other hand D. Vogt makes use of homological methods and splitting of certain exact sequences (and thereby using Komura-Komura type embedding theorems) to conclude that every linear continuous map between certain spaces can be lifted and he obtains the required embedding or the quotient map. E. Dubinsky's method is suitable for sub-, quotient spaces with basis whereas D. Vogt's method also works for sub-, quotient spaces without basis but is restrictive in the sense that certain exact sequences must be constructable.

In the case of stable power series spaces the complete characterizations are known. For an extensive survey on this topic we refer to D. Vogt [57].

2.4. GENERAL CASE FOR POWER SERIES SPACES

Here we mention some results concerning unstable or weakly stable or general power series spaces:

(a) (E. Dubinsky [16]) If $\Lambda_1(a)$ is unstable then it has no subspace isomorphic to a power series space of infinite type.

(b) (E. Dubinsky [17]) If $\Lambda_\infty(a)$ is unstable then a subspace of $\Lambda_\infty(a)$ is isomorphic to a power series space if and only if it is isomorphic to the subspace of $\Lambda_\infty(a)$ generated by a subsequence of the coordinate basis (e_n).

(c) (E. Dubinsky [20]) If $\Lambda_1(a)$ (resp. $\Lambda_\infty(a)$) is unstable then $\Lambda_1(b)$ (resp. $\Lambda_\infty(b)$) is isomorphic to a subspace of $\Lambda_1(a)$ (resp. $\Lambda_\infty(a)$) if and only if there is a subsequence (j_n) of indices such that

$$0 < \inf \frac{a_{j_n}}{b_n} \leq \sup \frac{a_{j_n}}{b_n} < \infty.$$

(d) (E. Dubinsky [20]) If $a_n = A^n$, $A > 1$ then $\Lambda_\infty(a)$ is isomorphic to a subspace of $\Lambda_1(a)$.

(e) (E. Dubinsky [20]) $\Lambda_1(b)$ is isomorphic to a subspace of $\Lambda_1((2^n))$ if and only if there is a subsequence (j_n) of indices such that

$$0 < \inf \frac{2^{j_n}}{b_n} \leq \sup \frac{2^{j_n}}{b_n} < \infty.$$

(f) (Z. Nurlu [49]) $\Lambda_\infty(b)$ is isomorphic to a subspace of $\Lambda_\infty(a)$ if and only if there is a subsequence (j_n) of indices such that

$$0 < \inf \frac{a_{j_n}}{b_n} \leq \sup \frac{a_{j_n}}{b_n} < \infty.$$

(g) (Z. Nurlu [49]) A regular Köthe space E is isomorphic to a subspace of $\Lambda_\infty(a)$ if and only if
i) E is (d_1)
ii) for any (some) normalized (d_1) representation (b_{kn}) of E,
$\forall k \; \exists p, k', n_k$ and $\exists$ a subsequence (j_n) of indices such that

$$b_{kn} \leq \exp(pa_{j_n}) \leq b_{k'n}, \; n \geq n_k.$$

(h) (Z. Nurlu [50]) If $\Lambda_\infty(a)$ is embeddable into $\Lambda_1(a)$ then $\Lambda_1(a)$ is weakly stable.

(i) (Z. Nurlu [50]) If $\Lambda_\infty(a)$ is isomorphic to a subspace of $\Lambda_1(a)$ spanned by a block basic sequence of the coordinate basis then $\Lambda_1(a)$ is stable.

(j) (D. Vogt [59]) If $\Lambda_1(a)$ is stable but not necessarily nuclear then a Fréchet space E is isomorphic to a quotient space (resp. subspace) of $\Lambda_1(a)$ if and only if E is hilbertizable, $\Lambda_1(a)$-nuclear and has $(\overline{\Omega})$ (resp. $(\underline{DN})$) where E is $\Lambda_1(a)$-nuclear if

$$\forall U \; \exists V \; \exists R > 1, \lim_n R^{a_n} d_n(V, U) = 0.$$

2.5. SUPERSPACES OF (s)

L. Holmström in [28] showed that if (s) is isomorphic to a subspace of a NFS E with a basis then (s) is isomorphic to a complemented subspace of E, and this property characterizes (s) among nuclear Köthe spaces with continuous norm. A. Aytuna, J. Krone, T. Terzioğlu in [6] generalized the first part of the above theorem by lifting the basis requirement on the NFS E.

3. Dragilev Spaces

$L_f(a,r)$-spaces (also called **Dragilev spaces**) were introduced by M.M. Dragilev [15] in 1965. These constitute the first concrete examples of Köthe spaces which are not necessarily power series spaces.

3.1. DEFINITION AND EXAMPLES

Let $f : \mathbf{R} \mapsto \mathbf{R}$ be an odd, increasing and logarithmically convex function (i.e. $\log \circ f \circ \exp$ is convex on $[0,\infty)$). Throughout this paper such a function will be called a **Dragilev function**. Let $a = (a_n)$ be an increasing sequence of positive numbers such that $\lim a_n = \infty$ and (r_k) a strictly increasing sequence of real numbers with $\lim r_k = r$ where $-\infty < r \leq +\infty$. Then the space $L_f(a,r)$ is defined as the Köthe space $K(a_{kn})$ where $a_{kn} = \exp f(r_k a_n)$.

From logarithmic convexity of f it follows that $\tau(a) = \lim_{x\to\infty} f(ax)/f(x)$ exists for every $a > 1$. Moreover either $\tau(a) < \infty$ for all $a > 1$ or $\tau(a) = \infty$ for all $a > 1$. In the first case f is called **slowly increasing**, in the second case f is called **rapidly increasing**.

The following example of R. Braun [8] shows that there are function spaces which are actually $L_f(a,r)$ spaces.

For the weight function $p(z) = (\log|z|)(\log\log|z|)$ consider the Hörmander algebras

$$A_p^0(\mathbf{C}) = \{f \in A(\mathbf{C}) : \forall \varepsilon > 0 \ \sup_{z\in\mathbf{C}} |f(z)| \exp(-\varepsilon p(z)) < \infty\}$$

and

$$A_p(\mathbf{C}) = \{f \in A(\mathbf{C}) : \exists A > 0 \ \sup_{z\in\mathbf{C}} |f(z)| \exp(-Ap(z)) < \infty\}.$$

Then we have that $A_p^0(\mathbf{C}) \cong L_f((n),\infty)$ and $A_p(\mathbf{C})'_b \cong L_f((n),0)$ where $f(x) = \exp(x-1)$ for large x and f is modified so that it is odd and logarithmically convex.

3.2. BASIC PROPERTIES

The following properties of Dragilev spaces are either immediate or have been observed by M.M. Dragilev [15].

(a) $L_f(a,r)$ is regular.

(b) $L_f(a,r)$ is isomorphic to a power series space if and only if f is slowly increasing in which case without loss of generality we can take f as the identity function. When f is slowly increasing we have that $L_f(a,\infty) \cong \Lambda_\infty(\alpha)$ and when $r < \infty$, $L_f(a,r) \cong \Lambda_1(\alpha)$.

(c) For a Dragilev function f,

$$L_f(a,r) \cong L_f(b,1) \text{ if } 0 < r < \infty, \text{ and } L_f(a,r) \cong L_f(b,-1) \text{ if } r < 0.$$

Hence basically there are four types of Dragilev spaces: $r = -1$, $r = 0$, $r = 1$, $r = \infty$. If the function is rapidly increasing, then the first two are of type (d_2) and the last two are of type (d_1).

(d) If f is rapidly increasing, then
$L_f(a,-1)$ and $L_f(a,1)$ are stable if and only if $\lim a_{2n}/a_n = 1$,
$L_f(a,0)$ and $L_f(a,\infty)$ are stable if and only if $\sup a_{2n}/a_n < \infty$,
$L_f(a,-1)$ and $L_f(a,1)$ are weakly stable if and only if $\lim a_{n+1}/a_n = 1$,
$L_f(a,0)$ and $L_f(a,\infty)$ are weakly stable if and only if $\sup a_{n+1}/a_n < \infty$,
$L_f(a,-1)$ and $L_f(a,1)$ are unstable if and only if $\liminf a_{n+1}/a_n > 1$,
$L_f(a,0)$ and $L_f(a,\infty)$ are unstable if and only if $\lim a_{n+1}/a_n = \infty$.

3.3. SOME DEFINITIONS

As a preparation for the forthcoming sections we give some definitions. Let E be a Fréchet space and f be a Dragilev function.

E has property $D_3(f)$ if $\exists j\ \forall M \geq 1\ \forall k\ \exists \ell$

$$Mf^{-1}\left(\log\frac{\|x\|_k}{\|x\|_j}\right) \leq f^{-1}\left(\log\frac{\|x\|_\ell}{\|x\|_k}\right),\ x \in E\setminus\{0\}.$$

E has property $D_4(f)$ if $\forall j\ \exists k > j\ \forall \ell > k\ \exists M \geq 1$

$$f^{-1}\left(\log\frac{\|y\|_k^*}{\|y\|_\ell^*}\right) \leq f^{-1}\left(\log\frac{\|y\|_j^*}{\|y\|_k^*}\right),\ y \in E_j'\setminus\{0\}.$$

The above definitions are due to H. Apiola [5].

Let $I^- = (0,1/2)$, $I^+ = (1/2,1)$ and let

$$U_k^\circ \subset \bigcap_{r\in\mathbf{R}}\left(e^{f((1-\lambda)r)}U_j^\circ + Ce^{-f(\lambda r)}U_\ell^\circ\right) \tag{1}$$

E has property

$S_0^-(f)$ if $\exists j\ \forall k\ \exists \ell\ \forall \lambda \in I^-\ \exists C$ (1) holds,

$S_1^-(f)$ if $\exists j\ \forall \lambda \in I^-\ \forall k\ \exists\ \ell\ \exists C$ (1) holds,

$S_2^-(f)$ if $\forall \lambda \in I^-\ \exists j\ \forall k\ \exists \ell\ \exists C$ (1) holds,

$S_3^-(f)$ if $\exists \lambda \in I^-\ \exists j\ \forall k\ \exists \ell\ \exists C$ (1) holds,

$S_4^-(f)$ if $\exists j\ \forall k\ \exists \ell\ \exists \lambda \in I^-\ \exists C$ (1) holds.

E has property $S_i^+(f)$ if it has property $S_i^-(f)$ with I^- replaced by I^+, $i = 0,\ldots,4$.

With the same quantifiers (1) is equivalent to each of the following

$$\|x\|_k \leq e^{f((1-\lambda)r)}\|x\|_j + Ce^{-f(\lambda r)}\|x\|_\ell,\ r \in \mathbf{R},\ x \in E,$$

$$f^{-1}\left(\log\frac{\|x\|_k}{\|x\|_j}\right) \leq \frac{1-\lambda}{\lambda}f^{-1}\left(\log\frac{C\|x\|_\ell}{\|x\|_k}\right),\ x \in E,$$

$$\bigcup_{r\in\mathbf{R}}\left(e^{-f((1-\lambda)r)}U_j \cap C^{-1}e^{f(\lambda r)}U_\ell\right) \subset U_k,$$

and if $E = K(a_{kn})$ then (1) is equivalent to

$$f^{-1}\left(\log\frac{a_{kn}}{a_{jn}}\right) \leq \frac{1-\lambda}{\lambda}f^{-1}\left(\log\frac{Ca_{\ell n}}{a_{kn}}\right),\ n \in \mathbf{N}.$$

Let

$$U_k \subset \bigcap_{r\in\mathbf{R}}\left(Ce^{f(\lambda r)}U_\ell + e^{-f((1-\lambda)r)}U_j\right) \tag{2}$$

E has property

$Q_1^-(f)$ if $\forall j\ \exists k\ \forall \ell\ \exists \lambda \in I^-\ \exists C$ (2) holds,

$Q_2^-(f)$ if $\forall j\ \exists k\ \exists \lambda \in I^-\ \forall \ell\ \exists C$ (2) holds,

$Q_3^-(f)$ if $\exists \lambda \in I^-\ \forall j\ \exists k\ \forall \ell\ \exists C$ (2) holds,

$Q_4^-(f)$ if $\forall \lambda \in I^-\ \forall j\ \exists k\ \forall \ell\ \exists C$ (2) holds,

$Q_5^-(f)$ if $\forall j\ \exists k\ \forall \lambda \in I^-\ \forall \ell\ \exists C$ (2) holds.

E has property $Q_i^+(f)$ if it has property $Q_i^-(f)$ with I^- replaced by I^+, $i = 1, \ldots, 5$. With the same quantifiers (2) is equivalent to each of the following

$$\|y\|_k^* \leq Ce^{f(\lambda r)}\|y\|_\ell^* + e^{-f((1-\lambda)r)}\|y\|_j^*, \ r \in \mathbf{R},\ y \in E',$$

$$f^{-1}\left(\log \frac{\|y\|_k^*}{C\|y\|_\ell^*}\right) \leq \frac{\lambda}{1-\lambda} f^{-1}\left(\log \frac{\|y\|_j^*}{\|y\|_k^*}\right),\ y \in E',$$

$$\bigcup_{r\in\mathbf{R}} \left(C^{-1}e^{-f(\lambda r)}U_\ell^\circ \cap e^{f((1-\lambda)r)}U_j^\circ\right) \subset U_k^\circ,$$

and if $E = K(a_{kn})$ then (2) is equivalent to

$$f^{-1}\left(\log \frac{a_{\ell n}}{Ca_{kn}}\right) \leq \frac{\lambda}{1-\lambda} f^{-1}\left(\log \frac{a_{kn}}{a_{jn}}\right),\ n \in \mathbf{N}.$$

We note that the condition $S_1^+(f)$ is equivalent to $D_3(f)$ and the condition $Q_1^+(f)$ is equivalent to $D_4(f)$.

We have the following relations:
$S_0^+(f) \Rightarrow \ldots \Rightarrow S_4^+(f) \Rightarrow S_0^-(f) \Rightarrow \ldots \Rightarrow S_4^-(f)$ and also $S_3^+(f) \Rightarrow S_1^+(f)$.
$Q_5^-(f) \Rightarrow \ldots \Rightarrow Q_1^-(f) \Rightarrow Q_5^+(f) \Rightarrow \ldots \Rightarrow Q_1^+(f)$, also $Q_3^-(f) \Rightarrow Q_4^-(f)$ and $Q_1^-(f) \Rightarrow Q_2^-(f)$.
Properties $S_i^\mp(f)$ and $Q_i^\mp(f)$ were defined by H. Ahonen [1].

3.4. SUBSPACES AND QUOTIENT SPACES

Since $L_f(a,r)$ is isomorphic to a power series space when f is slowly increasing, in this section (unless otherwise stated) we assume that f is rapidly increasing. In this case subspaces and quotient spaces have been studied by various researchers among whom we can cite E. Dubinsky, M. Alpseymen, N. DeGrande-DeKimpe, V.V. Kashirin, Z. Nurlu, H. Apiola and H. Ahonen.

In general there are partial results. Complete characterizations exist for nuclear and stable $L_f(a,\infty)$ and $L_f(a,0)$ spaces.

3.4.1. *Stable Case Characterizations.* In this subsection we assume that all Dragilev spaces are nuclear and stable. We have the following results.

(a) (H. Apiola [5]) A Fréchet space E is isomorphic to a subspace (resp. a quotient space) of $L_f(a,\infty)$ if and only if

$$\forall U\ \forall \rho > 0\ \exists V,\ \ d_n(V,U) \leq e^{-f(\rho a_n)},\ \text{n large}$$

and E has the property $D_3(f)$ (resp. $D_4(f)$).

We note that the subspace part of this result was shown earlier in [2] and [22] under the assumption that E has a basis.

(b) (H. Apiola [5]) A Fréchet space E is isomorphic to a complemented subspace of $L_f(a,\infty)$ if and only if E is isomorpic to a subspace and to a quotient space of $L_f(a,\infty)$.

(c) (H. Ahonen [1]) A Fréchet space E is isomorphic to a subspace (resp. a quotient space) of $L_f(a,0)$ if and only if

$$\forall U\ \exists V\ \exists \rho > 0\ \ d_n(V,U) \leq e^{-f(\rho a_n)},\ \text{n large}$$

and E has the property $S_4^-(f)$ (resp. $Q_4^-(f)$).

We note that the subspace part of this result was shown earlier in [22] under the assumption that E has a basis. In the proofs of (c), H. Ahonen constructs a particular exact sequence with precisely chosen maps and neighborhood bases following the splitting method of D. Vogt and his proof of the part concerning quotient spaces also uses the dead-end space method of B.S. Mitiagin adapted for the $L_f(a,0)$ case.

3.4.2. *The General Case.* In this subsection we assume that Dragilev functions are rapidly increasing (unless otherwise stated) and all Dragilev spaces are nuclear, but not necessarily stable. Here we summarize the results of V.V. Kashirin and Z. Nurlu.

V.V. Kashirin gave necessary and sufficient conditions on $a = (a_n)$ so that $L_f(a,r)$ has a subspace isomorphic to $L_g(b,s)$ for some b under the assumptions that $f = g$ or $g^{-1} \circ f$ is rapidly increasing or $f^{-1} \circ g$ is rapidly increasing.

The definition of the properties $(\alpha), \dots (\eta)$ as well as the Theorems (a),(b) and (c) below are due to V.V. Kashirin [29] (see also [31]).

Definition. The sequence $a = (a_n)$ is said to satisfy the condition (α) (resp. (β)) if there exist subsequences $(n(i))$ and $(k_j(i))$, $j \in \mathbf{N}$ of $\mathbf{N}$ such that for all $i \geq j$ the corresponding estimate is fulfilled:

(α) $j \leq a_{k_j(i)}/a_{n(i)} \leq C(j) < \infty,\ j \in \mathbf{N}$,

(β) $1 - (1/j) < a_{k_j(i)}/a_{n(i)} \leq C(j) < 1,\ j \in \mathbf{N}$.

(a) The space $L_f(a,r)$ contains a subspace isomorphic to some $L_f(b,s)$ if and only if the sequence (a_n) satisfies the corresponding condition in the table below:

	$s=0$	$s=-1$	$s=1$	$s=\infty$
$r=0$	always	(α)	(α)	(α)
$r=-1$	never	always	(β)	(α)
$r=1$	never	never	always	(α)
$r=\infty$	never	never	never	always

Definition. The sequence $a = (a_n)$ is said to satisfy the condition (γ) (resp. (δ), (ξ), (η)) if there exist subsequences $(n(i))$ and $(k_j(i))$, $j \in \mathbf{N}$ of $\mathbf{N}$ such that for all $i \geq j$ the corresponding estimate is fulfilled:

$(\gamma)\ 1 - (1/j) \leq a_{k_j(i)}/\varphi^{-1}(a_{n(i)}) \leq C(j) < 1,\ j \in \mathbf{N}$,

$(\delta)\ j \leq a_{k_j(i)}/\varphi^{-1}(a_{n(i)}) \leq C(j) < \infty,\ j \in \mathbf{N}$,

$(\xi)\ 1 - (1/j) \leq \varphi^{-1}(a_{k_j(i)})/a_{n(i)} \leq C(j) < 1,\ j \in \mathbf{N}$,

$(\eta)\ j \leq \varphi^{-1}(a_{k_j(i)})/a_{n(i)} \leq C(j) < \infty,\ j \in \mathbf{N}$

where depending on the context $f = g \circ \varphi$ or $g = f \circ \varphi$ and φ is rapidly increasing.

(b) Assume $f = g \circ \varphi$. Then the space $L_f(a, r)$ contains a subspace isomorphic to $L_g(b, s)$ for some b if and only if the sequence (a_n) satisfies the corresponding condition in the table below:

	$s = -1$	$s = 0$	$s = 1$	$s = \infty$
$r = -1$	(γ)	(γ)	(γ)	(γ)
$r = 0$	(δ)	(δ)	(δ)	(δ)

(c) Assume $g = f \circ \varphi$. Then the space $L_f(a, r)$ contains a subspace isomorphic to $L_g(b, s)$ for some b if and only if the sequence (a_n) satisfies the corresponding condition in the table below:

	$r = -1$	$r = 0$	$r = 1$	$r = \infty$
$s = 1$	(ξ)	(ξ)	(ξ)	(ξ)
$s = \infty$	(η)	(η)	(η)	(η)

Finally we summarize the results of Z. Nurlu (d) ...(h) which are taken from his thesis [49].

(d) Let f be a not necessarily rapidly increasing Dragilev fuction and assume $K(a_{kn}) = L_f(a, r)$ is (d_1) and unstable. Then a Köthe space E is isomorphic to a subspace of $L_f(a, r)$ if and only if

i) E is (d_1),

ii) for any (some) representation (b_{kn}) of E the following holds:

$\forall q\ \exists k_0 \geq q\ \forall k\ \exists p\ \exists k'\ \exists n_0\ \exists$ a sequence (m_n) of natural numbers such that $m_{n_1} \neq m_{n_2}$ for $n_1 \neq n_2$ such that:

$$\frac{b_{kn}}{b_{k_0 n}} \leq a_{p,m_n} \leq \frac{b_{k'n}}{b_{k_0 n}},\ n \geq n_0.$$

(e) Let f be a not necessarily rapidly increasing Dragilev function. A regular Köthe

space E is isomorphic to a subspace of $L_f(a,\infty)$ if and only if
i) for any (some) representation (b_{kn}) of E

$$\exists k_0\ \exists M > 1\ \forall k\ \exists k'\ \ \limsup_n \frac{f^{-1}(\log(b_{kn}/b_{k_0n}))}{f^{-1}(\log(b_{k'n}/b_{k_0n}))} < \frac{1}{M}$$

(hence in particular E is (d_1)),
ii) for any (some) regular and normalized (d_1) representation (b_{kn}) of E
$\forall k\ \exists p, k', n_k\ \exists$ a subsequence (m_n) of $\mathbf{N}$ such that

$$b_{kn} \le e^{f(pa_{m_n})} \le b_{k'n},\ n \ge n_k.$$

(f) Let f be a not necessarily rapidly increasing Dragilev function and assume $L_f(a,r)$ is (d_1). Then $L_f(b,r)$ is isomorphic to a subspace of $L_f(a,r)$ if and only if there is a subsequence (m_n) of $\mathbf{N}$ such that

$$0 < \inf b_n/a_{m_n} \le \sup b_n/a_{m_n} < \infty.$$

We note that this result was proved in [37] by V.P. Kondakov without the assumption of (d_1).

(g) Let f and g be not necessarily rapidly increasing Dragilev functions such that $f^{-1} \circ g$ is rapidly increasing. Then a (d_1) space $L_g(b,s)$ is isomorphic to a subspace of a (d_1) space $L_f(a,r)$ if and only if
$\exists q_0 \in (0,r)\ \forall p \in (0,s)\ \exists$ a subsequence (m_n) of $\mathbf{N}$, $\exists p' \in (0,s)\ \exists n_0$ such that

$$p \le \frac{g^{-1} \circ f(q_0 a_{m_n})}{b_n} \le p',\ n \ge n_0.$$

(h) Let f be rapidly increasing. Then
i) $L_f(b,\infty)$ is isomorphic to a subspace of $\Lambda_\infty(a)$ if and only if
$\forall k\ \exists$ a subsequence (m_n) of $\mathbf{N}$ and $M_k > 0$ such that

$$k \le \frac{f^{-1}(a_{m_n})}{b_n} \le M_k,\ \text{large } n.$$

ii) $L_f(b,1)$ is isomorphic to a subspace of $\Lambda_\infty(a)$ if and only if
$\forall p \in (0,1)\ \exists$ a subsequence (m_n) of $\mathbf{N}$ and $\exists p' \in (0,1)$ such that

$$p \le \frac{f^{-1}(a_{m_n})}{b_n} \le p',\ \text{large } n.$$

3.5. SOME OTHER FACTS ON DRAGILEV SPACES

In this section we summarize some results on Dragilev spaces which do not belong in the previous sections.

(a) While searching the frequency of occurence of power series subspaces of NFS, E. Dubinsky [18] constructed in every NFS not isomorphic to ω a subspace which does not contain (up to isomorphism) any power series subspace. These constructed subspaces turn out to be $L_f(a,\infty)$ spaces where f is rapidly increasing. H. Ahonen [1] proved an analogous result for quotient spaces, namely that every NFS not isomorphic to ω has a quotient space isomorphic to some $L_f(a,0)$ where f is rapidly increasing.

(b) Let $L_f(a,\infty)$ be stable and $E = K(b_{kn})$ satisfy

$$\exists M > 1 : \quad Mf^{-1}\left(\log \frac{b_{k+1,n}}{b_{kn}}\right) \leq f^{-1}\left(\log \frac{b_{k+2,n}}{b_{k+1,n}}\right), \quad n,k \in \mathbf{N}.$$

Then E has a complemented subspace isomorphic to a subspace of $L_f(a,\infty)$. (M. Alpseymen, M.S. Ramanujan, T. Terzioğlu [3].)

(c) H. Apiola has shown that Schwartz $L_f(a,\infty)$ spaces are quotient universal for NFS [4] in the sense that for every NFS E there is a stable Schwartz $L_f(a,\infty)$ such that E is isomorphic to a quotient space of this $L_f(a,\infty)$.

(d) H. Ahonen [1] has obtained the following result by using a Mitiagin-Henkin type interpolation argument:

Let f be a not necessarily rapidly increasing Dragilev function. Then a NFS E is isomorphic to $L_f(a,0)$ for some a if and only if E has properties $S_4^-(f)$ and $Q_4^-(f)$.

By combining this result with the fact that $S_4^-(f)$ (resp. $Q_4^-(f)$) passes to subspaces (resp. quotient spaces) we get that if a NFS E is isomorphic to a subspace of some $L_f(a,0)$ and a quotient space of some $L_f(b,0)$ then E has a basis. E. Dubinsky obtained this result in [20] but with some restrictions on f.

(e) H. Ahonen gives necessary and sufficient conditions in [1] in terms of the properties $S_i^{\mp}(f)$ and $Q_i^{\mp}(f)$ for a NFS with a basis to be isomorphic to some $L_f(a,r)$. We state these results below:

Theorem. Let E be a NFS with basis. Then

i) E is isomorphic to $L_f(a,\infty)$ for some a if and only if E has $S_1^+(f)$ and $Q_1^+(f)$ (where f is rapidly increasing or f=id.)

ii) E is isomorphic to $L_f(a,1)$ for some a if and only if E has $S_4^+(f)$ and $Q_4^+(f)$ (where f is rapidly increasing.)

iii) E is isomorphic to $L_f(a,-1)$ for some a if and only if E has $S_2^-(f)$ and $Q_2^-(f)$ (where f is rapidly increasing.)

The case of $L_f(a,0)$ is omitted in this list on purpose since in this case we had already mentioned a stronger result in (d), i.e. without assuming a basis.

(f) In [15], M.M. Dragilev has claimed that for rapidly increasing Dragilev functions the intersection (up to isomorphism) of classes of Dragilev spaces corresponding to different values of $r \in \{-1,0,1,\infty\}$ is empty. This claim is shown to be false by M. Kocatepe (Alpseymen) in [32] and [33] where she shows that there are spaces $L_f(a,1)$ which are isomorphic to some $L_g(b,\infty)$ and respectevily there are spaces $L_f(a,-1)$ which are isomorphic to some $L_g(b,0)$.

Below we state her results on characterizations of these intersection classes:

(g) (M. Kocatepe [34]) Let f be a rapidly increasing Dragilev function. Then $L_f(a,1)$ is isomorphic to some $L_g(b,\infty)$ if and only if
i) $\mathrm{Ext}(L_f(a,1), L_f(a,1)) = 0$,
ii) there is a strictly increasing sequence (r_k) with $\lim r_k = 1$ and a subsequence (n_k) of N such that

$$\frac{f(r_{k+1}a_n)}{f(r_k a_n)} \leq \frac{f(r_{k+2}a_n)}{f(r_{k+1}a_n)}, \; k \in \mathbf{N}, n \geq n_k.$$

Note that there is no nuclearity assumption here.

(h) (M. Kocatepe [35]) Let f be a rapidly increasing Dragilev function and assume that $L_f(a,-1)$ is nuclear. Then $L_f(a,-1)$ is isomorphic to some $L_g(b,0)$ if and only if the following conditions hold: there is a subsequence (a_{i_n}) of (a_n) and a sequence (s_k) which strictly decreases to 1 such that

i) $\liminf \frac{a_{i_{n+1}}}{a_{i_n}} > 1$,

ii) $\lim \frac{a_{i_{n+1}-1}}{a_{i_n}} = 1$,

iii) $\frac{f(s_1 a_{i_n})}{f(s_2 a_{i_n})} \leq \frac{f(s_k a_{i_{n+1}})}{f(s_{k+1} a_{i_{n+1}})}$, $k \in \mathbf{N}$, $n \geq n_k$ for some sequence (n_k) of indices.

(i) (M. Kocatepe [35]) Let g be a rapidly increasing Dragilev function and assume that $L_g(b,0)$ is nuclear. Then $L_g(b,0)$ is isomorphic to some $L_f(a,-1)$ if and only if the following conditions hold:
i) $\mathrm{Ext}(L_g(b,0), L_g(b,0)) = 0$,
ii) there is a sequence (s_k) which strictly decreases to 0 and a collection $(n_{k,u})$ of indices such that

$$\frac{g(s_{k+1}b_n)}{g(s_u b_n)} \leq \frac{g(s_k b_n)}{g(s_{k+1}b_n)}, \; k \geq 1, u > k+1, n \geq n_{k,u}.$$

4. The Functor Ext

The extensive use of the functor Ext in characterizing subspaces and quotient spaces of Köthe spaces has been initiated by D. Vogt. In the category of Fréchet spaces and continuous linear maps $\mathrm{Ext}(E,F) = 0$ if and only if every short exact sequence

$$0 \mapsto F \mapsto G \mapsto E \mapsto 0$$

splits.

As the first result in this direction we have we have a theorem due to D. Vogt [53] which states that if E is a NFS with the property (DN) then $\mathrm{Ext}(E,(s)) = 0$. Then D. Vogt and M.J. Wagner [60] have shown that if E and F are NFS with E having the property (DN) and F having the property (Ω), then $\mathrm{Ext}(E,F) = 0$. Later in [5] H. Apiola has defined the conditon (S) and by using this relation has shown that $\mathrm{Ext}(L_f(a,\infty), L_f(b,\infty)) = 0$. In order to establish a unified theory for the functor Ext in [58] D. Vogt has defined the conditions (S_1^*) and (S_2^*) where (S_1^*) is essentially the same as the condition (S) of H. Apiola.

Definition. Let E and F be Fréchet spaces. We say that the pair (E, F) satisfies (S_1^*) (resp. (S_2^*)) if the corresponding condition below is fulfilled:

(S_1^*) $\exists n_0\ \forall \mu\ \exists k\ \forall K, m\ \exists n, S\ \forall x \in E_n, y \in F_k'$

$$\|x\|_m \|y\|_k^* \leq S(\|x\|_n \|y\|_K^* + \|x\|_{n_0} \|y\|_\mu^*)$$

(S_2^*) $\forall \mu\ \exists n_0, k\ \forall K, m\ \exists n, S\ \forall x \in E_n, y \in F_k'$

$$\|x\|_m \|y\|_k^* \leq S(\|x\|_n \|y\|_K^* + \|x\|_{n_0} \|y\|_\mu^*)$$

In the same paper D. Vogt has shown that if $E = K(a_{kn})$ or nuclear or $F = K^\infty(a_{kn})$ or nuclear then

$$(S_1^*) \Rightarrow \operatorname{Ext}(E, F) = 0 \Rightarrow (S_2^*).$$

When $E = K(a_{ki})$ and $F = K(b_{kj})$ are Köthe spaces, the conditions (S_1^*) and (S_2^*) take the following form:

(S_1^*) $\exists n_0\ \forall \mu\ \exists k\ \forall K, m\ \exists n, S\ \forall i, j$

$$\frac{a_{mi}}{b_{kj}} \leq S \max\left(\frac{a_{ni}}{b_{Kj}}, \frac{a_{n_0 i}}{b_{\mu j}}\right)$$

(S_2^*) $\forall \mu\ \exists n_0, k\ \forall K, m\ \exists n, S\ \forall i, j$

$$\frac{a_{mi}}{b_{kj}} \leq S \max\left(\frac{a_{ni}}{b_{Kj}}, \frac{a_{n_0 i}}{b_{\mu j}}\right)$$

In [39] J. Krone and D. Vogt have shown that if E and F are both Köthe spaces then $\operatorname{Ext}(E, F) = 0$ if and only if the pair (E, F) satisfies (S_2^*).

D. Vogt in [58] has applied the general characterization to the particular case in which E or F is a power series space and has obtained the following results:

(a) Assume $\sup(a_{n+1}/a_n) < \infty$. Then $\operatorname{Ext}(\Lambda_\infty(a), F) = 0$ if and only if F has the property (Ω).

(b) Assume $\lim(a_{n+1}/a_n) = 1$. Then $\operatorname{Ext}(\Lambda_1(a), F) = 0$ if and only if F has the property $(\overline{\overline{\Omega}})$,

where a Fréchet space F is said to have the property $(\overline{\overline{\Omega}})$ if

$$\forall p\ \exists q\ \forall k, \varepsilon > 0\ \exists C\ \forall y \in F' : (\|y\|_q^*)^{1+\varepsilon} \leq C \|y\|_k^* (\|y\|_p^*)^\varepsilon.$$

(c) Assume $\sup(a_{n+1}/a_n) < \infty$. Then $\operatorname{Ext}(E, \Lambda_r^\infty(a)) = 0$ if and only if E has the property (DN).

(d) Assume that either $\lim(a_{n+1}/a_n) = 1$ or $\sup(b_{n+1}/b_n) < \infty$. Then $\operatorname{Ext}(\Lambda_r(a), \Lambda_s(b)) = 0$ if and only if $r = \infty$.

(e) In (a) ...(d) the respective conditions are sufficient for $\operatorname{Ext}(E, F) = 0$ or $\operatorname{Ext}(\Lambda_r(a), \Lambda_s(b)) = 0$ without the assumptions on $a = (a_n)$ or $b = (b_n)$.

In his thesis [27], J. Hebbecker applies D. Vogt's general conditions to the particular case in which E and F are both power series spaces or both Dragilev spaces. His characterizations are given in terms of the following conditions.

Notation. Let $a = (a_n)$ and $b = (b_n)$ be two exponent sequences. We say

$(a,b) \in HP$ if the set of finite limit points of the set $\{a_m/b_n : m,n \in \mathbf{N}\}$ is bounded,

$(a,b) \in HP_\infty$ if the set of finite limit points of the set $\{a_m/b_n : m,n \in \mathbf{N}\}$ consists of 0,

$(a,b) \in HP \not\asymp 1$ if there is $c \in [0,1)$ such that the set of finite limit points of the set $\{a_m/b_n : m,n \in \mathbf{N}\}$ is contained in $[0,c] \cup [1,\infty)$,

$(a,b) \in HP \neq 1$ if 1 is not a limit point of the set $\{a_m/b_n : m,n \in \mathbf{N}\}$.

J. Hebbecker [27] obtains the following results:

(a) $(a,b) \in HP \iff (\Lambda_1(a), \Lambda_\infty(b))$ satisfies $(S_1^*) \iff (\Lambda_1(a), \Lambda_\infty(b))$ satisfies (S_2^*).

(b) $(a,b) \in HP_\infty \iff (\Lambda_1(a), \Lambda_1(b))$ satisfies (S_1^*).

(c) $(a,b) \in HP \iff (\Lambda_1(a), \Lambda_1(b))$ satisfies (S_2^*).

Next J. Hebbecker considers pairs of Dragilev spaces which are defined by rapidly increasing Dragilev functions.

Notation. Let f and g be rapidly increasing Dragilev functions. We write

$f \succ g$ if $g^{-1} \circ f$ is a rapidly increasing function,

$f \approx g$ if $g^{-1} \circ f$ and $f^{-1} \circ g$ are both logarithmically convex and slowly increasing.

The rest of the results in this chapter are due to J. Hebbecker [27].

(d) Let f and g be rapidly increasing Dragilev functions. Then the pair $(L_f(a,r), L_g(b,s))$ has (S_1^*) or (S_2^*) if and only if the pair $(g^{-1}f(a), b)$ of sequences satisfies the corresponding condition in the table below:

	$r=\infty, s=\infty$	$r=\infty, s=1$	$r=1, s=1$	$r=1, s=\infty$
$f \succ g$	always	always	always	always
$f \approx g$	always	always	$HP \not\asymp 1$	HP
$g \succ f$	HP	$HP \not\asymp 1$	$HP \not\asymp 1$	HP

(e) Let f and g be rapidly increasing Dragilev functions. Then the pair $(L_f(a,r), L_g(b,s))$ has (S_1^*) if and only if the pair $(a, f^{-1}g(b))$ of sequences satisfies the corresponding condition in the table below:

	$r=0,s=0$	$r=0,s=-1$	$r=-1,s=-1$	$r=-1,s=0$
$f \succ g$	HP	HP	$HP \not\times 1$	$HP \not\times 1$
$f \approx g$	HP_∞	HP	$HP \neq 1$	always
$g \succ f$	always	always	always	always

(f) Let f and g be rapidly increasing Dragilev functions. Then the pair $(L_f(a,r), L_g(b,s))$ has (S_2^*) if and only if the pair $(a, f^{-1}g(b))$ of sequences satisfies the corresponding condition in the table below:

	$r=0,s=0$	$r=0,s=-1$	$r=-1,s=-1$	$r=-1,s=0$
$f \succ g$	HP	HP	$HP \not\times 1$	$HP \not\times 1$
$f \approx g$	HP	HP	always	always
$g \succ f$	always	always	always	always

(g) Let f and g be rapidly increasing Dragilev functions. Then the pair $L_f(a,r), L_g(b,s))$ has (S_1^*) or (S_2^*) if and only if the corresponding condition in the table below is satisfied:

	$r=0,s=\infty$	$r=0,s=1$
$f \succ g$	$(a, f^{-1}g(b)) \in HP$	$(a, f^{-1}g(b)) \in HP$
$f \approx g$	$(a, f^{-1}g(b)) \in HP$	$(a, f^{-1}g(b)) \in HP$
$g \succ f$	$(g^{-1}f(a), b) \in HP$	$(g^{-1}f(a), b) \in HP \not\times 1$

	$r=-1,s=1$	$r=-1,s=\infty$
$f \succ g$	$(a, f^{-1}g(b)) \in HP \not\times 1$	$(a, f^{-1}g(b)) \in HP \not\times 1$
$f \approx g$	$(a, f^{-1}g(b)) \in HP \not\times 1$	$(g^{-1}f(a), b) \in HP$
$g \succ f$	$(g^{-1}f(a), b) \in HP \not\times 1$	$(g^{-1}f(a), b) \in HP$

5. Compact Operators

Let E and F be Fréchet spaces. We write $(E,F) \in \mathcal{K}$ if all continuous linear maps $T : E \mapsto F$ are compact. The relation $\mathcal{K}$ has been investigated by various researchers. As the first result in this area we have the celebrated theorem of V.P. Zahariuta [63] which states that if E is (d_2) and F is (d_1) and Montel then $(E,F) \in \mathcal{K}$.

5.1. CASE OF POWER SERIES SPACES

(a) By the above mentioned theorem of V.P. Zahariuta we have that $(\Lambda_1(a), \Lambda_\infty(b)) \in$

$\mathcal{K}$ for all exponent sequences a and b. But E. Dubinsky [16] showed that the converse of the above relation is not true. His result states that for any a there is b such that $(\Lambda_\infty(a), \Lambda_1(b)) \notin \mathcal{K}$.

(b) The relation $\mathcal{K}$ for pairs of nuclear power series spaces of the same type has been studied by L. Crone and W. Robinson [9]. They have proved the following result:

Let $r = 1$ or $r = \infty$ and assume $\Lambda_r(a)$ and $\Lambda_r(b)$ are nuclear. Then the following are equivalent:
i) $(\Lambda_r(a), \Lambda_r(b)) \in \mathcal{K}$
ii) $(\Lambda_r(b), \Lambda_r(a)) \in \mathcal{K}$
iii) $\lim(b_{m_i}/a_{n_i-1}) = \lim(a_{n_i}/b_{m_{i+1}-1}) = \infty$
where $\ldots \leq a_{n_i-1} < b_{m_i} \leq \ldots \leq b_{m_{i+1}-1} \leq a_{n_i} \leq \ldots \leq a_{n_{i+1}-1} < \ldots$

(c) Finally we have the following result of N. DeGrande-DeKimpe and W. Robinson [13].

Let $\Lambda_\infty(a)$ and $\Lambda_1(b)$ be nuclear. Then $(\Lambda_\infty(a), \Lambda_1(b)) \in \mathcal{K}$ if and only if there is a non-decreasing sequence (k_j) of positive integers such that

$$\lim(b_{k_j+1}/a_j) = \infty \text{ and } \sup(b_{k_j}/a_j) < \infty.$$

5.2. CASE OF DRAGILEV SPACES

(a) The following theorem of V.P. Zahariuta on the relation $\mathcal{K}$ for pairs of Dragilev spaces appeared in [63].

Let f and g be Dragilev functions and assume that $f^{-1} \circ g$ is rapidly increasing. Then
i) For $r, s \in (0, \infty]$, $(L_f(a, r), L_g(b, s)) \in \mathcal{K}$.
ii) For $r, s \in (-\infty, 0]$, $(L_g(b, s), L_f(a, r)) \in \mathcal{K}$.
iii) $(L_f(a, 1), L_f(b, \infty)) \in \mathcal{K}$.
iv) If f is rapidly increasing, then $(L_f(a, 0), L_f(b, -1)) \in \mathcal{K}$.

(b) N. DeGrande-DeKimpe in [11] and [12] considered pairs of Dragilev spaces of the same type for which the Dragilev functions have the same growth rate. She obtained the following results.

Let f and g be rapidly increasing and assume $f^{-1} \circ g$ and $g^{-1} \circ f$ are both slowly increasing. Assume also all of the spaces under consideration are nuclear. Let (p_i) and (q_i) be increasing sequences of indices such that

$$\ldots \leq f(a_{p_i-1}) < g(b_{q_i}) \leq \ldots \leq g(b_{q_{i+1}-1}) \leq f(a_{p_i}) \leq \ldots \leq f(a_{p_{i+1}-1}) < \ldots$$

(α) The following are equivalent:
i) $(L_f(a, \infty), L_g(b, \infty)) \in \mathcal{K}$
ii) $(L_g(b, \infty), L_f(a, \infty)) \in \mathcal{K}$
iii) $\exists k_1 \ \forall m \ \lim(g(mb_{q_{i+1}-1})/f(k_1 a_{p_i})) = 0$ and $\exists m_1 \ \forall k \ \lim(f(ka_{p_i-1})/g(m_1 b_{q_i})) = 0$.

(β) The following are equivalent:
i) $(L_f(a, 1), L_g(b, 1)) \in \mathcal{K}$
ii) $(L_g(b, 1), L_f(a, 1)) \in \mathcal{K}$
iii) $\exists r \in (0, 1) \ \lim(g(b_{q_{i+1}-1})/f(ra_{p_i})) = 0$ and $\exists s \in (0, 1) \ \lim(f(a_{p_i-1})/g(sb_{q_i})) = 0$.

(c) In [29] and [31] V.V. Kashirin considered pairs of Dragilev spaces which are defined either by the same function or by the functions with different growth rates and he gave necessary and sufficient conditions in terms of a and b so that $(L_f(a,r), L_g(b,s)) \in \mathcal{K}$. The following definitions and theorems appeared in [29] (see also [31]).

Definition. The ordered pair of sequences (a,b) is said to satisfy the condition (A) (resp. (B),(C),(D)) if the corresponding estimate is fulfilled:

(A) $\forall M > 0\ \exists (k_n), k_n \to \infty :\ M \le \liminf(b_{k_n}/a_n) < \infty$,

(B) $\forall \delta \in (0,1)\ \exists (k_n), k_n \to \infty :\ 1 - \delta < \liminf(b_{k_n}/a_n) < 1$,

(C) $\exists (k_n), k_n \to \infty,\ \exists (i_n), i_n \to \infty :\ 0 < \lim(b_{k_n}/a_{i_n}) < \infty$,

(D) $\exists (k_n), k_n \to \infty,\ \exists (i_n), i_n \to \infty :\ \lim(b_{k_n}/a_{i_n}) = 1$.

Theorem. Assume f is rapidly increasing. Then $(L_f(a,r), L_f(b,s)) \in \mathcal{K}$ if and only if the pair (a,b) satisfies the corresponding condition in the table below:

	$s = 0$	$s = -1$	$s = 1$	$s = \infty$
$r = 0$	non (C)	always	always	always
$r = -1$	non (A)	non (D)	always	always
$r = 1$	non (A)	non (B)	non (D)	always
$r = \infty$	non (A)	non (A)	non (A)	non (C)

Next he considers pairs of Dragilev spaces where the functions have different growth rates.

Definition. Let φ be an odd rapidly increasing function. The ordered pair (a,b) of sequences is said to satisfy the condition (α) (resp. $(\beta),(\gamma),(\delta)$) if the corresponding estimate is fulfilled (we note that V.V. Kashirin calls these conditions (a),(b),(c) and (d) respectively):

(α) $\forall \delta \in (0,1)\ \exists (k_n), k_n \to \infty :\ 1 - \delta < \liminf(b_{k_n}/\varphi^{-1}(a_n)) < 1$,

(β) $\forall M > 0\ \exists (k_n), k_n \to \infty :\ M < \liminf(b_{k_n}/\varphi^{-1}(a_n) < \infty$,

(γ) $\forall \delta \in (0,1)\ \exists (k_n), k_n \to \infty :\ 1 - \delta < \liminf(\varphi^{-1}(a_{k_n})/b_n) < 1$,

(δ) $\forall M > 0\ \exists (k_n), k_n \to \infty :\ M < \liminf(\varphi^{-1}(a_{k_n})/b_n) < \infty$.

Theorem. Assume f and g are rapidly increasing and $\varphi = f^{-1} \circ g$ is rapidly increasing. Then $(L_f(a,r), L_g(b,s)) \in \mathcal{K}$ if and only if the pair (a,b) of sequences satisfies the corresponding condition in the table below:

	$s = -1$	$s = 0$	$s = 1$	$s = \infty$
$r = -1$	non (α)	non (β)	always	always
$r = 0$	non (α)	non (β)	always	always
$r = 1$	non (α)	non (β)	always	always
$r = \infty$	non (α)	non (β)	always	always

Theorem. Assume f and g are rapidly increasing and $\varphi = f^{-1} \circ g$ is rapidly increasing. Then $L_g(b,s), L_f(a,r)) \in \mathcal{K}$ if and only if the pair (a,b) of sequences satisfies the corresponding condition in the table below:

	$r=-1$	$r=0$	$r=1$	$r=\infty$
$s=-1$	always	always	always	always
$s=0$	always	always	always	always
$s=1$	non (γ)	non (γ)	non (γ)	non (γ)
$s=\infty$	non (δ)	non (δ)	non (δ)	non (δ)

(d) Finally we summarize the work of H. Faeskorn [25] in which she characterized the relation $\mathcal{K}$ for pairs of spaces one of which is a Dragilev space. In doing so she applied the following general conditions due to D. Vogt to the particular cases.

For two Fréchet spaces E and F, $L(E,F)$ denotes the set of all continuous linear maps $E \mapsto F$ and $LB(E,F)$ denotes the set of all bounded linear maps $E \mapsto F$. If in particular F is a Montel space, then $L(E,F) = LB(E,F)$ if and only if $(E,F) \in \mathcal{K}$.

Theorem (D. Vogt [56]). Let E and F be Fréchet spaces.

(α) The following are equivalent:

i) $L(K(b_{kn}),F) = LB(K(b_{kn}),F)$

ii) $\forall L(K) \nearrow \ \exists k \ \forall \ell \ \exists K_0, C \ \forall x \in F, n \in \mathbb{N}$

$$\frac{\|x\|_\ell}{b_{kn}} \leq \max\left\{\frac{\|x\|_K}{b_{L(K),n}} : 1 \leq K \leq K_0\right\}.$$

(β) The following are equivalent:

i) $L(E, K^\infty(a_{kn})) = LB(E, K^\infty(a_{kn}))$

ii) $\forall L(K) \nearrow \ \exists k \ \forall \ell \ \exists K_0, C \ \forall n \in \mathbb{N}, y \in E'$

$$a_{\ell n}\|y\|_k^* \leq C \max\left\{a_{Kn}\|y\|_{L(K)}^* : 1 \leq K \leq K_0\right\}.$$

Now we list the characterizations of H. Faeskorn [25] (see also [26]).

Theorem. In the following assume that all Dragilev spaces are weakly stable and E and F are Fréchet spaces.

(α) The following are equivalent:

i) $L(L_f(a,\infty),F) = LB(L_f(a,\infty),F)$

ii) $\forall \rho_n \uparrow \infty \ \exists k \ \forall \ell \ \exists K_0, C \ \forall x \in F \ \exists K_2, \ \ell < K_2 \leq K_0$

$$f^{-1}\log\left(\frac{C\|x\|_{K_2}}{\|x\|_\ell}\right) \geq \rho_{K_2} f^{-1}\log\left(\frac{\|x\|_\ell}{C\|x\|_k}\right)$$

(β) The following are equivalent:
i) $L(L_f(a,1),F) = LB(L_f(a,1),F)$
ii) $\exists k\ \exists M > 1\ \forall \ell\ \exists K_0, C\ \forall x \in F$

$$f^{-1}\log\left(\frac{C\|x\|_{K_0}}{\|x\|_\ell}\right) \geq M f^{-1}\log\left(\frac{\|x\|_\ell}{C\|x\|_k}\right)$$

(γ) The following are equivalent:
i) $L(L_f(a,0),F) = LB(L_f(a,0),F)$
ii) $\exists k\ \exists M > 1\ \forall \ell\ \exists K_0, C\ \forall x \in F$

$$M f^{-1}\log\left(\frac{C\|x\|_{K_0}}{\|x\|_\ell}\right) \geq f^{-1}\log\left(\frac{\|x\|_\ell}{C\|x\|_k}\right)$$

(δ) The following are equivalent:
i) $L(L_f(a,-1),F) = LB(L_f(a,-1),F)$
ii) $\forall \rho_N \downarrow 1\ \exists k_0\ \forall \ell\ \exists K_0, C\ \forall x \in F\ \exists K_1,\ 1 \leq K_1 < k_0$

$$\rho_{K_1} f^{-1}\log\left(\frac{C\|x\|_{K_0}}{\|x\|_\ell}\right) \geq f^{-1}\log\left(\frac{\|x\|_\ell}{C\|x\|_{K_1}}\right)$$

(ε) The following are equivalent:
i) $L(E, L_f^\infty(a,\infty)) = LB(E, L_f^\infty(a,\infty))$
ii) $\forall \rho_N \uparrow \infty\ \forall p\ \exists q\ \forall k_0\ \exists K_0, C\ \forall y \in E'\ \exists K_2,\ k_0 < K_2 \leq K_0$

$$f^{-1}\log\left(\frac{\|y\|_q^*}{C\|y\|_{K_2}^*}\right) \leq \rho_{K_2} f^{-1}\log\left(\frac{C\|y\|_p^*}{\|y\|_q^*}\right)$$

(ζ) The following are equivalent:
i) $L(E, L_f^\infty(a,1)) = LB(E, L_f^\infty(a,1))$
ii) $\forall p\ \exists q\ \forall M > 1\ \forall k_0\ \exists C\ \forall y \in E'$

$$f^{-1}\log\left(\frac{\|y\|_q^*}{C\|y\|_{k_0}^*}\right) \leq M f^{-1}\log\left(\frac{C\|y\|_p^*}{\|y\|_q^*}\right)$$

(η) The following are equivalent:
i) $L(E, L_f^\infty(a,0)) = LB(E, L_f^\infty(a,0))$
ii) $\forall p\ \exists q\ \forall M > 1\ \forall k_0\ \exists C\ \forall y \in E'$

$$M f^{-1}\log\left(\frac{\|y\|_q^*}{C\|y\|_{k_0}^*}\right) \leq f^{-1}\log\left(\frac{C\|y\|_p^*}{\|y\|_q^*}\right)$$

(θ) The following are equivalent:
i) $L(E, L_f^\infty(a,-1)) = LB(E, L_f^\infty(a,-1))$
ii) $\forall \rho_N \downarrow 1\ \forall p\ \exists q\ \forall k_0\ \exists C\ \forall y \in E'\ \exists K_1,\ p \leq K_1 < q$

$$\rho_{K_1} f^{-1}\log\left(\frac{\|y\|_q^*}{C\|y\|_{k_0}^*}\right) \leq f^{-1}\log\left(\frac{C\|y\|_{K_1}^*}{\|y\|_q^*}\right)$$

Next by using conditions ii) in the previous theorem, H. Faeskorn gives a complete characterization of $L(L_f(a,r), L_g^\infty(b,s))=LB(L_f(a,r), L_g^\infty(b,s))$ in terms of f and g under the assumption that both spaces are weakly stable. For this purpose she uses a slightly weaker formulation of logarithmic convexity.

Definition. A strictly increasing, odd function $h : \mathbf{R} \mapsto \mathbf{R}$ is said to be **semi logarithmically convex** if there are $M > 1$, $B > 1$ such that $\liminf_{x\to\infty}(h(Bx)/h(x)) \geq M$.

Then she obtains the following result:

Theorem. Assume $L_f(a,r)$ and $L_g^\infty(b,s)$ are weakly stable. Then $L(L_f(a,r), L_g^\infty(b,s)) =LB(L_f(a,r), L_g^\infty(b,s))$ if and only if f and g satisfy the corresponding condition in the table below:

	$s=-1$	$s=0$	$s=1$	$s=\infty$
$r=-1$	$g^{-1}f$ rap. inc.	$g^{-1}f$ rap. inc.	always	always
$r=0$	$g^{-1}f$ semi log. conv.	$g^{-1}f$ rap. inc.	always	always
$r=1$	never	never	$f^{-1}g$ rap. inc.	$f^{-1}g$ semi log. conv.
$r=\infty$	never	never	$f^{-1}g$ rap. inc.	$f^{-1}g$ rap. inc.

6. Köthe Spaces Defined by Transition Functions

The Vogt method emphasized the importance of knowing when $\mathrm{Ext}(K(a_{kn}), K(a_{kn})) = 0$. In fact D. Vogt characterized this case in terms of concrete conditions (S_1^*) and (S_2^*), but these are too involved. J. Krone in his Diplomarbeit [38] attacked the problem and was able to get a partial answer, namely in the case when the space has (DN).

The Köthe space $\lambda_\Phi(a)$, where $a = (a_n)$ is an increasing sequence of positive numbers with $\lim a_n = \infty$, $\Phi = (\phi_k)$ is a sequence of increasing functions $\phi_k : \mathbf{R}^+ \mapsto \mathbf{R}^+$ satisfying

$$x^2 \leq \phi_1(x) \leq \ldots \leq \phi_k(x) \leq \phi_{k+1}(x) \leq \ldots$$

for all large enough x, denotes the space $K(a_{kn})$ where $a_{kn} = \phi_k \circ \ldots \circ \phi_1(a_n)$. It is easy to see that $\lambda_\Phi(a)$ is Schwartz, has (DN) and Φ can be modified so that $\lambda_\Phi(a)$ becomes regular.

J. Krone proves the following theorem:

Let E be a Schwartz Köthe space. Then the following are equivalent:
i) (E,E) has (S_1^*)
ii) (E,E) has (S_2^*) and E has (DN)
iii) $E \cong \lambda_\Phi(a)$ (up to a diagonal transform and permutation)

Constructing appropriate exact sequences and following D. Vogt's splitting method, J. Krone also characterizes sub- and quotient spaces of $\lambda_\Phi(a)$ in terms of the (DN_Φ) and (Ω_Φ) conditions (see [38]).

For an application and examples of $\lambda_\Phi(a)$ spaces see the section on $S_g(a,r)$-spaces.

7. $S_g(a,r)$-spaces

$S_g(a,r)$ spaces were introduced by V.V. Kashirin [30] in 1980 to refute a conjecture of M.M. Dragilev who had asked whether every regular (d_1) or (d_2) type nuclear Köthe space was isomorphic to some L_f-space.

An $S_g(a,r)$ space is defined similar to an $L_f(a,r)$ but convexity of $g(x)$ for $x \geq 0$ is required instead of logarithmic convexity. Hence "rapidly or slowly increasing" has no meaning anymore.

The following properties of $S_g(a,r)$ spaces are either immediate or their proofs can be found in Z. Nurlu [51].

Let $E = S_g(a,r)$. Then

(a) E is regular.

(b) E belongs to one of the four types:
i) if $r < 0$, then $E \cong S_g(b,-1)$ and E is (d_2)
ii) if $r = 0$, then E is (d_2)
iii) if $0 < r < \infty$, then $E \cong S_g(b,1)$ and E is (d_5)
iv) if $r = \infty$, then E is (d_1).

(c) Every $L_f(a,r)$ space is isomorphic to some $S_g(a,r)$.

(d) Converse of (c) is false, in fact there are $S_g(b,1)$ spaces of type $E_1 \times E_2$ where E_i is of type (d_i), i=1,2. So $S_g(b,1)$ spaces deserve special attention. If an $S_g(b,1)$ space is also (d_1), then g exhibits some rapidly increasing behaviour.

In [30] V.V. Kashirin showed that every unstable Köthe space of type (d_1) (resp. (d_2)) is isomorphic to some $S_g(b,\infty)$ (resp. $S_g(b,0)$).

In [51] Z. Nurlu studied $S_g(a,r)$ spaces in connection with the functor Ext and obtained the following results:

Theorem. Let E be a nuclear and (d_1) Köthe space. Then the following are equivalent:
i) E is a repeated form of an unstable Köthe space where each coordinate is repeated a finite number of times
ii) $E \cong S_g(a,\infty)$ and $E \cong S_f(b,1)$
iii) $E \cong S_f(b,1)$ and $\operatorname{Ext}(E,E) = 0$

Theorem. The class of $S_g(a,\infty)$ spaces coincides with the subclass of $\lambda_\Phi(b)$ spaces where Φ consists of a single function, i.e. $\Phi = (\phi,\phi,\ldots)$.

By means of an example Z. Nurlu shows that the subclass of $\lambda_\Phi(b)$ spaces as described in the previous theorem is actually proper, namely he constructs a (d_1) Köthe space E such that $\operatorname{Ext}(E,E) = 0$, but E is not isomorphic to any $S_g(b,\infty)$ space.

Finally Z. Nurlu shows that the popular Köthe space $K(a_{kn})$ where $a_{kn} = \exp \circ \ldots \circ \exp(a_n)$ (k-times composition) is not only an $S_g(b,\infty)$ space but also an $L_f(a,\infty)$ space.

8. Quasi-equivalence Property and Bessaga's Conjecture

A NFS E with basis is said to have the **quasi-equivalence property** (QEP) if for any two bases (x_n) and (y_n) of E there exists a sequence (t_n) of non-zero scalars and a permutation $\pi : \mathbf{N} \mapsto \mathbf{N}$ such that the map $T : E \mapsto E$, $Tx_n = t_n y_{\pi(n)}$ is an automorphism. In this case one also says that the bases (x_n) and (y_n) are quasi-equivalent.

Whether every Köthe has the QEP is a question that goes back to G. Köthe himself. The answer is known only in some particular cases which we list below:

(a) (M.M. Dragilev [14]) $A(\Delta) = \Lambda_1((n))$ has QEP.

(b) (B.S. Mitiagin [45]) All power series spaces have QEP.

(c) (M.M. Dragilev [15]) All nuclear regular (d_1) and (d_2) spaces have QEP.

(d) (independently by V.P. Kondakov [36] and by L. Crone and W. Robinson [10]) All nuclear and regular Köthe spaces have QEP.

Hence it follows that all power series spaces, $L_f(a,r)$ spaces and $S_g(a,r)$ spaces have the quasi-equivalence property.

Finally V.P. Zahariuta and P.A. Chalov got an equivalent intrinsic characterization of QEP for nuclear spaces in [64], but it is expressed in terms of combinatorial inequalities and is not very practical in helping to settle the question of QEP.

In [7] C. Bessaga proved the following result which he called the Crucial Lemma.

Theorem. Let E be a NFS with basis (x_n). If F is a complemented subspace of E with basis (y_n) then there exists a non-decreasing sequence (k_n) of indices with $\lim k_n = \infty$ such that (y_n) is quasi-equivalent to (x_{k_n}).

This result comes close to showing that all complemented basic sequences in E are obtained this way and he was led to the following problem which is known as **Bessaga's conjecture.**

Conjecture. (k_n) can be chosen so that $k_n \neq k_m$ if $n \neq m$.

The conjecture is proved only in some particular cases which we list below:

(a) In [46] B.S. Mitiagin proved for $\Lambda_1(a)$.

(b) In [17] E. Dubinsky proved for stable and unstable $\Lambda_\infty(a)$.

(c) In [47] B.S. Mitiagin proved for $\Lambda_\infty(a)$.

(d) In [10] L. Crone and W. Robinson proved for stable and regular Köthe spaces.

(e) Finally the conjecture holds for all $L_f(a,r)$ spaces, since H. Ahonen [1] shows that a complemented subspace with basis of an $L_f(a,r)$ space is an $L_f(b,r)$ space, and in [37] V.P. Kondakov shows that if $L_f(b,r)$ is isomorphic to a subspace of $L_f(a,r)$ (not necessarily complemented) then there is a subsequence (k_n) of indices such that

$$0 < \inf \frac{a_{k_n}}{b_n} \leq \sup \frac{a_{k_n}}{b_n} < \infty.$$

REFERENCES

[1] H. Ahonen, 'On nuclear Köthe spaces defined by Dragilev functions', *Ann. Acad. Scient. Fenn.*, Series A, I. Mathematica, Dissertationes **38**, Helsinki, 1981.

[2] M. Alpseymen, 'Basic sequences in some nuclear Köthe sequence spaces', Thesis, University of Michigan, 1978.

[3] M. Alpseymen, M.S. Ramanujan and T. Terzioğlu, 'Subspaces of some nuclear sequence spaces', *Indag. Math.* **82** (1979), 217-224.

[4] H. Apiola, 'Every nuclear Fréchet space is a quotient of a Köthe Schwartz space', *Arch. Math.* **35** (1980), 559-573.

[5] H. Apiola, 'Characterization of subspaces and quotients of nuclear $L_f(\alpha, \infty)$-spaces', *Compositio Math.* **50** (1983), 65-81.

[6] A. Aytuna, J. Krone and T. Terzioğlu, 'Complemented infinite type power series subspaces of nuclear Fréchet spaces', to appear in *Math. Ann.*

[7] C. Bessaga, 'Some remarks on Dragilev's theorem', *Studia Math.* **31** (1968), 307-318.

[8] R.W. Braun, 'Linear topological structure of closed ideals in weighted algebras of entire functions', *Arch. Math.* **50** (1988), 251-258.

[9] L. Crone and W. Robinson, 'Diagonal maps and diameters in Köthe spaces', *Israel J. Math.* **20** (1975), 13-22.

[10] L. Crone and W. Robinson, 'Every nuclear Fréchet space with a regular basis has the quasi-equivalence property', *Studia Math.* **52** (1975), 203-207.

[11] N. DeGrande-DeKimpe, '$L_f(a, r)$-spaces between which all the operators are compact I', *Comment. Math. Univ. Carolinae* **18** (1977), 659-674.

[12] N. DeGrande-DeKimpe, '$L_f(a, r)$-spaces between which all the operators are compact II', *Comment. Math. Univ. Carolinae* **19** (1978), 1-12.

[13] N. DeGrande-DeKimpe and W. Robinson, 'Compact maps and embeddings from an infinite type power series space to a finite type power series space', *J. reine angew. Math.* **293/294** (1977), 52-61.

[14] M.M. Dragilev, 'Standard form of basis for the space of analytic functions' (Russian), *Usp. Mat. Nauk.* **15** (1960), 181-188.

[15] M.M. Dragilev, 'On regular bases in nuclear spaces', *Math. Sb.* **68** (1965), 153-173. (Amer. Math. Soc. transl. **93** (1970), 61-82.)

[16] E. Dubinsky, 'Infinite type power series subspaces of finite type power series spaces', *Israel J. Math.* **15** (1973), 257-281.

[17] E. Dubinsky, 'Infinite type power series subspaces of infinite type power series spaces', *Israel J. Math.* **20** (1975), 359-368.

[18] E. Dubinsky, 'Concrete subspaces of nuclear Fréchet spaces', *Studia Math.* **52** (1975), 209-219.

[19] E. Dubinsky, 'Basic sequences in (s)', *Studia Math.* **59** (1977), 283-293.

[20] E. Dubinsky, *The structure of nuclear Fréchet spaces*, Lecture Notes in Mathematics **720**, 1979.

[21] E. Dubinsky, 'Basic sequences in a stable finite type power series space', *Studia Math.* **68** (1980), 117-130.

[22] E. Dubinsky, M. Alpseymen and N. DeGrande-DeKimpe, 'Basic sequences in some stable, nuclear $L_f(b,r)$-spaces', *Indag. Math.* **82** (1979), 203-215.

[23] E. Dubinsky and W. Robinson, 'Quotient spaces of (s) with basis', *Studia Math.* **63** (1977), 39-53.

[24] A.S. Dynin and B.S. Mitiagin, 'Criterion for nuclearity in terms of approximative dimension', *Bull. Acad. Polon. Sci.* **3** (1960), 535-540.

[25] H. Faeskorn, 'Bedingungen für die Übereinstimmung der Klassen der stetigen und der beschränkten linearen Abbildungen zwischen Fréchetträumen, von denen einer ein L_f-Raum ist', Diplomarbeit, Wuppertal, 1982.

[26] H. Faeskorn, 'Some Fréchet spaces with corresponding classes of linear continuous and bounded operators', *Doğa, Tr. J. Math.* **10** (1986), 109-124.

[27] J. Hebbecker, 'Auswertung der Splittingbedingungen (S_1^*) und (S_2^*) für Potenzreihenräume und L_f-Räume', Diplomarbeit, Wuppertal, 1984.

[28] L. Holmström, 'Superspaces of (s) with basis', *Studia Math.* **75** (1983), 139-152.

[29] V.V. Kashirin, 'Isomorphic embeddings of some generalized power series spaces', Uniwersytet Warszawski, Instytut Matemtyki, Warszawa, 1979.

[30] V.V. Kashirin, 'On the representation of Köthe spaces of class d in the form of generalized power series spaces and their Cartesian products' (Russian), *Bull. Acad. Pol. Sci.* **28** (1980), 27-32.

[31] V.V. Kashirin, 'Isomorphic embeddings of some generalized power series spaces', *Studia Math.* **71** (1981), 169-178.

[32] M. Kocatepe (Alpseymen), 'On Dragilev spaces and the functor Ext', *Arch. Math.* **44** (1985), 438-445.

[33] M. Kocatepe, 'Some counterexamples on Dragilev spaces', *Doğa, Tr. J. Math.* **10** (1986), 136-142.

[34] M. Kocatepe, 'An isomorphism theorem for Dragilev spaces', *Arch. Math.* **50** (1988), 281-286.

[35] M. Kocatepe, 'Classification of Dragilev spaces of types −1 and 0', to appear in *Math. Balkanica.*

[36] V.P. Kondakov, 'On linear *b*-dimension and quasi-equivalence of bases of Köthe spaces' (Russian), *Matem. Analiz i Ego Prilozh.* **5** (1973), 210-213.

[37] V.P. Kondakov, 'On a certain generalization of power series spaces' (Russian), *Aktyal'n. Vopros. Mat. Analiža*, Rostov (1978), 92-98.

[38] J. Krone, 'Zur topologischen Charakterisierung von Unter- und Quotientenräumen spezieller nuklearer Kötheräume mit der Splittingmethode', Diplomarbeit, Wuppertal, 1984.

[39] J. Krone and D. Vogt, 'The splitting relation for Köthe spaces', *Math. Z.* **180** (1985), 387-400.

[40] R. Meise, 'Sequence space representations for (DFN)-algebras of entire functions modulo closed ideals', *J. reine angew. Math.* **363** (1985), 59-95.

[41] R. Meise, 'Structure of closed linear translation invariant subspaces of $A(\mathbf{C})$ and kernels of analytic convolution operators', *Functional Analysis: Surveys and Recent Results III*, (Eds: K.-D. Bierstedt and B. Fuchssteiner), North Holland Math. Studies **90** (1984), 331-347.

[42] R. Meise, K. Schwerdtfeger and B.A. Taylor, 'On kernels of slowly decreasing convolution operators', *Doğa, Tr. J. Math.* **10** (1986), 176-197.

[43] R. Meise and B.A. Taylor, 'Splitting of closed ideals in (DFN)-algebras of entire functions and the property (DN)', *Trans. Amer. Math. Soc.* **302** (1987), 341-370.

[44] R. Meise and B.A. Taylor, 'Sequence space representations for (FN)-algebras of entire functions modulo closed ideals', *Studia Math.* **85** (1987), 203-227.

[45] B.S. Mitiagin, 'Approximative dimension and bases in nuclear spaces', *Usp. Mat. Nauk.* **16** (1961), 73-132. (English translation, *Russian Math. Surveys* **16** (1961), 59-127.)

[46] B.S. Mitiagin, 'Equivalence of bases in Hilbert scales' (Russian), *Studia Math.* **37** (1971), 111-137.

[47] B.S. Mitiagin, 'Structure of subspaces of infinite Hilbert scales' (Russian), Trudy 7 Simney Szkoly, Drogovic (1974), 127-133.

[48] B.S. Mitiagin and G. Henkin, 'Linear problems of complex analysis' (Russian), *Usp. Mat. Nauk.* **26** (1972), 93-152.

[49] Z. Nurlu, 'On basic sequences in some Köthe spaces and existence of non-compact operators', Thesis, Clarkson College of Technology, Potsdam NY, 1981.

[50] Z. Nurlu, 'Embeddings of $\Lambda_\infty(\alpha)$ into $\Lambda_1(\alpha)$ and some consequences', *Math. Balkanica*, New series, **1** (1987), 14-24.

[51] Z. Nurlu, 'S_g spaces and the vanishing of the functor Ext', *Studia Math.* **88** (1988), 13-21.

[52] S. Rolewicz, 'On spaces of holomorphic functions', *Studia Math.* **21** (1962), 135-160.

[53] D. Vogt, 'Charakterisierung der Unterräume von s', *Math. Z.* **155** (1977), 109-117.

[54] D. Vogt, 'Charakterisierung der Unterräume eines nuklearen stabilen Potenzreihenraumes von endlichem Typ', *Studia Math.* **71** (1982), 251-270.

[55] D. Vogt, 'Eine Charakterisierung der Potenzreihenräume von endlichem Typ und ihre Folgerungen', *Manuscripta Math.* **37** (1982), 269-301.

[56] D. Vogt, 'Frécheträume, zwischen denen jede lineare stetige Abbildung beschränkt ist', *J. reine angew. Math.* **345** (1983), 182-200.

[57] D. Vogt, 'Structure theory of nuclear stable power series spaces', *Vorlesungen aus dem Fachbereich Math. der Universität Essen* **10** (1983), 319-329.

[58] D. Vogt, 'On the functors $\mathrm{Ext}^1(E, F)$ for Fréchet spaces', *Studia Math.* **85** (1987), 163-197.

[59] D. Vogt, 'On the characterization of subspaces and quotient spaces of stable power series spaces of finite type', *Arch. Math.* **50** (1988), 463-469.

[60] D. Vogt and M.J. Wagner, 'Charakterisierung der Quotientenräume von s und eine Vermutung von Martineau', *Studia Math.* **67** (1980), 225-240.

[61] D. Vogt and M.J. Wagner, 'Charakterisierung der Unterräume und Quotientenräume der nuklearen stabilen Potenzreihenräume von unendlichem Typ', *Studia Math.* **70** (1981), 63-80.

[62] M.J. Wagner, 'Quotientenräume von stabilen Potenzreihenräumen endlichen Typs', *Manuscripta Math.* **31** (1980), 97-109.

[63] V.P. Zahariuta, 'On the isomorphism of Cartesian products of locally convex spaces', *Studia Math.* **46** (1973), 201-221.

[64] V.P. Zahariuta and P.A. Chalov, 'A quasi-equivalence criterion for absolute bases in an arbitrary F-space' (Russian), *Izv. Sev-Kav. Nauc. Ts. Vyss. Šk. Estest. Nauko-Izd. Rostovs. Univ. Rostov* **42** (1983) No:2.

ON PELCZYNSKI'S PROBLEM

J. KRONE
Universität-Gesamthochschule
Wuppertal
Fachbereich Mathematik
Gaussstr. 20
F.R. Germany

In 1970 Studia Mathematica published unsolved problems posed by some of the most distinguished mathematicians. Pelczynski asked, whether complemented subspaces of a nuclear Fréchet space with basis have a basis [18]. In view of Dynin-Mitiagin's absolute basis theorem [19] this problem is equivalent to the question, whether complemented subspaces of nuclear Köthe spaces are again Köthe spaces.

Although this problem is still open, there are positive solutions in many special cases. Most of these results can be proved with one of the following two methods: First is the interpolation method of Mitiagin-Henkin [16] based on the spectral representation of compact Hilbert space operators. Second is the decomposition method of Vogt [22] which was motivated by Pelczynski's decomposition result for Banach spaces.

Both methods have many applications to function spaces. In case of analytic functions the most interesting results are contained in Aytuna's report [3] for this conference and many other cases are solved by Vogt [23]. The present article does not focus on these applications but on the methods and their relation to Vogt's splitting and structure theory, which is in form of the more general concept of the projective limit functor one of the main issues of this conference [26].

§1. Notations and Basic Results:

See also Dubinsky [7]. E and F are always locally convex spaces. For a subset M of E the gauge functional is defined by $p_M(x) := \inf\{c \geq 0 : x \in cM\}$, $\inf\{\} := +\infty$. M is a Hilbert ball if M is an absolutely convex, closed set and if p_M is a Hilbert norm on $E_M := U_{c \geq 0} cM$;

T. Terzioğlu (ed.), Advances in the Theory of Fréchet Spaces, 297–304.

that means the completion $(E_M, p_M)\hat{}$ is a Hilbert space.

For a fundamental system of absolutely convex and closed neighborhoods of zero $(U_a)_{a\in A}$ in E and $(V_b)_{b\in B}$ in F the $\|\ \ \|_a := p_{U_a}$ and the $\|\ \ \|_b := p_{V_b}$ are the corresponding fundamental systems of seminorms. If E is a Fréchet space ((F)-space) we can choose $A = \mathbf{N}$ and $U_n \supset U_{n+1}$.

For a continuous, linear mapping $T \in L(E, F)$ we define:

$$\| T \|_{N,M} := \inf\{c \geq 0 : T(M) \subset cN\} \text{ and } \| T \|_{b,a} := \| T \|_{V_b, U_a}.$$

The range of T is always endowed with the subspace topology of F.

$$A = (a_{j,k})_{j,k\in\mathbf{N}} \text{ with } 0 \leq a_{j,k} \leq a_{j,k+1} \text{ and } \sup_{k\in\mathbf{N}} a_{j,k} > 0 \text{ for all } j, k \in \mathbf{N}$$

is a Köthe matrix. A Köthe space (Köthe sequence space) is defined by

$$\lambda(A) := \{(x_j)_{j\in\mathbf{N}} \subset \mathbf{K} : \| x \|_k := \sum_{j=1}^{\infty} | x_j a_{j,k} | < \infty \text{ for all } k \in \mathbf{N}\}$$

This space is nuclear if and only if for all k there is an m so that

$$\sum_{j\in J_m} \frac{a_{j,k}}{a_{j,m}} < \infty \quad \text{where} \quad J_m := \{j \in N : a_{j,m} > 0\}$$

If A has the form $a_{j,k} = exp(r_k \alpha_j)$, where α is a sequence with $0 \leq \alpha_j \leq \alpha_{j+1} \to +\infty$ and $r_k < r_{k+1} \to r$, $r \in \mathbf{R}$ or $r = +\infty$, then $\lambda(A)$ is called power series space $\lambda_r(\alpha)$. If $a_{j,k} = exp(f(r_k\alpha_j))$ with an odd, rapidly increasing function f, we call $\lambda(A)$ a Dragilev space $L_f(\alpha, r)$ (see Dragilev [6] and Ahonen [1]). In the following we always assume that $\lambda(A)$ and $\lambda(B)$ are nuclear Fréchet spaces although some results are true under weaker assumptions.

Now we want to recall the splitting condition S^* for Köthe spaces and its use.

Theorem: (Krone-Vogt [13]). For a pair of nuclear Köthe spaces $\lambda(A)$, $\lambda(B)$ the following is equivalent:

(a) Condition S^*: $\forall u\ \exists p, k\ \forall v, m\ \exists n, S\ \forall j, i$

$$\frac{a_{i,m}}{b_{j,k}} \leq S \max\left(\frac{a_{i,n}}{b_{j,v}}, \frac{a_{j,p}}{b_{j,u}}\right).$$

(b) Every exact sequence of Fréchet spaces $0 \to \lambda(B) \to F \to \lambda(A) \to 0$ splits.

(c) $L_b(\lambda(A), \lambda(B))$ is bornological.

(d) $L_b(\lambda(A), \lambda(B))$ is barrelled (quasi-barrelled).

As it was developed during this conference by Vogt [26] and Meise [14], the above result holds in a much more general setting and it has also in this general frame very interesting applications.

For the present article we need the stronger version of S^* which is due to Apiola [2] and will be called here according to Vogt's terminology [24] S_1^*. The dual version of S^* called DS^* was introduced in [12].

Notation:

(1) The pair of Köthe spaces $\lambda(A)$, $\lambda(B)$ satisfies S_1^* iff

$$\exists p \ \forall \mu \ \exists k \ \forall K, m, \ R > 0 \ \exists n, S : \ \frac{a_{i,m}}{a_{j,k}} \leq \max\left(S\frac{a_{i,n}}{a_{j,K}}, \ \frac{1}{R}\ \frac{a_{i,p}}{a_{j,\mu}}\right)$$

for all i, j.

(2) The pair of Köthe spaces $\lambda(A)$, $\lambda(B)$ satisfies DS^* iff

$$\exists \mu \ \forall p, k \ \exists m, v \ \forall n \ \exists S : \ \frac{b_{j,k}}{a_{i,m}} \leq S \max\left(\frac{b_{j,v}}{a_{i,n}}, \frac{b_{j,\mu}}{a_{i,p}}\right)$$

for all i, j.

(3) The Köthe space $\lambda(A)$ satisfies S^*, (DS^* resp. S_1^*) iff the pair $\lambda(A)$, $\lambda(A)$ satisfies S^*, (DS^* resp. S_1^*).

In case of power series and Dragilev spaces the condition S^* and its variations are completely understood (cf. e.g. Hebbecker [10], Krone [11], [12]).

Remark:

(1) A nuclear Köthe space satisfies S_1^* iff it satisfies S^* and condition DN of Vogt [21].

(2) A pair of power series spaces satisfies S^* resp. DS^* iff

	$(\lambda_r(\alpha),\ \lambda_\rho(\beta)) \in (DS^*)$	$(\lambda_r(\beta),\ \lambda_\rho(\alpha)) \in (S^*)$
$r = +\infty$	M bounded	always
$r = 0$	always	M bounded

Here M is the set of all finite limit points of $\{\frac{\beta_j}{\alpha_i} : i, j \in N\}$.

(3) A pair of Dragilev spaces satisfies S^* resp. DS^* iff

	$(L_f(\alpha,r),\ L_f(\beta,r)) \in (DS^*)$	$(L_f(\beta,r),\ L_f(\alpha,r)) \in (S^*)$
$r=+\infty$	M bounded	always
$r=1$	always	N bounded
$r=0$	always	M bounded
$r=-1$	N bounded	always

Here $N := \{\frac{1}{1-x} : x \in M \text{ and } x < 1\}$.

§2. Interpolation Method

First we present the interpolation method due to Mitiagin-Henkin [16] in the setting of [12].

Lemma: Let E and F be complete, locally convex spaces and $T \in L(E,F)$. If there are Hilbert balls K and U so that E_K is dense in E, K is compact, U a neighborhood of zero in F and $\{L \in L(E,F) : L(K) \subset T(K) \text{ and } T^{-1}(U) \subset L^{-1}(U)\}$ is equicontinuous, then the range of T has a basis.

Proof: If we endow E_K with the topology of p_K and if $F_U := (F, p_U)^\wedge$, then $T : E_K \to F_U$ is a compact mapping between Hilbert spaces. Because of the spectral theorem for compact operators we can write T in the form

$$Tx = \sum_{j=1}^{\infty} \alpha_j \langle x, e_j \rangle_K f_j \quad \text{for all } x \in E_K,$$

where $\{e_j\}$ and $\{f_j\}$ are orthonormal systems in E_K and F_U and where $\alpha_j \geq \alpha_{j+1} \to 0$. We may assume that $\alpha_j > 0$ since otherwise the image of T is finite dimensional.

Since $T(e_j) = \alpha_j f_j$, we have $f_j \in range(T)$. From $\sum_{j=1}^{n} a_j f_j \to 0$ as $n \to +\infty$ in $range(T)$ follows that the series is also convergent in F_U, hence the $a_j = 0$ for all j and the set $\{f_j\}$ is linearly independent in the range of T.

For $x \in E, n \in \mathbb{N}$ we define $T_n(x) := \sum_{j=1}^{n} \langle Tx, f_j \rangle_U f_j$ and we get

$$T_n\left(T^{-1}(U)\right) \subset \{\sum_{j=1}^{n} \langle z, f_j \rangle_U f_j : \quad z \in U\} \subset U$$

$T_n(x) = T\left(\sum_{j=1}^{n} x_j e_j\right)$ if $x = \sum_{j=1}^{\infty} x_j e_j$ in E_K and

$$T_n(K) = \{\sum_{j=1}^{n} \alpha_j x_j f_j : \sum_{j=1}^{\infty} | x_j | \leq 1\} \subset T(K).$$

Hence the sequence $\{T_n\}_{n\in \mathbb{N}}$ is equicontinuous. Since $T_n(x) \to T(x)$ for $x \in E_K$, which is dense in E, we obtain the convergence on the whole of E by a standard argument e.g. used in Banach Steinhaus theorem.

From this result it follows:

Theorem: (Krone [12]). For a pair of nuclear Köthe spaces $\lambda(A)$, $\lambda(B)$ with DS^* the range of every operator from $L(\lambda(A), \lambda(B))$, has a basis.

In view of our Remark 2 in the first section this theorem implies for power series spaces the well known results. Namely Pelczynski's problem has a positive solution for power series spaces of finite type (Mitiagin [15], Mitiagin-Henkin [16]) and for power series spaces of infinite type with the limit point condition (Dubinsky-Vogt [8]). The latter result has very interesting generalisations:

Theorem: (Dubinsky-Vogt [8]). The range of every tame mapping in a power series space has a basis.

For this theorem one needs a new construction of the crucial set K in the interpolation method explained above. In fact, the construction of K depends not only on the space but also on the tame operator. Recently Vogt gave a very interesting application of this result.

Theorem: (Vogt [25]). Every nuclear, tame Fréchet space is tame isomorphic to a power series space.

It is not the purpose of the present note to discuss the various concepts of tame spaces, but we want to point out, that Vogt's result holds regardless whether one uses the notations of Moser [17], Hamilton [7] or Sergeraert [20].

§3. Decomposition Method

The following two theorems are well known for power series spaces, which is the most important case. In order to show the underlying structure we prefer to give here the more general version, which indicates that the splitting conditions are crucial for this method. This more general version of the following two theorems is proved in a brief note [4] contained in this proceedings.

Theorem: (Vogt-Pelczynski decomposition). Let E be a stable nuclear Köthe space which satisfies the condition (S_1). If F is isomorphic to a complemented subspace of E and E is isomorphic to a complemented subspace of F, then F is isomorphic to E.

Theorem: (Aytuna-Krone-Terzioğlu [5]). Let E be a nuclear Köthe space which satisfies the condition (S_1). Then every complemented subspace of E with stable diametral dimension has a basis.

Since the proof of both results is contained in this proceedings [4], I only want to make a final remark. Vogt's decomposition method is a very, very powerful tool to obtain power series space representations (cf. Vogt [22], [23]), but there was one interesting unsolved question. Zaharjuta [27] conjectured that for a Stein manifold M of dimension of the following are equivalent:

(1) The space $\mathcal{O}(M)$ of analytic functions on M is isomorphic to $\mathcal{O}(\mathbf{C}^d)$. (By Taylor expansion $\mathcal{O}(\mathbf{C}^d)$ is isomorphic to the power series space of infinite type with exponent sequence $\alpha_n = n^{1/d}$).

(2) $\mathcal{O}(M)$ has Vogt's property DN [21].

(3) Every bounded, plurisubharmonic function on M is constant.

Our above stated improvement of Vogt's decomposition result proves Zaharjuta's conjecture. For more information see Aytuna's report [3] for this conference.

REFERENCES

[1] Ahonen, H., 'On nuclear Köthe spaces defined by Dragilev functions', Dissertation, Helsinki 1981.

[2] Apiola, H., 'Characterization of subspaces and quotients of nuclear $L_f(\alpha, \infty)$-spaces', *Compositio Math.* **50** (1983), 65-81.

[3] Aytuna, A., 'Stein spaces M for which $\mathcal{O}(M)$ is isomorphic to a power series space'. This volume.

[4] Aytuna, A., Krone, J., Terzioğlu, T., 'On complemented subspaces of certain nuclear Köthe spaces'. This volume.

[5] Aytuna, A., Krone, J., Terzioğlu, T., 'Complemented infinite type power series subspaces of nuclear Fréchet spaces', to appear in *Math. Ann.*.

[6] Dragilev, M.M., 'On regular bases in nuclear spaces', *Amer. Math. Soc. Transl.* (2) **93** (1970), 61-82.

[7] Dubinsky, E., *The Structure of Nuclear Fréchet spaces.* Lecture Notes in Mathematics 720, 1979.

[8] Dubinsky, E., 'Bases in complemented subspaces of power series spaces', *Bull. Acad. Polon. Sci.* **34** (1986), 65-67.

[9] Hamilton, R.S., 'The inverse function theorem of Nash and Moser', *Bull. Am. Math. Soc.* **7**, (1982) 65-222.

[10] Hebbecker, J., 'Auswertung der Splittingbedingungen (S_1^*) und (S_2^*) für Potenzreihenräume und L_f-Räume', Diplomarbeit, Wuppertal 1984.

[11] Krone, J., 'Zur topologischen Charakterisierung von Unter - und Quotientenräumen spezieller nuklearer Kötheräume mit der Splittingmethode', Diplomarbeit, Wuppertal 1984.

[12] Krone, J., 'Existence of Bases and the Dual Splitting Relation for Frêchet Spaces', to appear in *Studia Math.*

[13] Krone, J., Vogt, D., 'The Splitting Relation for Köthe Spaces', *Math. Z.* **190** (1985), 387-400.

[14] Meise, R., 'Applications of the projective limit functor to convolution and linear partial differential equations on Gevrey classes'. This volume.

[15] Mitiagin, B.S., 'The equivalence of bases in Hilbert scales', *Studia Math.* **37** (1970), 111-137 (Russian).

[16] Mitiagin, B.S., Henkin, G.M., 'Linear problems of complex analysis', *Russian Math. Surveys* **26** (1971), 99-164.

[17] Moser, J., 'A new technique for the construction of solutions of nonlinear differential equations', *Proc. Nat. Acad. Sci. USA* **47** (1961), 1824-1831.

[18] Pelczynski, A., 'Problem 37', *Studia Math.* **38**, (1970), 476.

[19] Pietsch, A., *Nuclear Locally Convex Spaces*, Berlin- Heidelberg-New York, 1972.

[20] Sergeraert, F., 'Un théorème de fonctions implicites sur certains espaces de Fréchet et quelques applications', *Ann. Sci. Ec. Norm. Super* 4e séries, **5**, (1972), 599-660.

[21] Vogt, D., 'Charakterisierung der Unterräume von s', *Math 7*, **155** (1977), 109-117.

[22] Vogt, D., 'Ein Isomorphiesatz für Potenzreihenräume', *Arch. Math.* **38** (1982), 540-548.

[23] Vogt, D., 'Sequence Space Representations of Spaces of Test Functions and Distributions', *Funct. Anal., Holomorphy and Approximation Theory* (ed. G. Zapata), Lecture Notes Pure Appl. Math. **83**, 405-443, New York 1983.

[24] Vogt, D., 'On the functors $Ext'(E, F)$ for Fréchet spaces', *Studia Math.* **85** (1987), 163-197.

[25] Vogt, D., 'Tame Spaces and Power Series Spaces', *Math. Z.* **196** (1987), 523-536.

[26] Vogt, D., 'On the projective limit functor'. This volume.

[27] Zaharjuta, V.P., 'Isomorphism of spaces of analytic functions', *Sov. Math. Dokl.* **22** (1980), 631-634.

SOME INVARIANTS OF FRECHET SPACES AND IMBEDDINGS OF SMOOTH SEQUENCE SPACES

T. TERZİOĞLU

Department of Mathematics

Middle East Technical University

06531 Ankara - Turkey

Introduction

The purpose of this article is to give an exposition of some recent results related to imbedding smooth sequence spaces into nuclear Fréchet spaces. Although some new results are stated and proved in the present article, essentially they are modifications of the results of Aytuna, Krone and the author which are contained in [6] and [7].

In this article we consider the problem of the existence of imbeddings of nuclear smooth sequence spaces which are not necessarily metrizable. Definitions of these spaces and the duality between the two types are given in §1. One advantage in using this more general set-up is that we can apply our results not only to power series spaces but also to their duals. In §2 we describe the procedure for constructing a complemented smooth sequence space in a nuclear space. Although the result stated here (2.1. Theorem) is more general than the original one in [6] in several directions, the proof is a slight modification of the original one. We also indicate how we can dualize this result.

A major application of our theorem is to the decomposition method of Vogt [36], which is one of the main tools we have if we want to show that a certain nuclear Fréchet space is isomorphic to a stable power series space of infinite type. In this method it is required that the space under consideration should contain a complemented copy of the power series space. To get this we only need to know that there is what we call a "local imbedding" from the power series space and thus generalize the decomposition method (2.2. Corollary).

The existence of local imbeddings plays a major role in all our results. Although this concept is defined here in a more general fashion than in [6], we show in §3 that these two different definitions really boil down to the same thing. This gives us a certain degree of freedom which is exploited in the applications.

T. Terzioğlu (ed.), Advances in the Theory of Fréchet Spaces, 305–324.

The existence of imbeddings of nuclear Fréchet spaces into product spaces of the form Λ_∞^N is related to the diametral dimension of the space. Results obtained in this direction are almost always modelled after the well-known Komura imbedding theorem [16]. In §3, we show how we can also relate the existence of local imbeddings to some invariants like the diametral dimension, conditions ($\underline{DN}$) and (Ω) of Vogt and Wagner [37] [44]. In this context, the concept of the "associated exponent sequence" introduced in [7] is also briefly discussed.

The results of §2 are then combined in §4 with what we know about the existence of local imbeddings in terms of the invariants. In this way we obtain a positive solution to Pelczynski's problem for complemented subspaces of s which have stable diametral dimension. For a further contribution to this problem using the our approach and the decomposition method of Vogt we refer to [8]. Krone's article [19] in this volume summarizes different approaches to this problem and describes the present state. One result which yields a true imbedding instead of a local one is 4.1. Theorem whose proof is given in [7].

The final section is devoted to some applications of our results to some Fréchet spaces of functions. We begin by describing briefly what we have obtained in [6] and [7] about the linear-topological structure of the space of holomorphic functions on a Stein manifold. Some new results obtained by combining our results with the isomorphism theorem of Djakov [11] are also included. For further developments in this direction we refer to the articles by Aytuna [3], [4].

Another class of spaces we consider are the kernels of a linear elliptic partial differential operator. Here the works of Vogt [40] and Wiechert [46] deliver us readily what we need in order to apply our results. Finally we consider spaces of functions holomorphic on the duals of some nuclear Fréchet spaces, where the results of Börgens, Meise and Vogt ([9], [10], [22]) are essential in constructing our set-up.

§1. In this section we give some definitions, establish terminology and notation and describe the duals of smooth sequence spaces. For the standard theory of locally convex spaces we refer to [17] and [15], for the theory of nuclear spaces and the structure of nuclear Fréchet spaces to [13] and [27].

Let A be a Köthe set of sequences with the following additional properties:

(1) each $a \epsilon A$ is a positive non-decreasing sequence.

(2) for each $a \epsilon A$ there is a $b \epsilon A$ with $(a_n)^2 = 0(b_n)$.

The Köthe space $\lambda(A)$ is then called a G_∞-space or a *smooth sequence space of infinite type* [32]. Well-known examples of such spaces are the power series spaces of infinite-type or $L_f(\alpha;\infty)$- spaces of Dragilev [12] (cf. also [13]). Here and throughout $a_n = 0(b_n)$ means $a_n \leq \rho b_n$ for some constant $\rho > 0$. Those G_∞- spaces which are metrizable are exactly the (D_1)-spaces of Dragilev [12]. This was observed by Robinson [30]. A G_∞-space $\lambda(A)$ is nuclear if and only if $(1/a_n)\epsilon\ell_1$ for some $a\epsilon A$. We call a locally convex space E *stable* if it is isomorphic to $E \times E$. Stability of a G_∞-space $\lambda(A)$ is equivalent to: for every $a\epsilon A$ there is a $b\epsilon B$ with $a_{2n} = 0(b_n)$ [34].

A Köthe space $\lambda(Q)$ is called a G_1-space or a *smooth sequence space of finite type* [32] if Q satisfies the following properties:

(1) each $q\epsilon Q$ is positive and non-increasing.

(2) for each $q\epsilon Q$ there is a $p\epsilon Q$ with $\sqrt{q_n} = 0(p_n)$.

A power series of finite type is certainly a G_1-space. The class of Fréchet G_1-spaces coincides with the class (D_2) of Dragilev [30]. $\lambda(Q)$ is nuclear if and only if $Q \subset s$, where s is as usual the space of rapidly decreasing sequences. $\lambda(Q)$ is stable if and only if for each $q\epsilon Q$ there is a $p\epsilon Q$ with $q_n = 0(p_{2n})$. For further properties of smooth sequence spaces we refer to [34].

Let $\lambda(Q)$ be a nuclear G_1-space and

$$A_Q = \{x\epsilon\lambda(Q) : 1 \leq x_1 \leq x_2 \leq \ldots\}.$$

It is simple to prove $\lambda(A_Q)$ is a G_∞-space. The dual $\lambda(Q)'$ is densely contained in $\lambda(A_Q)$. For $y\epsilon\lambda(Q)$ we let $B_y = \{x\epsilon\lambda(Q) : \mid x_n \mid\leq\mid y_n \mid\}$. The sets B_y, $y\epsilon\lambda(Q)$, form a fundamental system of bounded subsets of $\lambda(Q)$ [18]. For each $y\epsilon\lambda(Q)$ we can find $x\epsilon A_Q$ with $\mid y_n \mid\leq x_n$. So $\lambda(Q)'_b$ is topologically isomorphic to a subspace of $\lambda(A_Q)$. By the Banach- Steinhaus theorem we have $\lambda(A_Q) = \lambda(Q)'_b$ if and only if $\lambda(Q)$ is barrelled. $\lambda(A_Q)$ is nuclear, and stable if $\lambda(Q)$ is stable. This is a summary of the proof of the following result. For further details we refer to [33] and [34].

1.1. Proposition: The dual $\lambda(Q)'_b$ of a nuclear G_1- space $\lambda(Q)$ is isomorphic to a dense subspace of the nuclear G_∞-space $\lambda(A_Q)$. If $\lambda(Q)$ is stable then $\lambda(A_Q)$ is also stable. $\lambda(Q)$ is barrelled if and only if $\lambda(Q)'_b = \lambda(A_Q)$. In particular, the dual of a nuclear power series space of finite type is a nuclear G_∞-space.

For the duals of smooth sequence spaces of infinite type we have a result of the same nature. However there is a minor point which requires additional consideration. If we let A be the set of all positive non-decreasing sequences then as the G_∞-space $\lambda(A)$ we get

φ, the space of all sequences with only a finite number of non-zero terms. The dual of φ, which is ω, the space of all sequences, cannot be represented as a G_1-space. If $\lambda(A)$ is a nuclear G_∞-space and $\lambda(A) \neq \varphi$ we let

$$Q_A = \{x \epsilon \lambda(A) : 0 < x_{n+1} \leq x_n\}.$$

Again $\lambda(Q_A)$ is a nuclear G_1-space which contains $\lambda(A)'$ as a dense subspace. We have the following result [34].

1.2. Proposition: Let $\lambda(A)$ be a nuclear G_∞-space and $\varphi \neq \lambda(A)$. $\lambda(A)'_b$ is isomorphic to a dense subspace of the nuclear G_1-space $\lambda(Q_A)$. If $\lambda(A)$ is stable, then $\lambda(Q_A)$ is also stable. $\lambda(A)$ is barrelled if and only if $\lambda(A)'_b = \lambda(Q_A)$. In particular the dual of a nuclear power series space of infinite type is a nuclear G_1-space.

On a nuclear sequence space $\lambda(A)$ we have various equivalent systems of semi-norms defining the topology [27]. We reserve the symbol $|\ |_a$, $a \epsilon A$, for

$$| x |_a = \sum_{n=1}^{\infty} | x_n | a_n.$$

The diametral dimension $\Delta(\lambda(A))$ of a nuclear G_∞-space $\lambda(A)$ is exactly $\lambda(A)'$ and $\Delta(\lambda(Q))$ is $\lambda(Q)$ where $\lambda(Q)$ is a nuclear G_1-space [32]. The diametral dimensions of those Fréchet spaces which satisfy the invariants (DN) or $(\overline{\Omega})$ of Vogt are also related to diametral dimensions of some Fréchet smooth sequence spaces [35]. In particular, if E is a nuclear Fréchet space which satisfies (DN), then there is a specific nuclear G_∞-space $\lambda(A)$ with $\Delta(\mathrm{E}) = \Delta(\lambda(A)) = \lambda(A)'$ [35].

§**2.** Throughout this section we denote by $\lambda(A)$ a nuclear G_∞-space which is always assumed to be *stable*. We call a continuous linear map i from $\lambda(A)$ into a locally convex space E a *local imbedding* if there is a sequence $\sigma = (\sigma_n)$ and a continuous semi-norm $\|\ \|$ on E with

$$| x |_\sigma = \sum | x_n | \sigma_n \leq \| ix \| .$$

where we assume $\sigma = (\sigma_n)$ satisfies the following condition:

$$(\ell_i): \quad \sigma_n > 0 \text{ and } \sigma_n = 0(a_n), 1/\sigma_n = 0(b_n) \text{ for some a and } b \epsilon A.$$

We postpone the discussion of local imbeddings to the next section. The definition given here is different than the one in [6], but as we shall observe, we are justified in using the same terminology. It should be noted at this point that an imbedding is certainly a local imbedding and a local imbedding is one to one.

Our aim is this section is to describe the procedure given in [6] for constructing a complemented basic sequence in E, provided there is a local imbedding $i : \lambda(A) \to E$ and E in turn is isomorphic to a subspace or a quotient space of $\lambda(A)$. The theorem we will prove here is more general than the one in [6]. For example, we do not assume $\lambda(A)$ is metrizable. However the essential construction is the same as in [6] and we modify the proof given there only slightly.

For the moment let us assume that E is also nuclear. Then we may assume $|x|_\sigma \leq ||| ix |||$ where the semi-norm $||| \cdot |||$ is given by a semi-inner product (,). This is in fact an inner product on the subspace $i(\lambda(A))$. Let (f_ν) be a sequence of vectors in E which will be specified later. Since $sp\{ie_1, \ldots, ie_{2n}\}$ is a 2n-dimensional subspace, by induction we can find $g_n \epsilon i(\lambda(A))$ with the following properties:

$$g_n \,\epsilon\, sp\{ie_1, \ldots, ie_{2n}\} \tag{1}$$

$$\begin{aligned} (g_n, f_\nu) &= 0 \text{ for } \quad \nu = 1, \ldots, n \text{ and} \\ (g_n, g_j) &= 0 \text{ for } \quad j = 1, \ldots, n-1 \end{aligned} \tag{2}$$

$$||| g_n ||| = 1 \tag{3}$$

Now if $g_n = i\left(\sum^{2n} \mu_j^n e_j\right)$, we have $\sum^{2n} | \mu_j^n | \sigma_j = | \sum \mu_j^n e_j |_\sigma \leq ||| g_n ||| = 1$. Let $\| \ \|$ be a continuous semi-norm on E. By continuity of i we have $c \epsilon A$ and $\rho > 0$ with

$$\| g_n \| \leq \rho \sum_{j=1}^{2n} | \mu_j^n | c_j \leq \rho \max \left\{ \frac{c_j}{\sigma_j} : j = 1, \ldots, 2n \right\} \leq \tilde{\rho} c_{2n} b_{2n}$$

where $b \epsilon A$ is obtained from the condition (ℓ_i). If we use the stability of $\lambda(A)$, we obtain that for any continuous semi-norm $\| \ \|$ on E there is some $d \epsilon A$ with

$$\| g_n \| = 0(d_n) \tag{4}$$

Let $j : \lambda(A) \to E$ be defined by $j(e_n) = g_n$. So far we have proved that j is well-defined and continuous. Our aim is to prove that under some additional assumptions $G = \overline{j(\lambda(A))}$ is a complemented subspace of E, (g_n) is a basis of G and j is in fact an isomorphism of $\lambda(A)$ onto G. This approach is somewhat more direct than the one in [6] in that we avoid the argument involving the calculation of diametral dimensions.

2.1. Theorem: Let $\lambda(A)$ be a nuclear stable G_∞-space and suppose there is a local imbedding of $\lambda(A)$ into E. Each one of the following assumptions implies that E has a complemented subspace isomorphic to $\lambda(A)$.

(1) $\lambda(A)$ is a Fréchet space and E is isomorphic to a closed subspace of $\lambda(A)^N$.

(2) E is barrelled and isomorphic to a closed subspace of $\lambda(A)$.

(3) E is sequentially complete and isomorphic to a quotient space of $\lambda(A)$.

Proof: In each of the three cases, E is automatically nuclear and so the construction of (g_n) is valid. We now select (f_ν) in each case differently. In case (1) we set $f_\nu = e_n^k$, where (e_n^k) is the canonical basis of $\lambda(A)^N$, $\nu = 2^{k-1}(2n-1)$ (cf.[6]). In case (2), we simply let $f_\nu = e_\nu$ and in (3), $f_\nu = qe_\nu$ where $q : \lambda(A) \to E$ is a quotient map. In all the cases, we have $||| f_\nu ||| = 0(\tilde{a_\nu})$ for some $\tilde{a}\epsilon A$. Now if $x\epsilon E$, then $x = \sum \xi_\nu f_\nu$ and we have for any $a\epsilon A$

$$|(g_n, x)| a_n \leq \sum_{\nu>n} \tilde{a}_\nu |\xi_\nu| a_n \leq \sum_{\nu>n} |\xi_\nu| \tilde{a}_\nu a_\nu.$$

From the nuclearity of $\lambda(A)$ we have $\tilde{a}_\nu a_\nu = 0(\frac{1}{\nu^2} b_\nu)$ for some $b\epsilon A$. So we have proved that for every $a\epsilon A$ there is a $b\epsilon A$ and $\rho > 0$ such that

$$|(g_n, x)| a_n \leq \frac{\rho}{n^2} \| \xi \|_b \tag{5}$$

where ξ is x or $q\xi = x$ in cases (1), (2) and (3) respectively. Since E is sequentially complete, for each $x\epsilon E$ we have $Px\epsilon E$ where

$$Px = \sum_{n=1}^{\infty} (g_n, x) g_n.$$

Note that $Pg_n = g_n$. In cases (1) and (2), continuity of $P : E \to E$ is immediate from (4) and (5). In case (3), we have that $Pq : \lambda(A) \to E$ is a continuous linear map by (4) and (5). However q is open and so again P is continuous. Hence P is a projection of E onto $G = \overline{sp(g_n)}$ and (g_n) is a basis of G.

To conclude the proof it remains to show that the map $j : \lambda(A) \to G$ defined by $j(e_n) = g_n$ is an isomorphism. We already know that j is one-to-one and continuous. From (5) we have $\sum |(g_n, x)| a_n < \infty$ for any $a\epsilon A$ and $x\epsilon G$. So $\eta = ((g_n, x))\epsilon\lambda(A)$ and $j\eta = x$. Hence j is a surjection. Further for each $a\epsilon A$

$$\| x \|_a = \sum |(g_n, x)| a_n$$

defines a continuous semi-norm on the barrelled space G. Therefore j is an imbedding.

The decomposition method of Vogt [36] states that if E is isomorphic to a complemented subspace of a stable nuclear power series space $\Lambda_\infty(\alpha)$ and $\Lambda_\infty(\alpha)$ in turn isomorphic to a complemented subspace of E, then E and $\Lambda_\infty(\alpha)$ are isomorphic. This simple method has been used very effectively to establish the isomorphism of various spaces to a power series space of infinite type. As only a sample list we mention [36], [39], [21] and [23].

To prove that a Fréchet space is isomorphic to a complemented subspace of some $\Lambda_\infty(\alpha)$ we have the invariants (DN) and (Ω) at our disposal [45]. Our theorem relaxes quite considerably the requirement that $\Lambda_\infty(\alpha)$ should be isomorphic to a complemented subspace of E. We state here, as a corollary, this generalization of the decomposition method of Vogt.

2.2. Corollary: [6]. Let E be a complemented subspace of a nuclear stable power series space $\Lambda_\infty(\alpha)$. If there is a local imbedding of $\Lambda_\infty(\alpha)$ into E, then E is isomorphic to $\Lambda_\infty(\alpha)$.

We mention that in [8] the class of power series spaces in the decomposition method is replaced by a more general class of nuclear G_∞-spaces.

Let E be a Fréchet space which is $\Lambda_N(\alpha)$-nuclear. This is equivalent to assuming that E is isomorphic to a subspace of $\Lambda_\infty(\alpha)^N$ [29]. In case E is simply nuclear, then the celebrated Komura imbedding theorem [16] says that E is isomorphic to a subspace of $(s)^N$. So another immediate consequence of our theorem is the following which generalizes a result of Holmström [14].

2.3. Corollary: [6]. If there is a local imbedding of s into a nuclear Fréchet space E, then E has a complemented subspace isomorphic to s. If there is a local imbedding of $\Lambda_\infty(\alpha)$ into a $\Lambda_N(\alpha)$-nuclear Fréchet space E, then E has a complemented subspace isomorphic to $\Lambda_\infty(\alpha)$.

Another class of stable G_∞-spaces important in applications is the class of duals of stable nuclear finite-type power series spaces. We would like to extend our result to cover this class by dualizing. We have noted in §1, (1.1. Proposition) the dual of a stable, barrelled nuclear G_1-space $\lambda(Q)$ is a stable nuclear G_∞- space $\lambda(A_Q)$. To dualize we need also the concept which is dual to the concept of local imbeddings.

Let $\lambda(Q)$ be a nuclear G_1-space. For $x = (x_j)\epsilon\lambda(Q), x_j \geq 0$, let

$$B_x = \{y : \mid y_j \mid\leq x_j\}.$$

A continuous linear map $h : E \to \lambda(Q)$ is called a *local quotient* if there is an $x\epsilon\lambda(Q)$ satisfying (ℓ_q) given below and $B \subset E$ bounded with $B_x \subset \overline{h(B)}$. The condition (ℓ_q) is

$$(\ell_q): \quad x_j > 0\,,\, x \text{ and } (1/x_j) \text{ belong to } \lambda(Q).$$

Note that $e = (1, 1, ...)$ satisfies (ℓ_q).

2.4. Proposition: Let $\lambda(Q)$ be a nuclear barrelled G_1-space and $h : E \to \lambda(Q)$ a local quotient. Then $h' : \lambda(Q)'_b \to E'_b$ is a local imbedding.

Proof: We identify $\lambda(Q)'_b$ with the nuclear G_∞- space $\lambda(A_Q)$ where $A_Q = \{x\epsilon\lambda(Q) : 0 < x_1 \leq x_2 \leq \ldots\}$. If $B_x \subset \overline{h(B)}$, where x satisfies (ℓ_q) and $B \subset E$ bounded, then we have

$$\| u \|_{B_x^0} \leq sup_{b\epsilon B} \ | u(hb) | = \| h' u \|_{B^0} .$$

Since x and $(1/x_j)$ both belong to $\lambda(Q)$, we can find y, $z\epsilon A_Q$ with $x_j = 0(y_j), (1/x_j) = 0(z_j)$. This means that x satisfies (ℓ_i) and since

$$B_x^0 = \left\{(u_j) : \sum | u_j | x_j \leq 1\right\}$$

we have $\| u \|_x = \| u \|_{B_x^0} \leq \| h' u \|_{B^0}$. Therefore h' is a local imbedding of $\lambda(A_Q)$ into E_b'.

With the duality between G_1 and G_∞-spaces, local imbeddings and local quotients at our disposal, we can dualize 2.1. Theorem in various ways. We give below one rather simple but important case as an example. Here the duality theorem for nuclear Fréchet spaces [27] and 2.1. Theorem gives an immediate proof.

2.5. Proposition: Let $\Lambda_1(\alpha)$ be a stable nuclear finite type power series space. Suppose there is a local quotient from E into $\Lambda_1(\alpha)$. If E is either isomorphic to a closed subspace of $\Lambda_1(\alpha)$ or to a quotient space of $\Lambda_1(\alpha)$, then E has a complemented subspace isomorphic to $\Lambda_1(\alpha)$.

§3. In this section we shall consider the problem of existence of a local imbedding and relate it to some invariants like the diametral dimension as well as $(\underline{DN})$ and (Ω). First let us start with some simple observations.

The map which sends each $x = (x_n)$ to $(x_n 2^{\alpha_n})$ is a local imbedding of $\Lambda_\infty(\alpha)$ into $\Lambda_1(\alpha)$ and it is also compact. We also note that in general $\Lambda_\infty(\alpha)$ need not be isomorphic to a subspace of $\Lambda_1(\alpha)$ [13] (cf. also [25]).

If $\lambda(B)$ is such that there is some $b\epsilon B$ with $inf_n b_n > 0$ and $\lambda(B)' \subset \lambda(A)'$ where $\lambda(A)$ is a G_∞- space, then the map $(x_n) \to (x_n)$ is a local imbedding of $\lambda(A)$ into $\lambda(B)$. Also if $i : \lambda(A) \to E$ is a local imbedding and $j : E \to F$ is an imbedding then ji is certainly a local imbedding of $\lambda(A)$ into F.

3.1. Proposition: Let $\lambda(A)$ be a nuclear G_∞-space. Suppose there is a local imbedding of $\lambda(A)$ into E. Then we have

(1) there is a continuous semi-norm $\| \ \ \|$ on E such that for every $a\epsilon A$ there is a local imbedding $i_a : \lambda(A) \to E$ such that $| x |_a \leq \| i_a x \|$.

(2) there is an imbedding of $\lambda(A)$ into E^A.

Proof: Let $D_a(x) = (x_n a_n)$ for $a_n \geq 0$. For $a \epsilon A$ we have $| D_a x |_e = | x |_a$. Suppose $i : \lambda(A) \to E$ is a local imbedding. That is $| x |_\sigma \leq \| ix \|$ for some continuous semi-norm $\| \ \|$ on E and $\sigma = (\sigma_n)$ satisfying (ℓ_i). For $a \epsilon A$ we set $c_n = a_n/\sigma_n$ and find by (ℓ_i) a $d \epsilon A$ with $c_n = 0(d_n)$. Then the diagonal map $D_c : \lambda(A) \to \lambda(A)$ is continuous and we simply define $i_a = iD_c$.

By the first part we have a continuous semi-norm $\| \ \|$ with the property that for each $a \epsilon A$, there is a local imbedding i_a of $\lambda(A)$ into E with $| x |_a \leq \| i_a x \|$. We now define $j : \lambda(A) \to E^A$ by $jx = (i_a x)$. j is continuous and one to one. For $b \epsilon A$ we let $V_b = \left\{ (y_a) \epsilon E^A : \| y_b \| \leq 1 \right\}$ and obtain

$$j(\lambda(A)) \cap V_b \subset j(U_b)$$

where $U_b = \{ x \epsilon \lambda(A) : \ | x |_b \ \leq 1 \}$.

In [6] we have called $i : \lambda(A) \to E$ a "local imbedding" if $| x |_e \ \leq \ \| ix \|$. By the first part of the proposition just proved, we are justified in using the same terminology if we are only interested in the existence of such mappings. It also tells us that if we have a local imbedding, we are free to "control" any semi-norm we choose on $\lambda(A)$.

For a Fréchet space $\lambda(A)$ the existence of a local imbedding from $\lambda(A)$ into E implies that $\lambda(A)$ is isomorphic to a subspace of the product space E^N. Suppose $j : \lambda(A) \to E^I$ is a local imbedding for some index set I. So we have $| x |_\sigma \leq \| jx \|$ where we can assume that the continuous semi-norm $\| \ \|$ on E^I satisfies $\| y \| = \| Py \|$ for some canonical projection of E^I onto some finite product $E \times \cdots \times E$. Then $Pj : \lambda(A) \to E \times \cdots \times E$ is a local imbedding.

3.2. Proposition: If there is a local imbedding of $\lambda(A)$ into E^I, then there is a local imbedding of $\lambda(A)$ into some finite product $E \times \cdots \times E$.

In our next proposition we give a necessary condition for the existence of a local imbedding in terms of the diametral dimension. This result will also shed some light onto the limitations of our theorem.

3.3. Proposition: [6]. If there is a local imbedding of a nuclear G_∞-space $\lambda(A)$ into E, then $\Delta(E) \subset \Delta(\lambda(A))$.

Proof: By the previous proposition we have a local imbedding $i : \lambda(A) \to E$ which satisfies $| x |_e \leq \| ix \|$. Let $(\xi_n) \epsilon \Delta(E)$ and $U = \{ y \epsilon i(\lambda(A)) : \| y \| \leq 1 \}$. Since $\Delta(E) \subset \Delta(i(\lambda(A))$, there is an absolutely convex, closed neighborhood V in $i(\lambda(A))$ with $lim \xi_n d_n(V, U) = 0$. By continuity of i we have an $a \epsilon A$ with $i(U_a) \subset \rho V$ for some $\rho > 0$. Now if $V \subset dU + L$, for some finite-dimensional subspace L of $i(\lambda(A))$ and $d > 0$, since i is one-to-one, we can find a subspace $\tilde{L}$ of $\lambda(A)$ with the same dimension as L and $i(\tilde{L}) = L$. So we get

$$U_a \subset \rho\, d\, U_e + \tilde{L}$$

and therefore for the Kolmogorov diameters we have $\frac{1}{a_n} = d_n(U_a, U_e) = 0(d_n(V, U))$ which implies $\xi \epsilon \lambda(A)' = \Delta(\lambda(A))$ ([32]).

We will now give a partial converse to the above result in the context of power series spaces. Let E be a nuclear Fréchet space which satisfies the conditions $(\underline{DN})$ and (Ω). Then it was proved in [7], that there is an exponent sequence $\epsilon = (\epsilon_n)$ such that

$$\Lambda_1(\epsilon) \subset \Delta(E) \subset \Lambda_\infty(\epsilon)'.$$

These inclusions in fact determine ϵ up to equivalence and we call ϵ the *associated exponent sequence* of E. We would like to mention that Langenbruch also makes use of inclusions of the form $\Lambda_1(\alpha) \subset \Delta(E) \subset \Lambda_\infty(\alpha)'$ in examining the structure of some solution spaces of partial differential operators ([20], [21]).

Some properties of the diametral dimension of E are reflected in the behaviour of ϵ. For example, if $\Delta(E \times E) = \Delta(E)$, then ϵ is stable. Also much more important is the fact that there is a local imbedding of $\Lambda_\infty(\epsilon)$ into E [7]. This is a partial converse of the previous proposition and so the existence of a local imbedding can be formulated in this context in terms of the diametral dimension.

We shall now state and prove a more general result than the one in [7] (cf. also [6]; 2.3. Lemma). This result will be technical in nature. Its proof is almost the same as in [7] but we repeat it with the aim to highlight the essential ingredients.

We recall that E satisfies (Ω) if for every p there is a $q(p)$ such that for every $k \geq q(p)$ there is a j and $C > 0$ so that

$$U_{q(p)} \subset C r^j U_k + \frac{1}{r} U_p$$

holds for all $r > 0$ ([44], [45]). We assume E is nuclear and so each U_k is a Hilbert ball.

3.4. Proposition: Let E be a nuclear Fréchet space with (Ω) and $\Lambda_\infty(\alpha)$ a nuclear power series space. Suppose there is a p such that $d_n(U_{q(p)}, U_p) > 0$ for all n and $-log d_n(U_{q(p)}, U_p) = 0(\alpha_n)$. If $\lambda(A)$ is a G_∞-space with $\Lambda_\infty(\alpha)' \subset \lambda(A)'$ then there is a local imbedding of $\lambda(A)$ into E.

Proof: The condition (Ω) gives

$$d_n(U_{q(p)}, U_p) \leq \rho\, d_n(U_k, U_p)^{\frac{1}{j+1}}$$

where $p, q(p), k$ and j are as in the condition and $\rho > 0$ is some constant ([35]; §3, (1)). In particular $lim\; d_n(U_{q(p)}, U_p) = 0$. $(\Lambda_\infty(\alpha), E)$ satisfies the splitting condition $(\tilde{S}_1)$, of Vogt

([42], [43]), which means that for every μ there is an $m(\mu)$ such that given an operator $T : \Lambda_\infty(\alpha) \to E_{m(\mu)}$, k and $r > 0$ there is an operator $S : \Lambda_\infty(\alpha) \to E_k$ with

$$\| Tx - Sx \|_\mu \leq r \mid x \mid_0 .$$

Here E_k is the Hilbert space obtained by completing E equipped with the norm $\| \ \|_k$. For $\mu = p$, we can choose $m_1 = m(p) > q(p)$ according to $(\tilde{S}_1)$. Using our assumption and setting $k = m_1$ in the above we get

$$-log\, d_n(U_{m_1}, U_p) = 0(\alpha_n)$$

By nuclearity of $\Lambda_\infty(\alpha)$ we find j_0 and $C_0 > 0$ with $\mid x \mid_e \leq C_0 \| x \|_{j_0}$, $x \epsilon \Lambda_\infty(\alpha)$. Hence for some j_1 we have

$$e^{j_0 \alpha_n} \leq e^{j_1 \alpha_n} d_n(U_{m_1}, U_p).$$

Let $\rho : E_{m_1} \to E_p$ be the canonical linking map. Since $lim d_n(U_{q(p)}, U_p) = 0$ and $m_1 \geq q(p)$, ρ is compact. Hence there are orthonormal sequences (f_n) and (g_n) in the Hilbert spaces E_{m_1} and E_p respectively so that

$$\rho(y) = \sum d_n(U_{m_1}, U_p)(y \mid f_n) g_n .$$

We now define $T_1 : \Lambda_\infty(\alpha) \to E_{m_1}$ by setting

$$T_1 x = \sum e^{j_0 \alpha_n} (d_n(U_{m_1}, U_p))^{-1} x_n f_n .$$

We have $\| T_1 x \|_{m_1} \leq \| x \|_{j_1}$ and

$$\mid x \mid_e \leq C_0 \| x \|_{j_0} = C_0 \left(\sum \mid x_n \mid^2 e^{2 j_0 \alpha_n} \right)^{\frac{1}{2}} = C_0 \| T_1 x \|_p .$$

Starting with T_1 which is already constructed, we will use $(\tilde{S}_1)$ to obtain a local imbedding of $\Lambda_\infty(\alpha)$ into E. To simplify notation, we pass to a subsequence (m_k) with $m_0 = p$, $m_1 = m(p)$, $m_k = m(m_{k-1})$ so that for each $T : \Lambda_\infty(\alpha) \to E_{m_k}$ and $r > 0$ there is an $S : \Lambda_\infty(\alpha) \to E_{m_{k+1}}$ with

$$\| Tx - Sx \|_{m_{k-1}} \leq r \mid x \mid_e .$$

We choose $r_i > 0$ with $\sum r_i < (\frac{1}{2C_0})$ and then find $T_2 : \Lambda_\infty(\alpha) \to E_{m_2}$ such that

$$\| T_1 x - T_2 x \|_{m_0} \leq r_1 \mid x \mid_e .$$

We have

$$(C_0^{-1} - r_1) \mid x \mid_e \leq \| T_2 x \|_{m_0} .$$

We apply $(\tilde{S}_1)$ repeatedly to find $T_k : \Lambda_\infty \to E_{m_k}$ such that

$$\| T_k x - T_{k+1} \|_{m_{k-1}} \leq r_k \mid x \mid_e$$

and so we have

$$\left(C_0^{-1} - \sum_{i=1}^{k-1} r_i\right) \mid x \mid_e \leq \| T_k x \|_{m_0} .$$

Hence from the representation of E as the projection limit of the spectrum (E_{m_k}), we obtain a continuous linear map $T : \Lambda_\infty(\alpha) \to E$ which satisfies $\mid x \mid_e \leq 2C_0 \| Tx \|_{m_0}$.

If $\Lambda_\infty(\alpha)' \subset \lambda(A)'$, the inclusion of $\lambda(A)$ into $\Lambda_\infty(\alpha)$ is obviously a local imbedding. Combining this with the local imbedding already constructed, we get a local imbedding of $\lambda(A)$ into E.

In case E satisfies $(\underline{DN})$ and (Ω), we can adjust things so that $-logd_n(U_{q(1)}, U_1) = 0(\epsilon_n)$ where ϵ is the associated exponent sequence [7]. So in this special case we get a local imbedding of $\Lambda_\infty(\epsilon)$ into E.

§4. We start this section with a discussion of an imbedding theorem which is stated and proved in [7]. Let E be a Fréchet space with $(\underline{DN})$ and (Ω) and ϵ the associated exponent sequence. We can represent E as the projective limit of Hilbert spaces (E_k) where each $(-logd_n(U_{k+1}, U_k))$ is equivalent to the associated exponent sequence ϵ (cf. [7]). We replace this projective spectrum by an equivalent one (F_k) where each F_k is in fact isomorphic to $\Lambda_1(\epsilon)$ itself. Suppose Y is isomorphic to a closed subspace of $\Lambda_1(\epsilon)$ and Y has (DN). Since (Y, E) satisfies the splitting condition $(\tilde{S}_1)$ of Vogt, ([42], [43]) we modify the imbeddings of Y into $\Lambda_1(\epsilon)$ by compact perturbations, and obtain an imbedding of Y into E. Although the method used has the same flavour as the proof of 3.4. Proposition, the argument here is more elaborate. For the full proof we refer to [7].

4.1. Theorem: [7]. Let E be a Fréchet space which satisfies $(\underline{DN})$ and (Ω), and ϵ the associated exponent sequence. Assume $\Lambda_1(\epsilon)$ is nuclear. Then any subspace Y of $\Lambda_1(\epsilon)$ which has (DN) is also isomorphic to a subspace of E.

A power series space of infinite-type $\Lambda_\infty(\alpha)$ need not be isomorphic to a subspace of $\Lambda_1(\alpha)$ [13]. In [25], the problem of imbedding $\Lambda_\infty(\alpha)$ into $\Lambda_1(\alpha)$ is discussed in detail. For this question from our theorem we have the following result.

4.2. Corollary: Let $\Lambda_1(\alpha)$ be nuclear. Every subspace of Y of $\Lambda_1(\alpha)$ which has (DN) is also isomorphic to a subspace of $\Lambda_\infty(\alpha)$.

An important case is when the associated exponent sequence is stable. In this case, using Vogt's characterization of subspaces of a nuclear finite type power series spaces [37], [38] and our theorem we get

4.3. Corollary: [7]. Let E be a Fréchet space ($\underline{DN}$) and (Ω) whose associated exponent sequence ϵ is stable and assume $\Lambda_1(\epsilon)$ is nuclear. Then $\Lambda_\infty(\epsilon)$ is isomorphic to a subspace of E and E itself is isomorphic to a subspace of $\Lambda_1(\epsilon)$.

The assumption of the nuclearity of the finite type power series space $\Lambda_1(\epsilon)$ in the preceeding results is used in the proof of the theorem. If we assume that E is nuclear, we automatically get the nuclearity of $\Lambda_\infty(\epsilon)$, but $\Lambda_1(\epsilon)$ may fail to be nuclear.

Suppose $\lambda(A)$ is a nuclear G_∞-space and there is a local imbedding of $\lambda(A)$ into E. We know then $\Delta(E) \subset \lambda(A)' = \Delta(\lambda(A))$ by 3.3. Proposition. Now suppose we want to apply 2.1. Theorem to E. Each one of the assumptions which lead to the conclusion in that theorem implies $\lambda(A)' = \Delta(\lambda(A)) \subset \Delta(E)$. Hence we can apply the theorem only in case E and $\lambda(A)$ have equal diametral dimension. For example, if $E = \Lambda_1(n) \times \Lambda_\infty(n)$, then $\Delta(E) = \Delta(\Lambda_1(n))$ and this is not equal to $\Lambda_\infty(n)'$. Therefore in this case we cannot use our result, although $\Lambda_\infty(n)$ is a complemented subspace of E.

In the rest of this section we shall apply the results of the previous section to our Theorem 2.1. and its consequences.

4.4. Proposition: [7]. Let E be a nuclear Fréchet space with ($\underline{DN}$) and (Ω). If $\Delta(E) = \Lambda_\infty(\alpha)'$ for some α stable, then E has a complemented subspace isomorphic to $\Lambda_\infty(\alpha)$.

For the proof we simply observe that α is the associated exponent sequence of E, which is isomorphic to a quotient space of $\Lambda_\infty(\alpha)$ by the result of Vogt and Wagner [45]. From 3.4. Proposition, we have a local imbedding of $\Lambda_\infty(\alpha)$ into E and so the result follows from 2.1. Theorem.

After the negative answers to many of the conjectures related to the existence of bases, and various approximation properties [13], there is still a problem posed by Pelczynski; that is, whether every complemented subspace of a nuclear Köthe space has a basis. This problem is open in this generality. However one can select a class of nuclear Köthe spaces and consider the problem for complemented subspaces of a space from that class. In the case of finite type power series space, the dead-end method of Mitiagin and Henkin [24] (cf. also [38]) gives a satisfactory positive answer. For the complemented subspaces of the space s of rapidly decreasing sequences there are certain positive answers. For a survey of this problem we refer to the article by Krone [19] in this volume and simply state our contribution for the sake of completeness. We refer to [8] for a generalization.

4.5. Theorem: [6]. Let E be a nuclear Fréchet space which satisfies (DN) and (Ω). If $\Delta(E \times E) = \Delta(E)$ then E is isomorphic to a power series space $\Lambda_\infty(\epsilon)$.

The ϵ here is the associated exponent sequence. The assumptions (DN) and (Ω) give easily $\Delta(E) = \Lambda_\infty(\epsilon)'$ [35], and so ϵ is stable. By the theorem of Vogt and Wagner [45] E is isomorphic to a complemented subspace of $\Lambda_\infty(\epsilon)$ and by 3.4. Proposition there is a local imbedding of $\Lambda_\infty(\epsilon)$ into E. So our generalized decomposition method 2.2. Corollary gives the result.

§5. In this final section we consider several spaces of functions and illustrate how our results can be applied. Whenever relevant, we determine the associated exponent sequence. The examples we treat here are neither the most general nor do we seek the full consequences of our results.

(A) As the first example we consider the space of analytic functions $O(M)$ on a connected k-dimensional Stein manifold M. In the special cases we have that $O(C^k)$ is isomorphic to $\Lambda_\infty(n^{\frac{1}{k}})$ and $O(\Delta^k)$ to $\Lambda_1(n^{\frac{1}{k}})$ where Δ^k denotes the k-dimensional unit polydisc [31]. We also know that $O(M)$ is isomorphic to a subspace of $O(\Delta^k)$ [5]. Hence $O(M)$ has ($\underline{DN}$) and (cf. also [37])

$$\Lambda_1(n^{\frac{1}{k}}) = \Delta(O(\Delta^k)) \subset \Delta(O(M)).$$

By the Oka-Cartan theorem, $O(M)$ is isomorphic to a quotient space of some $O(C^j)$ and so it has property (Ω). In [6] we have shown how one can construct directly a local imbedding of $\Lambda_\infty(n^{\frac{1}{k}})$ into $O(M)$. Hence we have

$$\Delta(O(M)) \subset \Lambda_\infty(n^{\frac{1}{k}})'$$

by 3.3. Proposition. We can also imbed $O(M)$ into the product $(O(C^k))^N$ [5], [24] and use 3.2. Proposition to get this inclusion (see **(B)**). Therefore $(n^{\frac{1}{k}})$ is the exponent sequence associated to $O(M)$.

From 4.3. Corollary we have that $O(C^k)$ is isomorphic to a subspace of $O(M)$. Further, $O(M)$ is isomorphic to $O(C^k)$ if and only if $O(M)$ has property (DN) [6] (2.2. Corollary or 4.5. Theorem.) We would like to mention that intrinsic characterizations in terms of plurisubharmonic functions of those M for which $O(M)$ has (DN) are also known ([1], [2], and [3]).

In order to apply 4.4. Proposition to the space $O(M)$, we need to know when the equality $\Delta(O(M)) = \Delta(O(C^k))$ holds. Using a result of Zaharjuta [47], complete Reinhardt domains D for which $\Delta(O(D)) = \Delta(O(C^k))$ can be characterized [7]. However for $O(\Delta^i \times C^j)$, $i, j \geq 1$ one can also show by more elementary methods that $\Delta(O(\Delta^i \times C^j)) = \Delta(O(C^{i+j}))$ (cf. also [31]). Hence by 4.4. Proposition the space

$O(\Delta^i \times C^j)$ has a complemented subspace which is isomorphic to $O(C^{i+j})$ [7]. We now generalize this in the following.

5.1. Proposition: $O(M \times C^j), j \geq 1$, has a complemented subspace isomorphic to $O(C^{k+j})$.

Proof: By the tensor product representation $O(M \times C^j)$ is isomorphic to a subspace of $O(\Delta^k \times C^j)$. Hence

$$\Delta(O(C^{k+j})) = \Delta(O(\Delta^k \times C^j)) \subset \Delta(O(M \times C^j)).$$

Since the associated exponent sequence of $O(M \times C^j)$ is $(n^{\frac{1}{k+j}})$ we have therefore

$$\Delta(O(M \times C^j)) = \Delta(O(C^{k+j}))$$

and we conclude by 4.4. Proposition.

Let us consider $M \times \Delta^i \times C^j$, $i \geq 1, j \geq 1$. We have that $O(M \times \Delta^i \times C^j)$ is isomorphic to the tensor product $O(M)\hat{\otimes}O(\Delta^i \times C^j)$. Now $O(\Delta^i \times C^j)$ is isomorphic to $O(\Delta^{i+j-1} \times C)$ by a result of Djakov [11]. Hence $O(M \times \Delta^i \times C^j)$ is isomorphic to $O(M \times C)\hat{\otimes}O(\Delta^{i+j-1})$. By what we have shown $O(M \times C)$ has a complemented subspace isomorphic to $O(C^{k+1})$; that is $O(M \times C) = O(C^{k+1}) \oplus F$ for some F. Using the above tensor product representation and the result of Djakov once more, we obtain the following

5.2. Proposition: $O(M \times \Delta^i \times C^j)$ has a complemented subspace which is isomorphic to $O(\Delta^{k+i} \times C^j)$.

(B) Let X be an analytic subvariety of C^k, J_x the space of all functions in $O(C^k)$ which vanish on X. We follow the discussion given by Vogt in [43] (7.C.) (cf. also [41]). J_x is a subspace of $O(C^k)$. Hence it has (DN) and $\Lambda_\infty(n^{\frac{1}{k}})' \subset \Delta(J_x)$. Further J_x also has (Ω). Therefore $\Delta(J_x) = \Lambda_\infty(\beta)'$ for some β [35]. $O(C^k)$ is isomorphic to a subspace of the product $(J_x)^N$ [43]. Hence, by 3.2. Proposition, there is a local imbedding of $O(C^k)$ into $(J_x)^s$ for some integer s. So by 3.3. Proposition we have $\Delta((J_x)^s) \subset \Lambda_\infty(n^{\frac{1}{k}})'$. Stability of $O(C^k)$ yields

$$\Lambda_\infty(n^{\frac{1}{k}})' = \Delta((J_x)^s) \subset \Lambda_\infty(\beta)'.$$

However $\Delta((J_x)^s) = \Lambda_\infty(\beta^{(s)})'$ where $\beta^{(s)}$ is that sequence obtained from β by repeating each term s-times. Since $\beta_n^{(s)} = O(n^{\frac{1}{k}})$ and $(n^{\frac{1}{k}})$ is stable, we have $\beta_n = O(n^{\frac{1}{k}})$. (cf. also [5]; Proof of Theorem 2.). So we have $\Delta(J_x) = \Lambda_\infty(n^{\frac{1}{k}})'$. From 4.5. Theorem we have

5.3. Proposition: J_x is isomorphic to $O(C^k)$.

This delivers a more direct proof of 7.3. Theorem in [43]. For further developments in this direction we refer to [4].

(C) Let $P(D)$ be an elliptic linear partial differential operator with constant coefficients on R^k, $k \geq 2$. Let O be an open subset of R^k. The space of null solutions $N(O) = \{f \epsilon C^\infty(O) : P(D)f = 0\}$ has property (Ω) [40] (cf. also [26]). If O is connected, then Vogt [40] has proved that $N(O)$ has $(\underline{DN})$.

Wiechert [46] has shown that $N(R^k)$ is isomorphic to the power series space $\Lambda_\infty(n^{\frac{1}{k-1}})$. If O is a bounded convex subset of R^k, then $N(O)$ is isomorphic to $\Lambda_1(n^{\frac{1}{k-1}})$ [46]. We also know [37] for any open subset O

$$\Lambda_1(n^{\frac{1}{k-1}}) \subset \Delta(N(O)).$$

Also, we have an imbedding of $N(R^k)$ into the product $(N(O))^N$. Hence, as in the previous example we get $\Delta(N(O)) \subset \Lambda_\infty(n^{\frac{1}{k-1}})'$. Therefore for any open connected subset O of R^k the space $N(O)$ has the properties $(\underline{DN}) and (\Omega)$ and its associated exponent sequence is $(n^{\frac{1}{k-1}})$. Hence as a corollary of 4.3. Corollary and 4.5. Theorem we have the following result.

5.4. Proposition: Let O be an open connected subset of R^k. Then $N(R^k)$ is isomorphic to a subspace of $N(O)$ and $N(O)$ isomorphic to a subspace of $N(B)$ where B is a bounded convex subset of R^k. Further $N(O)$ has (DN) if and only if it is isomorphic to $N(R^k)$.

(D) Let E be a nuclear Fréchet space and $H(E'_b)$ the space of holomorphic functions on the dual E'_b with the compact-open topology. For a nuclear power series space $\Lambda_\infty(\alpha)$, Börgens, Meise and Vogt [9] have proved that the space of holomorphic functions $H(\Lambda_\infty(\alpha)'_b)$ is isomorphic to the power series space $\Lambda_\infty(\beta(\alpha))$. We refer to [9] and [10] for the definition and properties of the sequence $\beta(\alpha)$. One remarkable property is that $\beta(\alpha)$ is always stable. Also $\beta(log(n+1))$ is equivalent to $(log(n+1))$ and so $H(s'_b)$ is isomorphic to s itself [9].

5.5. Proposition: If there is a local imbedding of $\Lambda_\infty(\alpha)$ into a nuclear Fréchet space E, then there is a local imbedding of $\Lambda_\infty(\beta(\alpha))$ into $H(E'_b)$.

Proof: We let $p_m(f) = sup\{| f(u) | : u \epsilon U_m^0\}$ for $f \epsilon H(E'_b)$ and U_m a neighborhood of E. Let $T : \Lambda_\infty(\beta(\alpha)) \rightarrow H(\Lambda_\infty(\alpha)'_b)$ be an isomorphism. We determine m so that $| x |_e \leq p_m(Tx)$ holds. For this m we find by 3.4. Proposition, and 3.1. Proposition a local imbedding $L : \Lambda_\infty(\alpha) \rightarrow E$ with $| y |_m \leq \| Ly \|_k$ for some k. Then $V_m^0 \subset \overline{L'(U_k^0)} = L'(U_k^0)$. We define

$h : \Lambda_\infty(\beta(\alpha)) \to H(E_b')$ by

$$h(x)(u) = Tx(L'u), \quad x\epsilon\Lambda_\infty(\beta(\alpha)), \quad u\epsilon E'.$$

Then we have

$$| x |_e \le p_m(Tx) \le sup\left\{| h(x)(u) |: u\epsilon U_k^0\right\} = p_k(h(x)).$$

Hence h is a local imbedding.

With the existence of the local imbedding of $\Lambda_\infty(\beta(\alpha))$ into $H(E_b')$ at our disposal we can now apply our results to spaces of holomorphic functions.

5.6. Proposition: Let E be a complemented subspace of s. Then $H(E_b')$ is isomorphic to a nuclear stable power series space of infinite type.

Proof: We know $\Delta(E) = \Lambda_\infty(\alpha)'$ for some α [35]. By a result of Meise and Vogt [22] (2.5. Lemma) $H(E_b')$ is a complemented subspace of $H(s_b')$ which in turn is isomorphic to s. So $H(E_b')$ has (DN) and (Ω). Further, $H(E_b')$ is $\Lambda_N(\beta(\alpha))$-nuclear and so $\Lambda_\infty(\beta(\alpha))' \subset \Delta(H(E_b'))$ by [10]; 4.1. Theorem. There is a local imbedding of $\Lambda_\infty(\alpha)$ into E [6]. So by 5.5. Proposition we have a local imbedding of $\Lambda_\infty(\beta(\alpha))$ into $H(E_b')$ and therefore by 3.3. Proposition we get

$$\Delta(H(E_b')) = \Lambda_\infty(\beta(\alpha))'.$$

Since $\beta(\alpha)$ is stable, by 4.5. Theorem we have that $H(E_b')$ is isomorphic to $\Lambda_\infty(\beta(\alpha))$. We note that here we do not know whether E has a basis, although $H(E_b')$ does have one.

Let E be a nuclear Fréchet space with $(\underline{DN})$ and (Ω) and ϵ its associated exponent sequence. Let us also assume that E is isomorphic to a subspace of $\Lambda_1(\epsilon)$ and $\Lambda_1(\epsilon)$ is nuclear. Then $H(E_b')$ is isomorphic to a subspace of $\Lambda_1(\beta(\epsilon))$ ([22]; 3.7. Corollary). So $H(E_b')$ has $(\underline{DN})$, (Ω) and by 5.5. Proposition we have

$$\Lambda_1(\beta(\epsilon)) \subset \Delta(H(E_b')) \subset \Lambda_\infty(\beta(\epsilon))'.$$

Therefore $\beta(\epsilon)$ is the associated exponent sequence of $H(E_b')$. Further, from the nuclearity of $\Lambda_1(\epsilon)$, we have that $\Lambda_1(\beta(\epsilon))$ is also nuclear ([10]; 2.5. Proposition). Hence, with these assumptions on E we have that $H(\Lambda_\infty(\epsilon)_b')$ is isomorphic to a subspace of $H(E_b')$ by 4.3. Corollary.

Also, if O is a non-empty open subset of E_b', then covering E_b' by countably many translates of O, we can get an imbedding of $H(E_b')$ into $H(O)^N$. Combining this with the local imbedding of $\Lambda_\infty(\beta(\epsilon))$ into $H(E_b')$, we obtain a local imbedding of $\Lambda_\infty(\beta(\epsilon))$ into $H(O)^k$ for some $k \ge 1$ by 3.2. Proposition.

Let $\lambda(P)$ be a quotient space of $\Lambda_\infty(\epsilon)$ which is nuclear and assume $\Delta(\lambda(P)) = \Lambda_\infty(\epsilon)'$. Let $a \epsilon \lambda(P)$ be a sequence with non-negative terms and

$$D_a = \{(x_j) : sup \mid x_j \mid a_j < 1\}.$$

Suppose $\lambda(P)$ is also isomorphic to a subspace of some nuclear $\Lambda_1(\delta)$. Then $H(D_a)$ is isomorphic to a subspace of $\Lambda_1(\beta(\delta))$ and so it has $(\underline{DN})$ ([22]; 3.6. Proposition). By another result of Meise and Vogt ([22]; 4.4. Corollary), $H(D_a)$ is a quotient space of $\Lambda_\infty(\beta(\epsilon))$. So it has (Ω) and also $\Lambda_\infty(\beta(\epsilon))' \subset \Delta(H(D_a))$. We already noted that there is a local imbedding of $\Lambda_\infty(\beta(\epsilon))$ into $H(D_a)^k$ for some $k \geq 1$. Using now the stability of $\beta(\epsilon)$, and 3.3. Proposition we get $\Delta(H(D_a)) = \Lambda_\infty(\beta(\epsilon))'$.

Our final result now follows immediately from 4.4. Proposition.

5.7. Proposition: Let $\lambda(P)$ be quotient space of some $\Lambda_\infty(\epsilon)$ with $\Delta(\lambda(P)) = \Lambda_\infty(\epsilon)'$ and assume $\lambda(P)$ is isomorphic to a subspace of some nuclear power series space of finite type. Let D_a be an open polydisc in $\lambda(P)'_b$. Then $H(D_a)$ has a complemented subspace which is isomorphic to $H(\Lambda_\infty(\epsilon)'_b)$.

Polydiscs $D_a \subset \Lambda_1(\alpha)'$ for which $H(D_a)$ is isomorphic to $\Lambda_1(\beta(\alpha))$ have been characterized by Meise and Vogt [22]. We also note that $H(\Lambda_1(\alpha)'_b)$ can never be isomorphic to $\Lambda_1(\beta(\alpha))$, ([9], [22]). In case $\Lambda_\infty(\alpha)$ is isomorphic to a subspace of $\Lambda_1(\alpha)$ and D_a a polydisc in $\Lambda_\infty(\alpha)'_b$, then our proposition gives that $H(D_a)$, has a complemented subspace which is isomorphic to $H(\Lambda_\infty(\alpha)'_b)$. Further, $H(D_a)$ is isomorphic to $H(\Lambda_\infty(\alpha)'_b)$ if and only if $H(D_a)$ has (DN), since $\Delta(H(D_a))$ is equal to $\Delta(\Lambda_\infty(\beta(\alpha))$.

REFERENCES

[1] A. Aytuna:'On the linear topological structure of spaces of analytic functions'. *Doga Tr. J. Math.* **10**, 46-49 (1986).

[2] A. Aytuna: 'On Stein manifolds M for which $O(M)$ is isomorphic to $O(\Delta^n)$ as Fréchet spaces'. *manus. math.*, **62** 297-315 (1988).

[3] A. Aytuna: 'Stein spaces M for which $O(M)$ is isomorphic to a power series space', these Proceedings.

[4] A. Aytuna: 'The Fréchet space structure of global sections of coherent analytic sheaves'. To appear in *Math. Balkanica*.

[5] A. Aytuna and T. Terzioğlu: 'On certain subspaces of a nuclear power series space of finite type'. *Studia Math.* **69**, 79-86 (1980).

[6] A. Aytuna, J. Krone and T. Terzioğlu: 'Complemented infinite type power series subspaces of nuclear Fréchet spaces'. *Math. Ann.* **283**, 193-202 (1989).

[7] A. Aytuna, J. Krone and T. Terzioğlu: 'Imbedding of power series spaces and spaces of analytic functions'. Preprint.

[8] A. Aytuna, J. Krone and T. Terzioğlu: 'On complemented subspaces of certain nuclear Köthe spaces', these Proceedings.

[9] J. Börgens, R. Meise, D. Vogt: 'Entire functions on nuclear sequence spaces'. *J. reine angew. Math.* **322**, 196-220 (1981).

[10] J. Börgens, R. Meise, D. Vogt: '$\Lambda_\infty(\alpha)$ -nuclearity in infinite-dimensional holomorphy'. *Math. Nachr.* **106**, 129-146 (1982)

[11] P.B. Djakov: 'On isomorphisms of some spaces of holomorphic functions' (Russian). *F. Analysis and appl.* (Trans. of 6. Winter School in Progobych (1973); ed. B. Mityagin 1975, 129-135.

[12] M. M. Dragilev: 'On regular bases in nuclear spaces'. *AMS Trans. (2)* 61-82 (1970).

[13] E. Dubinsky: *The structure of nuclear Fréchet spaces,* Lecture Notes in Mathematics 720, 1979.

[14] L. Holmström: 'Superspaces of (s) with basis'. *Studia Math.* **75**, 139-152 (1983).

[15] H. Jarchow: *Locally convex spaces.* Teubner Verlag, 1981.

[16] T. Komura und Y. Komura: 'Über die Einbettung der nuklearen Räume in $(s)^A$'. *Math. Ann.* **162**, 284-288 (1966).

[17] G. Köthe: *Topological Vector Spaces I, II.* Berlin-Heidelberg-New York, 1969 and 1979.

[18] G. Köthe: 'Über nukleare Folgenräume'. *Studia Math.* **31**, 267-271 (1968).

[19] J. Krone: 'On Pelczynski's Problem'. This volume.

[20] M. Langenbruch: 'Kolmogorov diameters in solution spaces of systems of partial differential equations'. *manus. math.* **53**, 35-64 (1985).

[21] M. Langenbruch: 'Isomorphieklassen von Losungsräumen partieller Differentialgleichungssysteme'. Habilitationsschrift, Münster 1984.

[22] R. Meise and D. Vogt: 'Structure of spaces of functions on infinite- dimensional polydiscs'. *Studia Math.* **75**, 235-252 (1983).

[23] R. Meise and B.A. Taylor: 'Linear extension operators for ultradifferentiable functions of Beurling type on compact sets'. To appear in *Amer. J. Math.*

[24] B.S. Mitiagin and G.M. Henkin: 'Linear problems of complex analysis'. *Russian Math. Surveys* **26**, 99-164 (1971).

[25] Z. Nurlu: 'Imbedding $\Lambda_\infty(\alpha)$ into $\Lambda_1(\alpha)$ and some consequences'. *Math. Balkanica.* New Series **1**, 14-24 (1987).

[26] H.J. Petzsche: 'Some results of Mittag-Leffler type for vector valued functions and spaces of class A'. *Func. Analysis: Surveys and Recent Results II.* (K.D. Bierstedt and B. Fuchssteiner, eds.) North-Holland Math. Studies 38, 183-204 (1979).

[27] A. Pietsch: *Nuclear locally convex spaces.* Berlin-Heidelberg-New York, 1972.

[28] M.S. Ramanujan and B. Rosenberger: 'On $\lambda(\phi, P)$ nuclearity'. *Compositio Math.* **37**, 113-125 (1975).

[29] M.S. Ramanujan and T. Terzioglu: 'Power series spaces $\Lambda_k(\alpha)$ of finite type and related nuclearities'. *Studia Math.* **53**, 1-13 (1975).

[30] W. Robinson: 'Some equivalent classes of Köthe spaces'. *Rocz. Pol. Tow. Math. Ser.I, Mat. Prace*, **20**, 449-451 (1978).

[31] S. Rolewicz: 'On spaces of holomorphic functions'. *Studia Math.* **21**, 135-160 (1962).

[32] T. Terzioğlu: 'Die diametrale Dimension von lokalkonvexen Räumen'. *Collect. Math.* **20**, 49-99 (1969).

[33] T. Terzioğlu: 'Smooth sequence spaces and associated nuclearity'. *Proc. Amer. Math. Soc.* **37**, 497-504 (1973).

[34] T. Terzioğlu: 'Smooth sequence spaces'. *Proceedings of Symp. on Func. Analysis, Silivri*, 31-41 (1974).

[35] T. Terzioğlu: 'On the diametral dimension of some classes of F-spaces'. *J. Karadeniz Uni. Ser. Math.-Physics* **8**, 1-13 (1985).

[36] D. Vogt: 'Ein Isomorphiesatz für Potenzreihenräume'. *Arch. Math.* **38**, 540-548 (1982).

[37] D. Vogt: 'Charakterisierung der Unterräume eines nuklearen stabilen Potenzreihenräumes von endlichem Typ'. *Studia Math.* **71**, 251-270 (1982).

[38] D. Vogt: 'Eine Charakterisierung der Potenzreihenräume von endlichem Typ und ihre Folgerungen'. *manus. math.* **37**, 269-301 (1982).

[39] D. Vogt: 'Sequence space representations of spaces of test functions and distributions'. *Advances in Func. Analysis, Holomorphy and Approximation Theory* (ed: G.I. Zapata) 405-443, New York - Basel (1983).

[40] D. Vogt: 'On the solvability of P(D)f = g for vector valued functions'. *RIMS Kokyoroku* **508**, 168-182 (1983).

[41] D. Vogt: 'Über die Existenz von Ausdehnungsoperatoren für holomorphe Funktionen auf analytischen Mengen in C^n'. Preprint.

[42] D. Vogt: 'Some results on continuous linear maps between Fréchet spaces'. *Functional Analysis: Surveys and Recent Results III.* (Eds.: K.D. Bierstedt and B. Fuchssteiner) 349-381 North-Holland Math. Studies 90 (1984).

[43] D. Vogt: 'On the functor $Ext^1(E, F)$ for Fréchet spaces'. *Studia Math.* **85**, 163-197 (1987).

[44] D. Vogt and M.J. Wagner: 'Charakterisierung der Quotientenräume von s und eine Vermutung von Martineau'. *Studia Math.* **67**, 225-240 (1980).

[45] D. Vogt and M.J. Wagner: 'Charakterisierung der Unterräume und Quotientenräume der nuklearen stabilen Potenzreihenräume von unendlichem Typ'. *Studia Math.* **70**, 65-80 (1981).

[46] G. Wiechert: 'Dualitäts - und Strukturtheorie der Kerne von linearen Differentialoperatoren'. Dissertation, Wuppertal 1982.

[47] V.P. Zaharjuta: 'Generalized Mityagin invariants and a continuum of pairwise nonisomorphic spaces of analytic functions'. *Func. Anal. and App.*, **11**, 182-188 (1978).

ON COMPLEMENTED SUBSPACES OF CERTAIN NUCLEAR KÖTHE SPACES

A. AYTUNA
Department of Mathematics
M. E. T. U.
06531 Ankara-Turkey

J. KRONE
FB Mathematik
Bergische Universität.
Wuppertal - F.R. Germany

T. TERZİOĞLU
Department of Mathematics
M. E. T. U.
06531 Ankara-Turkey

INTRODUCTION

Report [6] of this conference contains a brief description of the state of the art concerning Pelczynski's problem [10]. An important concept mentioned there is Vogt's decomposition method [15], also called Pelczynski's trick, and our improvement which is given in [2]. This method implies a power series space representation for many function spaces, but it is not limited to power series spaces of infinite type. In fact, it is the purpose of the present note to prove the analogous result for a more general case of nuclear Köthe spaces that satisfy the splitting condition S_1 of Apiola [1] and Vogt [16].

The structure of the spaces in this class, which includes infinite type power series spaces, $L_f(\alpha;\infty)$ spaces of Dragilev [3], (cf. also [9]) and some weighted spaces of analytic functions [7], [8] has been investigated by Krone [5]. For more explanations and further relevant literature we refer to [6].

§1. We use the standard terminology and notation of the theory of locally convex spaces as in [4]. For nuclear spaces we refer the reader to [3], [11]. By a subspace of a locally convex space we shall mean a closed subspace. For a fundamental sequence $\{\|\ \|_k\}_k$ of seminorms of E, the set $\{x \in E : \| x \|_k \leq 1\}$ will be denoted by U_k. Also we let E_k to denote the completion of the normed space $\{E/Ker \|\ \|_k ; \|\ \|_k\}$. If E, F are locally convex spaces, L(E, F) is the space of continuous linear maps from E into F. In case E and F are Banach spaces, the unit ball of L(E, F) will be denoted by B(E, F).

T. Terzioğlu (ed.), Advances in the Theory of Fréchet Spaces, 325–332.

The following conditions on a pair of Fréchet spaces have been introduced by Vogt [16].

$$(S_1): \qquad \exists n\ \forall \mu\ \exists k\ \forall K, m, R>0\ \exists n, S;$$
$$B(E_m, F_k) \subset S B(E_n, F_K) + \tfrac{1}{R} B(E_{n_0}, F_\mu).$$

$$(\tilde{S}_1): \qquad \exists n_0\ \forall \mu\ \exists k\ \forall K, R>0;$$
$$L(E, F_k) \subset L(E, F_K) + \tfrac{1}{R} B(E_{n_0}, F_\mu).$$

We will write $(E, F) \in S_1$, $(\tilde{S}_1)$ in case E and F satisfy the condition S_1 $(\tilde{S}_1)$. A Fréchet space E is said to satisfy the condition $(\tilde{S}_1)$ in case $(E, E) \in S_1$. If $(E, F) \in S_1$ then any exact sequence of Fréchet spaces of the form

$$0 \to F \to G \to E \to 0$$

splits. Clearly the condition (S_1) implies $(\tilde{S}_1)$ and under certain conditions they coincide. We refer the reader to [16] for more information on these matters.

A Fréchet space E is said to have the property (DN) in case there exists a fundamental sequence of seminorms $\{\|\ \|_k\}_{k=0}^{\infty})$ such that

$$\| x \|_k^2 \leq \| x \|_{k+1} \| x \|_0, \quad \forall x \in E.$$

In the category of nuclear Fréchet spaces, this condition characterizes the subspaces of the space of rapidly decreasing sequences [14]. A Fréchet space E is called *stable* in case it is isomorphic to $E \times E$.

Let $\phi = \{\phi_1, \phi_2, \cdots\}$ be a sequence of monotone increasing functions from R^+ into R^+ with the following property (*): $\exists r_0 > 0$ such that for $r > r_0$;

$$r^2 \leq \phi_1(r) \leq \phi_2(r) \cdots \leq \phi_k(r) \leq \phi_{k+1}(r) \leq \cdots.$$

For a given positive increasing, unbounded sequence (α_n) of real numbers the Köthe space $\lambda(A)$ where $A = \left\{a_n^k\right\}, a_n^k = \phi^k(\alpha_n) = \phi_k \circ \cdots \circ \phi_1(\alpha_n)$ will be denoted by $\Lambda_\phi(\alpha)$. The class of spaces $\{\Lambda_\phi(\alpha)\}$ will be refered to as G_∞- spaces defined by transition functions [5]. For G_∞-spaces we refer the reader to [12]. A space $\Lambda_\phi(\alpha)$ is nuclear if there exists a $k > 0$ such that $\sum \alpha_i^{-k} < \infty$.

In this paper we will be interested in nuclear Köthe spaces which satisfy the condition (S_1). The structure of these spaces has been investigated by Krone [5]. It turns out that this class coincides with the class of G_∞-spaces defined by transition functions. In fact it is shown in [5] that a nuclear Fréchet space E with basis $\{x_n\}$ satisfies the property (S_1) if and only if there exists a space $\Lambda_\phi(\alpha)$, scalars $\lambda_n \neq 0$, and a permutation π of natural numbers such that the operator that sends $x_{\pi(n)}$ to $\lambda_n e_n$, n=1,2,..., induces an isomorphism between E and $\Lambda_\phi(\alpha)$.

Let $\phi = (\phi_k)$ be a sequence of functions satisfying (*) above. The following conditions on a Fréchet space E play an important role in the structure theory of G_∞-spaces defined by transition functions [5]:

$$\overline{DN}_\phi : \qquad \exists\, p,\ \forall\, m, K, R > 0,\ \exists\, n, S : \ U_m^0 \subset \tfrac{r}{R} U_p^0 + \tfrac{S}{\phi_K(r)} U_n^0 \ \forall\, r > 0$$

$$\Omega_\phi : \qquad \forall\, \mu\ \exists\, k\ \forall\, K\ \exists\, n, S : \ U_k \subset S\phi^n(r) U_K + \tfrac{S}{r} U_\mu \ \forall\, r > 0.$$

The condition $\overline{DN}_\phi$ is stronger than the condition DN introduced above. It is shown in [5] (2.1.6. Folgerung, 2.2.4 Satz) that a nuclear Fréchet space E has the property $\overline{DN}_\phi$ if and only if it is isomorphic to a subspace of a $\Lambda_\phi(\alpha)$ space and it has the property Ω_ϕ if and only if it is isomorphic to a quotient space of a $\Lambda_\phi(\alpha)$ space.

The diametral dimension $\Delta(E)$ of E is the set of all sequences (ξ_n) such that for every k there is a p with $\lim\ \xi_n\, d_n(U_p, U_k) = 0$, where $d_n(U_p, U_k)$ denotes the n-th Kolmogorov diameter of U_p with respect to U_k ([12], [13]). A Fréchet space E is said to have stable diametral dimension in case $\Delta(E \times E) = \Delta(E)$.

§2. Let E be a nuclear Köthe space which satisfies (S_1). In this section we will prove that every complemented subspace of E with stable diametral dimension has a basis. We will follow the technique given in [15] (cf. [2]). This strategy can be summarized as follows. Suppose a complemented subspace F of E is given. First one tries to find a space $\Lambda_\phi(\alpha)$ with the following properties:

(i) F has a complemented subspace isomorphic to $\Lambda_\phi(\alpha)$ and

(ii) $\Lambda_\phi(\alpha)$ has a complemented subspace isomorphic to E.

Once this is done, one can then use Vogt's decomposition method and prove that F and $\Lambda_\phi(\alpha)$ are in fact isomorphic to each other. We recall that a continuous linear mapping T from a G_∞-space $\lambda(A)$ into a Fréchet space X is called a *local imbedding* in case there is a continuous seminorm $\|\ \|$ on X such that

$$\sum_{n=1}^{\infty} | x_n | = | x |_0 \le \| Tx \|$$

holds for every $x \in \lambda(A)$ [2]. Our first lemma, which is a corollary of the proof of 2.3. Lemma of [2], combined with 1.3. Theorem of [2] and 2.3.1. Satz [5], will allow us to reduce the task of finding a space with (i) and (ii) above to diametral dimension considerations. These are contained in 2.2., and 2.3. Lemmas which generalize some of results of [13] to our setting.

2.1. Lemma: Let $\lambda(A)$ be a nuclear G_∞-space. Suppose that E is a nuclear Fréchet space with a fundamental system of seminorms $\{\| \ \|_k\}$ which satisfies,

(i) $(\lambda(A), E) \in \tilde{S}_1$

(ii) $\exists\, k_0$ such that $\forall\, m\ \exists\, \ell$ with $\inf_n\, a_n^\ell\, d_n(U_m, U_{k_0}) > 0$

where U_k is the unit ball corresponding to $\| \ \|_k$. Then there exists a local imbedding of $\lambda(A)$ into E.

Proof: In view of (i), we can choose an increasing sequence of integers (m_k), tending to infinity such that $m_0 = k_0$ of the condition (ii) and

$$L(\lambda(A), E_{m_k}) \subset L(\lambda(A), E_{m_{k+1}}) + \varepsilon B(\lambda(A)_0, E_{m_{k-1}})$$

holds for every k and $\epsilon > 0$. From condition (ii) we fix an ℓ with $\inf\ a_n^\ell\, d_n(U_{m_1}, U_{m_0}) > 0$. To conclude the proof we simply repeat the argument given in 2.3. Lemma of [2].

Although we will not use it in this paper we would like to point out that the condition (ii) in 2.1. Lemma can be replaced with a weaker condition; namely that there exists a k_0 and ℓ such that $\inf_n\ a_n^\ell\, d_n(U_{k_1}, U_{k_0}) > 0$, where k_1 is the index corresponding to k_0 in the condition $(\lambda(A), E) \in S_1$. A deeper and more detailed discussion of results of this type is given by Vogt in [17].

In the next two lemmas we aim to identify the diametral dimension of a complemented subspace of a $\Lambda_\phi(\alpha)$-space with that of a suitable $\Lambda_\phi(\beta)$ space. In the case of power series spaces this was done in [13]. Our first lemma is rather technical in nature and it is in fact implicitly contained in the proof of 2.1.5 Satz [5].

2.2. Lemma: Let F be a nuclear Fréchet space with the property $\overline{DN}_\phi$. There exists a p such that for every k and $R > 0$ there exists an n and $C > 0$ with

$$\phi^k\left(\frac{R}{d_i(U_k, U_p)}\right) \leq \frac{C}{d_i(U_n, U_k)}.$$

Proof: According to 2.1.2 Lemma of [5], F also satisfies the condition $\overline{DN}_{\tilde{\phi}}$ where $\tilde{\phi} = \left\{\phi^k\right\}$. We choose a fundamental sequence of norms $\{\| \ \|_k\}$ defined by inner products. The condition $\overline{DN}_{\tilde{\phi}}$ implies that there exists a p such that for every k and $R > 0$ there exists an n and $S > 0$ such that

$$U_k^0 \subset \frac{r}{3R} U_p^0 + \frac{S}{\phi^k(r)} U_n^0 \ \forall r > 0.$$

We fix a k and $R > 0$. Choosing $r = \frac{R}{d_i(U_k,U_p)}$ the above expression takes the form,

$$U_k^0 \subset \frac{1}{3d_i(U_k,U_p)} U_p^0 + \frac{S}{\phi^k\left(\frac{R}{d_i(U_k,U_p)}\right)} U_n^0.$$

Hence there exists a subspace F_i of dimension at most i such that

$$U_k^0 \subset \frac{1}{2} U_k^0 + \frac{S}{\phi^k\left(\frac{R}{d_i(U_k,U_p)}\right)} U_n^0 + F_i.$$

Repeated applications of the above expression yields, for any j,

$$U_k^0 \subset (\frac{1}{2})^j\ U_k^0 + \frac{S\sum_{\ell=0}^{j-1}(\frac{1}{2})^\ell}{\phi^k\left(\frac{R}{d_i(U_k,U_p)}\right)} U_n^0 + F_i.$$

So, by taking j sufficiently large, we obtain;

$$U_k^0 \subset \frac{3S}{\phi^k\left(\frac{R}{d_i(U_k,U_p)}\right)} U_n^0 + F_i.$$

Since the subspace F_i has dimension at most i, the above expression yields;

$$d_i(U_n,U_k) \leq 3S\left(\phi^k\left(\frac{R}{d_i(U_k,U_p)}\right)\right)^{-1}.$$

This finishes the proof of 2.2. Lemma.

2.3. Lemma: Let F be a nuclear Fréchet space with the properties $\overline{DN}_\phi$ and Ω_ϕ. Then there exists an increasing, unbounded, sequence $\{\beta_i\}_i$ of positive real numbers such that,

(i) $\Delta(F) = \Delta(\Lambda_\phi(\beta))$,

(ii) $\Lambda_\phi(\beta)$ can be locally imbedded into F.

Proof: We choose a p according to 2.2. Lemma and a corresponding k_1 according to Ω_ϕ. Since

$$\phi^k(2Sr) \leq S'\phi^k(r^2) \leq S''\phi^{k+1}(r) \quad r > 0$$

we obtain the following version of Ω_ϕ: (**) $\forall m\ \exists k, S > 0$ with

$$U_{k_1} \subset S\phi^k(r)U_m + U_p \quad r > 0.$$

We can assume moreover $(d_i(U_k,U_p)) \in \ell_1$. For $\beta_i = (d_i(U_{k_1},U_p))^{-1}$ 2.2. Lemma implies that for each $k \geq k_1$ there are n, C with

$$\phi^k(\beta_i) \leq \frac{C}{d_i(U_n,U_k)}.$$

Hence the sequence $(\phi^k(\beta_i))$ belongs to $\Delta(F)$ and so we obtain

$$\Delta(\Lambda_\phi(\beta)) = \Lambda_\phi(\beta)' \subset \Delta(F).$$

If we show that condition (ii) of 2.1. Lemma is satisfied then we have $\Delta(F) \subset \Delta(\Lambda_\phi(\beta))$ and that there is a local imbedding of $\Lambda_\phi(\beta)$ into F. Now, (**) with $r = \beta_i$ implies that for every m there are k and S with

$$U_{k_1} \subset \left(S\phi^k(\beta_i)\, d_i\,(U_m, U_p) + (2\beta_i)^{-1}\right) U_p + F_i$$

for some subspace F_i of F with dimension $\leq i$. Hence we conclude

$$\frac{1}{\beta_i} = \frac{1}{d_i\,(U_{k_1}, U_p)} \leq S\phi^k(\beta_i)\, d_i\,(U_m, U_p) + \frac{1}{2\beta_i}.$$

So

$$\frac{1}{2} \leq S\phi^k\,(\beta_i)\,\beta_i d_i\,(U_m, U_p) \leq S'\phi^{k+1}(\beta_i)\, d_i\,(U_m, U_p)$$

for all $i \in N$. Our next result is an adjustment of 1.3. Satz of [15] to our setting.

2.4. Proposition: (Vogt-Pelczynski decomposition method). Let E be a stable nuclear Köthe space which satisfies the condition (S_1). If F is isomorphic to a complemented subspace of E and E is isomorphic to a complemented subspace of F, then F is isomorphic to E.

Proof: By a theorem of Vogt and Wagner (2.3. Satz [18]), we have an exact sequence of the form

$$0 \to E \to E \to E^N \to 0.$$

Let G be a complemented subspace of E. Since $E^N \approx G \oplus E^N$, as in (1.2. Lemma, [14]), there is a Fréchet space H and a commutative diagram of exact sequences;

$$\begin{array}{ccccccccc}
 & & & & 0 & & 0 & & \\
 & & & & \uparrow & & \uparrow & & \\
0 & \to & E & \to & E & \to & E^N & \to & 0 \\
 & & & & \uparrow & & \uparrow & & \\
0 & \to & E & \to & H & \to & G \oplus E & \to & 0 \\
 & & & & \uparrow & & \uparrow & & \\
 & & & & E & & E & & \\
 & & & & \uparrow & & \uparrow & & \\
 & & & & 0 & & 0 & &
\end{array}$$

In view of the condition (S_1), the vertical exact sequence containing H splits. According to 2.1.3. Satz of [5] the horizontal exact sequence containing H splits also. Taking the stability of E into account, these observations imply that E is isomorphic to $G \oplus E$. So if $F \approx G \oplus E$, then G is a complemented subspace of E and therefore the above argument shows that F is isomorphic to E. This finishes the proof of 2.4. Proposition.

Now we are in a position to prove our main result. We note that it generalizes 2.2. Theorem in [2].

Theorem. *Let E be a nuclear Köthe space which satisfies the condition (S_1). Then every complemented subspace of E with stable diametral dimension has a basis.*

Proof: Let F be a complemented subspace of E with stable diametral dimension. Since E is isomorphic to a space $\Lambda_\phi(\alpha)$, F has the properties $\overline{DN}_\phi$ and Ω_ϕ. In view of 2.3. Lemma there exists a sequence (β_n) such that $\Delta(F) = \Delta(\Lambda_\phi(\beta))$. Since F has stable diametral dimension, the space $\Lambda_\phi(\beta)$ is stable. Now the fact that $\Lambda_\phi(\beta)$ can be locally imbedded into F implies, in view of 1.3 Theorem in [2], that $\Lambda_\phi(\beta)$ is isomorphic to a complemented subspace of F. On the other hand F is isomorphic to a complemented subspace of $\Lambda_\phi(\beta)$ according to 2.3.1. Satz [5]. Hence 2.4. Proposition implies that F is isomorphic to $\Lambda_\phi(\beta)$, and has, in particular, a basis.

REFERENCES

[1] Apiola, H.: 'Characterization of subspaces and quotients of nuclear $L_f(\alpha.\infty)$-spaces', *Compositio Math.* **50** (1983), 65-81.

[2] Aytuna, A., Krone, J., Terzioğlu, T.: 'Complemented infinite type power series subspaces of nuclear Fréchet spaces'. To appear in *Math. Ann.*

[3] Dubinsky, E.: *The structure of nuclear Fréchet spaces.* Lecture Notes in Mathematics 720 (1979).

[4] Köthe, G.: *Topological vector spaces I*, Berlin-Heidelberg-New York, 1969.

[5] Krone, J.: 'Zur topologischen Charakterisierung von Unter - und Quotientenräumen spezieller nuklearer Kötheräume mit der Splittingmethode', Diplomarbeit, Wuppertal 1984.

[6] Krone, J.: On Pelczynski's problem. This volume.

[7] Meise, R.: 'Sequence space representations for (DFN)-algebras of entire functions modulo closed ideals'. *J. Reine angew. Math* **363**, (1985), 59-95.

[8] Momm, S.: 'Ideale in gewichteten Algebren holomorpher Funktionen auf dem Einheitskreis', Dissertation. Düsseldorf 1988.

[9] Nurlu, Z.: 'S_g spaces and vanishing of the functor Ext'. *Studia Math.* **88**, (1988), 13-21.

[10] Pelczynski, A.: 'Problem 37', *Studia Math.* **38**, 476 (1970).

[11] Pietsch, A.: *Nuclear Locally Convex Spaces.* Berlin-Heidelberg-New York, 1972.

[12] Terzioğlu, T.: 'Die diametrale Dimension von lokalkonvexen Räumen'. *Collect. Math.* **20** (1969), 49-99.

[13] Terzioğlu, T.: 'On the diametral dimension of some classes of F-spaces', *J. Karadeniz Univ. Ser. Math.-Physics* **8**, (1985), 1-13.

[14] Vogt, D.: 'Charakterisierung der Unterräume von s', *Math. Z.* **155** (1977), 109-117.

[15] Vogt, D.: 'Ein Isomorphiesatz für Potenzreihenräume'. *Arch. Math.* **38** (1982), 540-548.

[16] Vogt, D.: 'On the functors $Ext^1(E,F)$ for Fréchet spaces'. *Studia Math.* **85**, (1987), 163-197.

[17] Vogt, D.: 'Lectures on interpolation theory'. Preprint.

[18] Vogt, D., Wagner, M.I.: 'Charakterisierung der Unterräume und Quotientenräume der nuklearen stabilen Potenzreihenräume von unendlichem Typ',*Studia Math.* **70** (1981), 65-80.

SOME NEW METHODS IN THE STRUCTURE THEORY OF NUCLEAR FRÉCHET SPACES

MAX–JOSEF WAGNER
Weinheimer Str. 54
6703 Limburgerhof
Fed. Rep. of Germany

Abstract. We develop two methods which can help to get new informations about the structure of nuclear Fréchet spaces E. The first one is to construct a basis in a Hilbert space $(E, \| \|_k)$ with growth estimates in all other norms of E. As an application we prove that each complemented subspace of s with r–FDD and each complemented subspace of s with the diametral dimension of a tame power series space has a basis.

The second method can be used as a tool to prove the existence of certain exact sequences $0 \to E \to F \to G \to 0$. As an application we show that each stable complemented subspace of s has a basis.

O. Introduction

We want to develop two methods which are helpful in the structure theory of nuclear (F)–spaces.

The growth of the seminorms of the unit vectors is very important for the structure of a nuclear (F)–space with basis, for example $\lambda(A)$ is a subspace of s iff for a basis $(e_n)_{n=1,2,\dots}$ and suitable seminorms $\| \|_k$ we have

$$\frac{\| e_j \|_{k+1}}{\| e_j \|_k} \geq \frac{\| e_j \|_k}{\| e_j \|_{k-1}} .$$

We will develop a method for a special class of (F)–spaces, the (F)–spaces with property Ω, which make it possible to construct a lokal basis (a basis in the Banach space E_k) with a certain growth.

As an application of this method we prove a theorem about the existence of a basis in a special class of (F)–spaces and an imbedding theorem.

The existence of exact sequences of the form

$$0 \to \Lambda_\infty(\alpha) \to \Lambda_\infty(\alpha) \to \Lambda_\infty^{\mathbf{N}}(\alpha) \to 0$$

for stable $\Lambda_\infty(\alpha)$ plays an important role in the theory of nuclear (F)–spaces.

T. Terzioğlu (ed.), Advances in the Theory of Fréchet Spaces, 333–353.

We want to discuss a method which makes it possible to construct under general assumptions (E countably normed) exact sequences of the form

$$0 \to E \to F \to \lambda^{\mathbb{N}}(A) \to 0 .$$

We will show in an example (E has DN) how to construct F in such a way that the property DN from E is inherited to F. This special example is the main tool in proving that each stable complemented subspace of s has a basis.

1. Ω–sequences

In this paragraph we want to introduce the notation of an Ω–sequence. An Ω–sequence is a local basis (this means a basis in a Banach space E_k) with a fixed growth of the seminorms. In the next paragraph we will show how to use this concept in different applications.

We start this paragraph in proving a theorem which shows that in spaces with property Ω there exists a system of seminorms with some special properties.

Definition 1. A Fréchet space $(F, \| \|_k)$ has property Ω iff for each p there exists a q such that for all k we have an n_k and C_k with

$$\mathcal{U}_q \subset C_k r^{n_k} \mathcal{U}_k + \frac{1}{r}\mathcal{U}_p \text{ for all } r > 0$$

where the $\mathcal{U}_k$ are the unit–balls associated with the seminorms $\| \|_k$.

Theorem 1. *Let $(E, \| \|_{\tilde{k}})_{k=1,2,\dots}$ be a Fréchet space with property Ω. Then there exists a system of seminorms $(\| \|_k)_{k=1,2,\dots}$ defining the topology with the following property:*
For $x \in \mathcal{U}_k$ and all $r > 0$ there exists an $x_1(r)$ and $x_2(r)$ with

a) $x = x_1(r) + x_2(r)$

b) $\| x_1(r) \|_s \leq C_s \cdot r^{n_s}$ *for all* $s \in \mathbb{N}$, C_s, n_s *suitable*

c) $\| x_2(r) \|_{k-1} \leq \frac{1}{r}$

d) $\| x_2(r) \|_k \leq 1$.

Proof. Without loss of generality we can assume that

$$\tilde{\mathcal{U}}_{k+1} \subset C_s r^{n_s} \tilde{\mathcal{U}}_s + \frac{1}{r}\tilde{\mathcal{U}}_k .$$

It is sufficient to show: For each seminorm $\| \|_k$ exists a seminorm $\| \|_{\bar{q}}$ with $\| \|_k \leq \| \|_{\bar{q}}$ and for $x \in \mathcal{U}_{\bar{q}}$ we have

$$\begin{aligned} x &= x_1(r) + x_2(r) \\ \| x_1(r) \|_s &\leq C_s r^{n_s} \qquad \| x_2(r) \|_{\bar{q}} \leq 1 \qquad \| x_2(r) \|_k \leq \frac{1}{r} . \end{aligned}$$

Let $(y_j^k)_{j=1,2,\ldots}$ be the joint basis of E_{k+1}, E_k with $\| y_j^k \|_k = 1$ and $\| y_j^k \|_{k+1}$, in j increasing. We define a new norm $\| \; \|_q^{\sim}$:

$$\| x \|_q^{\sim 2} = \| \sum_j \lambda_j y_j^k \|_q^{\sim 2} = \sum_j |\lambda_j^2| \, \| y_j^k \|_{k+1}^{\left(\frac{2}{n_{k+2}+1}\right)} .$$

It is obvious that $\| \; \|_k \leq \| \; \|_q^{\sim} \leq \| \; \|_{k+1}$. We can decompose an $x \in \widetilde{\mathcal{U}}_q$, $x = x_1 + x_2$ with $x_2 \in \frac{1}{r}\mathcal{U}_k$, $x_1 \in r^{\tilde{n}_{k+1}}\mathcal{U}_{k+1}$ and $\| x_2 \|_q^{\sim} \leq 1$. In the next step we will show

$$\mathcal{U}_{k+1} \subset C \cdot r^{\bar{n}_{k+2}} \mathcal{U}_{k+2} + \frac{1}{r} \widetilde{\mathcal{U}}_q .$$

Let $z \in \mathcal{U}_{k+1} \Longrightarrow z = z_1 + z_2$, $\| z_1 \|_{k+2} \leq C \cdot r^{n_{k+2}}$ and $\| z_2 \|_k \leq \frac{2}{r^2}$. It is obvious that $\| z_2 \|_{k+1} \leq 2Cr^{n_{k+2}}$. Let N_r be the set of indices with $\| y_j^k \|_{k+1}^{\frac{1}{n_{k+2}+1}} < r$.

$$z_2 = \sum_j \lambda_j y_j^k = \sum_{j \in N_r} \lambda_j y_j^k + \sum_{j \notin N_r} \lambda_j y_j^k = \tilde{z}_1 + \tilde{z}_2$$

We have

$$\begin{aligned} \| z_2 \|_q^{\sim} &\leq \| \tilde{z}_1 \|_q^{\sim} + \| \tilde{z}_2 \|_q^{\sim} \leq \| \tilde{z}_1 \|_k \cdot r + \| \tilde{z}_2 \|_{k+1} \cdot \frac{1}{r^{n_{k+2}+1}} \\ &\leq \| z_2 \|_k \cdot r + \| z_2 \|_{k+1} \cdot \frac{1}{r^{n_{k+2}+1}} \leq \frac{\bar{C}}{r} . \end{aligned}$$

Now it is to prove the theorem:

$$x \in \mathcal{U}_{\tilde{q}} \Longrightarrow x = z_1 + z_2 \quad \| z_2 \|_q^{\sim} \leq 1 \quad \| z_2 \|_k \leq \frac{1}{r} \quad \| z_1 \|_{k+1} \leq r^{n_{k+1}} .$$

We can decompose z_1 $z_2 = z_3 + z_4$ with $\| z_4 \|_{\tilde{q}} \leq \frac{1}{r^2}$ and $\| z_3 \|_{k+2} \leq C_{k+2} r^{\tilde{n}_{k+2}}$. We can write $z_3 = z_5 + z_6$ with $\| z_6 \|_{k+1} \leq \frac{1}{r^4}$ and $\| z_5 \|_{k+3} \leq C_{k+3} r^{\tilde{n}_{k+3}}$. An iterative use of this decomposition gives us a sequence $(z_i)_{i=1,2,\ldots}$ with:

$$\begin{aligned} \| \sum_i z_{2i} \|_q^{\sim} &\leq \| z_2 \|_q^{\sim} + \| \sum_{i=2}^{\infty} z_{2i} \|_q^{\sim} \leq 1 + \sum_{i=2}^{\infty} \frac{1}{r^{2i}} \leq 2 \qquad r \geq 2 \\ \| \sum_i z_{2i} \|_k &\leq \sum_{i=1}^{\infty} \| z_{2i} \|_k \leq \frac{1}{r} + \sum_{i=1}^{\infty} \frac{1}{r^{2i}} \leq \frac{2}{r} . \end{aligned}$$

It remains to be seen that

$$\| x - \sum_i z_{2i} \|_s \leq \widetilde{C}_s r^{\widetilde{k}_s} \text{ for all } s .$$

We have: $\| x - \sum_i z_{2i} \|_s = \| x - \sum_i z_{2i} \|_{k+\bar{s}} = \| z_{2\bar{s}-1} - \sum_{i=\bar{s}}^{\infty} z_{2i} \|_{k+\bar{s}}$ $\leq \| z_{2\bar{s}-1} \|_{k+\bar{s}} + \sum_{i=\bar{s}}^{\infty} \| z_{2i} \|_{k+\bar{s}} \leq C_s r^{\widetilde{k}_s} + \sum_{i=\bar{s}}^{\infty} \| z_{2i} \|_{k+i} \leq C \cdot r^{\bar{k}_s}$.

Now we define the notation of an Ω–sequence.

Definition 2. A Fréchet space has property DN iff there exists a norm $\| \ \|$ and for each k a p and a $C \in \mathbb{N}$ with

$$\| \ \|_k \leq Cr \| \ \| + \frac{1}{r} \| \ \|_{k+p} \text{ for all } r > 0 .$$

Definition 3. Let E be an (F)–space and $\| \ \|_m \leq \| \ \|_{\bar{m}}$ two continuous norms on E. A sequence $(x_i)_{i=1,2,\dots} \in E$ is an Ω sequence with respect to $m, \bar{m}$ iff

1) $(x_i)_{i=1,2,\dots}$ is a basis in $(\widehat{E}, \| \ \|_m)$ and $(\widehat{E}, \| \ \|_{\bar{m}})$
2) For all $l \in \mathbb{N}$ there exists C_l, n_l such that

$$\| x_i \|_l \leq C_l \| x_i \|_m \left(\frac{\| x_i \|_{\bar{m}}}{\| x_i \|_m} \right)^{n_l} \text{ for all } i .$$

Theorem 2. *Let E be a nuclear (F)–space with property Ω and $\| \ \|_{k_1} \leq \| \ \|_{\tilde{k}_2}$ two continuous Hilbertnorms on E. Then there exists a continuous Hilbertnorm $\| \ \|_{k_2}$ on E with $\| \ \|_{k_1} \leq \| \ \|_{k_2} \leq \| \ \|_{\tilde{k}_2}$ and for each $n \in \mathbb{N}$ an Ω–sequence with respect to k_1, k_2 and*

$$\| x_i \|_{k_2}^{n+1} \geq \| x_i \|_{\tilde{k}_2}^{n} \cdot \| x_i \|_{k_1} .$$

PROOF. Let $(y_i)_{i=1,2,\dots}$ be the joint basis of $\widehat{E}_{k_1}$ and $\widehat{E}_{\tilde{k}_2}$. We can assume that $\| y_i \|_{k_1} = 1$ and $\| y_i \|_{\tilde{k}_2}$ is increasing in i. Choose δ such that $n \leq \frac{\delta}{1-\delta}$ and $\delta \geq \frac{2}{3}$. Define

$$\| x \|_{k_2}^2 = \| \sum_i x_i y_i \|_{k_2}^2 := \sum_i |x_i|^2 \| y_i \|_{\tilde{k}_2}^{\delta} .$$

$\| \ \|_{k_2}$ is a continuous Hilbertnorm and we have the inequality

$$\| x_i \|_{k_2}^{n+1} \geq \| x_i \|_{\tilde{k}_2}^{n} \cdot \| x_i \|_{k_1} .$$

Using Theorem 1 we can conclude

$$x \in \mathcal{U}_{\tilde{k}} \Longrightarrow x = x_1(r) + x_2(r) \text{ for all } r \text{ and } x_2(r) \in \frac{1}{r^s} \mathcal{U}_{k_1} \cap \mathcal{U}_{\tilde{k}_2},\ x_1(r) \in Cpr^{n(p)} \mathcal{U}_{k+p} \cap \mathcal{U}_{\tilde{k}_2}$$

$$x_2(r) = \sum_{\substack{i \\ \| y_i \|_{\tilde{k}_2} \leq r^s}} \lambda_i y_i + \sum_{\substack{i \\ \| y_i \|_{\tilde{k}_2} > r^s}} \lambda_i y_i = z_1 + z_2 .$$

From $\| z_1 \|_{k_1} \leq \frac{1}{r^s}$ we can conclude that $\| z_1 \|_{k_2} \leq \frac{1}{r}$ and from $\| z_2 \|_{\tilde{k}_2} \leq 1$ we can conclude $\| z_2 \|_{k_2} \leq \frac{1}{r}$. Summing up this we have

$$x_2(r) \in \mathcal{U}_{\tilde{k}_2} \cap \frac{2}{r} \mathcal{U}_{k_2} .$$

We choose $r = \left(\frac{1}{\|y_i\|_{k_2}}\right)^m$ with m such that $\left(\sum \frac{1}{\|y_i\|_{k_2}}\right)^{2m} < 1$ and get

$$y_i = y_i^1 + y_i^2 \text{ with } \| y_i^2 \|_{k_2} \leq \left(\frac{1}{\| y_i \|_{k_2}}\right)^m \text{ and } \| y_i^1 \|_{k+p} \leq C \cdot \| y_i \|_{k_2}^{n(p)} \text{ for all } p .$$

From $\sum_i \| y_i - y_i^1 \|_{k_2} = \sum_i \| y_i^2 \|_{k_2} < 1$ it is obvious that $(y_i^1)_{i=1,2,\ldots}$ is a basis in $(\widehat{E}, \| \|_{k_1})$ and $(\widehat{E}, \| \|_{k_2})$.

In the case that E has also property DN we show the following lemma.

Lemma 3. *Let E be a nuclear (F)-space with properties DN and Ω and $\| \|_k \leq \| \|_{\tilde{k}_2}$ be two continuous Hilbertnorms on E. Then there exists a continuous seminorm $\| \|_{k_2}$ on E with $\| \|_{k_2} \geq \| \|_{\tilde{k}_2}$ and an Ω-sequence with respect to $\| \|_{k_1}$ and $\| \|_{k_2}$.*

PROOF. Property DN gives us the existence of a seminorm $\| \|_{\bar{k}_2}$ on E with

$$\| \|_{\tilde{k}_2}^{2} \leq \| \|_{\bar{k}_2} \| \|_{k_1} \ .$$

Using theorem 2 we have a seminorm $\| \|_{k_2}$ and an Ω-sequence $(x_i)_{i=1,2,\ldots}$ with respect to $\| \|_{k_1}$ and $\| \|_{k_2}$ and

$$\| x_i \|_{k_2}^3 \geq \| x_i \|_{\tilde{k}_2}^2 \| x_i \|_{k_1} \ .$$

Using property DN and the nuclearity of E it is easy to see that $(x_i)_{i=1,2,\ldots}$ fulfills all requirements.

2. Applications

In this paragraph we want to demonstrate some applications of the concept of Ω-sequences. We start with the following definition.

Definition 1. Let E be a locally convex vector space and $V \subset U$ two neighbourhoods of zero. We define

$$\delta_r(U, V) = \{\inf \delta \ : \ V \subset \delta U + F \ \dim F \leq r\} \ .$$

The diametral dimension $\Delta(E)$ of E is the set of all sequences $(\delta_r)_{r=1,2,\ldots}$ such that for each zero neighbourhood V there exists a zero neighbourhood $U \subset V$ in E with $\delta_r(V, U) \leq \delta_r$ for all r.

From [7] we know:

Theorem 1. *The diametral dimension $\Delta(\Lambda_\infty(\lambda))$ is the set of all sequences $(\delta_r)_{r=1,2,\ldots}$ such that there exists $C, s \in \mathbb{N}$ with $\delta_r \geq C \cdot s^{\alpha_r}$.*

Now we are able to prove:

Theorem 2. *Let E be a nuclear, stable (F)–space with properties DN, Ω. If $\Delta(E) = \Delta(\Lambda_\infty(\alpha))$ then $\Lambda_\infty(\alpha)$ is isomorphic to a subspace of E.*

PROOF. Let E be equipped with the seminorms $\| \ \|_{\tilde{k}}$ $k = 1, 2, \ldots$. E is stable, therefore we can associate to each x a sequence $(x_i)_{i=1,2,\ldots}$ with $x_i \in E_i \cong E$ and the system of seminorms $\| x \|_k = \sum_{i=1}^{k} \| x_i \|_{\tilde{k}} + \| x - \sum_{i=1}^{k} x_i \|_{\tilde{k}}$ defines the topology of E.

Let $(y_i)_{i=1,2,\ldots}$ be an Ω–sequence in E with respect to the norm $\| \ \|_k$ and a norm $\| \ \|_{\tilde{k}}$, with $\tilde{k}$ choosen in such a way that

$$\frac{\| y_j \|_{\tilde{k}}}{\| y_j \|_k} \geq k^2 \cdot k^{\alpha_j}.$$

Because E and $\Lambda_\infty(\alpha)$ have the same diametral dimension, we will have also an inequality in the other direction,

$$\frac{\| y_j^k \|_{\tilde{k}}}{\| y_j^k \|_k} \leq C \cdot s^{\alpha_j} \text{ with } C, s \text{ suitable}.$$

We define a mapping T on $\Lambda_\infty(\alpha)$ by defining the images of the unit vectors $T : e_j \to \sum_{n=1}^{\infty} C_{k,j} y_j^k$ with $C_{k,j}$ choosen such that $\| C_{k,j} y_j^k \|_{\tilde{k}} = k^{\alpha_j}$. For the n-th norm we can conclude that

$$\| C_{k,j} y_j^k \|_k \leq \frac{1}{k^2}$$

We have to show that T is an isomorphism from $\Lambda_\infty(\alpha)$ to a subspace of E.

We estimate for $x = \sum_{j=1}^{\infty} \lambda_j e_j \in \Lambda_\infty(\alpha)$:

$$\begin{aligned}
\| Tx \|_l &\leq \sum_{k=1}^{\infty} \| \sum_{j=1}^{\infty} \lambda_j y_j^k \|_l \\
&\leq \sum_{k=1}^{l} \| \sum_{j=1}^{\infty} \lambda_j C_{k,j} y_j^k \|_l + \sum_{k=l+1}^{\infty} \| \sum_{j=1}^{\infty} \lambda_j C_{k,j} y_j^k \|_l \\
&\leq \sum_{k=1}^{l} \sum_{j=1}^{\infty} |\lambda_j| \, \| C_{k,j} y_j^k \|_l + \sum_{k=l+1}^{\infty} \sum_{j=1}^{\infty} |\lambda_j| \, \| C_{k,j} y_j^k \|_k \\
&\leq \sum_{k=1}^{l} \sum_{j=1}^{\infty} |\lambda_j| s(k)^{\alpha_j} + \sum_{k=l+1}^{\infty} \sum_{j=1}^{\infty} |\lambda_j| \frac{1}{k^2} \leq C \| x \|^{\sim}
\end{aligned}$$

In the other direction we have

$$\begin{aligned}
\| Tx \|_{\tilde{l}} &= \| \sum_{j=1}^{\infty} \lambda_j \cdot \sum_{n=1}^{\infty} C_{k,j} y_j^k \|_{\tilde{l}} \geq \| \sum_{j=1}^{\infty} \lambda_j C_{l,j} y_j^l \|_{\tilde{l}} \\
&= \left(\sum_{j=1}^{\infty} |\lambda_j|^2 \cdot e^{2\alpha_j} \right)^{\frac{1}{2}} = \| x \|_l
\end{aligned}$$

Therefore $\Lambda_\infty(\alpha)$ is isomorphic to a subspace of E.

In our next theorem we show that each nuclear (F)–space with properties DN and Ω and a r–FDD has a basis (cf. [1] for a more special result).

We start with a definition (see [2]).

Definition 2. A nuclear (F)–space E has a r–FDD iff there are continuous maps A_n, $n \in \mathbb{N}$, such that $A_n A_m = \delta_{n,m} A_m$, $\dim \operatorname{im} A_n \leq r$ and $x = \sum_n A_n x$ for all $x \in E$.

Theorem 3. *A nuclear (F)–space E with properties DN and Ω and an r–FDD has a basis.*

PROOF. Let $\| \|_k$ be a system of seminorms on E which defines the topology and has the property that $\| \|_k^2 \leq \| \|_{k-1} \| \|_{k+1}$. We prove our assumption using induction.

Let $r = 2$.
If $z = (z_1, z_2, \ldots)$ is an Ω–sequence with respect to $\| \|_1$ and $\| \|_k$ it is obvious that we can choose z such that $z_i = (z_{i,1}, z_{i,2})$ with $z_{i,1}, z_{i,2} \in E_i$.

We can assume that $\| z_{i,1} \|_1 = 1$ and $\| z_{i,2} \|_k \geq \| z_{i,1} \|_k$. Let $x_i = \lambda_{i,1} z_{i,1} + \lambda_{i,2} z_{i,2}$ with $\| x_i \|_1 = 1$. We have to show for any l that

$$a := \sup(\| \lambda_{i,1} z_{i,1} \|_l, \| \lambda_{i,2} z_{i,2} \|_l) \leq \| x_i \|_{s(l)} .$$

Case 1: $\| \lambda_{i,1} z_{i,1} \|_l \leq \frac{1}{2} \| \lambda_{i,2} z_{i,2} \|_l$
$\Longrightarrow \| \lambda_{i,2} z_{i,2} \|_l \leq 2(\| \lambda_{i,2} z_{i,2} \|_l - \| \lambda_{i,1} z_{i,1} \|_l) \leq 2 \| x \|_l$
Case 2: $\| \lambda_{i,1} z_{i,1} \|_l \geq \frac{1}{2} \| \lambda_{i,2} z_{i,2} \|_l$
$\Longrightarrow a \leq 2 \| \lambda_{i,1} z_{i,1} \|_l \leq 2C \| z_{i,1} \|_k^{n(l)} \leq 2C \| x_i \|_k^{n(l)} \leq C' \| x_i \|_{s(l)}$
The last inequality is true because E has property DN. Now we can carry out the induction step $s \to s+1$. Let $z = (z_i)_{i=1,2,\ldots}$ with $z_i = (z_{i,1}, z_{i,2}, \ldots, z_{i,s+1})$ an Ω–sequence with respect to $\| \|_1$ and $\| \|_k$. We can assume $\| z_{i,1} \|_1 = \ldots = \| z_{i,s+1} \|_1 = 1$ and $\| z_{i,1} \|_k \leq \ldots \leq \| z_{i,s+1} \|_k$. With $x = \sum_i c_i x_i$ and $x_i = \sum_{l=1}^{k+1} \lambda_{i,l} z_{i,l}$ we have to show that

$$a = \sup(\| \lambda_{i,1} z_{i,1} \|_l, \ldots, \| \lambda_{i,s+1} z_{i,s+1} \|_l) \leq C \cdot \| x_i \|_{s(l)} .$$

We can assume $\| x_i \|_l = 1$ and we get
Case 1: $a = \| \lambda_{i,1} z_{i,1} \|_l \Longrightarrow$ The conclusion is trivial.
Case 2: $a \neq \| \lambda_{i,1} z_{i,1} \|_l$
From the induction we know that

$$a \leq C \| \sum_{i=2}^{k+1} \lambda_{i,k+1} z_{i,k+1} \|_{n(l)} \text{ with suitable } n(l).$$

If $\| \lambda_{i,1} z_{i,1} \|_{s(l)} \leq \frac{1}{2} \| \sum_{i=2}^{k+1} \lambda_{i,k+1} z_{i,k+1} \|_{n(l)}$ our conclusion is obvious.
But in the other case we have

$$a \leq 2 \| \lambda_{i,1} z_{i,1} \|_{n(l)} \leq C \| x_i \|_{s(l)} .$$

As a last application we will demonstrate that nuclear (F)–spaces with properties DN, Ω and the diametral dimension of a tame power series space $\Lambda_\infty(\alpha)$ have a basis.

We quote from [10] the following theorem.

Theorem 4. *A power series space $\Lambda_\infty(\alpha)$ is tame iff $\Lambda_\infty(\alpha) = \Lambda_\infty(\beta)$ with $\beta = (\beta_j)_{j=1,2,\ldots}$, $\beta_j = (\beta_{j,1},\ldots,\beta_{j,n(j)})$ and $\beta_{i,1} = \ldots = \beta_{i,n(i)}$,*

$$\lim_{i\to\infty} \frac{\beta_{i,1}}{\beta^s_{i-1,n(i)}} = \infty \text{ for all } s\,.$$

Theorem 5. *Let E a Fréchet space with properties DN, Ω and $\Delta(E) = \Delta(\Lambda_\infty(\alpha))$, with $\Lambda_\infty(\alpha)$ tame then E has a basis.*

PROOF. Let $\|\ \|_j$ be a system of seminorms which defines the topology of E and has the property $\|\ \|_k^2 \leq \|\ \|_{k+1}\|\ \|_{k-1}$. The canonical mapping $E_2 \to E_1$ can be assumed as nuclear.

If $z = (z_1, z_2, \ldots)$ is an Ω–sequence with respect to $\|\ \|_1$ and $\|\ \|_2$, we can start out from

a) $\| z_i \|_1 = 1$

b) $\| z_i \|_2$ is increasing in i

c) $(z_i) = (y_i)$ with $y_i = (y_{i,1},\ldots,y_{i,n(i)})$ and $\lim_{n\to\infty} \frac{\|y_{i+1,1}\|_l}{\|y^s_{i,n(i)}\|_l} = \infty$ for all s (since $\Delta(E) = \Delta(\Lambda_\infty(\alpha))$).

We want to show that $(z_i)_{i=1,2,\ldots}$ is a basis of E. We have $x = \sum_i \lambda_i z_i = \sum_{j=1}^{\infty}\sum_{n=1}^{n(j)} \lambda_{j,k} y_{j,k}$ in the $\|\ \|_2$ norm. It remains to be seen that $\| \lambda_{j,k} y_{j,k} \|_l \leq C_l \| x \|_{s(l)}$. Let i, k fixed and $\| x \|_1 = 1$

$$x = \sum_{j=1}^{i-1}\sum_{n=1}^{n(j)} \lambda_{j,k} y_{j,k} + \sum_{j=i}^{\infty}\sum_{n=1}^{n(j)} \lambda_{j,k} y_{j,k} = x_1 + x_2\,.$$

Using the fact that $(y_{j,k})$ is an Ω–sequence we get

$$|\lambda_{j,k}|\, \| y_{j,k} \|_l \leq C |\lambda_{j,k}|\, \| y_{j,k} \|_2^p \leq C \| x_2 \|_2 \left(\frac{\| x_2 \|_2}{\| x_2 \|_1}\right)^p .$$

Because E has property DN we can estimate

$$|\lambda_{j,k}|\, \| y_{j,k} \|_l \leq C \| x_2 \|_s \,.$$

Case 1: $\| x_1 \|_s \leq \frac{1}{2} \| x_2 \|_s$ or $\| x_2 \|_s \leq \frac{1}{2} \| x_1 \|_s$. It is obvious that $|\lambda_{j,k}|\, \| y_{j,k} \|_l \leq C \| x \|_s$.

Case 2: $\| x_1 \|_s = C \| x_2 \|_s$ with $\frac{1}{2} \leq C \leq 2$. $(z_i)_{i=1,2,\ldots}$ is an Ω–sequence therefore we can estimate

$$\begin{aligned} \| x_1 \|_{s+p} &\leq \sum_{k,j<i} \| \lambda_{j,k} y_{j,k} \|_{s+p} \leq \sum_{k,j<i} |\lambda_{j,k}|\, \| y_{j,k} \|_2^m \\ &\leq C \cdot \| x_1 \|_2 \| y_{i-1,n(i)} \|_2^m \leq C \| x_1 \|_s \| y_{i-1,n(i)} \|_2^m \,. \end{aligned}$$

Using property DN we have for p suitable $\| x_2 \|_{s+p} \geq \| x_2 \|_s \cdot \| y_{i,1} \|_2$ which results in

$$\lim_{i\to\infty} \frac{\| x_2 \|_{s+p}}{\| x_1 \|_{s+p}} = \infty\,.$$

For $j \geq J_0$, J_0 suitable, we have

$$|\lambda_{j,k}| \, \| y_{j,k} \|_l \leq C \| x \|_{s(l)}\,.$$

For $j < J_0$ we can use the fact that $\| x \|_1 = 1$, which means $\sum_{j,k} |\lambda_{j,k}|^2 = 1$, and estimate

$$\begin{aligned} |\lambda_{j,k}| \, \| y_{j,k} \|_l &\leq C |\lambda_{j,k}| \, \| y_{j,k} \|_2^p \leq C \| x \|_2 \| y_{1,1} \|_2^{p-1} \leq C \| x \|_2^p \\ &\leq C \| x \|_{s(l)} \end{aligned}$$

3. Exact Sequences

The (F)–spaces $\lambda(A)^{\mathbf{N}}$ are important because often (F)–spaces with a common property can be imbedded in $\lambda(A)^{\mathbf{N}}$. For example the theorem of Komura–Komura shows that each nuclear (F)–space is isomorphic to a subspace of $s^{\mathbb{N}}$. Let E, Q two (F)–spaces and $(*)$ a (F)–space property which is inherited to subspaces. If you are interested in the existence of an exact sequence

$$0 \to E \to F \to Q \to 0$$

(F having property $*$ and Q is isomorphic to a subspace of $\lambda(A)^{\mathbf{N}}$), it is sufficient to have a sequence $0 \to E \to \widetilde{F} \to \lambda(A)^{\mathbf{N}} \to 0$ with $\widetilde{F}$ having property $(*)$. Therefore exact sequences

$$0 \to G \to H \to \lambda(A)^{\mathbf{N}} \to 0$$

are of importance.

We want to prove that for each stable, nuclear, countably normed space E and for each nuclear $\lambda(A)$ there exists a sequence

$$0 \to E \to F \to \lambda(A)^{\mathbf{N}} \to 0$$

with F having a continuous norm. In the next paragraph we will give a special example which illustrates how properties of E can be transformed to properties of F.
We start with a definition.

Definition 1. Let $\pi : \mathbb{N}^2 \to \mathbb{N}$, $\pi(i,j) = 2^{i-1} + (j-1)2^j$ and $x = (x^i_{j,k})_{i,j,k=1,2,3,\ldots}$ a triple sequence with $\sum_{i=1}^{\infty} x^i_{j,k}$ exists for all j,k. Then $\widetilde{x} = (\widetilde{x}_{j,k})_{j,k=1,2,\ldots}$ is the double sequence with $\widetilde{x}_{\pi(i,j),k} := \sum_{m=1}^{i} x^m_{j,k}$. $\bar{x} = (\bar{x}_{j,k})_{j,k=1,2,\ldots}$ is the double seqence with $\bar{x}_{\pi(i,j),k} := -\sum_{m=i+1}^{\infty} x^m_{j,k}$.
Let $x = (x_{j,k})_{j,k=1,2,\ldots}$ a double sequence then $x^{(m)} = (x^{(m)}_{j,k})_{j,k=1,2,\ldots}$ is defined by

$$x^{(m)}_{j,k} = \begin{cases} x_{j,k} & k > m \\ 0 & k \leq m \end{cases}$$

and $x^{[m]} = (x^{[m]}_{j,k})_{j,k=1,2,\dots}$ is defined by

$$x^{[m]}_{j,k} = \begin{cases} 0 & k > M \\ x_{j,k} & k \le m\,. \end{cases}$$

Now we can state the main theorem of this paragraph.

Theorem 1. *Let E be a stable, nuclear, countably normed (F)–space and $\lambda(A)$ a nuclear Köthe space with continuous norm, then there exists a nuclear (F)–space F with a continuous norm and an exact sequence*

$$0 \to E \to F \to \lambda(A)^{\mathbf{N}} \to 0\,.$$

Proof. Let $(\| \;\|_{\tilde{k}})_{k=1,2,\dots}$ a system of Hilbertnorms defining the topology of E. E is stable, therefore $E = E_1 \oplus \widetilde{E_2}$, with $E_1, \widetilde{E_2} \subset E$ and $E_1, \widetilde{E_2}$ isomorphic to E. Because $\widetilde{E_2}$ is isomorphic to E, we have $\widetilde{E_2} = E_2 \oplus \widetilde{E_3}$ with $E_2, \widetilde{E_3} \subset \widetilde{E_2}$ and $E_2, \widetilde{E_3}$ isomorphic to E. This mechanism assigns to each $x \in E$ a well defined sequence $(x_j)_{j=1,2,\dots}$ with $x_j \in E_j \subset E$ and $E_j \cong E$.

The system of seminorms

$$\| x \|_k := \sum_{l=1}^{k} \| x_l \|_{\tilde{k}} + \| x - \sum_{l=1}^{k} x_l \|_{\tilde{k}}$$

defines also the topology of E. Without loss of generality we assume $a_{j,1} = 1$. We are now ready to define F. Using the spectral theorem we get $(y_j^k)_{j=1,2,\dots}$ the joint basis of $(E_k, \| \;\|_{k+1})$ and $(E_k, \| \;\|_k)$ with

$$(*) \quad \| y_j^k \|_k = \tfrac{1}{j^2} \quad \text{and} \quad \frac{\|y_j^k\|_{k+1}}{\|y_j^k\|_k} \le \frac{\|y_{j+1}^k\|_{k+1}}{\|y_{j+1}^k\|_k}.$$

The first condition fixes only the k-th norm of y_j^k and the second condition is easy to fulfill because E is nuclear.

Without loss of generality we can assume that all $y_j^k \in E$. So we can assign to each $x \in E$ a double sequence $(x_{j,k})_{j,k=1,2,\dots}$ of skalars because each $x_k \in E_k$ has in E_k the expansion

$$x_k = \sum_{j=1}^{\infty} x_{j,k} y_j^k\,.$$

Let $F = \left\{ x = \left(x^i_{j,k} \right)_{i,j,k=1,2,\dots} \right\}$ the space of triple sequences with following properties

1) $\bar{x}^{[m]} \in E$ for all $m \in \mathbb{N}$, which is equivalent to say that $z_k = \sum_{k=1}^{m} \sum_{j=1}^{\infty} \bar{x}_{j,k} y_j^k \in E$ for all $m \in \mathbb{N}$.

2) $\tilde{x}^{(m)} \in (E, \| \, \|_{m+1})$ for all $m \in \mathbb{N}$, which is equivalent to say that $\sum_{j=1}^{\infty} \tilde{x}_{j,k} y_j^k$ converges in $(E_k, \| \, \|_{m+1})$ $k > m$ to y_k and $\sum_{k=m+1}^{\infty} y_k$ converges in $(E, \| \, \|_{m+1})$ to y.

3) $\| x \|_l := \sum_{n=1}^{l} \sum_{j=1}^{\infty} | \sum_{i=1}^{\infty} x_{j,k}^i | a_{\pi(k,j),l-k+1} + \| \bar{x}^{[l]} \|_l + \| \tilde{x}^{(l)} \|_l < \infty$

We want to prove that $\| \, \|_1$ is a continuous norm on F. So assume that $\| \left(x_{j,k}^i \right) \|_1 = 0$. We have $\| \bar{x}^{[1]} \|_1 = 0$ and therefore $\sum_{m=i+1}^{\infty} x_{j,1}^m = 0$ for $i = 1, 2, \ldots$. Together with the fact that $\sum_{m=1}^{\infty} x_{j,1}^m$ is equal zero we can conclude that $x_{j,1}^i = 0$ for all i, j.

$\sum_{k=2}^{\infty} y_k$ converges to zero in $(E, \| \, \|_1)$ because $\| \tilde{x}^{(1)} \|_1 = 0$. Together with the fact that E is countably normed and the definition of F this forces the convergence of $\sum_{k=2}^{\infty} y_k$ in $(E, \| \, \|_2)$ to zero in $(E, \| \, \|_2)$.

From the definition of the $\| \, \|_2$-norm it is obvious that $y_2 = 0$, and therefore $\tilde{x}_{j,2} = 0$ for all j and $x_{j,2}^i = 0$ for all i, j. Repeating this argument we get $y_k = 0$ for all $k \geq 2$ and finally $x_{j,k}^i = 0$ for all i, j, k, which proves that $\| \, \|_1$ is in fact a norm.

To show that F is a Fréchet–space with a continuous norm we prove that $\| \, \|_k \leq C_k \| \, \|_{k+1}$ with a suitable constant C_k. It is sufficient to have the existence of the inequality

$$\| \bar{x}^{[k]} \|_k + \| \tilde{x}^{(k)} \|_k \leq \left(\sum_{j=1}^{\infty} \| \sum_{i=1}^{\infty} x_{j,l+1}^i \|_{k+1} \right) + \| \bar{x}^{[k+1]} \|_{k+1} + \| \tilde{x}^{(k+1)} \|_{k+1} \, .$$

The definition of the $\| \, \|_k$-norm gives

$$\| \bar{x}^{[k+1]} \|_{k+1} \leq \| \bar{x}^{[k]} \|_k + \| \sum_{i,j=1}^{\infty} \left(- \sum_{j=l+1}^{i} x_{j,l+1}^m \right) y_{\pi(i,j)}^{k+1} \|_{k+1}$$

and

$$\| \tilde{x}^{(k)} \|_k \leq \| \tilde{x}^{(k+1)} \|_k + \| \sum_{i,j=1}^{\infty} \left(\sum_{m=1}^{i} x_{j,l+1}^m \right) y_{\pi(i,j)}^{k+1} \|_k$$

so it is sufficient to show

$$\| \sum_{i,j=1}^{\infty} \left(\sum_{m=1}^{i} x_{j,k+1}^m \right) y_{\pi(i,j)}^{k+1} \|_k \leq C_k \left(\sum_{j=1}^{\infty} \sum_{i=1}^{\infty} x_{j,k+1}^i \right) + \| \sum_{i,j=1}^{\infty} \left(- \sum_{m=i+1}^{\infty} x_{j,l+1}^m \right) y_{\pi(i,j)}^{k+1} \|_{k+1} \, .$$

We have

$$\begin{aligned}
&\| \sum_{i,j=1}^{\infty} \left(\sum_{m=1}^{i} x_{j,k+1}^{m} \right) y_{\pi(i,j)}^{k+1} \|_k = \| \sum_{i,j=1}^{\infty} \left(\sum_{m=1}^{\infty} x_{j,k+1}^{m} - \sum_{m=i+1}^{\infty} x_{j,k+1}^{m} \right) y_{\pi(i,j)}^{k+1} \|_k \\
\leq\ &\| \sum_{i,j=1}^{\infty} \left(\sum_{m=1}^{\infty} x_{j,k+1}^{m} \right) y_{\pi(i,j)}^{k+1} \|_k + \| \sum_{i,j=1}^{\infty} \left(- \sum_{m=i+1}^{\infty} x_{j,k+1}^{m} \right) y_{\pi(i,j)}^{k+1} \|_k \\
\leq &\sum_{j=1}^{\infty} | \sum_{m=1}^{\infty} x_{j,k+1}^{m} | \cdot \| \sum_{i=1}^{\infty} y_{\pi(i,j)}^{k+1} \|_k + \| \sum_{i,j=1}^{\infty} \left(- \sum_{m=i+1}^{\infty} x_{j,k+1}^{m} \right) y_{\pi(i,j)}^{k+1} \|_k \\
\leq\ &\sum_{j=1}^{\infty} | \sum_{m=1}^{\infty} x_{j,k+1}^{m} | \cdot \sum_{i=1}^{\infty} \frac{1}{\pi(i,j)^2} + \| \sum_{i,j=1}^{\infty} \left(- \sum_{m=i+1}^{\infty} x_{j,k+1}^{m} \right) y_{\pi(i,j)}^{k+1} \|_k
\end{aligned}$$

which proves that F is a Fréchet space.

Our next goal is to define a mapping from F to $\lambda(A)^{\mathbf{N}}$. Let e_j^k the j-th unit vector in the k-th factor of $\lambda(A)^{\mathbf{N}}$. We define q

$$\mathrm{q}(x) = \mathrm{q}((x_{j,k}^{i})_{i,j,k=1,2,\ldots}) := \sum_{j,k=1}^{\infty} \left(\sum_{i=1}^{\infty} x_{j,k}^{i} \right) e_j^k .$$

Since $\| \mathrm{q}(x) \|_m = \sum_{k=1}^{m} \sum_{j=1}^{\infty} | \sum_{i=1}^{\infty} x_{j,k}^{i} | a_{j,m-k+1} \leq \| x \|_m$ we have that q is a continuous mapping from F to $\lambda(A)^{\mathbf{N}}$.

Before showing the surjectivity of q, we prove that the kernel of q is isomorphic to a dense subspace of F. From the definition of q it is clear that

$$\mathrm{kernel(q)} = \left\{ x \in F : \sum_{i=1}^{\infty} x_{j,k}^{i} = 0 \text{ for all } j, k \right\} .$$

So if x is an element of kernel(q) we have $\| x \|_l = \| \bar{x}^{[l]} \|_1 + \| \tilde{x}^{(l)} \|_l$, $\bar{x} = \tilde{x}$ and $\bar{x} = \bar{x}^{[l]} + \tilde{x}^{(l)} \in (E, \| \|_l)$ for all l. So the mapping i : $x \rightarrow \bar{x}$ is an isomorphism from kernel(q) to E, because

$$\| x \|_l = \| \bar{x}^{[l]} \|_l + \| \tilde{x}^{(l)} \|_l = \| \bar{x}^{[l]} \|_l + \| x^{(l)} \|_l = \| \bar{x} \|_l$$

Now we have to answer the question whether i is dense.

Let $\hat{x} \in E$, and $(z_1, z_2, \ldots)$ the orthonormal basis in $(E, \| \|_m)$ which you get by orthonormalising the sequences $\left(y_j^k\right)_{j,k=1,2,\ldots}$. For each $m \in \mathbb{N}$ and $\varepsilon > 0$ we find an $\tilde{n} \in \mathbb{N}$ with $\| \hat{x} - \sum_{i=1}^{\tilde{n}} \lambda_i z_i \|_m < \varepsilon$, because $\hat{x} = \sum_{i=1}^{\infty} \lambda_i z_i$ is in $(E, \| \|_m)$. Obviously is $z = \sum_{i=1}^{\tilde{n}} \lambda_i z_i \in \bigoplus_{i=1}^{n} E_i$ with a suitable $n \in \mathbb{N}$. Therefore i is dense if for each $z \in \bigoplus_{i=1}^{k} E_i$ there exists a $y \in$ kernel q with $\mathrm{q}(y) = z$. If $(z_{j,k})_{j,k=1,2,\ldots}$ is a double sequence which belongs to z, we have $z_{j,k} = 0$ for $k > n$. As the first step we want to show that $\lim_{j \rightarrow \infty} z_{j,k} = 0$ for $k \leq n$. E is nuclear and $\| y_j^k \|_k = \frac{1}{j^2}$, so we can without loss of generality assume that

(**) $\quad \| y_j^k \|_{k+1} \geq 1.$

The series $\sum_{j=1}^{\infty} z_{j,k} y_j^k$ converges in $(E, \| \ \|_{k+1})$, so together with (**) it is obvious that $(z_{j,k})_{j=1,2,\ldots}$ converges to zero.

We define

$$x = \left(x_{j,k}^i\right)_{i,j,k=1,2,\ldots} \quad \text{with } x_{j,k}^i = \begin{cases} \sum_{m=1}^{i} z_{\pi(m,j),k} - z_{\pi(m-1,j),k} & k \le n \\ 0 & k > n \end{cases}$$

where $\pi(0,j) = 0$.

From $\lim z_{j,k} = 0$ we can conclude that $\sum_{i=1}^{\infty} x_{j,k}^i = 0$ and this results in the fact that the triple sequence $(x_{j,k}^i)$ is an element of kernel q. On the other hand it's obvious that $ix = z$.

We have seen that q is a continuous mapping from F to $\lambda(A)^{\mathbb{N}}$. Now we want to prove that q is also surjective.

Take an $x \in \lambda(A)^{\mathbb{N}} \Longrightarrow x = (x_1, x_2, \ldots)$ $x_i \in \lambda(A)$ and each x_k has an expansion $x_k = \sum_{j=1}^{\infty} \lambda_{j,k} e_j^k$ with suitable $\lambda_{j,k}$.

Let $\varepsilon > 0$, $s \in \mathbb{N}$ and $n(j,k) \in \mathbb{N}$ fixed, and choose n_k that $\| \sum_{j=n_k}^{\infty} \lambda_{j,k} e_j^k \|_k < \frac{\varepsilon}{2^k}$.

We define

$$y_{j,k}^i = \begin{cases} \lambda_{j,k} & i = 1 & k \le s & \\ 0 & i \ne 1 & k \le s & \\ \lambda_{j,k} & i = 1 & k > s & j \ge n_k \\ 0 & i \ne 1 & k > s & j \ge n_k \\ \lambda_{j,k} & i = n(j,k) & k > s & j < n_k \\ 0 & i \ne n(j,k) & k > s & j < n_k \end{cases}$$

Now we can compute $\bar{y}_{j,k}$.

$$\bar{y}_{\pi(i,j),k} = \begin{cases} 0 & k \le s & & \\ 0 & k > s & j \ge n_k & \\ -\lambda_{j,k} & k > s & j < n_k & i < n(j,k) \\ 0 & k > s & j < n_k & i \ge n(j,k) \end{cases}$$

So $\bar{y}^{[m]}$ has only a finite number of nonzero components and therefore $\bar{y}^{[m]}$ is an element of E for all m.

If s is equal m, we have $\| \bar{y}^{[m]} \|_m = 0$.

$$\widetilde{y}_{\pi(i,j),k} = \begin{cases} \lambda_{j,k} & k \le s & & \\ \lambda_{j,k} & k > s & j \ge n_k & \\ 0 & k > s & j < n_k & i < n(j,k) \\ \lambda_{j,k} & k > s & j < n_k & i \ge n(j,k) \end{cases}$$

We have

$$\| y_k \|_k = \| \sum_{j=1}^{\infty} \widetilde{y}_{j,k} y_j^k \|_k \le \sum_{j=1}^{\infty} |\lambda_{j,k}| \cdot \sum_{j=1}^{\infty} \| y_j^k \|_k \le C \cdot \sum_{j=1}^{\infty} |\lambda_{j,k}| < \infty .$$

Because of that y_k is an element of $(E_k, \| \|_k)$. Furthermore we can conclude

$$\sum_{n=m+1}^{\infty} \| y_k \|_{m+1} \leq \sum_{k=m+1}^{s} \| y_k \|_{m+1} + \sum_{k=s+1}^{\infty} \sum_{j=n_k}^{\infty} |\lambda_{j,k}| \sum_{s=1}^{\infty} \| y_s^k \|_k +$$
$$+ \sum_{k=s+1}^{\infty} \sum_{j=1}^{n_k} |\lambda_{j,k}| \sum_{s=2^{n(j,k)-1}}^{\infty} \| y_s^k \|_k < \varepsilon$$

with n_k, $n(j,k)$ suitable and $s = m$.

Therefore $y \in F$ and as we have $\mathrm{q}(y) = x$ it is proved that q is surjective.

Let us also mention the fact that

$$\inf_{\substack{y \in F \\ \mathrm{q}(y) = x}} \| y \|_m = \| x \|_m \text{ for all } m.$$

Let $\bar{F}$ the completition of F and $\bar{\mathrm{q}}$ the expansion from q to $\bar{F}$, then we have the exact sequence

$$0 \to \operatorname{kernel} q \to \bar{F} \to \lambda(A)^{\mathbf{N}} \to 0$$

with $\bar{F}$ an (F)–space with continuous norms.

It remains to be seen that kernel $\bar{\mathrm{q}}$ is isomorphic to E. We know that kernel q is isomorphic to a dense subspace of E, so it is sufficient to show that

$$\operatorname{kernel} \bar{\mathrm{q}} = \overline{\operatorname{kernel} \mathrm{q}}$$

Obviously we have $\overline{\operatorname{kernel} \mathrm{q}} \subset \operatorname{kernel} \bar{\mathrm{q}}$.

On the other hand let $x \in \operatorname{kernel} \bar{\mathrm{q}}$, then there exists a sequence $(x_n)_{n=1,2,\ldots} \in F$ with $x_n \to x$ and $\mathrm{q}(x_n) = y_n \to 0$ in $\lambda(A)^{\mathbf{N}}$.

We have shown that $\inf_{\substack{y \in F \\ \mathrm{q}(y) = x}} \| y \|_m = \| x \|_m$ for all m and so we find to each y_n a $z_n \in F$ with $\mathrm{q}(z_n) = -y_n$ and $\| z_n \|_n < \frac{2}{n}$.

As we can assume that $\| y_n \|_n < \frac{1}{n}$ we have $x_n + z_n \to x$ and $x_n + z_n \in$ kernel q which results in the fact that $x \in \overline{\operatorname{kernel} \mathrm{q}}$.

4. A special exact sequence

We want to discuss the result from paragraph 3 in the special case of a space E having property DN and $\lambda(A) = \Lambda_\infty(\alpha)$. We start with the following simple lemma.

Lemma 1. *Let $(E, \| \|_k)_{k=1,2,\ldots}$ an (F)–space. If for each k there exists a p, such that for all $r > 0$*

$$\| \|_{k+1} \leq r^{\left(\frac{1}{2}\right)^{k+1}} \| \|_k + \frac{1}{r} \| \|_{k+p}$$

then E has property DN.

Proof. We have to show: For each k there exits a $\tilde{p}$ and C with $\| \ \|_{k+1} \leq r \| \ \|_1 + \frac{C}{r} \| \ \|_{k+\tilde{p}}$ for each $r > 1$.

Starting from our assumptions we get

$$\begin{aligned}
\| \ \|_{k+1} &\leq r^{\left(\frac{1}{2}\right)^{k+1}} \| \ \|_k + \frac{1}{r} \| \ \|_{k+p_{k+1}} \\
&\leq r^{\left(\frac{1}{2}\right)^{k+1}} r^{2\cdot\left(\frac{1}{2}\right)^k} \| \ \|_{k-1} + r^{\left(\frac{1}{2}\right)^{k+1}-2} \| \ \|_{k+p_k} + \frac{1}{r} \| \ \|_{k+p_{k+1}} \\
&\leq r^{\left(\frac{1}{2}\right)^{k+1}+2\cdot\left(\frac{1}{2}\right)^k} \| \ \|_{k-1} + \frac{2}{r} \| \ \|_{k+\bar{p}} \\
&\leq r^{\sum_{m=2}^{k+1} 2\cdot\left(\frac{1}{2}\right)^m} \| \ \|_1 + \frac{\tilde{C}}{r} \| \ \|_{k+\tilde{p}} \\
&\leq r \| \ \|_1 + \frac{\tilde{C}}{r} \| \ \|_{k+\tilde{p}}
\end{aligned}$$

Theorem 2. *Let $E, \Lambda_\infty(\alpha)$ two nuclear, stable (F)–spaces with the same diametral dimension.*

If E has property DN, then there exists an (F)–space F with property DN and an exact sequence

$$0 \to E \to F \to \Lambda_\infty^{\mathbf{N}}(\alpha) \to 0 \,.$$

Proof: The existence of an exact sequence of the described type is proved in Theorem 1 paragraph 3. It remains to be seen, that the space F has property DN.

We have to show (see Lemma 1):
For each l there exists C, p with

$$\| \ \|_{l+1} \leq r^{\left(\frac{1}{2}\right)^{l+1}} \| + \|_1 \frac{C}{r} \| \ \|_{l+p} \text{ for all } r \,.$$

We use all definitions of Theorem 1 paragraph 3, especially let

$$\begin{aligned}
z_k &= \sum_{j=1}^{\infty} \bar{x}_{j,k} y_j^k \text{ and} \\
y_k &= \sum_{j=1}^{\infty} \tilde{x}_{j,k} y_j^k
\end{aligned}$$

In the proof of Theorem 1 the $(y_j^k)_{j,k=1,2,\dots}$ are the joint orthogonal basis of $(E_k, \| \ \|_{k+1})$ and $(E_k, \| \ \|_k)$. The property of being a basis in $(E_k, \| \ \|_k)$ is used in $(*)$ and $(**)$. It is easy to see that the joint basis of $(E_k, \| \ \|_{k+1})$ and $(E_k, \| \ \|_1)$ in a space with property DN also fulfills $(*)$ and $(**)$. Therefore we can assume that $(y_j^k)_{j=1,2,\dots}$ is the joint orthogonal basis of $(E_k, \| \ \|_{k+1})$ and $(E, \| \ \|_1)$. We estimate the $(l+1)$–norm.

$$\| x \|_{l+1} = \sum_{k=1}^{l+1} \sum_{j=1}^{\infty} \left| \sum_{i=1}^{\infty} x_{j,k}^i \right| a_{\pi(k,j),l-k+2} + \| \bar{x}^{[l+1]} \|_{l+1} + \| \tilde{x}^{(l+1)} \|_{l+1}$$

$$\leq \sum_{k=1}^{l}\sum_{j=1}^{\infty}|\sum_{i=1}^{\infty}x_{j,k}^{i}|a_{\pi(k,j),l-k+2}+\sum_{k=1}^{l}\| z_k \|_{l+1}+\| \tilde{x}^{(l+2)} \|_{l+1}$$
$$+\sum_{j=1}^{\infty}|\sum_{i=1}^{\infty}x_{j,l+1}^{i}|+\| z_{l+1} \|_{l+1}+\| y_{l+2} \|_{l+1}\leq$$
$$\leq \sum_{k=1}^{l}\sum_{j=1}^{\infty}|\sum_{i=1}^{\infty}x_{j,k}^{i}|a_{\pi(k,j),l-k+2}+\sum_{k=1}^{l}\| z_k \|_{l+1}+\| \tilde{x}^{(l+2)} \|_{l+1}$$
$$+\sum_{j=1}^{\infty}|\sum_{i=1}^{\infty}x_{j,l+1}^{i}|+\| z_l \|_{l+1}+\sum_{j=1}^{\infty}|\sum_{i=1}^{\infty}x_{j,l+2}^{i}|+\| z_{l+2} \|_{l+2}$$

The last inequality is true because we have a monotone increasing system of seminorms in F. E is stable and so we can assume that

$$\| \tilde{x}^{(l)} \|_l \geq \| y_{l+1} \|_1 + \| y_{l+2} \|_1 + \| \tilde{x}^{(l+2)} \|_1$$

which gives us

$$\| x \|_l \geq \sum_{k=1}^{l}\sum_{j=1}^{\infty}|\sum_{i=1}^{\infty}x_{j,k}^{i}|a_{\pi(k,j),l-k+1}+\sum_{k=1}^{l}\| z_k \|_l$$
$$+\| \tilde{x}^{(l+2)} \|_1+\| y_{l+1} \|_1+\| y_{l+2} \|_1 .$$

For the $\| \ \|_{l+p}$–norm we get the following estimate

$$\| x \|_{l+p} \geq \sum_{k=1}^{l}\sum_{j=1}^{\infty}|\sum_{i=1}^{\infty}x_{j,k}^{i}|a_{\pi(k,j),l+p-k+1}+\sum_{k=1}^{l}\| z_k \|_{l+p}$$
$$+\sum_{j=1}^{\infty}|\sum_{i=1}^{\infty}x_{j,l+1}^{i}|a_{\pi(l+1,j),p}+\sum_{j=1}^{\infty}|\sum_{i=1}^{\infty}x_{j,l+2}^{i}|a_{\pi(l+2,j),p-1}$$
$$+\| z_{l+1} \|_{l+p}+\| z_{l+2} \|_{l+p}+\| \tilde{x}^{(l+2)} \|_{l+p} .$$

In stating this inequality we have used the fact that the system of seminorms in F is monotone increasing.

If we use Lemma 1, we still have to prove

$$\| z_{l+1} \|_{l+1}+\| z_{l+2} \|_{l+2}+\sum_{j=1}^{\infty}|\sum_{i=1}^{\infty}x_{j,l+1}^{i}|+\sum_{j=1}^{\infty}|\sum_{i=1}^{\infty}x_{j,l+2}^{i}|\leq$$
$$\leq r^{\left(\frac{1}{2}\right)^{l+1}}(\| y_{l+1} \|_1+\| y_{l+2} \|_1)+\frac{C}{r}\Bigg(\| z_{l+1} \|_{l+p}+\| z_{l+2} \|_{l+p}$$
$$+\sum_{j=1}^{\infty}|\sum_{i=1}^{\infty}x_{j,l+1}^{i}|a_{\pi(l+1,j),p}+\sum_{j=1}^{\infty}|\sum_{i=1}^{\infty}x_{j,l+2}^{i}|a_{\pi(l+2,j),p-1}\Bigg)$$

for all r with suitable p, C.

To do this it is sufficient to show

$$\| z_l \|_l + \sum_{j=1}^{\infty} |\sum_{i=1}^{\infty} x_{j,l}^i| \leq r^{\left(\frac{1}{2}\right)^{l+1}} \| y_l \|_1 + \frac{C}{r} \left(\| z_l \|_{l+p} + \sum_{j=1}^{\infty} |\sum_{i=1}^{\infty} x_{j,l}^i| a_{\pi(l,j),p} \right) .$$

We start with an estimate for $\| z_l \|_l$.

Let $C = \sum_{i,j=1}^{\infty} \frac{1}{\pi(i,j)^2}$.

$$\begin{aligned} \| z_l \|_l &\leq \| \sum_{i,j=1}^{\infty} \left(- \sum_{m=i+1}^{\infty} x_{j,l}^m y_{\pi(i,j)}^l \right) \|_l \leq C \sum_{i,j=1}^{\infty} | \sum_{m=i+1}^{\infty} x_{j,l}^m | \frac{1}{C\pi(i,j)^2} \\ &\leq C \sum_{j\neq 1}^{\infty} \sup_i | \sum_{m=i+1}^{\infty} x_{j,l}^m | \leq C \sum_{i,j=1}^{\infty} |x_{j,l}^i| . \end{aligned}$$

So we have proved that

$$\| z_l \|_l + \sum_{j=1}^{\infty} |\sum_{i=1}^{\infty} x_{j,l}| \leq (C+1) \sum_{i,j=1}^{\infty} |x_{j,l}^i| .$$

Our next goal is to get an estimate for $\| y_l \|_1$.

$$\begin{aligned} \| y_l \|_1 &= \sum_{i,j=1}^{\infty} \left(|\sum_{m=1}^{i} x_{j,l}^m|^2 \| y_{\pi(i,j)}^l \|_1^2 \right)^{\frac{1}{2}} \\ &\geq \bar{C} \sum_{i,j=1}^{\infty} | \sum_{m=1}^{i} x_{j,l}^m | \cdot \| y_{\pi(i,j)}^l \|_1 \cdot \frac{1}{\pi(i,j)^2} \\ &= \bar{C} \sum_{i,j=1}^{\infty} | \sum_{m=1}^{i} x_{j,l}^m | \cdot \frac{\| y_{\pi(i,j)}^l \|_1}{\| y_{\pi(i,j)}^l \|_l} \cdot \frac{1}{\pi(i,j)^4} \\ &\geq \bar{C} \sum_{i,j=1}^{\infty} | \sum_{m=1}^{i} x_{j,l}^m | \cdot \left(\frac{\| y_{\pi(i,j)}^l \|_l}{\| y_{\pi(i,j)}^l \|_{l+1}} \right)^{\frac{1}{2C^2}\left(\frac{1}{2}\right)^{l+2}} . \end{aligned}$$

The last inequality is true because we can assume that (E is nuclear and has property DN)

$$\frac{\| y_{\pi(i,j)}^l \|_{l+1}}{\| y_{\pi(i,j)}^l \|_l} \geq \pi(i,j)^4 \left(\frac{\| y_{\pi(i,j)}^l \|_l}{\| y_{\pi(i,j)}^l \|_1} \right)^{2C^2 \cdot 2^{l+2}} .$$

Let $x = (x_1, x_2, \ldots)$ be a sequence of real numbers with $\sum_{k=1}^{\infty} |x_k| < \infty$, $\alpha = (\alpha_1, \alpha_2, \ldots)$ a monotone decreasing null sequence, and e_j the canonic unit vectors, then we have using $\| x \| = \sum_j |x_j| a_j$

$$x = \sum_{i=1}^{\infty} \left(\sum_{j=1}^{i} x_j \right) (e_i - e_{i+1}) \text{ and } \| x \| \leq 2 \cdot \sum_{i=1}^{\infty} |\sum_{j=1}^{i} x_j| \| e_i \| .$$

Applying this results to the sequence $\frac{\|y^l_{\pi(i,j)}\|_l}{\|y^l_{\pi(i,j)}\|_{l+1}}$ we get

$$\| y_l \|_1 \geq \frac{\bar{C}}{2} \sum_{i,j=1}^{\infty} |x^i_{j,l}| \left(\frac{\| y^l_{\pi(i,j)} \|_l}{\| y^l_{\pi(i,j)} \|_{l+1}} \right)^{\frac{1}{2C^2}\left(\frac{1}{2}\right)^{l+2}} .$$

Now we can find an estimate for $\| z_l \|_{l+p} + \sum_{i,j=1}^{\infty} | \sum_{i=1}^{\infty} x^i_{j,l} | a_{\pi(l,j),p}$. We start with

$$\begin{aligned}
\| z_l \|_{l+1} &= \sum_{i,j=1}^{\infty} \left(| \sum_{m=1}^{\infty} x^m_{j,l} |^2 \, \| y^l_{\pi(i,j)} \|^2_{l+1} \right)^{\frac{1}{2}} \\
&\geq \bar{C} \sum_{i,j=1}^{\infty} | \sum_{m=i+1}^{\infty} x^m_{j,l} | \cdot \| y^l_{\pi(i,j)} \|_{l+1} \cdot \frac{1}{\pi(i,j)^2} \\
&\geq \bar{C} \sum_{i,j=1}^{\infty} | \sum_{m=i+1}^{\infty} x^m_{j,l} | \cdot \frac{1}{\pi(i,j)^4} \cdot \frac{\| y^l_{\pi(i,j)} \|_{l+1}}{\| y^l_{\pi(i,j)} \|_l} \\
&\geq \bar{C} \sum_{i,j=1}^{\infty} | \sum_{m=i+1}^{\infty} x^m_{j,l} | \cdot \left(\frac{\| y^l_{\pi(i,j)} \|_{l+1}}{\| y^l_{\pi(i,j)} \|_l} \right)^{\frac{1}{2}} .
\end{aligned}$$

The sequence $\frac{\|y^l_{\pi(i,j)}\|_{l+1}}{\|y^l_{\pi(i,j)}\|_l}$ is in i monotone increasing therefore we can use the following remark.

Let $\| x \| = \| (x_1, x_2, \ldots) \| = \sum_j \alpha_j |x_j|$, where α_j is an increasing sequence with $\alpha_j \geq 1$, then we have with the unit vectors e_i

$$\begin{aligned}
& x = \sum_{i=1}^{\infty} \left(- \sum_{j=i+1}^{\infty} x_j \right) (e_i - e_{i+1}) + \left(\sum_{i=1}^{\infty} x_i \right) e_1 \\
\Longrightarrow \quad & \| x \| \leq 2 \cdot \sum_{i=1}^{\infty} | \sum_{j=i+1}^{\infty} x_j | \, \| e_{i+1} \| + | \sum_{i=1}^{\infty} x_i | \, \| e_1 \| .
\end{aligned}$$

So we can conclude

$$\begin{aligned}
& \| z_l \|_{l+p} + \sum_{j=1}^{\infty} | \sum_{m=1}^{\infty} x^m_{j,l} | a_{\pi(l,j),p} \geq \\
\geq \; & \bar{C} \sum_{i,j=1}^{\infty} | \sum_{m=i+1}^{\infty} x^m_{j,l} | \cdot \left(\frac{\| y^l_{\pi(i,j)} \|_{l+1}}{\| y^l_{\pi(i,j)} \|_l} \right)^{\frac{1}{2}} + \sum_{j=1}^{\infty} | \sum_{m=1}^{\infty} x^m_{j,l} | a_{\pi(l,j),p} \\
\geq \; & \bar{C} \sum_{i,j=1}^{\infty} | \sum_{m=i+1}^{\infty} x^m_{j,l} | \cdot \left(\frac{\| y^l_{\pi(i,j)} \|_{l+1}}{\| y^l_{\pi(i,j)} \|_1} \right)^{\frac{1}{4}} + \sum_{j=1}^{\infty} \sum_{m=1}^{\infty} | x^m_{j,l} | \frac{\| y^l_{\pi(i,j)} \|_{l+1}}{\| y^l_{\pi(i,j)} \|_l} .
\end{aligned}$$

The last inequality holds, because $\lambda(A)$ has the same diametral dimension as E, and as we can assume that $a_{j,1} = 1$ and p as great as we want.

So we get

$$\| z_l \|_{l+p} + \sum_{j=1}^{\infty} | \sum_{m=1}^{\infty} x_{j,l}^{m} | a_{\pi(l,j),p} \geq$$

$$\geq \frac{\bar{C}}{2} \sum_{j=1}^{\infty} |x_{j,l}^{1}| \left(\frac{\| y_{\pi(i,j)}^{l} \|_{l+1}}{\| y_{\pi(i,j)}^{l} \|_1} \right)^{\frac{1}{4}} + \sum_{j=1}^{\infty} \sum_{i=2}^{\infty} |x_{j,l}^{i}| \left(\frac{\| y_{\pi(i-1,j)}^{l} \|_{l+1}}{\| y_{\pi(i-1,j)}^{l} \|_1} \right)^{\frac{1}{4}} .$$

We have

$$s^{\alpha_{\pi(i,j)}} \leq C \frac{\| y_{\pi(i,j)}^{l} \|_{l+1}}{\| y_{\pi(i,j)}^{l} \|_1} \leq s'^{\alpha_{\pi(i,j)}}$$

with suitable s, s', C, because E and $\Lambda_\infty(\alpha)$ have the same diametral dimension.

The sequence α is stable i.e. $\alpha_{2j} \leq C\alpha_j$ and the sequence $\frac{\|y_{\pi(i,j)}^{l}\|_{l+1}}{\|y_{\pi(i,j)}^{l}\|_1}$ is in i increasing, so we can conclude:

If i is the smallest index with $\frac{\|y_{\pi(i,j)}^{l}\|_{l+1}}{\|y_{\pi(i,j)}^{l}\|_1} \geq s'^{\frac{1}{2}\alpha_{\pi(i,j)}}$ then we have for all $i_0 \geq i$

$$\frac{\| y_{\pi(i_0,j)}^{l} \|_{l+1}}{\| y_{\pi(i_0,j)}^{l} \|_1} \geq s'^{\frac{1}{2C}\alpha_{\pi(i_0,j)}}$$

because otherwise it would be true that

$$\frac{\| y_{\pi(i_0,j)}^{l} \|_{l+1}}{\| y_{\pi(i_0,j)}^{l} \|_1} < s'^{\frac{1}{2C}\alpha_{\pi(i_0,j)}} \leq s'^{\frac{1}{2}\alpha_{\pi(i_0,j)}} .$$

So we have with two suitable constants C_1, C_2

$$C_1 \cdot s'^{\frac{1}{2C}\alpha_{\pi(i,j)}} \leq \frac{\| y_{\pi(i,j)}^{l} \|_{l+1}}{\| y_{\pi(i,j)}^{l} \|_1} \leq C_2 s'^{\frac{1}{2}\alpha_{\pi(i,j)}} .$$

Using this we can estimate

$$\| z_l \|_{l+p} + \sum_{j=1}^{\infty} | \sum_{m=1}^{\infty} x_{j,l}^{m} | a_{\pi(l,j),p} \geq$$

$$\geq \tilde{C} \sum_{j=1}^{\infty} |x_{j,l}^{1}| \left(\frac{\| y_{\pi(1,j)}^{l} \|_{l+1}}{\| y_{\pi(1,j)}^{l} \|_1} \right)^{\frac{1}{4}} + \sum_{j=1}^{\infty} \sum_{i=2}^{\infty} |x_{j,l}^{i}| s'^{\frac{1}{8C}\alpha_{\pi(i-1,j)}}$$

$$\geq \tilde{C} \sum_{j=1}^{\infty} |x_{j,l}^{1}| \left(\frac{\| y_{\pi(1,j)}^{l} \|_{l+1}}{\| y_{\pi(1,j)}^{l} \|_1} \right)^{\frac{1}{4}} + \sum_{j=1}^{\infty} \sum_{i=2}^{\infty} |x_{j,l}^{i}| s'^{\frac{1}{8C^2}\alpha_{\pi(i,j)}}$$

$$\geq \tilde{\tilde{C}} \sum_{j=1}^{\infty} |x_{j,l}^{1}| \left(\frac{\| y_{\pi(1,j)}^{l} \|_{l+1}}{\| y_{\pi(1,j)}^{l} \|_1} \right)^{\frac{1}{4}} + \sum_{j=1}^{\infty} \sum_{i=2}^{\infty} |x_{j,l}^{i}| \left(\frac{\| y_{\pi(i,j)}^{l} \|_{l+1}}{\| y_{\pi(i,j)}^{l} \|_1} \right)^{\frac{1}{8C^2}}$$

$$\geq \tilde{\tilde{C}} \sum_{i,j=1}^{\infty} |x_{j,l}^{i}| \left(\frac{\| y_{\pi(i,j)}^{l} \|_{l+1}}{\| y_{\pi(i,j)}^{l} \|_1} \right)^{\frac{1}{8C^2}} .$$

It remains to be seen

$$1 \leq r^{\left(\frac{1}{2}\right)^{l+1}} \left(\frac{\| y_{\pi(i,j)}^{l+1} \|_l}{\| y_{\pi(i,j)}^{l} \|_1} \right)^{\frac{1}{2C^2}\left(\frac{1}{2}\right)^{l+2}} + \frac{1}{r} \left(\frac{\| y_{\pi(i,j)}^{l} \|_{l+1}}{\| y_{\pi(i,j)}^{l} \|_l} \right)^{\frac{1}{8C^2}} .$$

If we put $a = \left(\frac{\| y_{\pi(i,j)}^{l} \|_{l+1}}{\| y_{\pi(i,j)}^{l} \|_l} \right)^{\frac{1}{8C^2}}$ we have to show

$$1 \leq r^{\left(\frac{1}{2}\right)^{l+1}} \cdot a^{\left(\frac{1}{2}\right)^{l}} + \frac{1}{ar} \quad \text{for each } r > 0 .$$

Case 1: $r \leq \frac{1}{a} \Longrightarrow \frac{1}{ar} \geq 1$

Case 2: $r > \frac{1}{a} \Longrightarrow r^{\left(\frac{1}{2}\right)^{l+1}} \cdot a^{\left(\frac{1}{2}\right)^{l}} \geq \frac{1}{a} > 1$

5. Applications

In this paragraph we want to give an application of our yet proved results. We prove that each nuclear, stable (F)–space with properties DN and Ω has a basis. We start with the following theorem.

Theorem 1. *If a nuclear, stable (F)–space has the properties DN and* Ω *and* $\Lambda_\infty(\alpha)$ *has the same diametral dimension as* E*, then* $\Lambda_\infty(\alpha)$ *is isomorphic to a complemented subspace of* E*.*

PROOF. We know from our previous results

a) $\Lambda_\infty(\alpha)$ is isomorphic to a subspace of E.

b) There exists a (F)–space F with property DN and an exact sequence

$$0 \to E \to F \xrightarrow{q} \Lambda_\infty(\alpha)^{\mathbf{N}} \to 0 .$$

$\Lambda_\infty(\alpha)$ is a subspace of E, therefore we have an exact sequence

$$0 \to \Lambda_\infty(\alpha) \to E \to Q \to 0 .$$

Q is isomorphic to a subspace of $\Lambda_\infty^{\mathbf{N}}(\alpha)$, so we have

$$\begin{array}{ccccccc} & & & & 0 & & \\ & & & & \uparrow & & \\ 0 \to E \to & \mathrm{q}^{-1}(Q) & \to & & Q & \to & 0 \\ & & & & \uparrow & & \\ & & & & E & & \\ & & & & \uparrow & & \\ & & & & \Lambda_\infty(\alpha) & & \\ & & & & \uparrow & & \\ & & & & 0 & & \end{array}$$

In [9] it is shown that under this circumstances there exists an exact sequence

$$0 \to \Lambda_\infty(\alpha) \to E \oplus E \to \mathrm{q}^{-1}(Q) \to 0\,.$$

$\Lambda_\infty(\alpha)$ has property Ω and $\mathrm{q}^{-1}(Q)$ has property DN, so we can use the splitting theorem of [10], which means that $\Lambda_\infty(\alpha)$ is complemented in $E \oplus E$, which is isomorphic to E.

Now it is easy to prove our main result:

Theorem 2. *If a nuclear, stable (F)-space has properties DN and Ω then E has a basis.*

Proof From [10] we know that F is complemented in $\Lambda_\infty(\alpha)$, on the other hand we have shown that $\Lambda_\infty(\alpha)$ is complemented in E.

Using [8] we get the result.

References

[1] Djakov P. B., Dubinsky E., 'Complemented block subspaces of Köthe spaces', *Serdica* **14** , (1988), 278-282.

[2] Dubinsky E., 'The structure of nuclear Fréchet spaces', *Lecture notes in Math.* **720** Berlin–Heidelberg–New York 1979.

[3] Dubinsky E., Ramanujan M. S., 'On λ–nuclearity', *Mem. Am. Soc.* **128** (1972).

[4] Dubinsky E., Vogt D., 'Bases in complemented subspaces of power series spaces', *Bull. Acad. Polon. Sci.* **34** (1986) 65-67.

[5] Dubinsky E., Vogt D., 'Complemented subspaces in tame power series spaces', to appear in *Studia Math.*.

[6] Komura T., Komura Y., 'Über die Einbettung der nuklearen Räume nach $(s)^A$', *Math. Ann.* **162** (1966) 284-288.

[7] Pietsch A., 'Nukleare lokalkonvexe Räume', *Akademie Verlag,* Berlin 1969.

[8] Vogt D., 'Ein Isomorphiesatz für Potenzreihenräume', *Arch. Math.* **38** (1982), 540-548.

[9] Vogt D., 'Tame spaces and power series spaces', *Math. Z.* **196** (1987), 523-536.

[10] Vogt D., Wagner M., 'Charakterisierung der Unterräume und Quotientenräume der nuklearen, stabilen Pottenzreihenräume vom unendlichen Typ', *Studia Math.* **70** (1981), 63-80.

EVERY QUOJECTION IS THE QUOTIENT OF A COUNTABLE PRODUCT OF BANACH SPACES

J. BONET
Departamento de Matematica Aplicada
E.T.S. Arquitectura
Universidad Politécnica
46071 Valencia - (Spain)

M. MAESTRE
Departamento de Analisis Matematico
Facultad de Farmacia
Universidad de Valencia
46010 Valencia - (Spain)

G. METAFUNE
Università della Basilicata
Potenza - (Italy)

V.B. MOSCATELLI
Dipartimento di Matematica
Università di Lecce - (Italy)

D. VOGT
FB Mathematik
Universität -GHS- Wuppertal
D-5600 Wuppertal 1 - (F.R. Germany)

Abstract: It is proved that every quojection in the sense of Bellenot and Dubinsky [1] is the quotient of a countable product of copies of $\ell_1(I)$ for a suitable index set I.

During the NATO Workshop "Advances in the Theory of Fréchet Spaces" the above authors realized they had obtained the same result. Here we present a short direct proof of it.

A Fréchet space E is called a quojection if for every continuous seminorm p on E the space $E/ker p$ is normable or, equivalent, if E is the surjective limit of a sequence of Banach spaces. Quojections occur frequently in the problem of determining pairs (E, F) of Fréchet spaces such that $Ext^1(E, F) = 0$ (cf. [4], [5]). Clearly if E is nuclear or a Köthe echelon space $\lambda_1(A)$ and F is a quotient of a countable product of Banach spaces, then every exact sequence

$$O \to F \to G \to E \to O$$

splits. Here we show that quojections are exactly quotients of countable products of Banach spaces. Moscatelli [3] gave examples of quojections not isomorphic to a countable product of Banach spaces.

Let E be a quojection, $\|\cdot\|_1 \leq \|\cdot\|_2 \leq \cdots$ a fundamental system of seminorms, $E_k = E/ker\ \|\cdot\|_k, q_k : E \to E_k$ the quotient map, $q_{k+1,k} : E_{k+1} \to E_k$ the connecting (=quotient) map, U_k the open unit ball in E_k.

1. Lemma (cf. [1, Cor. 4], [2, Prop. 1]). For every k there is a bounded set $B_k \subset E$ such that $q_k B_k \supset U_k$.

T. Terzioğlu (ed.), Advances in the Theory of Fréchet Spaces, 355–356.

2. Corollary. For every index set $I, k \in \mathbf{N}$ and $\phi \in L(\ell_1(I), E_k)$ there is $\psi \in L(\ell_1(I), E)$ such that $q_k \circ \psi = \phi$.

3. Theorem. There exists I such that E is isomorphic to a quotient of $\ell_1(I)^{\mathbf{N}}$.

Proof: We put $F_0 := E/ker \, \|\cdot\|_1, F_k := ker \, \|\cdot\|_k / ker \, \|\cdot\|_{k+1}$ for $k = 1, 2, \cdots$. So $F_0 := ker \, \|\cdot\|_0 / ker \, \|\cdot\|_1$ with $\|\cdot\|_0 := 0$. The spaces F_k are Banach spaces. We choose I such that for every k we have a surjective continuous linear map $\phi_k : \ell_1(I) \to F_k$. We obtain $\psi_k \in L(\ell_1(I), ker \, \|\cdot\|_k)$ according to Corollary 2 applied to the quojection $ker \, \|\cdot\|_k$. We set

$$\psi(x) = \sum_{k=0}^{\infty} \psi_k(x_k) \quad \text{for} \quad x = (x_k) \in \ell_1(I)^{\mathbf{N}_0}.$$

Since

$$\sum_{k=0}^{\infty} \|\psi_k(x_k)\|_m = \sum_{k=0}^{m-1} \|\psi_k(x_k)\|_m$$

the series converges in E and defines a continuous linear map $\ell_1(I)^{\mathbf{N}_0} \to E$.

We show that ψ is surjective. For $x \in E$, we put $x_0 = x$ and proceed inductively. Given $x_k \in ker \, \|\cdot\|_k$ we choose $\xi_k \in \ell_1(I)$ such that

$$q_{k+1} \, x_k = \phi_k \, \xi_k, \quad \text{i.e.} \quad x_k = \psi_k \, \xi_k + x_{k+1}, \quad x_{k+1} \in ker \, \|\cdot\|_{k+1}.$$

This yields a sequence $\xi = (\xi_k) \in \ell_1(I)^{\mathbf{N}_0}$ such that

$$\psi(\xi) = \sum_{k=0}^{\infty} \psi_k(\xi_k) = x.$$

REFERENCES

[1] S.F. Bellenot, E. Dubinsky: 'Fréchet spaces with nuclear Köthe quotients'. *Trans. Amer. Math. Soc.* **273** (1982) 579-594.

[2] S. Dierolf, D.N. Zarnadze: 'A note on strictly regular Fréchet spaces'. *Arch. Math.* **42** (1984) 549-556.

[3] V.B. Moscatelli: 'Fréchet spaces without continuous norms and without bases'. *Bull. London Math. Soc.* **12** (1980] 63-66.

[4] D. Vogt: 'Some results on continuous linear maps between Fréchet spaces'. pp.349-381 in *Functional Analysis: Surveys and Recent Results III*, K.D. Bierstedt, B. Fuchssteiner (eds). *North Holland Math. Studies* **90**.

[5] D. Vogt: 'On the functor $Ext^1(E, F)$ for Fréchet spaces'. *Studia Math.* **85** (1987) 163-197.

DUAL KŌMURA SPACES

Hikosaburo Komatsu
Department of Mathematics
Faculty of Science
University of Tokyo
Hongo, Tokyo, 113 Japan

ABSTRACT. A dual Kōmura space is by definition a locally convex space E in which for any absolutely convex bounded set A there is another $B \supset A$ such that the natural mapping $\hat{E}_A \to \hat{E}_B$ is weakly compact. Fundamental properties and hereditarity are discussed together with motivations and applications.

1. INTRODUCTION

Everybody would agree that the theory of Fréchet spaces was founded by the three papers, Dieudonné-Schwartz [1], Grothendieck [2] and [3].

The first paper is easy to understand. The second one is moderately difficult. However, it is very hard to read through the third paper and, in particular, the part from Theorem 3 to Theorem 10 of Chapter I in which Grothendieck proves essentially that the product of an integral linear mapping and a weakly compact linear mapping is nuclear.

This is very fundamental in his theory but later expositions, Schwartz [4], Trèves [5] and Pietsch [6], all evade this part. The author's [7] is not an exception either. The main difficulty is in the Phillips theorem [8] of Radon-Nikodým type for vector measures. Here again Phillips' original proof is very difficult to read. Grothendieck quotes an earlier version [9] and reconstructs the proof based on Bourbaki's definition of measures. It seems, however, that the lack of adequate terminology causes more confusion of readers.

In this paper we introduce new names for certain properties of locally convex spaces, which Grothendieck [2] and [3] tacitly employed, and we hope that those concepts will be useful for paraphrasing his deep theory.

According to our exposition [7] we call a nuclear space a Grothendieck space because a class of spaces is usually named after the founder of its theory.

2. KŌMURA SPACES, SCHWARTZ SPACES AND GROTHENDIECK SPACES

Before we go into the theory of dual Kōmura spaces, which is our main subject, we recall the theory of Kōmura spaces, Schwartz spaces and Grothendieck spaces as developed by Grothendieck [2] for Schwartz spaces, by Grothendieck [3] for Grothendieck spaces and by Grothendieck [2], Kōmura [10] and Komatsu [7] for Kōmura spaces.

DEFINITION 1. A locally convex space E is said to be a *Kōmura space* (resp. a *Schwartz space*, resp a *Grothendieck space*), if for each absolutely convex closed neighborhood U of zero in E there

T. Terzioğlu (ed.), Advances in the Theory of Fréchet Spaces, 357.

is an absolutely convex closed neighborhood V of zero included in U such that the natural mapping $\hat{E}_V \to \hat{E}_U$ is weakly compact (resp. compact, resp. nuclear).

Here $\hat{E}_U$ denotes the Banach space obtained as the completion of the normed space $E_U = E/\sim$ having U as the unit ball. A linear mapping u from a locally convex space E into a locally convex space F is said to be ***weakly compact*** (resp. ***compact***) if it maps a neighborhood of zero in E into a relatively weakly compact (resp. relatively compact) set in F. A linear mapping $u : E \to F$ is said to be ***nuclear*** if it admits the factorization

$$E \longrightarrow l^\infty \xrightarrow{\lambda} l^1 \longrightarrow F$$

into continuous linear mappings in which λ is the multiplication by a sequence (λ_n) in l^1.

The following Proposition follows easily from the definition.

PROPOSITION 1. *A locally convex space E is a Kōmura space (resp. a Schwartz space, resp. a Grothendieck space) if and only if every continuous linear mapping u from E into an arbitrary Banach space N is weakly compact (resp. compact, resp. nuclear).*

These classes of locally convex spaces enjoy unusual stability under standard operations on locally convex spaces.

THEOREM 1. *Subspaces, arbitrary direct products, arbitrary projective limits, quotient spaces, countable direct sum and countable inductive limits of Kōmura spaces (resp. Schwartz spaces, resp. Grothendieck spaces) are Kōmura spaces (resp. Schwartz spaces, resp. Grothendieck space).*

A locally convex space E is a Kōmura space (resp. a Schwartz space, resp. a Grothendieck space) if and only if so is its completion $\hat{E}$.

Grothendieck [2] (resp. [3]) gives a proof for Schwartz spaces (resp. Grothendieck spaces). Grothendieck [2, pp. 115–116] discusses on Kōmura spaces without giving their names and most statements of Theorem are proved there. We gave a proof in [7] based on our Definition 1.

Davis-Figiel-Johnson-Pełczyński [11] has proved that a linear mapping $u : E \to F$ from a locally convex space E into another F is weakly compact if and only if it can be factorized through a reflexive Banach space R as

$$E \longrightarrow R \longrightarrow F$$

with continuous linear mappings.

Hence it follows that a locally convex space E is a Kōmura space (resp. a complete Kōmura space) if and only if it is isomorphic to a subspace (resp. a closed subspace) of a direct product of reflexive Banach spaces. (For Grothendieck spaces E we can take Hilbert spaces l^2 in this representation.)

Hence follows the following important theorem due to Kōmura [10].

THEOREM 2. *Every complete Kōmura space E is semi-reflexive and its strong dual E' is ultrabornologic (i.e. the inductive limit of Banach spaces).*

The completeness is not necessary for the semi-reflexivity. We have

PROPOSITION 2. *A Kōmura space (resp. a Schwartz space) is semi-reflexive (resp. semi-Montel) if and only if it is boundedly complete (i.e. every bounded closed set is complete).*

A Fréchet space E is a Kōmura space (resp. a Schwartz space, resp. a Grothendieck space) if and only if it is represented as the projective limit $\varprojlim E_n$ of a sequence E_n of Banach spaces such that the mappings $u_n : E_n \to E_{n-1}$ are weakly compact (resp. compact, resp. nuclear). Here Banach spaces E_n can be chosen so that the image of E under the canonical mapping $E \to E_n$ is dense for all n.

As we remarked earlier, we can choose reflexive Banach spaces (resp. Hilbert spaces) E_n for Kōmura spaces (resp. Grothendieck space). We call a Fréchet Kōmura space (resp. a Fréchet

Schwartz space, resp. a Fréchet Grothendieck space) an (FK) space (resp. an (FS) space, resp. an (FG) space) for short.

3. COMPLETING BOUNDED SETS AND COMPLETING LOCALLY CONVEX SPACES

In defining the dual concepts we employ the normed space E_B generated by an absolutely convex closed bounded set B in a locally convex space E having B as the unit ball. We also employ its completion $\hat{E}_B$.

DEFINITION 2. A bounded set B in a locally convex space E is said to be *completing* if it is absolutely convex and closed and if the normed space E_B generated by B is a Banach space.

A locally convex space E is said to be *completing* if every absolutely convex closed bounded set in E is completing.

THEOREM 3. *The following conditions are equivalent for a locally convex space E :*
(a) *E is completing ;*
(b) *Every bounded set A in E is included in a completing bounded set B in E ;*
(c) *Let (e_n) be a sequence in E. If there is an absolutely convex closed set B in E such that*

$$e_n \in E_B \quad \text{and that} \quad \sum_{n=1}^{\infty} \|e_n\|_{E_B} < \infty,$$

then the series $\sum_{n=1}^{\infty} e_n$ converges weakly in E.

PROOF. The implications (a) $\Rightarrow$ (b) $\Rightarrow$ (c) are trivial.

To prove that (c) implies (a), let B be an absolutely convex closed bounded set in E. We have to show that every absolutely convergent series $\sum e_n$ in E_B converges. By the assumption the series converges weakly in E. Let $e = \sum e_n$ be the sum. For any $\varepsilon > 0$ there is an N_ε such that for any $M, N \geq N_\varepsilon$ we have

$$\sum_{n=M}^{N} \|e_n\|_{E_B} \leq \varepsilon.$$

Hence we have for any l in the polar B° of B in E'

$$\left|\left\langle \sum_{n=M}^{N} e_n, \ l \right\rangle\right| \leq \varepsilon.$$

Let N tend to infinity. Then we have

$$\left|\left\langle e - \sum_{n=1}^{M-1} e_n, \ l \right\rangle\right| \leq \varepsilon.$$

Hence it follows that $e \in E_B$ and that

$$\left\| e - \sum_{n=1}^{M-1} e_n \right\|_{E_B} \leq \varepsilon.$$

In particular, every sequentially complete locally convex space is completing. The converse holds for metrizable locally convex spaces.

THEOREM 4. *Every completing metrizable locally convex space E is complete.*

PROOF. Let (e_n) be a Cauchy sequence in E. Since the metrizable locally convex space E has the strict Mackey condition of convergence [2, p.106] there is an absolutely convex closed bounded set B containing the bounded set $\{e_n\}$ and such that the uniform topology of E coincides with the norm topology of E_B on the convex hull A of $\{e_n\}$. Since E_B is complete, (e_n) converges in E_B and hence in E.

THEOREM 5. *Closed subspaces, arbitrary direct products, arbitrary projective limits and arbitrary direct sums of completing locally convex spaces are completing.*

PROOF. The proof is trivial for closed subspaces.

Let A be a bounded set in the direct product $\prod E_\alpha$ of completing locally convex spaces E_α. The projection $\pi_\alpha(A)$ of A into E_α is a bounded set. Hence it is included in a completing bounded set B_α. Then $B = \prod B_\alpha$ is a completing bounded set in $\prod E_\alpha$ and clearly includes A.

Since every bounded set A in a direct sum $\sum E_\alpha$ of locally convex space E_α is included in a finite direct product of E_α's, the same proof holds for direct sums. A projective limit is a closed subspace of the direct product.

Let E be a locally convex space. Its completion and even its sequential completion [12] are completing. On the other hand, it follows from the criterion (c) of Theorem 2 that the intersection of a family of completing subspaces of a locally convex space is completing. Hence there is the smallest completing locally convex space E^C including E, which we call the *completing hull* of E.

PROPOSITION 3. *The completing hull of a metrizable locally convex space E coincides with its completion.*

The completing hull of a (DF) space E is a (DF) space.

PROOF. The first part follows immediately from Theorem 4.

The completion $\hat{E}$ of a (DF) space E is a (DF) space in which every bounded set is included in the closure in $\hat{E}$ of a bounded set B in E [2, p.77]. Therefore the dual E' has the same bounded sets and the same equicontinuous sets for any space between E and $\hat{E}$. In particular, E^C is a (DF) space.

4. DUAL KŌMURA SPACES, DUAL SCHWARTZ SPACES AND DUAL GROTHENDIECK SPACES

DEFINITION 3. A locally convex space E is said to be a *dual Kōmura space* or *(DK) space* (resp. a *dual Schwartz space, or (DS) space, resp. a dual Grothendieck space or (DG) space*) if for each absolutely convex closed bounded set A in E there is an absolutely convex closed bounded set B including A such that the natural mapping $\hat{E}_A \to \hat{E}_B$ is weakly compact (resp. compact, resp. nuclear).

If E is completing, then we may replace $\hat{E}_A \to \hat{E}_B$ by $E_A \to E_B$. We remark, however, that the dual of Proposition 1 does not hold even in this case. In fact, the property that every continuous linear mapping v from an arbitrary normed space N into E is weakly compact (resp. compact) means that every bounded set B in E is relatively weakly compact (resp. relatively compact), that is, E is semi-reflexive (resp. semi-Montel). As shown later, completing (DK) spaces (resp. completing (DS) spaces) are semi-reflexive (resp. semi-Montel), but the converse does not hold in general.

We have, however, the following.

PROPOSITION 4. *A locally convex space E is a completing (DG) space if and only if every continuous linear mapping $v : N \to E$ from an arbitrary normed space N into E is nuclear.*

PROOF. Let $v : N \to E$ be a continuous linear mapping from a normed space N into a completing (DG) space E. Let A be the closure of the image of the unit ball in N under v. Then v is factorized into $N \to E_A \to E_B \to E$, where B is a completing bounded set so that $E_A \to E_B$ is nuclear. Hence v is nuclear.

Conversely suppose that every $v : N \to E$ is nuclear. Let A be an absolutely convex closed bounded set in E. Then the natural mapping $E_A \to E$ is nuclear and hence is factorized as a nuclear mapping $E_A \to L$ into a Banach space L followed by a continuous linear mapping $L \to E$. An absolutely convergent series in E_A has a convergent image in L and hence is convergent in E. Let B be the closure of the image of a sufficiently large ball of L. Then B includes A and the natural mapping $E_A \to E_B$ is nuclear.

Furthermore, if we restrict ourselves to Fréchet spaces, then the dual characterization holds for (DK) and (DS) spaces.

PROPOSITION 5. *Every reflexive (resp. Montel) Fréchet space is a (DK) space (resp. a (DS) space).*

PROOF. Grothendieck [3, Chap. I, p.105] (resp. [3, Chap. I, p.112]) has shown that a weakly compact set (resp. a compact set) A in a Fréchet space E is a weakly compact set (resp. a compact set) in E_B for a suitable absolutely convex closed bounded set B containing A.

Definition 3 is justified by the following.

PROPOSITION 6. (i) *A locally convex space E is a (DK) space (resp. a (DS) space, resp. a (DG) space) if and only if its strong dual E' is a Kōmura space (resp. a Schwartz space, resp. a Grothendieck space).*

(ii) *A quasi-barrelled space E is a Kōmura space (resp. a Schwartz space, resp. a Grothendieck space) if and only if its strong dual E' is a (DK) space (resp. a (DS) space, resp. a (DG) space).*

PROOF. Let $A \subset B$ be a pair of absolutely convex closed bounded sets in E. Then the dual mapping of the natural mapping $E_A \to E_B$ is identified with the natural mapping $E'_{B^\circ} \to E'_{A^\circ}$. Hence the first part follows from the Gantmacher-Nakamura theorem (resp. Riesz-Schauder theorem, resp. the corresponding theorem for nuclear and integral linear mappings).

The second part is proved similarly. Actually a countably quasi-barrelled space E (i.e. a locally convex space satisfying half of the conditions of (DF) spaces) is a Grothendieck space if its strong dual E' is a (DG) space.

COROLLARY. *Every reflexive (resp. Montel) (DF) space is a Kōmura space (resp. a Schwartz space).*

THEOREM 6. *Subspaces, countable direct products, countable projective limits and arbitrary direct sums of (DK) spaces (resp. (DS) spaces, resp. (DG) spaces) are (DK) spaces (resp. (DS) spaces, resp. (DG) spaces).*

PROOF. Let F be a subspace of a (DK) space (resp. a (DS) space, resp. a (DG) space) E. For each absolutely convex closed bounded set A in F there is another $B \supset A$ in E such that the natural mapping $\hat{E}_A \to \hat{E}_B$ is weakly compact (resp. compact, resp. the product of two nuclear linear mappings). Then $B_1 = B \cap F$ is an absolutely convex closed bound set in F including A, and the natural mapping $\hat{F}_A \to \hat{F}_{B_1}$ is weakly compact (resp. compact, resp. nuclear).

In view of Proposition 6 (i), the proof for countable direct products and arbitrary direct sums is reduced to Theorem 1. A direct proof similar to Theorem 1 is also possible. Then the statement for countable projective limits follows.

REMARK. Köthe and Grothendieck [2, p.95] have constructed a Fréchet Montel space E and its closed subspace F such that the quotient space E/F is isomorphic to the non-reflexive Banach space l^1. E is a (DS) space but its quotient space is not a (DK) space.

The dual E' of the space E is an example of (DF) Montel space which is not a (DK) space.

PROPOSITION 7. *A (DK) space (resp. a (DS) space) is semi-reflexive (resp. semi-Montel) if and only if it is completing.*

PROOF. A semi-reflexive space is always boundedly complete, and therefore is completing.

Let E be a completing (DK) space (resp. (DS) space). For every absolutely convex closed bound set A in E there is another $B \supset A$ such that the natural mapping $E_A \to E_B$ is weakly compact (resp. compact). Since the imbedding $E_B \to E$ is continuous, this implies that A is relatively weakly compact (resp. relatively compact) in E.

For the sake of completeness we quote the following theorem of Grothendieck [3, Chap. II, p.40].

THEOREM 7. *Let E be a Fréchet space or a (DF) space. Then E is a Grothendieck space if and only if it is a (DG) space.*

This is another difficult part of his theory. Pietsch [6] has simplified the proof by making use of his Property (B) which both Fréchet spaces and (DF) spaces satisfy. Actually Pietsch shows that a Grothendieck space is a (DG) space if and only if it has Property (B).

5. APPLICATIONS

A completing (DF) space E is a (DK) space (resp. a (DS) space, resp. a (DG) space) if and only if it is represented as the inductive limit $\varinjlim E_n$ of a sequence of Banach spaces E_n such that the mappings $u_n : E_n \to E_{n+1}$ are one-to-one and weakly compact (resp. compact, resp. nuclear). It is also characterized as the strong dual of an (FK) space (resp. an (FS) space, resp. an (FG) space). We call such a space a (DFK) space (resp. a (DFS) space, resp. a (DFG) space).

(DFK) spaces have been introduced by Makarov [13] and Komatsu [14] under different names. Earlier Sebastião e Silva [15] and Raikov [16], [17] discussed (DFS) spaces under different names. These spaces play very important roles in applications (see e.g. the series of [12]).

For practical purposes Grothendieck's Theorems 3,4 and 10 of [3] may be summarized as follows :

THEOREM 8. *Let E be a locally convex space and F a completing (DK) space. Then every integral linear mapping $u : E \to F$ is nuclear.*

Here $u : E \to F$ is called an *integral linear mapping* if its natural extension $u : E \to F''$ admits the factorization

$$E \longrightarrow L^\infty(M) \hookrightarrow L^1(M) \longrightarrow F''$$

for a suitable finite measure space M.

As Grothendieck remarks [3, Chap. I, p.138], the conclusion of Theorem 8 does not hold if F is the (DF) Montel space E' of Remark in §4.

REFERENCES

[1] J. Dieudonné - L. Schwartz, 'La dualité dans les espaces (F) et (LF)', *Ann. Inst. Fourier*, **1** (1949), 61–101.

[2] A. Grothendieck, 'Sur les espaces (F) et (DF)', *Summa Brasil. Math.*, **3** (1954), 57–122.

[3] A. Grothendieck, 'Produits Tensoriels Topologiques et Espaces Nucléaires', *Mem. Amer. Math. Soc.*, **16** (1955).

[4] L. Schwartz, ' Produits Tensoriels Topologiques d'Espaces Vectoriels Topologiques, Espaces Vectoriels Topologiques Nucléaires', *Sém. Schwartz*, 1953-54.

[5] F. Trèves, *Topological Vector Spaces, Distributions and Kernels*, Academic Press, New York, 1967.

[6] A. Pietsch, *Nuclear Locally Convex Spaces*, Springer, Berlin-Heidelberg-New York, 1972.

[7] H. Komatsu, *Theory of Locally Convex Spaces*, Dept. of Math. Univ. of Tokyo, 1974.

[8] R. S. Phillips, ' On weakly compact subsets of a Banach space', *Amer. J. Math.*, **65** (1943), 108–136.

[9] R. S. Phillips, ' On linear transformations', *Trans. Amer. Math. Soc.*, **48** (1940), 516–541.

[10] Y. Kōmura, ' On linear topological spaces', *Kumamoto J. Sci. A.*, **5** (1962), 148–157.

[11] W.J.Davis - T.Figiel - W.B.Johnson - A.Pełczyński, ' Factoring weakly compact operators', *J. Funct. Analysis*, **17** (1974), 311–327.

[12] H. Komatsu, ' Ultradistributions, III, Vector valued ultradistributions and theory of kernels', *J. Fac. Sci. Univ. Tokyo, Sec. IA*, **29** (1982), 653–718.

[13] B. M. Makarov, ' On inductive limits of normed spaces', *Dokl. Akad. Nauk. SSSR*, **119** (1958), 1092–1094; *Vestnik Leningrad Univ.*, **13** (1965), 50–58 (Russian).

[14] H. Komatsu, ' Projective and injective limits of weakly compact sequences of locally convex spaces', *J. Math. Soc. Japan*, **19** (1967), 366–383.

[15] J. Sebastião e Silva, ' Su certi classi spazi localmente convessi importanti per le applicazioni', *Rend Mat. Appl.*, **14** (1955), 388–410.

[16] D. A. Raikov, ' On two classes of locally convex spaces important in applications', *Trudy Sem. Funkt. Anal. Voronež*, **5** (1957), 22–34 (Russian).

[17] D. A. Raikov, ' Completely continuous spectra of locally convex spaces', *Trudy Moskov. Mat. Obšč*, **7** (1958), 413–438 (Russian).

GPSR Compliance

The European Union's (EU) General Product Safety Regulation (GPSR) is a set of rules that requires consumer products to be safe and our obligations to ensure this.

If you have any concerns about our products, you can contact us on ProductSafety@springernature.com

In case Publisher is established outside the EU, the EU authorized representative is:

Springer Nature Customer Service Center GmbH
Europaplatz 3
69115 Heidelberg, Germany

Batch number: 10815974

Printed by Printforce, the Netherlands